Essentials of Clinical Anatomy

Second Edition

Essentials of
Clinical Anatomy

Second Edition

The Parthenon Publishing Group

Essentials of Clinical Anatomy

Second Edition

Ralph Ger FRCS(Eng), FRCS(Ed), FACS
Professor of Anatomy, Albert Einstein College of Medicine, New York, USA
Professor of Surgery, State University of New York at Stony Brook, New York, USA
Associate Chairman, Department of Surgery, Nassau County Medical Center,
East Meadow, New York, USA

Peter Abrahams MB BS (London), FRCS(Ed), FRCR
Clinical Anatomist, University of Cambridge, Cambridge, UK

Todd R. Olson PhD, MA, BA
Associate Professor of Anatomy and Structural Biology, Albert Einstein College of
Medicine, New York, USA
Director, Medical Gross Anatomy, Albert Einstein College of Medicine, New York, USA

The Parthenon Publishing Group
International Publishers in Medicine, Science & Technology

NEW YORK   LONDON

Published in the USA by
The Parthenon Publishing Group Inc.
One Blue Hill Plaza, PO Box 1564, Pearl River, New York 10965, USA

Published in the UK by
The Parthenon Publishing Group Limited
Casterton Hall, Carnforth, Lancs LA6 2LA, UK

Library of Congress Cataloging-in-Publication Data
Ger. Ralph
    Essentials of clinical anatomy / Ralph Ger. Peter Abrahams. Todd
R. Olson. — 2nd ed.
      p.   cm.
    Includes index.
    ISBN 1-85070-635-2
    1. Human anatomy.   2. Anatomy, Pathological.   I. Abrahams, Peter
H. II. Olson, Todd R. III. Title.
    [DNLM: 1. Anatomy.   QS 4 G354e 1996]
    QM23.2.G47   1996
    6 — dc20
    DNLM/DLC
    for Library of Congress   96-33686
                                                       CIP

British Library Cataloguing-in-Publication Data
Ger, Ralph
    Essentials of clinical anatomy. – 2nd ed.
    1. Human anatomy
    I. Title II. Abrahams, Peter, 1947 Aug. 5. – III. Olson, Todd R.

ISBN 1-85070-635-2

First published 1986 (Pitman Publishing Ltd)
Reprinted 1989 (Churchill Livingstone)
Reprinted 1992
Second edition published 1996

Composition by H&H Graphics, Blackburn, UK
Printed and bound in Great Britain by
Butler & Tanner Ltd, Frome and London

This book is dedicated to
our wives, Dorrit and Lucy,
and children, Amanda, Michael, and Kevin;
and Tania, Nadia, and Bianca, whose
patience and forbearance allowed us
to spend countless hours in isolation,

and to

Alfred M Abrahams, FRCS,
without whom this book would never
have been written; he is partly responsible
for one of the authors' existence,
and was guide, mentor and
inspirational force for the academic
activities of the other.

Contents

Preface

It would appear difficult to justify yet another text, for the distribution of tissues is unchanging. However, we have been encouraged by both past and present medical students of several medical schools to do just that. The suggestion by these students was that the syllabus be based on a mostly unexpurgated version of lecture courses. This idea has been adopted and some material has in fact been taken verbatim from recorded lectures. The purpose of this approach is to attempt to re-capture the advantages of a live lecture. The live situation permits the use of particular words or phrases, the reference to current or historical situations, or humorous asides. While these vehicles are commonly used to enhance a speech or lecture, some of these 'peg-hangers' act in an underscoring capacity to reinforce significant facts, but they themselves should not be taken too seriously. In modern parlance, we hope you find the text 'user friendly'.

It should be stated that some aspects of the teaching of anatomy belong to a bygone era; there is surely little to be gained by insisting that the student be able to identify some obscure bony projection, or a minor muscle deep in the neck or the sole of the foot. It has often been said that the clinical disciplines are merely the application of the basic sciences. With respect to gross anatomy, never before has the knowledge of this subject been as vital to the modern clinician as at present. New diagnostic and therapeutic techniques have been rendered possible by the explosive advances in radiological imaging. The cardiovascular system is now a meeting place commonly used by physicians of many disciplines as catheters thread their way in diverse directions. Entrapment neuropathies have emerged as important causes of painful syndromes. Many more examples come to mind, but suffice it to say that the practicing physician of the late twentieth century needs to be part radiologist, part neurologist and even part anatomist, to mention only a few of the required skills. The practical use of these current advances requires a sound anatomical background; without this both the clinician and the patient will be at a severe disadvantage. For these reasons the emphasis of this text is heavily tilted toward the clinical setting as it unfolds before the budding physician.

This book is dedicated to the medical students of the University of Cape Town, South Africa, The Albert Einstein College of Medicine, New York, City University of New York, The Middlesex Hospital Medical School, London, Ben-Gurion University of the Negev, Israel, and the University of Iowa, USA. For without their intellectual curiosity, stimulation and encouragement this book might never have come to fruition.

RG
PA

Preface to the Second Edition

Second editions are required for several reasons. In this case, we wish to add new material and present an addition to the editorial staff. Deletions have been carried out to allow us to accommodate new material, resulting in a book that is modestly longer.

In view of the persistent and growing swing towards the clinical application in the study of anatomy, a new Part has been added which recognizes that the physical examination of the living patient is the ultimate application of anatomical knowledge. The distillation of the experience of the original authors, reflected in this new Part, offers the neophyte a method of developing a sound basis for the establishment of a reliable and lasting habitus for clinical application. Included in this new Part is the anatomic basis for procedures that often fall to the student, intern or resident and which frequently follow the physical examination.

Requests have been often made for expansion of the answers to the questions appearing at the end of each Part of the book. These requests have been met by referring the reader to a page where the explanatory data relevant to a question appears.

It is often advisable to inject new thoughts and ideas into a text; for this reason we welcome Todd R. Olson PhD as a co-author in the knowledge that his extensive teaching experience will complement the experience of the authors of the first edition.

RG
PA
TO

How to use this book

The reader will notice on a considerable number of pages either a symbol in the form of a stethoscope or a 'Q'. Both of these will occupy the margin opposite a portion of the text or a figure. The stethoscope symbol is to be used in conjunction with the page references in Chapter 36. These page numbers indicate the underlying anatomic basis that pertains to the clinical situation under discussion. On turning to the page indicated, the image of a stethoscope will immediately highlight the relevant information.

With regard to the 'Q' symbol, this is to be used in conjunction with the answer sheet on p. 529. The information on which the answer is based appears on the page mentioned, opposite the 'Q'.

Acknowledgements

We thank the following for their participation and assistance in the production of this publication.

Medical illustration

Ingram Chodorow has been innovative and has skilfully produced deceptively simple line diagrams of difficult subjects. His anatomical background has enabled him to present the material in an animated and educational manner. Additional help from John Hardie is acknowledged.

A number of line diagrams used for the second edition have been skilfully rendered by Scott Staton BA, first year Medical Class attendee, Albert Einstein College of Medicine, Bronx, New York.

Radiographs

Dr Robert Bernstein, Albert Einstein College of Medicine, Bronx, New York

Dr David Faegenburg, Nassau Hospital, Mineola, New York, and State University of New York at Stony Brook, Long Island

Dr I. Melbourne Greenberg, Nassau Hospital, Mineola, New York, and State University of New York at Stony Brook, Long Island

Dr Martin Goldman and Dr Paul Moh, Department of Radiology, Nassau County Medical Center, East Meadow, New York.

Dr Jamie Weir, University of Aberdeen Teaching Hospitals, Scotland. We are indebted to him for permission to use numerous plates from *An Atlas of Radiological Anatomy*, 1st edition (Weir and Abrahams, 1978, Pitman, London, and 2nd edition (Weir and Abrahams, 1986, Churchill Livingstone, Edinburgh).

Academic assistance

Dr France Baker-Cohen, Albert Einstein College of Medicine, has been an encouraging stimulus and an authoritative adviser

Prof. Bernard Wood, The Middlesex Hospital Medical School, assisted in the compilation of the glossary.

Secretarial

Mrs Gladys Scherrer, Albert Einstein College of Medicine, Bronx, New York

Mrs Elise Radichio and Mrs Lee Swartz, State University of New York at Stony Brook, Long Island

Miss Paula Smith, The Middlesex Hospital Medical School, University of London, England

Part I Introduction

1 Anatomical language – its origins and meanings

A, Arabic; A-S, Anglo-Saxon; Fr, French; G, Greek; L, Latin

ABDOMEN: (L) *abdere* = to hide, because it hides the viscera

ABDUCT: (L) *ab* = from, plus *ducere* = to draw; thus to draw or lead out from the midline

Abduction

ACCESSORY NERVE: (L) *accedere* = to be added to; thus the nerve (the cranial part) conducting the fibres which are 'added to' the vagus

ACETABULUM: (L) *acetum* = vinegar, plus *abulum*, diminutive of *abrum* = a holder

ACOUSTIC: (G) *akouein* = to hear

ACROMION: (G) *akros* = summit, plus *omos* = shoulder; hence highest point of the arm

ADDUCT: (L) *ad* = to, plus *ducere* = to draw; thus to draw or lead in to the midline

ADENOIDS: (G) *aden* = gland, plus *eidos* = like; thus gland-like

ADITUS: (L) from *adire* = to go to; an entrance

ADRENAL: (L) *ad* = to or toward, plus *ren* = kidney

ADVENTITIA: (L) *adventicius*, from *advenire* = to come to. Applied to an outer covering where this was derived from surrounding (or foreign) tissue

AFFERENT: (L) *ad* = toward, plus *fero* = to carry to; thus going towards the brain in sensory nerves

ALA: (L) = a wing

ALBA: (L) *albus* = white

Adduction

ALVEOLUS: (L) diminutive of *alveus* = a hollow; thus a small space or compartment

AMPULLA: (L) = a two-handled flask; hence a flask-like swelling; e.g. in vas, rectum

ANASTOMOSIS: (G) *ana* = through, plus *stoma* = a mouth; an opening or a coming together through a mouth

ANSA: (L) = handle (of a jug) or loop on a sandal; as in loops of nerve, e.g. ansa cervicalis

ANTEVERSION: (L) *ante* = forward, plus *vertere* = to turn

ANTRUM: (L) = a cave or cavity

3

ANULUS: (L) = a small ring

ANUS: (L) = a ring

AORTA: (G) *aorte* = to suspend, used because the aorta was thought to suspend the heart

APONEUROSIS: (G) *apo* = from, plus *neuron* = a tendon. An expansion from a tendon or any sheet of fibrous tissue associated with a muscle

APPENDIX: (L) = an appendage, from *appendere* = to hang to

AQUEDUCT: (L) *aqua* = water, plus *ducere* = to lead

ARACHNOID: (G) *arachne* = a spider or a spider's web, plus *eidos* = like

AREOLA: (L) *area* = a courtyard, or a little open space (adj. areolar)

ARTERY: (G) *aer* = air, plus *tereo* = I carry. Arteries were found to be empty after death and were thought only to carry air

ARTICULATION: (L) *articulus* = a joint

ATAVISTIC: (L) *atavus* = a remote ancestor. Anatomical characteristics thought to be a reversion to an earlier biological type

ATRESIA: (G) *a* = without, plus *tresis* = a hole; thus literally no hole, no perforation. The term is used for various conditions in which an opening or canal or duct is closed either from congenital or other causes

ATRIUM: (L) = court or hall, the central open entrance hall of a Roman house

AURICLE: (L) *auricula*, diminutive of *auris* = ear

AUTO-: (G) *autos* = self

AUTONOMIC: (G) *autos* = self, plus *nomos* = law; thus a 'self-controlled' system

AVULSE: (L) *avello* = to tear away, or separate by force

AXIS: (L) = axle, from *ago* = I carry

AXON: (G) see 'axis'. The nerve fibre, passing through its tubular sheath, forming the axis of the structure

AZYGOS: (G) *a* = without, plus *zygon* = yoke; thus unpaired

BASILIC: (L) *basis* = the base, also inner, thus the 'inner' vein. (G) *basilikos* = royal; the basilic vein is the largest vein of the arm

BI-: (L) *bis* = twice or double

BIO-: (G) *bios* = life

BRACHIUM: (L) = the upper part of the arm above the elbow

BREGMA: (G) = the upper or anterior part of the head. Probably from *brexo* = I moisten because this area is soft and feels moist in the newborn

BREVIS: (L) = brief or short

BRONCHUS: (G) *bronchos* = the windpipe. Said to be derived from *brexein* = to pour or moisten, from an ancient belief that solids were conveyed to the stomach by the esophagus, and fluids by the bronchi. Now applies to air passages beyond the trachea

BUCCA: (L) = mouth, cheek

BUCCINATOR: (L) = trumpeter: the buccinator muscle is the trumpeter's muscle

BULLA: (L) = bubble, or any swollen organ; e.g. bulla ethmoidalis

BURSA: (L) = a purse; e.g. a synovial sac

CADAVER: (L) *cadere* = to fall; hence a fallen man or corpse

CALCIFICATION: (L) *calx* = lime, plus *facio* = to make; thus the deposition of insoluble calcium salts

CALYX: (G) *kalyx* = a cup

CANCELLOUS: (L) *cancellus* = lattice; hence lattice-like, or spongy tissue

CANNULATION: (L) *canna* = reed, thin tube, and its insertion into a vessel or cavity

CAPILLARY: (L) *capillus* = a hair of the head; hence a vessel of hair-like size

CAPSULE: (L) *capsula*, diminutive of *capsa* = a box or case; from *capere*, to contain

CAPUT: (L) = the head, and *capitulum* = a small head; e.g. on the distal humerus

CARDIA: (G) *kardia* = the heart; also the cardiac portion of the stomach

CARINA: (L) = keel of a boat; thus a ridge dividing main bronchi

CAROTID: (G) *karoun* = a deep sleep or to stupefy. When both carotid arteries are pressed the person may become sleepy or even unconscious due to cerebral ischemia

CARPUS: (G) *karpos* = the wrist

CAUDA: (L) = tail

CAVERNOUS: (L) *caverna*; thus anything full of caverns or cavities

CECUM: (L) *caecus* = blind; hence the blind end of the large intestine

CELOM: (G) *koiloma* = a hollow, a body cavity

CEPHALIC: (G) *kephale* = head; thus pertaining to the head, e.g. cephalic vein, the vein directed towards the head

CERVIX: (L) = neck; i.e. neck of the uterus

CHIASMA: (G) the mark of the letter X

CHOLECYST: (G) *cole* = bile, plus *kystis* = bladder, thus gallbladder

CHONDRO-: (G) *chondros* = cartilage

CHORDA: (L) = a string of gut (used for musical instrument)

CHORION: (G) = any skin or leather, used as a membrane

CHYLE: (G) *chylos* = a juice or fluid

CILIUM: (L) = an eyelid, from *cillo* = to move

CIRCUMDUCTION: (L) *circum* = around, plus *ductus* = to move

CLAVICLE: (L) *clavicula*, diminutive of *clavis* = a key. Called this because it resembles a primitive key, or stick

CLIVUS: (L) = the slope of a hill, hence the slope of basioccipital and basisphenoid

CLOACA: (L) = a sewer or drain

COCCYX: (G) *kokkyx* = cuckoo, because of a supposed resemblance of the vertebrae to the bill of a cuckoo

COCHLEA: (G) *kochlias* = a snail with a spiral shell

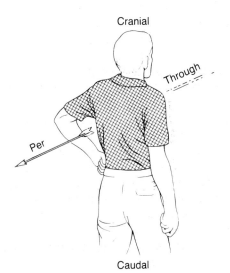

Cranial

Through

Per

Caudal

COLLAGEN: (G) *kolla* = glue, plus *gennao* = I produce

COLON: (G) *kolon* = the food passage, or large bowel

COMES: (L) = a companion

COMMITANS: (L) = a person accompanying

COMMUNICANS: (L) = to share with someone; thus connecting or joining

CONCHA: (L) = a shell

CONDYLE: (G) *kondylos* = knuckle, the knobs formed by the knuckle at any joint

Circumduction

CONJUNCTIVA: (L) *con* = together, plus *jungo* = I join, to connect or join together. The membrane which connects the globe of the eye with the lids

CORACOID: (G) *korax* = a crow or raven, plus *eidos* = like; like a raven's beak

CORNU: (L) = a horn

CORONA: (L) = a crown

CORPUS: (L) = body (pl. corpora)

CORTEX: (L) = peel, rind or bark

CRANIUM: (G) *kranion* = the skull

CREMASTER: (G) *kremaster* = a suspender; e.g. cremaster muscle 'suspends' the testes

CRIBRIFORM: (L) *cribrum* = a sieve, plus *forma* = form or shape

CRICOID: (G) *krikos* = a ring, plus *eidos* = like

CRISTA: (L) = a crest

CRUS: (L) = the shin; thus used for a 'leg-like' process, or processes (pl. crura)

CUBITUS: (L) = the elbow; from *cubo* = I lie down, because the Romans leant on the elbow when assuming a recumbent position for eating

CUNEIFORM: (L) *cuneus* = a wedge, plus *forma* = shape

CYST: (L, G) *kystis* = a bag or pouch

DARTOS: (G) = to be skinned or flayed

DELTOID: (G) Δ = delta, the fourth letter of the Greek alphabet; hence structures delta-shaped

DENS: (L) = a tooth

DERMATOME: (G) *derma* = skin, plus *temein* = to cut up

DERMIS: (G) *derma* = the skin

DIAPHRAGM: (G) *diaphragma* = a partition or wall

DIAPHYSIS: (G) = a growing through, from *dia* = through, plus *physis* = nature or growth

DIASTOLE: (G) = a pause, place apart, expansion

DIGASTRIC: (G) *di* = twice or double, plus *gaster* = belly. A muscle having two fleshy bellies with an intermediate tendon

DIGIT: (L) *digitus*, a finger or toe, from *dico* = I point out

DIPLOË: (G) = double; e.g. double layer of skull bone

DORSUM: (L) = the back; e.g. latissimus dorsi – the widest muscle of the back

DUCT: (L) *ductus*, from *ducere* = to lead

DUODENUM: (L) = twelve. Because 12 fingers' breadth in length

DURA MATER: (L) *dura* = hard, and *mater* = mother. The tough outer layer of the meninges

ECTOPIC: (G) *ektos* = outside, plus *topos* = place; hence something out of normal place

EFFERENT: (L) from *effero* = to bring out; thus outward from an organ or part

EMBRYO: (G) from *bryein* = to grow, *embryo* = the fruit of the womb before birth

ENTERIC: (G) *enterikos* = the intestine, from *entos* = within, because of its inner position

EPIDIDYMIS: (G) *epi* = upon, plus *didymos* = testis

EPIGASTRIUM: (G) *epi* = upon, plus *gaster* = belly or stomach

EPIGLOTTIS: (G) *epi* = upon, plus *glotta* = tongue. In this case probably applied to the epiglottis in animals, which does lie 'on' the tongue

EPIPHYSIS: (G) *epi* = upon, plus *physis* = growth. A portion of a bone which grows into another piece of bone

EPIPLOIC: (G) *epi* = upon or over, plus *ploon* = fold

ETHMOID: (G) *ethmos* = a sieve, plus *eidos* = like (see also 'cribriform')

EVERSION: (L) *e* = out, plus *versus* = to overturn; hence turning outward

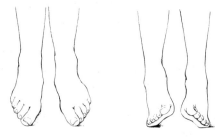

Eversion   Inversion

EXTENSOR: (L) *extendo* = I stretch out

FALCIFORM: (L) *falx* = a sickle, plus *forma* = shape

FASCIA: (L) = a band or bandage, or a fillet

FAUCES: (L) = a passage

FEMUR: (L) = the thigh, perhaps from *ferendum* = bearing

FENESTRA: (L) = a window, and hence opening

FETUS: (L) from *feo* = to bring forth; thus, offspring – that which is brought forth

FIBULA: (L) = a brooch or clasp, from *figo* = I fasten

FILIFORM: (L) *filum* = thread – a fine hair-like structure

FIMBRIA: (L) = a fringe or border

FISSURE: (L) *fissura* = fissure or a cleft, from *findere* = to cleave

FLAVUS: (L) = yellow – related to the color; e.g. ligamentum flava – yellow elastic ligament

FLEX: (L) *flecto* = bend; hence flexor – a muscle which bends

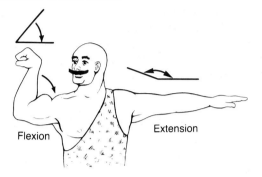

Flexion   Extension

FONTANEL(LE):(L) = a spring, or fountain – the gap between skull bones of infants

FORAMEN: (L) = an aperture or hole, from *forare* = to pierce

FORNIX: (L) = an arch, vault or gutter

FOSSA: (L) = a hole, ditch or pit; also a grave

FOURCHETTE: (Fr) = a small fork; applied to mucous membrane joining labia majora

FOVEA: (L) = a small pit, from *fodere* = to dig

FRENULUM: (L) diminutive of *frenum* = a bridle or curb

FRONTAL: (L) *frons* = the forehead or brow

FUNDUS: (L) = the bottom or base of anything. The bottom of a hollow structure or the part farthest removed from the external orifice; e.g. fundus of the eye and uterus, but why stomach?

FUNGIFORM: (L) *fungus* = a mushroom, plus *forma* = shape

FUSIFORM: (L) *fusus* = a spindle, plus *forma* = shape

GALEA: (L) = a helmet made of leather

GALLUS: (L) = a cock. With *crista* = crest, the crista galli is the cock's comb or elevation of the ethmoid bone into the anterior fossa

GANGLION: (G) = a knot. A swelling or tumor under the skin; also a collection of nerve cells

GASTRIC: (G) *gaster* = the stomach or belly

GASTROCNEMIUS: (G) *gaster* = belly, plus

kneme = leg; the calf of the leg

GEMELLUS: (L) diminutive of *geminus* = a twin; hence paired

GENU: (L) = bend; thus knee

GINGIVA: (L) the gums, from *gigno* = to beget, because the teeth are born in the gums

GINGLYMUS: (G) = a hinge; thus any hinge-like joint

GLABELLA: (L) *glaber* = smooth, without hair; thus the area between the eyebrows

GLAND: (L) from *glandula*, diminutive of *glans* = acorn

GLANS: (L) = acorn or pellet

GLENOID: (G) *glene* = socket, plus *eidos* = shape

GLOMERULUS: (L) diminutive of *glomus* = a skein, or ball, of wool or thread

GLOTTIS: (G) = larynx. Early anatomists compared the larynx to the tongue-shaped reed of a flute

GLUTEAL: (G) *gloutos* = the buttock or rump

GOMPHOSIS: (G) *gomphos* = a nail – hence the peg-in-socket fibrous joint of the teeth – plus *osis* = condition

GONAD: (G) *gonos* = seed, also offspring

GRACILIS: (L) = slender

GUBERNACULUM: (L) *gubernator* = governor or helmsman

HALLUX: (L) = the great toe

HAMATUS: (L) = hooked; e.g. the hamate bone with its small hook

HAUSTRUM: (L) = a machine for drawing water from a well; thus a sac or anything shaped like a bucket, pouch or saccule

HEMI-: (G) = half

HEPAR: (G) = the liver

HERNIA: (L) = a sprout or shoot, of a young plant; hence protrusion or rupture

HIATUS: (L) = a gap, aperture or opening

HILUM: (L) = a small thing. Term used by the Romans for the little spot on a seed which marked its point of attachment, especially the spot on a bean or seed

HUMERUS: (L) = the bone of the upper arm. (G) *omos* = the shoulder

HYMEN: (G) = a membrane; also the name of the Greek god of marriage

HYOID: (G) letter upsilon, *v*, plus *eidos* = like; hence shaped like a U

HYPER-: (G) = above, in excess of

HYPO-: (G) = under, deficient

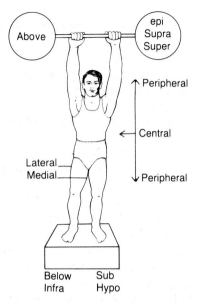

HYSTER-: (G) = uterus; e.g. hysterectomy – removal of the womb

ILEUM: (G) *eileos* = twisted, or roll around in pain

ILIUM: (L) = flank

IMA: (L) = lowest, as in thyroidea ima artery

INCUS: (L) = an anvil, from *incudere* = to strike upon

INFRA-: (L) = below or under

INFUNDIBULUM: (L) = a funnel, from *infundere* = to pour in

INGUINAL: (L) from *inguen* = the groin

INION: (G) = nape of neck, the back of the head or external occipital protuberance

INNOMINATE: (L) *innominatus* = without name; the innominate artery now has a name – the brachiocephalic!

INSULA: (L) = island

INTROITUS: (L) *intro* = within and, *ire* = to go; hence entrance to the female cavity (vagina)

INVERSION: (L) *in* = in, plus *versus* = to overturn; hence turning inward

IRIS: (G) = rainbow or halo

ISCHIUM: (G) *iskion* = hip; probably related to *iskus* = strength

ISTHMUS: (G) *isthmos* = narrow passage or entrance; also a neck of land separating two bodies of water

JEJUNUS: (L) = empty or hungry. At death the upper intestine is found empty

JUGULAR: (L) *jugulum* = the throat; from *jugum* = a yoke, because a yoke is carried on the neck

KYPHOSIS: (G) = bent or bowed

LABIUM: (L) = lip

LABRUM: (L) = lip, rim or margin; e.g. labrum glenoidale

LACRIMAL: (L) *lacrima* = a tear

LACUNA: (L) = a pit or hollow, a pond or pool, a little lake

LAMINA: (L) = a thin plate, a broad leaf or layer

LARYNX: (G) = the upper part of the windpipe

LEVATOR: (L) = one who lifts, from *levare* = to lift

LIEN: (L) = spleen

LIGAMENT: (L) *ligamentum*, from *ligare* = to bind, tie or bondage

LIGATURE: (L) *ligatus* = a bond or tie

LINEA: (L) = a linen thread, as in linea alba of the abdomen

LINGUA: (L) = the tongue, from *lingere* = to lick

LORDOSIS: (G) = to bend the body forward and inward

LUMBAR: (L) *lumbus* = the loin

LUMBRICAL: (L) *lumbricus* = a worm; the lumbrical muscles are worm-like

LUNG: (A-S) *lunge* = light – as the lungs float

LYMPH: (L) *lympha* = clear water from a spring

MACULA: (L) = a small spot or mark

MALLEUS: (L) = a hammer or mallet

MAMMA: (L) = the breast (female). (G) from the cry of the infant 'ma-ma'

MANDIBLE: (L) *mandibula* = the lower

jaw; from *mandere* = to chew

MANUBRIUM: (L) = a handle, from *manus* = a hand, plus *hibrium*, from *habeo* = I hold

MARROW: (A-S) = soft fatty substance filling the cavities of long bones

MASSETER: (G) from *masso* = I chew

MAST-: (G) *mastos* = breast; hence mastoid – like a breast

MAXILLA: (L) = upper jaw bone, from *macerare* = to chew

MEATUS: (L) = a passage or canal, from *meare* = to go, or pass

MEDIASTINUM: (G, L) *medius* = middle, plus *stare* = to stand in; thus the middle part between the lungs

MENINGES: (G) = membranes (pl. of meninx)

MENISCUS: (G) = a crescent, diminutive of *mene* = the moon

MESENTERY: (G) *mesos* = middle, plus *enteron* = intestine

MESODERM: (G) *mesos* = middle, plus *derma* = skin; hence the middle of the three primary germ layers which gives origin to connective tissues

METACARPUS: (G) *meta* = after, plus *karpos* = wrist

METOPIC: (G) *met* = between, plus *ops*, eye; the forehead suture between the eyes

MITRAL: (L) *mitra* = a kind of cap used by the bishops. (G) = a turban or headband worn in ancient Greece

MOLAR: (L) = a mill; hence a grinding tooth

MORPHOLOGY: (G) *morphos* = form or shape, plus *logos* = a discourse or study

MYOCARDIUM: (G) *mys* = muscle, plus *kardia* = heart

NARES: (L) = the nostrils (pl. of naris)

NAVICULAR: (L) *navicula* = a little boat (very little!)

NEPHRON: (G) *nephros* = kidney

NEURON: (G) = nerve, any structure with a white fibrous appearance

NOTOCHORD: (G) *notos* = back, plus

chorde = string of a lyre – hence cord of the developing back

NUCHAL: (A) = of spine or spinal cord, now used to refer to the nape of the neck

OBTURATOR: (L) *obturare* = to occlude, close or stop up

OCCIPITAL: (L) *occipito* = I begin, called thus because the back of the head is usually the first part to appear at birth

OCULAR: (L) *oculus* = eye

ODONTOID: (G) *odous* = tooth, plus *eidos* = like

OLECRANON: (G) *olene* = elbow, plus *kranion* = skull. The process at the superior (or 'skull') end of the ulna

OMENTUM: (L) *opimus* = fat, a fatty membrane around the internal organs

OMOS: (G) = shoulder

OMPHALOS: (G) = the navel. An island in the middle of the sea (see also 'Umbilicus')

OPHTHALMIC: (G) *ophthalmos* = the eye

ORBIT: (L) *orbita* = a track or circuit; thus the circle in which the eye sits

OSSIFICATION: (L) *os* = bone, plus *facio* = to make; the process of bone formation

OSTIUM: (L) = a door or opening

OTIC: (G) *otikos* = pertaining to the ear

OTORRHEA: (G) *ous* = ear, plus *rhoia* = flow; hence a discharge from the ear

OVARY: (L) *ovarium* = an egg receptacle

PALPEBRA: (L) *palpitare* = that which quivers

PAMPINIFORM: (L) *pampinus* = a vine tendril, plus *forma* = shape

PANCREAS: (G) *pas* = all, plus *kreas* = flesh; described thus because of its meaty character (i.e. no fat)

PANNICULUS: (L) diminutive of *pannus* = a piece of cloth. Panniculus carnosus is skeletal muscle in the superficial fascia, e.g. platysma

PAPILLA: (L) = a nipple or teat

PARA-: (G) = beside

Para - next to

PARAMETRIUM: (G) = beside the metrium (G) or uterus (L)

PARIETAL: (L) *parietalis* = belonging to a wall

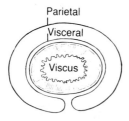

PAROTID: (G) *parotis*; from *para* = beside, plus *otos* = the ear; thus the gland beside the ear

PECTORAL: (L) = belonging to the breast, from *pectus* = the breast

PEDICLE: (L) *pediculus* = a little foot

PEDUNCLE: (L) variant of pedicle, from *pes* = a foot; used now to describe a stalk

PELVIS: (L) = a wide vessel, a basin or bowl

PENIS: (L) = a tail, from *pendere* = to hang down

PERI-: (G) = around, about

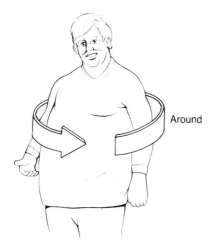

Around

PERICARDIUM: (G) *peri* = around, plus *kardia*, the heart. Around the heart there are both serous and fibrous layers

PERINEUM: (G) *perinaion*, probably from *perineo* = to swim around. Used by the Greeks because this part of the body is usually moist with sweat. The first descriptions referred to the space between the anal opening and the scrotum

PERITONEUM: (G) *peri* = around, plus *tonos* = tension; that which is stretched over the abdomen

PERONEAL: (G) *perone* = pin or brooch

PETROUS: (G) *petra* = a rock; hence the rock-like bone protecting the inner ear

PHALANX: (G) = line of battle. The bones of the fingers are arranged in ranks like a Greek phalanx going into battle (pl. phalanges)

PHARYNX: (G) = the throat, but later Galen used it for the joint respiratory and gastro-intestinal opening

PHRENIC: (G) = pertaining to the dia-phragm

PIA MATER: (L) *pia* = soft or kind, and *mater* = mother

PILOMOTOR: (L) *pilus* = hair, plus *motor* = mover; muscle or nerves causing contraction of the arrectores pilorum muscles, i.e. goose-pimples

PIRIFORM: (L) *pirum* = a pear, plus *forma* = shape

PISIFORM: (L) *pisum* = a pea, plus *forma* = shape; small sesamoid carpal bone

PITUITARY: (L) *pituita* = a mucous secretion. The hypophysis was thought to secrete mucus into the nose

PLACENTA: (G) = a flat cake

PLANTAR: (L) *planta* = the sole of the foot

PLATYSMA: (G) = a flat plate; now used to describe a thin flat superficial muscle (see also 'panniculus')

PLEURA: (G) = a rib; thus refers to the side of the body

PLEXUS: (L) *plectere* = to plait or braid; hence a network

PLICA: (L) *plicare* = to fold

POLLEX: (L) = the thumb. Perhaps from *polleo* = I am strong, because the thumb is stronger than the other fingers

PORTA: (L) = a gate; also *portare* = to

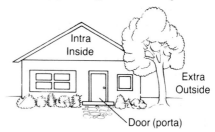

Intra Inside

Extra Outside

Door (porta)

carry, as in the hepatic portal venous system, i.e. the system that carries blood to the liver

PREPUCE: (L) *praeputium* = foreskin

PRONATOR: (L) from *pronus* = turned or inclined to face downwards

PROPRIOCEPTIVE: (L) *proprius* = one's own, plus *capio* = to take; thus capable of receiving stimuli from within

PROSTATE: (G) = one who stands before

PSOAS: (G) = the loins, hence the 'loin' muscle

PTERION: (G) *pteron* = a wing

PUBIS: (L) *pubes* = hair-covered portion of the pelvic or innominate bone

PUDENDAL: (L) *pudendus*, from *pudere* = to be ashamed; of which one should feel ashamed

PYLORUS: (G) *pyle* = a gate, plus *ouros* = a watcher or guard

QUAD: (L) = four

QUADRICEPS: (L) *quattor* = four, plus *caput* = head; thus four-headed muscle

RADIOLUCENT: (L) *radio* = ray, plus *lucens* = shining; hence partly or wholly penetrable (permeable) by x-rays or radiation

RADIUS: (L) = a rod, or spoke of a wheel

RAMUS: (L) = a branch

RAPHE: (G) *rhaphe* = a seam or suture

RECTUM: (L) = straight. Most early anatomical examinations were done on animals and, as this portion is straighter in quadrupeds than in humans, the term rectum was adopted. However, the human rectum is not straight!

RENAL: (L) *ren* = the kidney

RETE: (L) = a net or meshwork

RETINACULUM: (L) from *retinere* = to hold back or retain

RETRO-: (L) = behind or backwards

RHINORRHEA: (G) *rhis* = nose, plus *rhoia* = flow; thus a nasal discharge

RIMA: (L) = a slit, cleft or fissure

RUGA: (L) = a wrinkle; e.g. ruga gastrica, the creases or mucosal folds of the stomach

SACCULE: (L) *sacculus* = a small sac or pouch

SACRUM: (L) = holy or sacred; it looks like the pile of ashes after cremation (the sacred bone). It was supposed to resist decomposition longest and to be the seed from which the body was resurrected; i.e. because it 'enclosed' the sex organs

SAGITTAL: (L) *sagitta* = an arrow

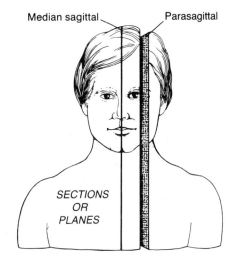

Median sagittal Parasagittal

SECTIONS OR PLANES

SALPINX: (G) = a trumpet or tube

SAPHENOUS: (G) *saphenes* = clear, easily visible, or (A) *al safin* = hidden or secret. The Arabic term applies to the healthy state of the vein; the Greek to the knotted appearance of a varicose vein

SARTORIUS: (L) *sartor* = a tailor. The tailor's muscle, so called because it flexes the knee and flexes and laterally rotates the thigh to bring leg and thigh into the traditional cross-legged sitting position of a tailor

SCALENUS: (G) *skalenos* = uneven or irregular. Applied to a triangle with unequal sides. The scalenus muscles together form a scalene triangle

SCAPULA: (L) = the shoulder blade (pl. scapulae)

SCLERA: (G) *skleros* = hard, firm

SCROTUM: (L) = a skin or a hide, various things made of leather, a bag of skin

SELLA TURCICA: (L) = Turkish saddle; thus the pituitary fossa

SEMEN: (L) = seed

SEPTUM: (L) from *saepire* = to fence in; hence a dividing wall, a partition or fence

SERRATUS: (L) = shaped like a saw, tooth-edged; e.g. the serratus anterior muscle has eight origins

SESAMOID: (G) *sesamoides* = like a sesame seed; thus small bones within tendons

SIGMOID: (G) letter sigma (σ), plus *eidos* = like

SINUS: (L) = anything hollowed out

SKELETON: (G) *skeletos* = dried up

SOLEUS: (L) *solea* = the sole. Flat like the sole of the foot or like the flatfish of that name

SPHENOID: (G) *sphen* = a wedge, plus *eidos* = like

SPHINCTER: (G) from *sphingein* = to bind tightly or strangle

SPLANCHNIC: (G) *splanchnikos* = relating to the viscera

SPLENIUS: (G) *splenion* = a bandage

SQUAMA: (L) = the scale of a fish or serpent; e.g. the 'flat' surface cells of a squamous epithelium

STAPES: (L) = a stirrup

STERNUM: (G) *sternon* = the chest or male breast

STRIATION: (L) *striatus* = furrowed; now striped or marked by lines

STYLOID: (G) *stylos* = a pillar, plus *eidos* = like

SUDOMOTOR: (L) *sudor* = sweat, plus *motor* = mover; thus nerves that cause sweating

SULCUS: (L) = a furrow or groove

SUPINATION: (L) *supinus* = lying on the back or to turn on to the back

SURAL: (L) *sura* = the calf of the leg, hence the sural nerve across the calf

SUSTENTACULUM: (L) from *sustentare* = to hold up or support

SUTURE: (L) = a seam, from *suo* = to sew

SYMPATHETIC: (G) *syn* = with, plus *pathas* = feeling, as this system has widespread connections

SYMPHYSIS: (G) = grown together

SYNOVIA: (G) *syn* = together, plus *ovum* = egg; thus the joint fluid which resembles white of egg

SYSTOLE: (G) = a contraction, from *sustellein* = to draw together or shorten

TABES: (L) = wasting away; e.g. tabes dor-

salis, a disease in which the dorsal white columns of the spinal cord degenerate or waste

TALUS: (L) = the ankle bone

TARSUS: (G) *tarsos* = a crate, a wickerwork mat; hence flat part of foot

TECTUM: (L) = a cover or roof

TEGMEN: (L) = a cover or roof; e.g. tegmen tympani, the roof of the middle ear

TEMPORAL: (L) *tempus* = time. The first white hairs appear in the temporal region; they indicate maturity, or that time has passed—*tempus fugit!*

TENDON: (L) from *tendere* = to stretch out

TENIA: (G) *tainia* = a fillet, tape or band

TENTORIUM: (L) = a tent

TESTIS: (L) = a witness – see 'testicle'

TESTICLE: (L) *testis* = a witness, one who testifies. Under Roman law no man was admissible as a witness unless he had both testes (but see 'pudendal'!)

THYROID: (G) *thyreos* = a large, oblong shield, plus *eidos* = form. The term was first applied to the thyroid cartilage and later used for the gland

TIBIA: (L) = a pipe or flute. Flutes were made with the tibias (shin-bones) of animals

TRABECULA: (L) = a little beam

TRACHEA: (G) from *trachys* = rough. Aristotle thought that both the arteries and the respiratory tract contained air; he distinguished the two by calling the arteries *arteria leia* (smooth) and the tract the *arteria trachia*

TRACHEOSTOMY: (G) trachea, plus *stoma* = mouth, an opening into the trachea

TRAGUS: (G) *tragos* = a goat, because of the hairy character of the skin which was likened to a goat's beard

TRICEPS: (L) *tri* = three, plus *caput* = head; thus having three heads

TRIGEMINAL: (L) *tri* = three, plus *geminus* = twins; thus three born at one time

TRIGONE: (G) *trigonon* = triangle

TRIQUETRAL: (L) *triquetrus* = having three corners

TROCHANTER: (G) = a runner or roller. First used to describe the head of the femur, which turns in its socket like a wheel. Later, and still, used to describe the bony processes of the femur

TROCHLEA: (L) = a pulley – the fixed portion

TUBERCLE: (L) *tuberculum* = a small swelling, bump or protuberance

TURBINATE: (L) from *turbin* = whirlwind, to spin; probably for the curled shape of the bones

TYMPANUM: (L) = drum

ULNA: (L) = elbow; it also is a measurement length, presumably equivalent to a forearm's length

UMBILICUS: (L) = the navel. Diminutive of *umbo* = a boss on a shield, or the bottom of a wine bottle; i.e. described as if there was a small umbilical hernia

UNCIFORM: (L) *uncus* = hook, plus *forma* = shape

URACHUS: (G) from *ouron* = urine, plus *-chlo* = to pour

URETER: (G) = urinary tract, from *ourein* = to make water

URETHRA: see 'ureter' – in early writings both words were interchangeable

UTERUS: (L) = a water bottle made of leather. At first the term was used only for a pregnant uterus

UVULA: (L) = a little grape

VAGINA: (L) = a sheath, a scabbard for the male phallic sword

VAGUS: (L) = wandering, from *vagari* = to wander

VAS: (L) = a vessel or duct; hence vascular relating to blood vessels

VEIN: (L) *vena* = a vein or blood vessel – also used for arteries originally. The vena cava the 'hollow' vein, because it was usually found to be empty after death

VENTRAL: (L) *ventralis* = of or relating to the belly; i.e. anterior

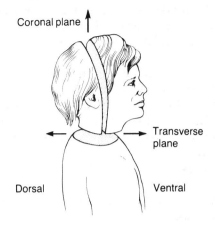

VENTRICLE: (L) diminutive of *venter* = the belly, or a cavity

VERMIS: (L) = a worm. Vermiculation is the worm-like movement made by the ureter when it is stimulated by touching

VERTEBRA: (L) = a joint, from *vertere* = to turn

VESICA: (L) = a bladder

VESTIBULE: (L) *vestibulum* = the space between the entrance of a house and the street

VISCERA: (L) = internal organs (pl. of viscus)

VISCOUS: (L) *viscus* = sticky fluid marked by high viscosity

VITELLINE: (L) from *vitellus* = the yolk of an egg

VOLAR: (L) *vola* = the palm of the hand, or the sole of the foot

VOMER: (L) = a ploughshare

VULVA: (L) = a wrapper or covering, from *volvere* = to wrap around

XIPHOID: (G) *xiphos* = a sword, plus *eidos* = like

ZONA: (L) = a belt, a girdle or zone

ZYGOMA: (G) = like a yoke

2 Introduction to basic tissues and organ systems

Basic tissues · Organ systems

Basic tissues

Everybody consists of five tissue types: epithelial, connective, skeletal, muscular and nervous. These will be studied in your histology/cell biology course, but perhaps now is the time to understand their general arrangement and function.

Epithelium

All surfaces – internal and external – and all tubes and ducts are lined by epithelial cells. Their functions vary. They form a protective envelope for the skin; in the gastrointestinal tract, they may be secretory or absorptive; in the serous cavities, they provide smooth, moist surfaces that facilitate movement. Epithelial cells lack blood vessels.

Connective tissue

Connective tissues are mesodermal structures, which tend to have a passive role, supporting or joining the more functional active tissues. This type of tissue is composed of cells and fibers; the cells are not as closely packed as those in epithelial structures.

The character of the cells and fibers in the connective tissue determines its type. In fat, the cells are modified and contain oil droplets. White fibrous tissue consists mostly of fibers. A preponderance of elastic fibers exists in yellow elastic tissue. The least highly specialized tissue is areolar tissue.

The blood supply of connective tissue is from vessels running through the tissues on their way elsewhere and is therefore sparse. Since combating infection and eventual healing depend on blood supply, infection of connective tissues often results in their destruction. Contrast the leg to the head and neck, where the prolific blood supply of the latter leads to rapid healing. This is a real advantage to the head and neck surgeon.

The connective tissue is well supplied with lymph vessels, and some of the nerves in this tissue have specialized functions.

Areolar tissue

This tissue contains intercommunicating small cavities, hence its name (see Chapter 1). Areolar tissue functions as packing between the different layers of the various tracts (respiratory, alimentary, genitourinary) and between organs and muscles.

Fat

Most parts of the body contain fat. Certain areas are partially or totally fat free for practical reasons: the subcutaneous tissues of the eyelids (accumulation of fat would partially close the eye), penis (a fatty penis has obvious disadvantages), scrotum (it will be mentioned later that the testes live outside the body cavity to be in a cool environment), beneath the nipples (smooth muscle maintains an erect nipple for feeding purposes), the intracranial cavity (accumulations of fat in the intracranial cavity will bulge the eyeballs), and the lungs (fatty lungs would make one very short of breath).

FUNCTIONS OF FAT

Fat is excellent packing material and helps to keep the organs in place (for example, perinephric fat supports the kidney, and orbital fat stabilizes the eyeball). Surprisingly, this seemingly soft substance can also act as a shock absorber. In certain circumstances, it is stiffened by being divided into lobules by fibrous tissue septa (for example: the sole of the foot, especially the heel; the buttock; and the palm).

Fat acts as a storehouse of energy, to be released by various hormones in times of need. The large buttocks of Namibians (an African tribe living under desert conditions with an extremely irregular food supply) are striking examples of fat storage.

With its relatively poor blood supply, fat is an insulator that prevents heat loss. The best long-distance swimmers are females, who usually have a more generous supply of subcutaneous fat and can therefore better withstand the effects of cold water. Conversely, the fat-free subcutaneous layer of the scrotum allows the normal testis to cool down.

White fibrous tissue

The purpose of this dense tissue is to provide strength without rigidity or elasticity. It forms ligaments, which unite bones and limits the range of joint movements, tendons for muscles, and protecting membranes or sheaths around many structures such as muscle (perimysium), bone (periosteum) and various visceral glands such as the liver capsule.

Elastic tissue

This highly specialized tissue is capable of being deformed and returning to its original shape. It is found in the ligamenta flava (ligaments uniting the vertebral laminae), in the true cords of the larynx which are deformed during speech, and in the coats of the arteries, which stretch at each heart beat.

Subcutaneous tissue (superficial fascia)

The subcutaneous tissue layer consists of areolar tissue with varying amounts of fat. Fibrous tissue bands may traverse this layer to tether the skin to the underlying deep fascia. The subcutaneous tissue layer also contains sheets of muscle, which are better developed in animals. In humans, these sheets of muscles exist primarily in the face and neck, although small remnants persist in the hand, around the anus and in the scrotum. This layer of muscle is sometimes called the panniculus carnosus; its noteworthy feature is that the muscle fibers insert into the skin, allowing it to be moved as in wrinkling the anus or facial grimacing, depending on your mood.

Deep fascia

This connective tissue layer wraps the limbs and body wall like a bandage or stocking. Of varying thickness, it may be substantial enough to be used as material for surgical repair; the fascia lata of the thigh may close a hernial defect or a defect in the dura mater. However, in other areas where considerable expansion is needed, such as the abdominal wall, or where the skin is moved by the direct insertion of muscles, as in the face, the deep fascia is thin or absent. The absence of deep fascia in the face leads to a big 'shiner' following a blow near the eye, whereas the same blow injuring the thigh muscles evokes little swelling, the bleeding from the ruptured vessels being contained by the firm fascial stocking. The latter injury may be just as painful, but the former victim elicits much more sympathy.

The deep fascia has several functions. In certain areas it serves as an attachment for the skin, which is tethered to it by fibrous strands. By this means, the skin in the palm of the hand is fixed and not wobbly. The deep fascia, particularly in the extremities, forms a sheath for the muscles, thereby shaping the enclosed part. A few decades ago, a common material used for repair of herniae was a portion of the fascia lata. Although the hernial repair was satisfactory, the muscles of the thigh bulged through the fascial defect, leading to such discontent that the operation was abandoned. Portions of muscles may arise from the deep (inner) surface of the fascia.

The elastic stocking around muscles not only contours the extremity but also plays an important role in the return of blood to the heart. The lower limb has to return blood against gravity; it does so mainly by compression of the veins as they lie in and between the muscles. Were the muscles to swell without restriction as they contracted, the leg would merely fill with blood. The limitation of the swelling by the fascia results in the blood being squeezed out of the muscle (this muscular pump is sometimes called the third heart). One of the reasons for severe changes in the leg following phlebitis (inflammation of the veins) is the failure of this muscle pump action. The function of the fibrous pericardium is another example of the limiting role of fascia; the fibrous pericardium limits the amount of blood that the heart accumulates during its resting phase (diastole), thereby preventing overdistension of its cavities. (See Chapter 9 for further discussion about the fibrous pericardium.)

Another function of fascia is the maintenance of position of structures by the formation of retaining straps (retinacula); for example, the tendons at the wrist and ankle are stabilized by these means. It is a golden rule that when fascia approaches bone, it becomes attached to it, blending with the covering periosteum.

Skeletal tissue

Skeletal tissues are modified connective tissues. The cells and fibers lie in the matrix, which has been solidified. In cartilage, chondroitin sulfate is mostly responsible for the relative rigidity; in bone, inorganic salts, mostly calcium phosphate, produce the absolute rigidity.

Cartilage

Cartilage is formed whenever strength, rigidity and some elasticity are required. In the early stage of fetal development, much of the skeleton is cartilaginous. Most of this cartilage is later replaced by bone (ossification); that which persists remains as permanent cartilage. Cartilage is also found in much of the respiratory system and in the ear. To a degree, cartilage is avascular; infections therefore almost always destroy the cartilage.

There are three types of cartilage: hyaline cartilage, white fibrocartilage and elastic fibrocartilage.

HYALINE CARTILAGE

The nonossified cartilage of the fetal skeleton is comprised primarily of hyaline cartilage. It coats the articular surface of the bone at joints (articular cartilage), where it provides both the elasticity to offset shocks and concussions and the smoothness to promote easy motion. The wear and tear in the joints is such that cell division occurs as an active continual process to replace the worn bearings. The avascular cartilage receives most of its nutrition from the joint fluid. In the respiratory system, the costal cartilages, xiphoid process, and cartilages of the nose, larynx and trachea appear to have a somewhat better vascular supply.

Hyaline cartilage tends to become calcified, and even ossified. The calcification of costal cartilages may show up on a chest x-ray film as a shadow and delude one into misdiagnosing a lung lesion, especially if x-ray examinations are given annually (e.g. during a routine examination of a worker in a tuberculosis sanatorium).

WHITE FIBROCARTILAGE

White fibrocartilage is the constituent of the large intervertebral discs and supplies great strength and rigidity with elasticity. It is present to a lesser degree in the discs inside joints and in the thickenings that deepen the glenoid and acetabular fossae of the shoulder and hip joints, respectively. There is some doubt as to whether the fibrocartilage is really fibrocartilage or merely white fibrous tissue, since the number of cells is small. This fibrocartilage may calcify and ossify, and has an adequate blood supply.

ELASTIC FIBROCARTILAGE

Yellow or elastic fibrocartilage is pervaded by a network of yellow elastic fibers. It is found in the corniculate cartilages of the larynx, epiglottis and ear. The constant movement of the epiglottis, preventing the inhalation of food on swallowing and facilitating speech, underlines the need for a mobile and elastic structure. The elasticity of this type of cartilage is so important that it does not calcify or ossify, and its blood supply is adequate. (Perhaps the ear developed this tissue as professional wrestling gained in popularity; the ear, bent into a 'Z' by a determined opponent, rapidly reverts to a gentle 'C' at the sound of the bell.)

Bone

Bone is the hardest tissue of the body. The fibrous connective tissue is impregnated with mineral salts. The connective tissue gives the bone its toughness and elasticity, and the mineral salts its hardness and rigidity. Removing the salts by immersing the bone in weak acids for a few days leaves the organic element in such a pliant condition that it may be tied in a knot – a doubtful advantage, and something that one is unlikely to inflict on a patient. On the other hand, when the organic matter is removed by heat and the fibrous connective tissue is left, the bone retains its original form but the slightest touch will crumble the bone. This mineral substance, which forms about two-thirds of the weight of bone, consists mostly of calcium salts, chiefly calcium phosphate. Nature has been good to little children – their bones are especially tough and resilient and even fractures are often incomplete (greenstick fractures). Nature is not so good to the aged; their bones are brittle and fracture from trivial causes; a common event is the fracture of a large bone, such as the femur, by tripping over the edge of a carpet. When the calcium salts are removed by disease (e.g. rickets in children and osteoporosis in adults), the bones become soft and pliable, and deformities such as bow legs and spinal curvatures occur.

Bone is surrounded by an important fibrous layer, the periosteum, an active bone-forming organ that is responsible for the growth in the width of the bone. When this layer is torn across, as in a fracture, it acts as the major producer of the new bone that unites the ends. Occasionally, some of its bone-forming cells escape into the surrounding tissue, and unnecessary bone is formed outside the regular bone (heterotopic ossification). When near a joint, this extra bone can act as a bony block to the movement of that joint.

The articular surfaces of the extremities of the bone are coated with a layer of hyaline articular cartilage. The periosteum continues beyond the extremities as a fibrous capsule that surrounds the joint between the bone and its succeeding bone.

TYPES OF BONE: LONG BONES, SHORT BONES, FLAT BONES

Long bones consist of a shaft, or diaphysis, and two ends that are usually articular. The shaft consists of dense ivory-like bone, known as compact bone, which surrounds cancellous bone (also known as spongy bone). At each end, cancellous bone contains trabeculae, running at angles to each other in an open spongy network. In weight-bearing areas, these bars are laid down in definitive patterns to counteract the stresses to which the bone is subjected in that particular area. In the middle of the shaft beneath the compact bone is a marrow cavity, which contains either red or yellow marrow, depending on the age of the person.

At birth, all the bones are filled with the blood-forming red marrow. With age, there is a progressive replacement of the red marrow by yellow marrow, which is fatty and non-blood producing. The adult retains red marrow in the central bones (ribs, vertebrae, skull, sternum and pelvis). The long peripheral limb bones have fatty marrow in their shafts except for the two proximal bones (femora and humeri), which have a small amount of red marrow in the upper parts of their shafts. In the elderly, even this red marrow disappears. In certain blood-destroying diseases, such as leukemia, the peripheral bones remember what they once could do, and the red marrow re-forms.

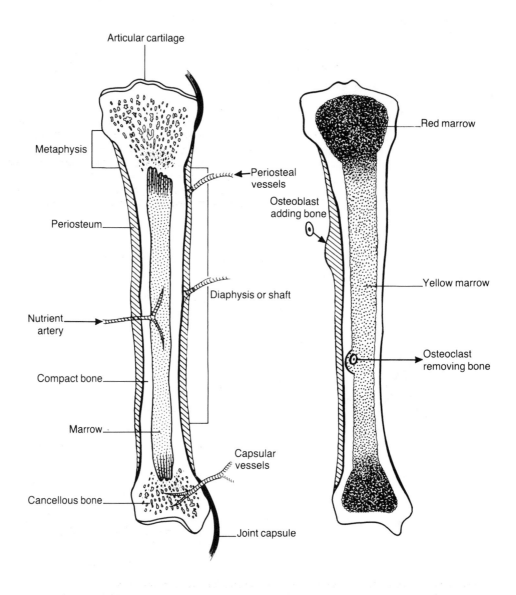

Fig. 2.1   A typical long bone. The metaphysis is that part of the shaft (diaphysis) close to the articular surface or epiphyseal disc before fusion.

NEUROVASCULAR SUPPLY

The dried-out bone that you are holding in your hand and staring at is anything but that in the living organism. The long bone is a highly vascular organ with a triple blood supply. The periosteum sends vessels from its deep surface into the bone. A nutrient artery, accompanied by veins, lymphatics and nerves, enters each bone through a foramen on the shaft of the bone, runs through a nutrient canal into the cavity and divides into branches which pass toward the ends of the bone. The direction of growth of a long bone can be deduced from the direction of the nutrient foramen in the growing bone. The third source of the blood supply is the capsular anastomosis around joints. Most bones have large holes near their extremities for the passage of these vessels (e.g. the neck of the femur).

Short bones (also known as irregular bones), such as vertebrae, consist of a thin

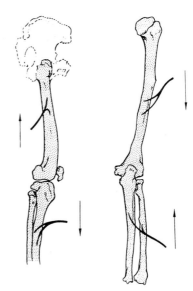

Fig. 2.2 Nutrient vessels to long bones of extremities. The direction of the nutrient vessels indicates the maximal growth site in the limbs. 'From the knee, I flee; to the elbow, I go' means that the most active growth occurs at the knee in the lower limb, and at the wrist and shoulder in the upper limb – the growing bone carries the vessels in the appropriate direction.

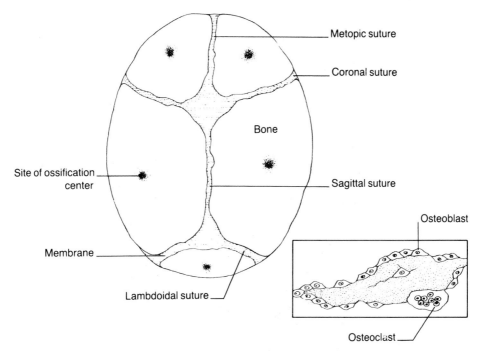

Fig. 2.3 Ossification in membrane. The original sites of the ossification centers in the membranous skull are seen. Bone has been laid down except for the stippled areas, which are still membrane. The inset depicts the osteoblasts involved in the laying down of bone, while on the surface a multinucleated osteoclast resorbs bone.

outline of compact bone surrounding cancellous tissue and red marrow. The blood supply is from the periosteum, which is also responsible for growth.

Flat bones have two layers of compact bone enclosing a layer of cancellous tissue. This is apparent in the skull, where the outer and inner tables enclose cancellous bone, which contains red marrow and a large number of venous channels known as the diploic veins. These veins are important because they are easily seen during x-ray examination, especially in older people, and may be confused with skull fractures. They also communicate with both intracranial and extracranial veins and may allow the dangerous spread of infections from outside the bone to the bone itself and to intracranial structures. The ribs and sternum are also good examples of flat bones.

BONE DEVELOPMENT (Figs. 2.2–2.5)

The earliest vestige of bone is a condensed mesoderm. This is replaced by cartilage, which forms a perfect model of the bone. The cartilage is then replaced by bone. In some instances there does not appear to be enough time for the mesoderm to pass through a cartilaginous phase; the mesoderm is transformed directly into bone. This certainly appears to be true in the case of the bones of the vault and sides of the skull and the face, where the rapidly developing brain needs more protection than that provided by cartilage. The clavicle and mandible (partially) appear to be in the same hurry. Ossification in these bones is rapid and simple, and they are often called membrane bones because they ossify in membrane and not cartilage.

Abnormal development of the membrane bones produces a broad skull, facial and dental anomalies, and the absence of part of the clavicles (cleidocranial dysostosis).

The formation of cartilage bone proceeds in the following manner. Initially, the membrane surrounding the cartilaginous model, the perichondrium, lays down bone from bone-forming cells (osteoblasts), which encircles the center of the shaft of the bone. At about the seventh or eighth week of fetal life, a center of ossification (primary center) appears in the center of the shaft of the bone and spreads toward each extremity. By birth, the shaft is all bone, except for the cartilaginous extremities. Secondary centers of ossification, called epiphyses, then appear in the cartilaginous extremities and grow toward the shaft until the shaft and epiphyses are separated only by plates of cartilage, known as the epiphyseal plates. The bone on the shaft side of this plate is often called the metaphysis. These cartilaginous epiphyseal plates cannot be seen on x-ray film because cartilage does not show on x-ray examination. It is very easy, therefore, for the unsuspecting doctor to confuse the epiphyseal plate with a fracture on x-ray examination. The epiphyseal plates ossify with the cessation of growth.

These secondary sites of ossification appear at definitive intervals throughout the skeleton. For those posing as gurus of anatomy, a knowledge of these dates is necessary and numerous mnemonics are for sale to achieve this feat of memory. This knowledge, however, is essential for orthopedic surgeons and pediatricians, so that variations of growth can be diagnosed and appropriate action taken (growth can be slowed by inhibiting the epiphyseal plate). A general rule of anatomy holds that the most active secondary centers of growth appear first and fuse last.

Living bone is continuously being laid down and resorbed. This remodeling, an ongoing process that is mostly under hormonal control, is partly responsible for stabilizing the serum calcium. Bone is very clever. Should a fractured bone angulate during repair, the stress is not carried along the angle, an undesirable state of affairs. New bone is laid down in the angle, which distributes the stress in the normal manner. Removal of one of two paired limb bones will lead to enlargement of the other. Conversely, disuse of a limb will lead to atrophy (like unused muscles); for example, the leg emerg-

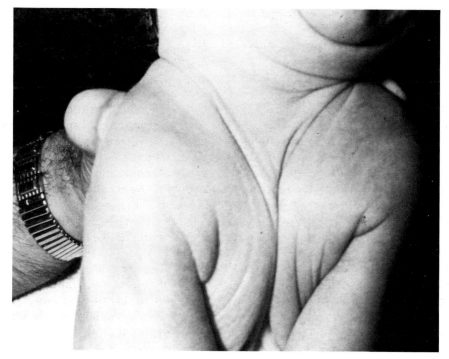

Fig. 2.4   An infant with absent clavicles.

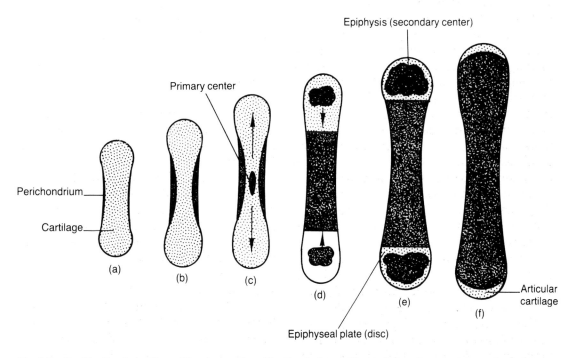

Fig. 2.5   Ossification in cartilage. The stages of ossification in the cartilage of a long bone are shown from the cartilage model (a) to the adult bone (f). The ends are covered by the remains of the cartilage, now subserving the function of the articular cartilage of a joint.

ing from a plaster cast is embarrassingly skinny.

Joints

There are three types of joints: fibrous, cartilaginous and synovial. At fibrous joints, the bones are joined by fibrous tissue; for example, the sutures of the skull or the teeth in the mandible. No movement occurs at these sites.

In cartilaginous joints, the bones are connected by a plate of cartilage and have only limited movement. Primary cartilaginous joints occur at the epiphyses while secondary joints appear solely in the midline; for example, the intervertebral discs and the symphysis pubis.

The synovial joint is the most commonly occurring type. The bony surfaces, covered by articular cartilage, are separated by a joint cavity and enclosed by a fibrous capsule lined by a synovial membrane. There are many varieties; for example, hinge as in the elbow joint, pivot as in the radioulnar joint, plane as in the carpal joints, and ball and socket as in the hip joint. Movement is always possible, and the range depends on the type of joint.

Synovial membrane

This vascular membrane lining the nonarticular components of the joint produces a rather viscous fluid that lubricates the joint and nourishes the articular cartilage. The articular cartilage is insensitive, mostly avascular, and deformable. Most of its nutrition is from the synovial fluid, but it does get some nourishment via diffusion from the vessels in the immediate vicinity. Cartilage is radiolucent, and joint space in radiologic language means the thickness of the articular cartilage of the two bone ends. Consequently, diminished joint space (abnormal approximation of the bones) means loss of cartilage, a frequent concomitant of arthritis.

Capsule

The two articular ends of the bone are enclosed by a fibrous capsule. Where stability is paramount (e.g. the hip) the capsule is taut, but where mobility is important (e.g. the shoulder) it is loose and baggy. The capsule has an excellent nerve supply, which conducts the sensations of pain and position.

Ligaments

Ligaments surround the joint and vary from mere thickenings of the capsule to substantial extracapsular or intracapsular structures with bony attachments. They not only serve to prevent mechanical disruption of the joints, but, like the capsule, also act as sense organs for the perception of movement and position.

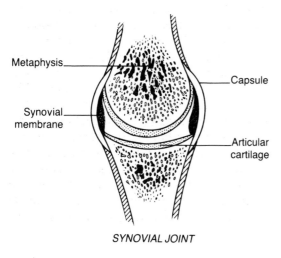

Metaphysis

Capsule

Synovial membrane

Articular cartilage

SYNOVIAL JOINT

CARTILAGINOUS JOINT

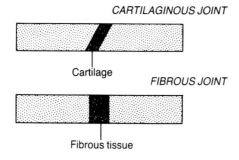

Cartilage

FIBROUS JOINT

Fibrous tissue

Fig. 2.6 The three main types of joints are shown: synovial, cartilaginous and fibrous.

Intra-articular structures

Intra-articular structures include the menisci or intra-articular discs, synovial folds and fat pads. The main function of these structures is to disperse synovial fluid throughout the joint, but, in some instances, they appear to act as shock absorbers. The fat pads are sucked toward the joint surfaces during movement to prevent the formation of a vacuum; sometimes a piece of the fat is pulled in excessively and nipped between the bony ends, causing severe pain.

NEUROVASCULAR SUPPLY

The good blood supply of the joint structures emanates from the periarticular anastomosis. The sensory nerves consist of pain fibers (anyone who has had his joints twisted knows how effective these are) and proprioceptive fibers. The latter dictate the position and movement of the joint and are essential in the reflex control of posture, walking and running. A patient with destruction of these fibers (e.g. in tabes dorsalis) wobbles all over the room while walking. A physician, while playfully twisting his young son's shoulder joint into odd positions, was heard to say, 'The nerves which supply the groups of muscles moving a joint also supply the skin over the insertion of these muscles.' The son cast a doleful look at him and said in simple terms, 'You mean the same nerve supplies the joint, the muscles moving the joint, and the skin over the joint?' The father responded with a savage twist, but agreed. This law was first enunciated by Hilton and is helpful clinically.

Tendons

Tendons are the fibrous ends of muscles, which often attach to bone. Sometimes, during their excursions, they play across a bony surface. To allay friction, they develop either a surrounding synovial sheath (inside which the tendon moves as if a piston in a cylinder) or a sesamoid bone, which articulates with the bone on which it is moving. The synovial sheath may become inflamed (tenosynovitis); the two layers of the sheath may then be heard to rub against each other to produce a creaking noise (crepitus).

Bursae

Bursae are synovium-lined sacs that may lie between tendons and bone, between ligaments, between two tendons, or where the bone lies against the skin (see Fig. 5.6). They facilitate movement between relatively rigid structures. In fact, a bursa will develop in response to circumstances where abnormal friction occurs. For example, a porter carrying heavy weights on his shoulder may develop a bursa over the acromion. The synovial lining secretes a small amount of fluid that diminishes any friction. Sometimes a large amount of fluid is secreted due to injury (acute traumatic bursitis) or, more commonly, chronic pressure causes a chronic bursitis (e.g. housemaid's knee, in which pressure occurs between the patella and the skin from scrubbing floors on one's knees).

Study outline

	Function	Neurovascular supply	Comments
Epithelial tissue (epithelium)	Protection, secretion, absorption, facilitation of movement in serous cavities	Epithelial cells have no blood vessels	Epithelial cells line all surfaces
Connective tissue	Supports or joins functional tissue	Blood supply sparse; well supplied with lymph vessels	Poor blood supply predisposes connective tissue to greater risk of destruction in presence of infection
Areolar tissue	Fills the space between organs, organ systems, muscles; acts as packing	Adequate blood supply	Amount of areolar tissue varies with the part
Fat	Supports and stabilizes organs, storehouse for energy; insulates and protects from heat loss	Relatively poor blood supply	Certain areas store fat; others dare not (see text). Poor blood supply of subcutaneous fat allows it to act as a heat insulator; the down side is its poor healing
White fibrous tissue	Provides strength without rigidity or elasticity; forms protective membranes or sheaths around certain structures	Reasonable blood supply	Forms ligaments and tendons
Elastic tissue	Provides flexibility to structures whose function requires stretching followed by a return to original shape	Adequate blood supply	Found in vocal cords, coats of arteries, ligamenta flava
Subcutaneous tissue	Attaches skin to deep fascia, loosely or firmly	Varies with the particular area	Consists of areolar tissue and fat; lies between skin and deep fascia

	Function	Neurovascular supply	Comments
Deep fascia	Protects limbs and body wall; partial origin of muscles; shapes extremities; plays a role in venous return in the extremities	Reasonable blood supply	Deep fascia is thin or absent where skin must be moved by direct insertion of muscles, such as the face or where organ expansion is necessary (e.g. abdominal wall)
Skeletal tissue	See specific types of skeletal tissue		
Cartilage	Provides tissues with strength, rigidity and elasticity	Relatively avascular	Chondroitin sulfate produces rigidity
1 Hyaline cartilage	Coats articular surface of joints; assists functions of respiratory system	Vascular supply varies; joint fluid supplies nutrition to avascular cartilage in joint; good supply in cartilage of respiratory system	Primary component of nonossified cartilage; tends to calcify and ossify
2 White fibrocartilage	Supplies great strength and rigidity with elasticity	Reasonable blood supply	Component of intervertebral discs, articular discs; may calcify or ossify
3 Elastic fibrocartilage	Provides flexibility and strength	Reasonable blood supply	Does not calcify or ossify

Types of bone

1 *Long bone*: composed of fibrous connective tissues and inorganic salts.

Shaft or diaphysis is the middle part of the bone. It consists of compact bone, enclosing a marrow cavity, which contains either red or yellow marrow (e.g. femur, ulna).

The periosteum, a layer of connective tissue surrounding the bone, has bone-forming properties.

Vascular supply: from the periosteum, through the nutrient foramen in the bone, and from the capsular anastomosis around joints.

2 *Short bone or irregular bone*: consists of thin layer of compact bone surrounding cancellous tissue and red marrow (e.g. vertebra).

Vascular supply: from the periosteum.

3 *Flat bone*: consists of two layers of compact bone enclosing cancellous tissue and red marrow (e.g. ribs, sternum, tables of the skull).

Development of bone

1 *Membrane bones*: mesoderm is transformed into bone directly (e.g. bones of the vault, and sides of skull and face).

2 *Cartilage bones*: mesoderm is replaced by cartilage, which is then replaced by bone – endochondral ossification.

Joints

Point of union of two or more bones. There may or may not be movement at this junction.

Types of joints

1 *Fibrous joint*: bones are joined by fibrous tissue (e.g. a skull suture). There is no movement.
2 *Primary and secondary cartilaginous joints*: bones are joined by cartilage plate. Movement is limited.
3 *Synovial joint*: bones are separated by a cavity and enclosed by a fibrous capsule, which is lined by a synovial membrane. Movement is variable.

Synovial membrane

Vascular membrane, which produces fluid that lubricates the joint and nourishes the articular cartilage.

Capsule

Fibrous capsule encloses the two articular ends of the bone. Well supplied with afferent nerves.

Ligaments

Surround and protect the joint; vary in thickness and structure. Act as sense organs, affecting movement and position.

Intra-articular structures

These include menisci, synovial folds and fat pads. Their functions are to disperse synovial fluid throughout the joint and to absorb shock.

Tendons

Fibrous ends of muscles usually attaching to bone. Where the tendon is subject to friction, it may be enclosed by a synovial sheath or may contain a sesamoid bone.

Bursae

Synovium-lined sacs that alleviate the friction between two structures.

Organ systems

Skin

Skin is more than a mere envelope covering the whole body. Skin protects the deeper tissues, controls the body temperature to a large degree, and acts as an absorptive and excretory organ. The terminal fibers of extremely sensitive nerves end in this layer. Modified skin is present wherever durable material is necessary. It lines the terminal parts of those ducts that open externally and are liable to trauma (e.g. the nose, ear, rectum, vagina and urethra).

The skin is composed of two layers: a superficial avascular epidermis covering a vascular dermis. The epidermis is extremely thick in certain areas, such as the palms of the hands and the soles of the feet. This is partly a response to intermittent pressure. However, since even a newborn infant has thickened epidermis in these parts, this is not the complete reason – tissues destined to sustain pressure are naturally adapted for this function.

The dermis, also known as the corium, is a thick vascular layer with a profuse nerve supply. It is inserted into the epidermis by a series of projections, called papillae, that act as pegs. These pegs prevent the epithelium from being stripped off the dermal layer by a shearing stress. Blood vessels of the dermis impart some of the color to the skin, and a refined nerve supply reaches its zenith in the highly sensitive fingertips or lips (think what happens to your lips when kissing on a very cold day).

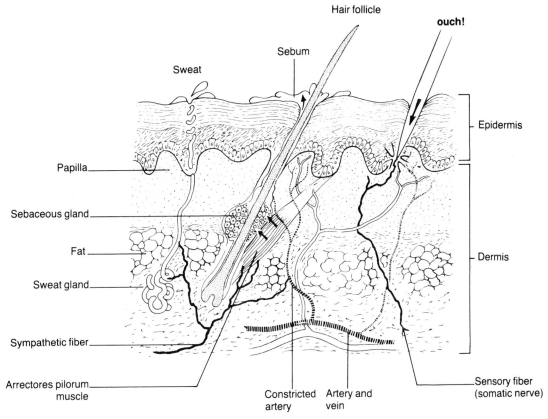

Fig. 2.7 The skin. The dark continuous lines represent nerves – on the left the sympathetic nerves supply the sweat gland (note sweat secretor), the arrectores pilorum and the blood vessels (note vasoconstrictor); on the right the sensory nerve is stimulated by a needle during an injection (ouch!). Contraction of the arrectores pilorum muscle straightens the hair (making it stand on end) and also squeezes and empties the sebaceous gland between the hair and the muscle (arrows) (note excretion of sebum).

The appendages of the skin are the nails, hairs, and sweat and sebaceous glands. The nail, a thickened and horny modification of the epidermis, is firmly attached to the underlying dermis. The nail grows from its proximal root forward – a removed distal part of the nail is replaced by proximal regrowth. This is surprisingly slow, with nail replacement usually taking about 3 months.

Hairs project into the dermis and are present all over the body, except for the palms, soles, the dorsal surfaces of the distal phalanges and the glans penis. Small smooth muscles, the arrectores pilorum, insert into the obliquely lying hair follicles and straighten them on contraction. These little muscles are under the control of a specialized part of the nervous system (the sympathetic division), and respond to numerous stimuli (e.g. a hair-raising moment in a horror movie or the goose-pimples of a freezing runner).

Sebaceous glands are small sacculated glands that lubricate the skin and are absent from the palms and soles (slippery grip and walking as if on banana skins are thereby avoided). These glands are especially numerous on the nose and face, where they may become large enough to become cosmetic disabilities. The glands are partly emptied by the action of the little arrectores pilorum muscles, which squeeze the glands against the hair follicles.

Sweat glands are found all over the body in the dermal layer. They usually are a single tube, although they may become coiled. Areas of maximal perspiration harbor the greatest number of glands (e.g. the axilla, soles and palms, where the moistened surfaces prevent friction). These are important excretory organs, especially with regard to the elimination of sodium chloride. (Tennis players replace the loss of the electrolyte-containing sweat with salt tablets, and people living in the tropics often add salt to their drinks.)

Collagen fibers of the dermis are arranged in parallel bundles; surface cleavage lines may indicate the direction of these fibers. These lines, also known as Langer's lines, are arranged in a definite manner. In general, the lines run transversely on the trunk and vertically on the limbs. Opposite the joints, the creases are extremely well formed. On the face, they can be best located by pinching the skin together. Cleavage lines are important to the surgeon because an incision that separates these parallel bundles of collagen fibers without rupturing them heals with a fine line, whereas an incision severing and disrupting them produces a broad scar.

Muscle

Muscles are classified as striated, cardiac and smooth. Striated muscle has cross-striations through its fibers; cardiac muscle has cross-striations and branching fibers that join to form indefinite cell boundaries (syncytium); and smooth muscle has no striations. Striated muscle is innervated by the somatic (voluntary) division of the nervous system, whereas cardiac and smooth muscle are served by the autonomic (involuntary) division. For this reason, striated muscle is often called voluntary, or under a person's control, whereas cardiac and smooth muscle are involuntary, or beyond the person's control. However, this is not entirely accurate because some striated muscles cannot be controlled (e.g. the striated muscle of the upper portion of the esophagus cannot be made to contract). It is rare for one muscle to carry out a particular action; other muscles assist (synergists),

relax the opposing muscles (inhibitors) or oppose the movement (antagonists). For example, to place the hand in a card-dealing position, muscles acting on the shoulder and elbow joints contract (synergists), while their antagonists relax. An antagonist may also contract to prevent an undesirable action of a multiaction muscle (e.g. the biceps brachii, see p. 79)

Striated muscle is found throughout the body and is popularly called flesh. Its contraction is characterized by a rapid response to a stimulus. Smooth muscle, found in the gastrointestinal tract and the vascular system, contracts in a more leisurely, sustained fashion. Cardiac muscle responds extremely rapidly and with considerable force. Cardiac and smooth muscles are able to contract spontaneously. The rate and degree of contraction is regulated by both the autonomic nervous system and various chemical substances.

Skeletal muscles

Skeletal muscles have fanciful names, which depend on their shape, position and action. The direction of the muscle fibers is related to the type of action of the muscle. Muscle fibers may run in parallel fashion along the length of the muscle, or they may insert into one side of a tendon (unipennate), into both sides of the tendon (bipennate) or consist of series of bipennate masses (multipennate). This may allow the surgeon to identify a particular muscle and to establish the anatomic location. The arrangement of the fibers influences the action of the muscle: parallel fibers pull in a longitudinal direction, and unipennate fibers pull obliquely. The latter movement is inefficient, but the larger number of oblique fibers, which fill the same volume of a parallel-fibered muscle, results in an overall increase in the power of the muscle. In parallel-fibered muscles, range of motion supersedes power; in oblique-fibered muscles, power is more important than range.

Origins and insertions

The terms 'origin' and 'insertion' are really inconsequential because a muscle can act from either attachment, depending on which end is fixed. Nevertheless, traditionally the proximal attachments are regarded as origins and the distal attachments as insertions. Musculotendinous insertions often cause bony tuberosities.

Blood supply

Muscles receive a substantial blood supply, varying from a single artery running throughout the muscle to several arterial branches entering along the length of the muscle. Since muscle transposition operations are common, a knowledge of the muscle's blood supply is important.

Nerve supply

Muscles are innervated by nerves that may arise from one or several spinal segments. Muscles having similar actions are often supplied by the same nerve. A nerve that runs through a muscle in the limbs always supplies the muscle. It is common but not invariable for the nerve to enter at a regular point known as a motor point. These motor points are dearly beloved by those who work in the field of electrical stimulation. By stimulating the nerve at the motor point, it can be determined whether the muscle is alive and well, degenerating or recovering. Almost half of the content of a nerve supplying a muscle is composed of sensory fibers that mediate proprioceptive impulses, which are vital to purposeful and coordinated muscular contractions. The nerve supplying the muscles also supplies the tendons, and often the capsule of the neighboring joint. Loss of nerve supply leads to atrophy and eventual replacement of muscle by fibrous tissue and/or fat. The motor nerve supplying a muscle may be stimulated by impulses re-

layed from the motor cortex of the brain to produce primary movement of the part, or by local action in which the stimulating impulse is initiated at a lower level without reaching higher controlling nervous centers.

Nervous system (Figs. 2.8–2.19)

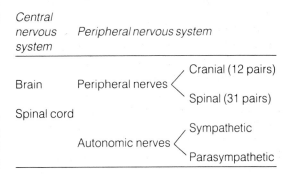

The nervous system is composed of central and peripheral portions. The central system is represented by the brain and its inferior extension, the spinal cord. The peripheral system is made up of branches from the brain and spinal cord. The branches from the brain are called cranial nerves, of which there are 12 pairs, and those from the cord are termed spinal nerves, of which there are 31 pairs. Both cranial and spinal nerves are partly controlled by fibers which originate at a higher level in the brain, the so-called higher centers.

Structurally, the nervous system is composed of impulse-conducting nerve cells, called neurons, and nonconducting supporting cells and ground substance that assist the neuron. The neuron consists of a cell body containing a nucleus and surrounding cytoplasm with cytoplasmic extensions. The longest extension is the axon, which conducts impulses away from the cell body. More numerous shorter processes, called dendrites, carry impulses toward the cell body. Axons communicate with one another at junctions known as synapses, through the release of chemical substances. Some axons are ensheathed by membranes derived from the supporting cells and are called myelinated fibers; those lacking this cell layer are called

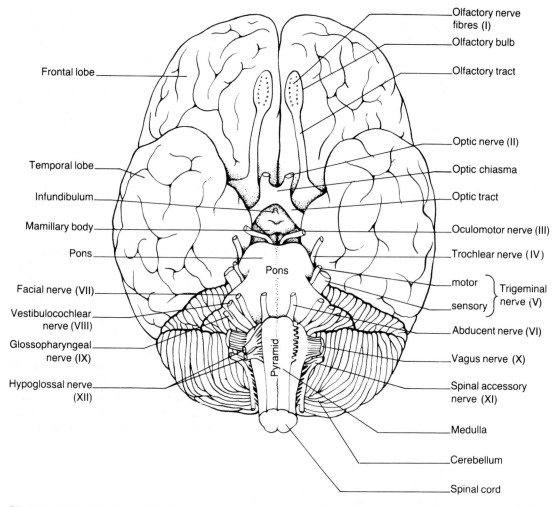

Olfactory nerve fibres (I)
Olfactory bulb
Olfactory tract
Frontal lobe
Optic nerve (II)
Temporal lobe
Optic chiasma
Infundibulum
Optic tract
Mamillary body
Oculomotor nerve (III)
Pons
Trochlear nerve (IV)
Pons
motor ⎱ Trigeminal
sensory ⎰ nerve (V)
Facial nerve (VII)
Vestibulocochlear nerve (VIII)
Abducent nerve (VI)
Glossopharyngeal nerve (IX)
Pyramid
Vagus nerve (X)
Hypoglossal nerve (XII)
Spinal accessory nerve (XI)
Medulla
Cerebellum
Spinal cord

Fig. 2.8 A view of the base of the brain and brain stem.

unmyelinated fibres. Fibers bundled together that have the same origin, function and termination are called tracts. Concentrations of neuronal cell bodies inside the central nervous sytem are called nuclei, whereas those outside are termed ganglia.

Spinal cord

The spinal cord is one of several structures 45 cm (18 in) long (the rest are to come). It extends from the lower end of the medulla oblongata (the lowermost portion of the brain stem), immediately below the foramen magnum of the skull, down the vertebral canal, to the lower border of the first lumbar vertebra. In the 3-month-old fetus, the spinal cord extends to the coccyx, but it becomes proportionately shortened with the more rapid growth of the vertebral column relative to the spinal cord. In the child, it reaches the level of the third lumbar vertebra, and, in the adult, the lower border of the first lumbar vertebra. It is important to remember this when performing a lumbar puncture on a small child.

The spinal cord has two enlargements caused by a profusion of cells. These enlargements indicate the origin of the large nerve trunks for the upper and lower extremities.

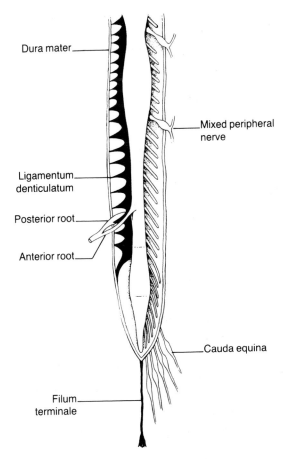

Dura mater

Mixed peripheral nerve

Ligamentum denticulatum

Posterior root

Anterior root

Cauda equina

Filum terminale

Fig. 2.9   Spinal cord, nerve roots and ligamenta denticulata. The spinal cord, with its cervical and lumbar enlargements, is attached by the ligamenta denticulata to its surrounding dural envelope. The filum terminale attaches the cord to the coccyx. The peripheral nerves are seen perforating the coverings of the cord.

The cervical enlargement lies opposite the lower part of the cervical spine, and the lumbar enlargement is related to the lower part of the thoracic spine.

Structurally the spinal cord is composed of an H-shaped arrangement of gray matter, consisting mostly of nerve cells, surrounded by white matter. Arising from the anterior and posterior ends of the 'H' (often called horns) by a row of rootlets are the anterior and posterior nerve roots. The cord is surrounded by the same three membranes as the brain: from within outwards, the flimsy pia mater, the spidery arachnoid mater and the firm dura

mater. The roots carry a sleeve of these membranes with them to the intervertebral canal (Fig. 2.10), where the dural envelope becomes continuous with the periosteum of the canal and the nerve divides into anterior and posterior primary rami. Because of this prolongation, the subarachnoid space, with its contained cerebrospinal fluid, reaches the intervertebal foramen (see p. 32). Injections to block and anesthetize the rami at this site sometimes penetrate the dura and enter the subarachnoid space with serious results, including damage to the spinal cord.

The lower portion of the spinal cord ends conically, with a prolongation of the enclosing pia mater, the filum terminale, extending distally to attach to the coccyx. This is one of the anchoring mechanisms of the spinal cord. Other anchors are thickenings of the pia mater, the denticulate ligaments, which project laterally like a series of teeth that fuse with the dura mater. The anterior and posterior roots pass laterally in front and behind these ligaments. Below the termination of the cord at the first lumbar vertebra, the vertically running roots and the central filum terminale constitute the cauda equina (horse's tail).

BLOOD SUPPLY

The blood supply of the spinal cord is the subject of much research. Basically, the cord is supplied by the anterior and posterior spinal arteries, which run downward along the length of the cord. These arteries arise from blood vessels near the base of the brain. The anterior and posterior spinal arteries are reinforced by branches of the segmental trunk arteries that pass through the intervertebral foramina, from the atlas to the sacrum. Those vessels supplying the cervical and lumbar enlargements are expectedly large. They may be seen on specialized radiographs.

Interference with the blood supply to the cord has disastrous results, and the vagaries of the anastomoses in each person prevent any prediction of the seriousness of an inter-

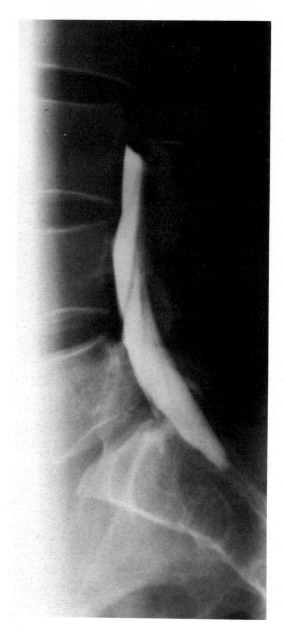 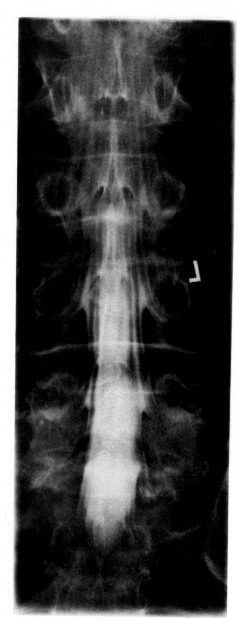

Fig. 2.10   Lumbar myelogram – erect lateral view and water-soluble radiculogram. (From Weir and Abrahams 1978)

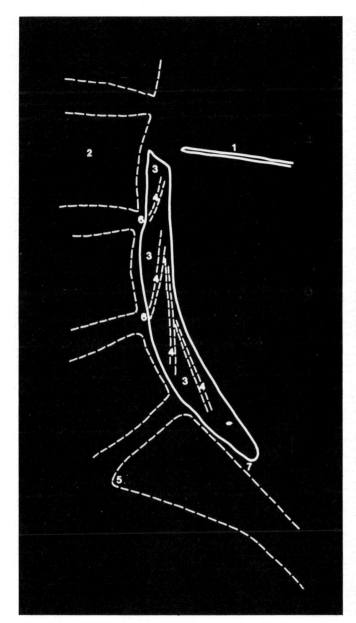
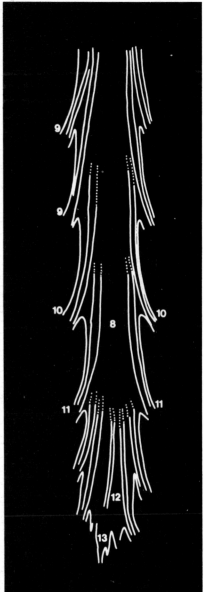

1 Lumbar puncture needle in the L2/L3 space.
2 Third lumbar vertebral body.
3 Contrast medium in subarachnoid space.
4 Spinal nerves within subarachnoid space (cauda equina).
5 Sacral promontory.
6 Normal intervertebral disc indentations on anterior thecal margin.

7 Terminal thecal sac at S1/S2.
8 Subarachnoid space at the level of L4.
9 Lateral extensions of subarachnoid space around spinal nerve roots.
10 Third lumbar spinal nerve.
11 Fourth lumbar spinal nerve.
12 Cauda equina.
13 Caudal thecal sac.

ruption. For example, the aorta is sometimes purposely occluded during surgery. Some of the blood supply to the spinal cord emanates from the aorta's segmental branches. If the anastomosis is heavily dependent on these branches, the results may be catastrophic, with loss of part of the cord's functions.

The spinal veins form plexuses, which drain to the segmental veins of the body and into the vertebral venous plexus which runs the length of the vertebral column (see Fig. 35.8).

Cranial and spinal nerves

CRANIAL NERVES

The first cranial nerve, the olfactory nerve, originates in the upper nasal cavity and passes intracranially to reach the olfactory tract of the brain (see Fig. 15.2). The second cranial nerve, the optic nerve, is really an extension of the brain which it reaches via the optic tract (see Fig. 18.4). Nerves III–XII arise in the brain stem which consists of the midbrain,

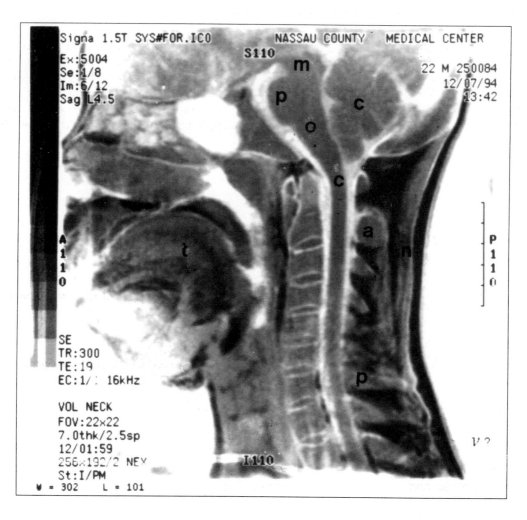

Fig. 2.11 MRI examination of the head and neck demonstrating the area of origin of cranial nerves III–XII. m, midbrain; p, pons; o, medulla oblongata; c, spinal cord; c, cerebellum; t, tongue; a, axis; p, vertebra prominens (the longest cervical spinous process); n, ligamentum nuchae.

pons, and the medulla oblongata (Fig. 2.11).

See Chapters 13 and 18 for discussion about the cranial nerves.

SPINAL NERVES

These nerves function through their terminal divisions, the anterior and posterior rami. The posterior primary rami run backward to supply all the extensor muscles of the vertebral column and most of those of the skull. Sensation is also supplied to the posterior skin from the vault of the skull to the upper part of the buttock. The area of skin served reaches as far forward as the posterior axillary line. The anterior primary rami are much more extensive. They supply all the sensation to the skin of the front and sides of the neck and trunk, and the extremities. Concerning motor activity, they innervate the flexor muscles of the spine as well as the body wall in a fairly segmental manner.

The motor and sensory innervation of the limbs are somewhat special (Fig. 2.14). Firstly, they are served by the anterior rami exclusively, and, secondly, the nerves anastomose with one another, forming networks called plexuses, which supply the extremities. This allows each segment of the cord to have a bigger say in its distribution of impulses. These plexuses divide into anterior and posterior divisions, the anterior divisions being responsible for flexor and the posterior for extensor functions.

Reflex arcs

When a baby picks up a very hot object, he quickly retracts his hand. When the same object is offered to the baby again, he withdraws his hand before or at the instant he sees or feels the hot object. The infant has developed a response that will be repeated

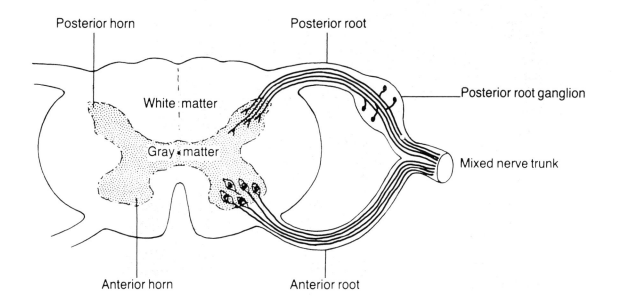

Fig. 2.12 Spinal cord and nerve roots. Cross-section of spinal cord, showing the H-shaped gray matter surrounded by the white matter. The anterior root, made of axons (fibers), arising from neurons in the anterior horn of the gray matter, joins with fibers of the posterior root, to form a mixed nerve trunk.

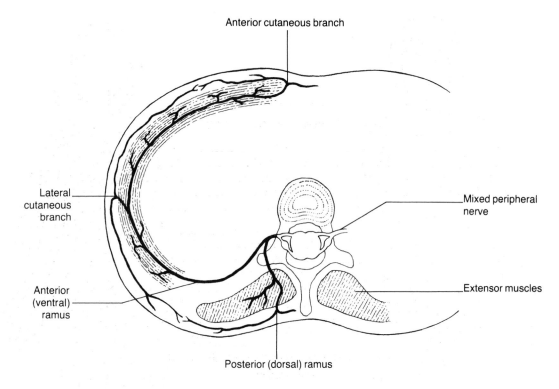

Anterior cutaneous branch

Lateral cutaneous branch

Anterior (ventral) ramus

Mixed peripheral nerve

Extensor muscles

Posterior (dorsal) ramus

Fig. 2.13 Distribution of primary rami. The posterior rami supply most of the extensor muscles of the skull, and all those of the spine and the overlying skin, while the anterior rami supply the anterolateral muscles and overlying skin.

under the same conditions. When an adult backs into something very hot, he or she will immediately jump away before the pain is even felt. These actions are too rapid for impulses to pass up to the higher centers and back again; it is a reflex action conducted at the local level in the spinal cord. The afferent axon (or sensory fiber) passes into the posterior root from its ganglion cell (the neuron of the sensory axon) and enters the posterior horn of the gray matter. Here the axon synapses with a cell whose axon (connecting axon) then passes to a motor cell in the anterior horn of the gray matter, which institutes the necessary action. This type of nervous system activity is called a local reflex arc, with an afferent axon, a connecting axon and an efferent axon as its primary components. If it were necessary for the higher centers to learn of the information coming in, the sensory

fiber would synapse and send another fiber superiorly.

Dermatomes

Dermatomes are areas of skin innervated by one segment of the spinal cord (see Fig. 2.14). Except for the limbs, dermatomes are arranged fairly segmentally. Neither the anterior or the posterior ramus of the first nerve (C1) usually reach the skin, expending themselves in supplying muscles. Both rami of the second, third and fourth cervical nerves (C2, 3, 4) supply that part of the head and neck not supplied by the fifth cranial nerve, and share that territory fairly equally. The upper limb is supplied by successive anterior rami, from the fifth cervical nerve to the first thoracic nerve (C5–T1). Posterior rami of C5–T1 end in muscles and fail to reach the

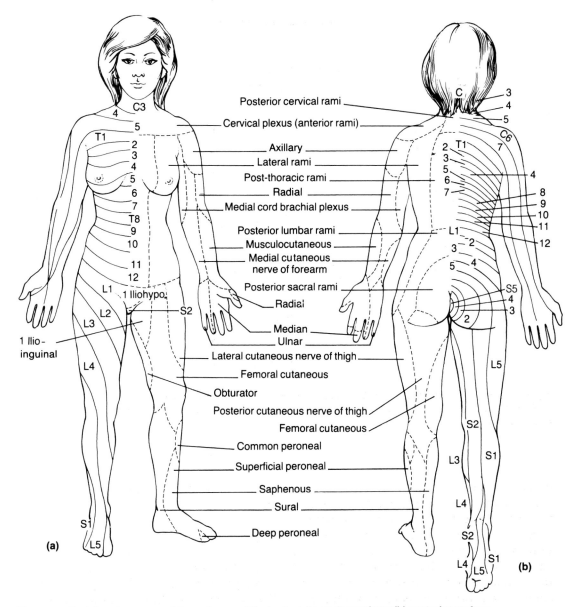

Fig. 2.14 Peripheral nerves and dermatomes of the body: (a) anterior surface; (b) posterior surface.

skin. The trunk is supplied on a fairly regular horizontally disposed sequential basis from T2 to T12. The lower limb is supplied by anterior rami of L2–S2 with the posterior primary rami confining themselves to the sacral and upper gluteal areas.

The overlap between dermatomes is considerable. Thus, there is no noticeable effect when small nerves are severed during surgery. A knowledge of these dermatomes is extremely useful to the clinician investigating pain.

Myotomes

A myotome is a group of muscles innervated by one segment of the spinal cord. As a single muscle seldom functions individually (see p. 28), myotomes are better thought of in terms of movements. Muscles are supplied by one or usually two segments, which is the same for muscles that share the same primary action on a joint. Muscles opposing that movement are supplied by segments which numerically follow those which supply the prime movers; for example, flexion of the elbow joint is mediated by spinal segments C5 and 6, while extension is the responsibility of C7 and 8. The more distal the joint, the lower the responsible segment of the cord; for example, the knee joint is extended by L3 and 4 segments and flexed by L5 and S1 segments. The ankle joint is extended (dorsiflexed) by L4 and 5 segments and flexed (plantar flexed) by S1 and 2 segments. These statements vastly oversimplify the position, but do indicate the underlying principles governing the supply of muscles by the spinal centers for joint movements. Summarized briefly, it is reasonably accurate to say that spinal centers for each large limb joint tend to occupy four continuous segments of the cord.

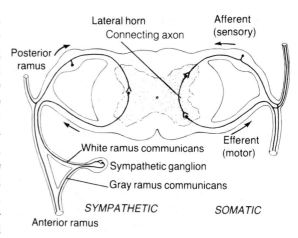

Fig. 2.16   Somatic and sympathetic systems. The right side of the cord shows the cells and fibers constituting a reflex arc. On the opposite side, the sympathetic efferent fiber originates in a cell in the lateral horn, and passes via the anterior root, mixed nerve, anterior ramus and white ramus to a sympathetic ganglion where it synapses; the succeeding fiber passes via a gray ramus communicans to rejoin the anterior ramus. The sensory sympathetic fiber enters the posterior root where its ganglion cell is located.

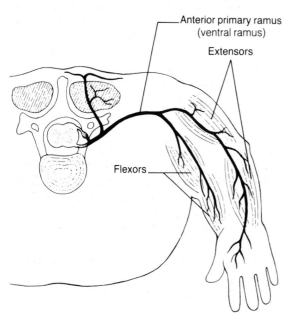

Fig. 2.15   Distribution of primary rami. The limb being totally supplied, both motor and sensory, by the anterior ramus. The anterior division of the latter supplies mainly the flexor portion, and the posterior division the extensor portion.

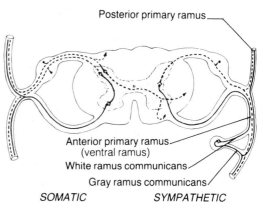

Fig. 2.17   Somatic and sympathetic systems. On the left-hand side, sensory fibers (broken lines) pass into the posterior horn where they (1) pass up in posterior columns, (2) synapse and cross to reach the opposite side where they pass into ascending tracts, (3) synapse with a connector fiber (solid line), which in turn synapses with a motor cell in the anterior horn, constituting part of the reflex arc. On the right side lie the sensory (broken line) and motor fibers (solid line) of the sympathetic system; the connector neuron lies in the lateral horn, and its fiber synapses in the sympathetic ganglion, reaching it via a white ramus communicans and returning to the anterior ramus via a gray ramus communicans.

Autonomic nervous system

The spinal and cranial nerves discussed so far are involved with common sensation, such as touch, pain, temperature, sense of position and vibration sense, and with motor activity, supplying the skeletal muscles, which are under voluntary control. All the nonstriated smooth muscle of the body – including that present in the walls of all the long tubes (e.g. the vascular tree and the alimentary, respiratory and genitourinary tracts), the striated involuntary muscle of the heart, and all the secreting glands present in the various tracts and in the skin – are supplied by the autonomic nervous system.

This system consists of two subgroups: the sympathetic and the parasympathetic systems, which are mostly antagonistic. The parasympathetic nervous system produces a sense of contentment. You are sitting in a comfortable armchair, listening to soothing music and munching a supply of never-ending delicacies. The parasympathetic system is looking after you – the pulse is slow, the glands are producing the necessary juices, the various viscera are obeying themselves. Suddenly the telephone rings and you are informed that the anatomy exam has been rescheduled for an earlier date. The pulse races, your hair stands on end, you sweat, and the viscera stop working. The sympathetic nervous system is responsible for this angry and frightened state – the state of 'flight or fight'.

The autonomic nervous system is often called the involuntary nervous system, and, to a large degree, this is true. However, there are controls at different levels. The highest level of control is the cerebral cortex. For example, a full bladder initiates a desire to micturate, and you start looking around for a toilet. However, when your companions describe a thrilling play in the World Series or World Cup, the cerebral cortex sends down impulses that inhibit the sensation of fullness and you go on merrily discussing the umpire's mistake and forget about your bladder.

The next level of control in the autonomic system is the hypothalamus, which is a portion of the base of the brain, that is particularly important in the control of body temperature. For example, when you step out of a warm house into the snow, the vessels of the skin immediately constrict to prevent heat loss; the skin becomes pale and cold. The hypothalamus also influences the next control level, the brain stem, which is concerned with the vital functions of respiration and circulation, although these functions often act reflexly through impulses that do not reach the higher centers.

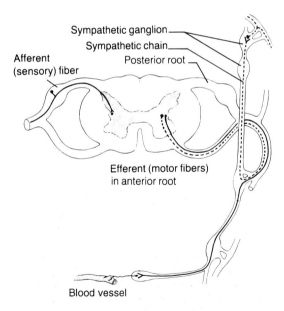

Fig. 2.18 Distribution of sympathetic fibers. The afferent fiber is seen on the left. On the right two fibers are seen originating in the lateral horn; one fiber (broken line) passes up the sympathetic chain to synapse at a ganglion higher up in the chain, while the other (solid line) passes down the chain and branches off to reach a ganglion outside the chain (e.g. celiac ganglion) where it synapses before being distributed to an artery.

SYMPATHETIC SYSTEM (Figs. 2.16–2.19)

Efferent outflow The anterior and posterior horns in the gray matter of the spinal cord have already been described. In the thoracic and upper lumbar regions (T1 to L1, 2), an

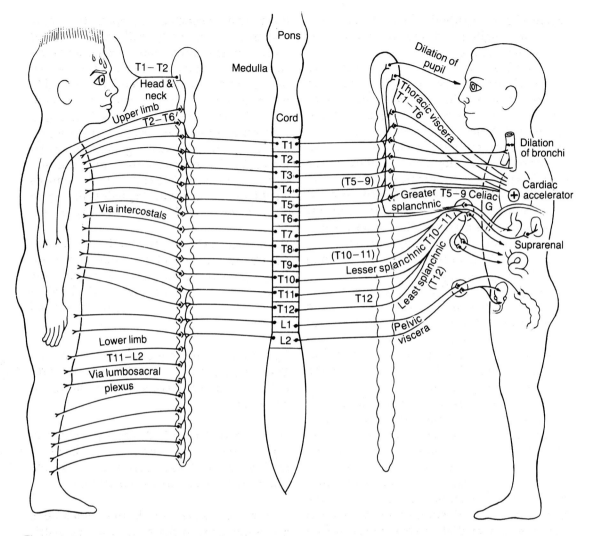

Fig. 2.19 The sympathetic system. The sympathetic outflow from T1–L2 segments of the cord is seen in the center. On the left, outgoing fibers synapse in the chain and supply the skin and its appendages. On the right, the outgoing fibers synapse partly in the chain (especially the cervical ganglia), but many pass through to synapse either in outlying ganglia (e.g. celiac ganglia) or even in the viscus itself (e.g. suprarenal).

additional horn, the lateral horn, represents the cell bodies for the sympathetic outflow.

Efferent fibers   Efferent axons pass through the anterior root, the mixed nerve and the commencement of the anterior ramus, before leaving the ramus for a ganglion situated anteromedial to the ramus. The ganglia lie in a paravertebral position in close relationship to the anterior primary rami as they emerge

from the intervertebral foramina. Reaching from the level of the atlas to the coccyx, the ganglia are strung together by fibers constituting a continuous sympathetic chain or trunk. Some of the fibers synapse in the ganglion of the same nerve and leave the ganglion to pass back to the anterior ramus for distribution through both anterior and posterior rami. The anterior ramus therefore has two connections with the ganglion – one where the sympathetic fiber passes from ramus to ganglion and one where it returns from ganglion to ramus.

Some fibers, however, do not synapse in the ganglion but pass through it, running superiorly or inferiorly to relay in another ganglion of the chain from which they are distributed. Finally, many fibers, especially those destined for the viscera, do not synapse in the chain at all. They pass from the chain as branches that relay in ganglia lying well away from the trunk and are distributed with the blood vessels. So each fiber has three choices: it can relay in its own ganglion of the chain; it can run up and down the chain and relay in another ganglion; or it can leave the chain to relay in outlying ganglia.

The ganglia usually correspond to the spinal nerves to which they are related. In the neck, however, the ganglia fuse – the superior cervical ganglion (C1–4), middle cervical ganglion (C5, 6), and an inferior cervical or stellate ganglion (C7, 8, and often T1). In the thoracic region, the chain is situated on the necks of the ribs proximally, but, with descent, it passes medially to lie on the sides of the bodies of the vertebrae. The thoracic chain usually has 12 ganglia. In the abdomen, where there are four or five ganglia, depending on possible fusions, they lie on the sides of the vertebral bodies. In the pelvis, the ganglia are placed anterior to the sacrum opposite and medial to each intervertebral foramen. The chains finally fuse at the coccyx to form the ganglion impar.

An efferent axon, arising from a cell in the lateral horn and passing to the ganglion, is myelinated, and is referred to as a preganglionic or white ramus communicans fiber. The axon leaving the ganglion (known as a postganglionic fiber) may be nonmyelinated and gray, after synapse, or white should it synapse more distally. The ganglia from T1–L2 receive white rami communicans only, whereas those ganglia above and below these levels receive fibers that begin as white rami but may have synapsed in various ganglia and therefore may be either white or gray.

Body wall branches Sympathetic nerve branches for the body wall are distributed mostly by the spinal nerves. Each nerve receives a gray ramus that has already synapsed in one or another ganglia and has returned to the spinal nerve for distribution by both rami. These efferent fibers reach the skin where they are vasoconstrictor for the vessels, sudomotor for the sweat glands and pilomotor for the muscles (arrectores pilorum). Sympathetic stimulation therefore produces a cold sweaty skin with the hair standing on end – the reaction of fear. (See also Fig. 2.7.)

Head and neck Preganglionic fibers arise from T1–2 segments of the spinal cord. After relaying in the three cervical ganglia, the fibers are distributed by the cervical spinal nerves and, to a lesser extent, by cranial nerves.

Upper limb Preganglionic fibers arise from T2–6 segments and ascend to join the brachial plexus, the nerve supply of the upper extremity.

Lower limb The spinal cord segments supplying the lower limb are T11–L2. The fibers arising from these segments join the lumbosacral plexus for their distribution.

Visceral branches While the sympathetic nerve branches to the body wall are distributed largely by the spinal nerves, those for the viscera are mostly distributed with the blood vessels. These nerves often pass through a synapse in secondary or subsidiary ganglia before reaching the vessels for their final distribution.

Head and neck After relaying mainly in the superior cervical ganglia, the fibers are distributed with the carotid and vertebral arteries to the dura mater, cerebral vessels and the pupil. The fibers acting on the pupil arise from the first thoracic segment of the spinal cord only.

Thorax Preganglionic fibers for the heart

and lungs arise from approximately T1–6 segments, and, for the esophagus, from T4–6 segments. The cardiac fibers take an apparently rather strange course in that they first ascend to the three cervical ganglia, where they relay, and then descend as cervical cardiac nerves into the mediastinum; as the heart is a cervical structure in the embryo, this is not too surprising. The esophageal and pulmonary plexuses also descend but only from the upper thoracic ganglia. All these fibers have synapsed in the sympathetic ganglia.

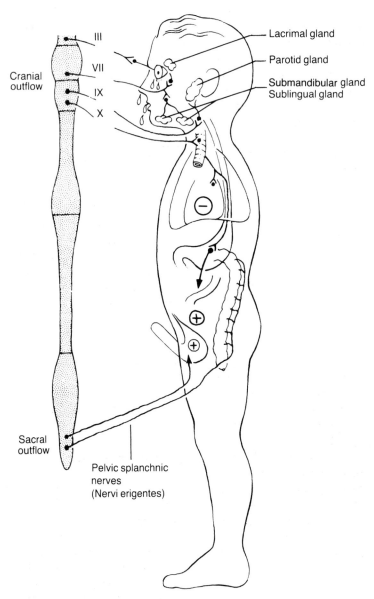

Cranial outflow

III

VII

IX

X

Lacrimal gland

Parotid gland

Submandibular gland
Sublingual gland

Sacral outflow

Pelvic splanchnic nerves
(Nervi erigentes)

Fig. 2.20   The parasympathetic system with its cranial and sacral outflow. The outflow from the brain stem passes via nerves III, VII, IX and X. The fibers in the first three synapse in outlying ganglia, but those in X, as well as those in the sacral outflow, synapse in or near viscera. Some of the actions are depicted: constriction of the pupil, lacrimation, salivation, slowing of the heart, stimulation of the gastrointestinal tract, penile erection and emptying of the bladder.

Abdomen The fibers subserving the abdomen arise from T5–12 segments and are known as the splanchnic nerves. The three splanchnic nerves run from the sympathetic trunk inferiorly and medially to pierce the diaphragm, and join intra-abdominal ganglia and plexuses. These splanchnic fibers are preganglionic (white) and they mostly relay in the celiac ganglia (see p. 419).

The spinal cord segments supplying the abdominal viscera are arranged approximately according to the anatomy of the abdomen: the structures of the upper abdomen are supplied by the higher nerves; the middle abdomen by the middle nerves; and the lower abdomen by the lower nerves. Specifically, the stomach is supplied by T7, 8, 9; the small bowel by T9, 10; the large bowel by T10, 11; and the rectum by T12–L2.

Pelvis Fibers arise from the lumbar ganglia and descend to a hypogastric plexus, which divides to enter the pelvis as the two pelvic plexuses (p. 473). The pelvic plexuses supply the structures in the pelvis and are innervated by segments in the lower spinal cord (approximately T12–L2). Some of these fibers are gray, and some are white, the latter relaying in the pelvic plexuses.

Afferent fibers The afferent sympathetic nerve fibers run approximately the same course as the efferent fibers until the fusion of the rami. They then pass into the posterior root, where the ganglion cell is situated, either to synapse with the cell body in the lateral horn of the cord for a sympathetic reflex or to ascend to higher centers by pathways presently not clear. The afferent fibres from the thoracic and abdominal structures travel with the sympathetic nerves, whereas the pelvic afferents travel with the parasympathetics.

PARASYMPATHETIC SYSTEM (Fig. 2.20)

The parasympathetic system has a double outflow: the cranial outflow, in which the fibers arise in the nuclei of the third, seventh, ninth and tenth cranial nerves; and a sacral outflow from the spinal segments S2, 3 and 4.

Cranial outflow Fibers arising from the third, seventh and ninth cranial nuclei pass to, and synapse with, parasympathetic ganglia in the head and neck that lie outside the central nervous system. The fibers of the tenth cranial nerve, however, synapse with minute ganglia that lie in or near the walls of the thoracic and alimentary viscera.

Sacral outflow The source of the sacral outflow of fibers is the second, third and fourth sacral nerves. The nerves consist of rami that pass forward to enter the pelvic plexuses through which they are distributed to all the derivatives of the original cloaca. Like the tenth cranial nerve, these fibers synapse around motor cell bodies in or near the walls of the viscera. This outflow is sometimes called the nervi erigentes for the part it plays in producing erection (see p. 491). Some branches from the pelvic plexus pass upward to reach the artery to the hindgut, through which they supply that part of the large intestine.

Afferent supply Afferent parasympathetic fibers run with cranial nerves VII, IX and X, the cell bodies being in the sensory ganglia related to those nerves (see Chapter 13), from which fibers pass up to the cranial nuclei and higher centers.

Chemical mediators

In both the sympathetic and parasympathetic nervous systems, the final mode of action is the liberation of chemical substances at the postganglionic terminals. Fibers of the sympathetic nervous system release norepinephrine (noradrenaline), with the exception of the sympathetic fibers supplying the hair muscles and sweat glands; these fibers secrete acetylcholine. At the parasympathetic nerve terminals, action is mediated by acetylcholine exclusively. Injections of these chemicals mimic the action of the two systems. Drugs that produce similar effects are therefore called mimetics (e.g. a parasympatheticomimetic drug will empty the bladder or produce gastric acid). Substances can also be given to block the action of these chemical transmitters and produce the opposite effect. For example, general anesthetic agents are usually bronchial irritants that induce increased bronchial secretion and produce a distinct tendency for the patient to drown in his own secretions. The administration of a drug to inhibit the parasympathetic system (e.g. atropine) will prevent the liberation of acetylcholine and abolish the bronchial secretion. Sufferers from peptic ulcers reduce gastric acid by a similar mechanism, but the nerve fibers to the salivary glands are also inhibited and the dry mouth accompanying the medication is often the price paid for relief from ulcer pain. Unfortunately, it is common for the benefits of a drug to be accompanied by unwanted side effects in other parts of the body.

Cardiovascular system (Fig. 2.21)

The cardiovascular system is comprised of the heart, various types of blood vessels and special vascular tissues.

Heart

The heart is a muscular pump divided into right and left sides. The right side receives deoxygenated venous blood from the whole body, except for the lungs, and pumps the blood through the pulmonary artery to the lungs. Here the blood is oxygenated and returned by the pulmonary veins to the left side of the heart, from where it is distributed throughout the body by the arterial system. The products of cellular metabolic activity are returned to the heart by either veins or a lymphatic vascular system. Veins return the blood to the right side of the heart, and lymphatic vessels open into the venous system

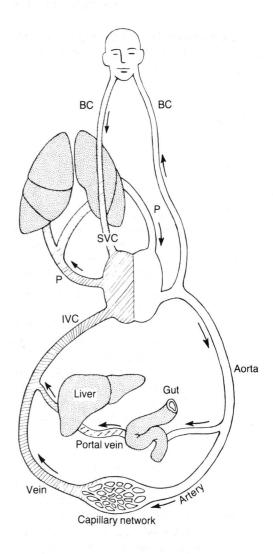

Fig. 2.21 The vascular system. The heart is depicted as having a right side receiving and discharging the venous blood (shaded conduits), while the left side receives and discharges arterial blood (clear conduits). P is the pulmonary circuit to and from the lungs. BC is the brachiocephalic circuit to and from the head, neck and upper limbs, the blood returning via the superior vena cava (SVC). Distally the blood passes via the aorta to the viscera, body wall and lower extremities, returning via the inferior vena cava (IVC). A large part of the digestive system passes the blood through the liver via the portal system before it reaches the inferior vena cava. The rest of the arterial system (apart from a few exceptions) passes the blood from the arterial to venous side via a capillary network.

by large ducts that drain the lymphatic channels (p. 164). (See Chapter 9 for in-depth coverage of the heart.)

Blood vessels

ARTERIES

The vessels leaving the heart, the arteries, have a large proportion of elastic tissue in their walls, enabling them to recover their original diameter after the expansion following the jet impulse delivered by the contracting heart. This recoil propels the blood during the resting phase of the heart, and is also responsible for filling the coronary arteries supplying the heart and closing some of the heart valves (see further discussion in Chapter 9). The large elastic arteries give rise to smaller arteries with less elastic tissue but more muscular tissue. Most vessels of the body are of the latter type; their caliber can be altered by contraction or inhibition of the smooth muscle.

Arteries give way to arterioles, the smallest divisions of the arteries, whose walls consist mostly of smooth muscle. Arterioles are largely responsible for maintaining blood pressure by providing peripheral resistance.

Arterioles diminish in size until they become so narrow as to be called capillaries. Capillaries are fine vessels with thin, semipermeable walls which allow the passage of substances essential for cellular metabolism and the return of the resulting waste products. The capillaries form an anastomotic network; active tissues such as the kidney or liver have vast capillary networks, whereas less active tissues such as tendons and ligaments are relatively poorly supplied. Certain structures are completely devoid of capillaries: the cornea of the eye, the epidermis and hyaline cartilage. In the case of the cornea, a corneal graft from a nonrelated person can be carried out with relative ease, since the lack of blood vessels prevents rejection by the defenders in the blood (the white cells), which normally make short work of transplants from another host. The collection of blood from the capillary plexus begins in small venules that join to form veins.

ANASTOMOSES

Throughout most of the body there are anastomoses, or communicating branches, between various arteries and veins. This defense mechanism exists to detour blood past a blocked or narrowed vessel. It also allows the surgeon to ligate a vessel without side effects. Anastomoses are particularly numerous in certain areas; for example, around joints, at the base of the brain, and in the hand and foot. However, in some parts of the body, a blocked artery is a disaster. These vessels are called end-arteries and cannot be bypassed; two examples are the central artery of the retina with occlusion resulting in blindness, and the intracerebral branches of the surface vessels of the brain with occlusion resulting in neurological deficits. Other examples of end-arteries are the renal and pulmonary arteries and, to a lesser degree, some of the vessels of the intestine.

The body also adapts itself to circumstances. For example, when a vein is placed in the arterial system to bypass a blocked artery, it becomes arterialized; with time, the vein thickens and closely resembles an artery. This may be seen when the long saphenous vein of the leg is used to bypass obstructed coronary vessels of the heart.

VEINS

Veins are more numerous than arteries, their walls are thinner and their diameters usually larger. Smaller arteries are usually accompanied by pairs of veins that often communicate around the artery and are called venae comitantes. Although most arteries are accompanied by veins of the same name, several venous systems are without any arterial counterpart. These are the azygos system of the chest (p. 137), the vertebral system of veins in and around the vertebral column (p. 481) and the intra-abdominal portal system (p. 409). Many veins have valves, in the form of infoldings of two or three cusps, to prevent reverse flow and to aid the return of blood to the heart. Remember that the veins return blood with relative difficulty, since they do not have the flow pressure that is imparted to the arteries by the heart. This is especially so when gravity is an additional burden, as in the legs. For this reason, valves are most numerous in the lower limbs. Most veins of the abdomen and thorax do not have valves; the return of blood in these cavities is aided by the positive pressure in the abdomen and the negative pressure in the thorax. Veins in the head and neck usually do not need valves because their flow is with gravity. However, some valves are present to prevent the reflux of blood when the pressure in the abdomen or thorax rises, such as occurs when coughing.

Special vascular arrangements

CAVERNOUS TISSUE

Cavernous tissue is present in the penis and clitoris where the corpora contain large spaces. When filled with blood, the pressure is sufficient to produce erection.

SINUSOIDS

Sinusoids are similar to, but wider and more tortuous than, capillaries. They substitute for capillaries in various parts of the body, especially the liver, spleen and endocrine glands (suprarenal and parathyroid). Many of the lining cells of these vessels are phagocytic.

ARTERIOVENOUS ANASTOMOSES

In some areas, arteries are not connected to the veins by capillaries but anastomose directly. This, of course, prevents the blood from being made available for tissue metab-olism. Arteriovenous anastomoses tend to occur in certain areas such as the skin of the nose, lips, hand and intestine; they are functionally useful. For example, on exposure to extreme cold, the arteriovenous anastomosis will open, reduce the amount of blood in the capillaries of the skin and so limit the heat loss. Arteriovenous anastomoses also conserve blood, diverting it to areas where it is needed. For example, between meals it is unnecessary for the intestine to be well supplied with blood so the anastomosis opens up, diverting the blood from the bowel. During periods of digestion, the anastomosis closes and the blood proceeds through the capillaries for use by the intestine.

Neurochemical control

The blood vessels play an exceedingly important part in the circulation, and are controlled by neural and chemical mechanisms. The neural mechanism consists of the sympathetic fibers, which release norepinephrine (noradrenaline) to constrict vessels and raise the blood pressure. The chemical control is by epinephrine (adrenaline) – closely related to norepinephrine – which is secreted by the suprarenal medulla and has a similar action. Both mechanisms are subject to control from higher neural centers, whose activity is partly dependent on incoming sensory information. Increase or decrease of the diameter of the vessels occurs by inhibition or stimulation, respectively, of the vasoconstrictor fibers. For example, the higher centers in the brain and brain stem may wish to divert blood to the intestine during a gastronomic banquet. Most of the participants will look fairly pale as the blood leaves the skin, where the vessels are constricted, and makes for the alimentary tract, where the vessels are dilated. Toward the end of the meal the pale gourmets start becoming red in the face as the increased metabolism required to dispose of the meal results in excessive heat production and dilatation of the skin vessels to dissipate this heat.

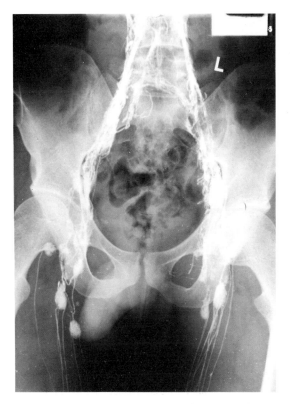 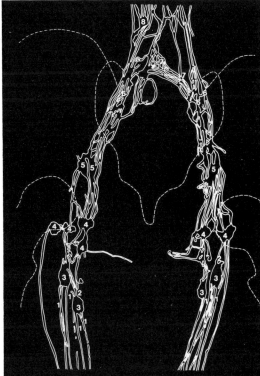

Fig. 2.22 Lymphangiogram – first day after injection of radio-opaque dye into lymphatic vessels of the foot. (From Weir and Abrahams 1978).

1 Numerous afferent lymph vessels in groin.
2 Efferent lymph vessels.
3 Superficial inguinal lymph nodes (lower group – subinguinal).

4 Superficial inguinal lymph nodes (upper group).
5 External iliac lymph vessels.
6 Common iliac lymph vessels.
7 Lumbar crossover of iliac vessels.
8 Lower para-aortic node (lateral aortic).

Lymphatic system

LYMPHATIC VESSELS

The vascular capillaries (described earlier) absorb most of the products of tissue activity. However, certain substances possessing large molecules cannot be so absorbed. For this reason, another system is necessary – the lymphatic system. The smallest vessels in the lymphatic system are similar to, and are also called, capillaries. They absorb substances and transport them as lymph. Like the vascular capillaries, the lymphatic capillaries form networks and eventually give way to larger collecting vessels that are usually valved to produce unidirectional flow. These valves are often seen on x-ray studies of the lymph vessels (lymphangiogram), giving them a beaded appearance. Some of these lymphatic capillaries are given special names; for example, those that project into the tips of the villi in the mucous membrane of the small intestine are called lacteals, due to their somewhat creamy appearance from the absorbed emulsified fat. Lymph vessels are present all over the body, except for the central nervous system and a few other tissues. From the collecting vessels, the lymph passes into larger trunks that even-

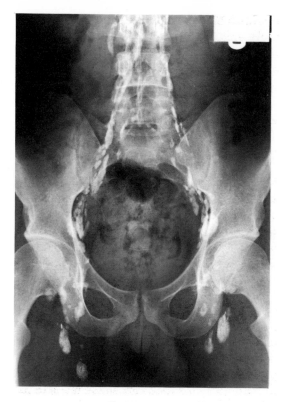

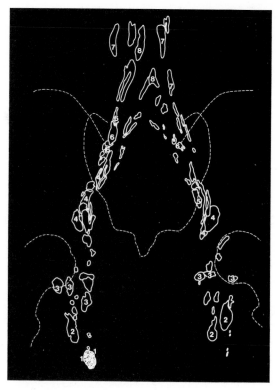

Fig. 2.23   Lymphangiogram – second day. (From Weir and Abrahams 1978).

1 Normal follicular appearance of lymph nodes.
2 Superficial inguinal lymph nodes (lower group).
3 Superficial inguinal lymph nodes (upper group).
4 External iliac lymph nodes.

5 Internal iliac lymph nodes.
6 Common iliac lymph nodes.
7 Lower para-aortic nodes (lateral aortic).
8 Pre-aortic lymph nodes.

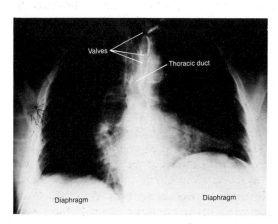

Fig. 2.24   The contrast medium finally reaches and outlines the thoracic duct – note the beading due to the valves.

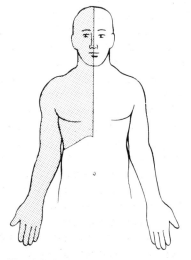

Fig. 2.25   The right half of the head and neck, right arm and right chest drain into the right lymphatic duct, while the rest of the body drains into the thoracic duct.

tually reach and empty into the large venous channels at the base of the neck. The right lymphatic ducts drain the area shown in Fig. 2.25 while the left (thoracic duct) drains the rest of the body. Lymphatic vessels also carry cells produced by the lymphatic tissue, the lymphocytes, which are important cells in pathologic conditions.

LYMPHATIC TISSUE

Certain collections of lymphocytes have a supporting stroma. They appear in various parts of the body: mucous membranes, lymph nodes, thymus, spleen and bone marrow. Collections of lymphatic tissue in the mucous membranes include the tonsils, the area in and around the oral cavity, and the intestines.

Lymph nodes are scattered throughout the body and tend to lie along the course of blood vessels. Fairly prominent at birth, they reach their maximal growth in adolescence, after which they atrophy. Lymph nodes are experienced as painful swellings in inflammatory conditions. For example, in a septic pinprick of the finger, organisms are transported by the lymph capillaries and collecting channels to the axillary lymph nodes, which enlarge to form tender lumps. Usually the lymph nodes win the battle and prevent the infection from passing further.

It is a truism that superficial lymphatics accompany veins while deep lymphatics accompany arteries. A possible reason for this relationship of the deep lymphatics appears to be the transmission of pulsation of the arteries to the lymphatics, with a milking effect that aids the flow. Therefore, if the arterial supply of an organ is known, so is the lymphatic drainage. Some confusion may arise because of the different accounts of lymphatic drainage in different textbooks. The probable reason for this discrepancy is that the lymph travels along the easiest of several available routes. In disease, certain lymphatic vessels become blocked, and the lymph passes in an unusual direction. Enough clinical examples exist, however, to establish patterns of organ disease and lymph node enlargement. Knowledge of these patterns is extremely important in the diagnosis and treatment of many diseases.

Genitourinary system

Urinary system

Most waste products are excreted through the kidneys, two organs situated in the upper abdomen (see Chapter 30). Having filtered the large volume of blood each receives, the filtrate is passed down a pair of muscular tubes, the ureters. These enter the pelvis and discharge into the bladder, a capacious hollow muscular organ that stores the urine until it causes discomfort. This initiates the act of micturition, at which time the contractions of the bladder, assisted by the contractions of the abdominal musculature, increase the pressure in the bladder, and urine passes into the urethra. This tube carries the urine to its ultimate destination, be it toilet or lamp post.

Genital system

FEMALE REPRODUCTIVE SYSTEM

The female is more sophisticated than the male – she has separated her reproductive and urinary systems. The reproductive apparatus lies in the pelvis. The ovum is produced by the ovaries. The ovaries are two solid structures that secrete hormones and produce and deliver an ovum to either of two fallopian tubes (salpinges). Each tube has an opening for receipt of the ovum. If fertilization occurs, it is by a single sperm in the outer part of the tube. The developing ovum travels along the tube to enter the uterus, a thick hollow muscular organ whose function is to receive and nurture the fertilized ovum. The ovum grows into a fetus. When the fetus is ready to face the outside world (about 40 weeks' gestation), the uterus, under hormonal influence, contracts and expels the fetus into the vagina, a hollow canal that opens exteriorly and through which the child is born.

In less dramatic times, the ovaries and uterus undergo cyclical changes, the stimuli for these changes originating through hormones produced by the ovary and other glands. This accounts for the proliferation of the uterine mucous membrane and its subsequent loss during the menstrual cycle. The vagina, when not acting as a birth canal, is the repository for the penis and the spermatozoa during the sexual act. (See Chapters 33 and 34 for detailed discussion about the contents of the female reproductive system.)

MALE REPRODUCTIVE SYSTEM

Spermatozoa are produced by a pair of testes that reside in a pouch, the scrotum, which lies outside the abdominal cavity. These sperm are transported through a much coiled canal, the epididymis, which ends in a thick muscular tube, the vas deferens. Shortly before each vas deferens empties into the urethra, the spermatozoa are bathed in seminal fluid, produced by a pair of seminal vesicles. From the urethra, which tunnels the penis, they pass to the exterior and into the vagina during coitus.

Urine also traverses the urethra, which is therefore a true genitourinary canal. The spermatozoa cannot function without the secretions of the seminal vesicles and prostate gland, the latter emptying directly into the urethra. (See Chapters 33–35 for detailed discussion about the contents of the male reproductive system.)

Respiratory system

The main functions of the respiratory system are to supply oxygen to the living organism and remove carbon dioxide. In the most simple unicellular organism respiration is carried out by a simple exchange between the cell and the surrounding fluid. At a higher level of development, fish carry out this gas interchange through their gills, extracting oxygen from the water. The human inhales air and absorbs the oxygen into the blood stream for distribution to the tissues; here the blood gives up the oxygen and receives the carbon dioxide that has accumulated as a result of metabolic activity. This carbon dioxide is then excreted through the expired air.

To carry out this blood/gas interchange, air is inhaled through the nasal cavities and/or mouth to reach the pharynx. It then passes through a continuous series of air-conducting tubes. The larynx gives way to the trachea, which bifurcates into two main bronchi that continually subdivide into bronchioles and, eventually, terminal bronchioles. All these ducts constitute the conducting airways. As small sacs start budding from the terminal bronchioles, they change to respiratory bronchioles. The sacs increase in number constituting the alveolar ducts and finally these ducts end in a large number of sacs (alveoli). The lung is made up mostly by the tissue between the terminal bronchioles and the most distal alveoli. This constitutes the respiratory zone and its volume is about 3000 ml. As the conducting airways do not contain alveoli and can take no part in gas exchange, they constitute the anatomical dead space, a volume of about 150 ml. Each alveolus, of which there are about 300 000 000 in each lung, measures about 0.3 mm (0.01 in) in diameter and is composed mainly of simple squamous epithelium surrounded by a network of elastic tissue and a capillary plexus. It is the extremely thin walls of both the alveoli and the capillaries that allow gaseous interchange between air and blood to occur. Some larger cuboidal cells are also present in the alveolar epithelium, which produce a substance, surfactant, that reduces surface tension. This material starts being produced at about the seventh month of fetal life and an absence or reduced amount of this substance may lead to hyaline membrane disease, a major cause of death in premature infants.

The respiratory system must protect itself

from a very hostile environment as the surface area of each lung is some 50–100 m² (60–120 yd). From above down, the protective mechanisms are as follows. The nasal hairs are able to block fairly large particles. The bronchial walls, richly supplied with goblet cells and mucous glands, continually produce mucus. The pseudostratified columnar epithelium is covered by a large number of tiny cilia which are able to move the stream of mucus in a continuous upward direction. Particles caught in this rising stream reach the epiglottis where both the mucus and the particles are swallowed. Paralysis of the cilia by various diseases may therefore lead to respiratory problems. Phagocytic cells also exist in the alveolar epithelium, and are important in engulfing particulate matter which has escaped the defence mechanisms at a higher level; since the alveoli have no cilia, they rely on these phagocytic cells.

It is important that the inspired air which reaches the alveoli be warm and moist. Warmth is provided by the rich network of blood vessels that lie in the connective tissue beneath the epithelium of the nose. Should these vessels become dilated, as in the common cold, nasal congestion will occur, leading to difficulty in breathing. It is also

this rich vascular network which may produce a nosebleed. The moistness is provided by the numerous goblet cells and seromucous glands which abound in the respiratory tract. This secretion is under neural control, and patients who have the nerve supply temporarily abolished, such as those receiving medication prior to anesthesia, will complain bitterly of a very dry nose and mouth. Both the diminished secretion and the paralysis of ciliary movement induced by preoperative medication play a part in postoperative respiratory problems.

Other functions of the respiratory tract consist of maintaining a proper acid–base balance of the body tissues by being able to excrete larger or smaller quantities of carbon dioxide as required, excreting water in small amounts and producing vocal sounds (phonation). For the last-named function, expired air passes cranially through the trachea and larynx, causing vibration of a pair of vocal folds (cords). The pharyngeal spaces, nasal and oral, as well as the paranasal sinuses, serve as resonating chambers. Congestion of these cavities produces the typical gravel voice of an upper respiratory infection. The expired air is moulded into human speech with the aid of the tongue, soft palate and lips.

Review questions

(answers on p. 529)

1 Fat
a) has a poor blood supply
b) is present in large quantities in the intracranial cavity
c) is absent in the buttocks
d) cannot be readily mobilized for energy purposes
e) is of little use in helping to support viscera

2 Elastic tissue can be found in
a) the walls of arteries
b) large quantities in the heart
c) the outer coverings of nerves
d) many muscles in the extremities
e) the skin of the lips

3 Subcutaneous tissue (superficial fascia)
a) contains no fat
b) contains no areolar tissue
c) may contain smooth muscle fibers
d) lies below the deep fascia
e) rarely has any innervation

4 Deep fascia
a) is a strong layer in the face
b) is a weak layer in the lower limb
c) may afford origin for muscles
d) is not attached to the skin of the hand
e) can resist abdominal distension

5 In joints
a) a cartilaginous joint exhibits more movement than a synovial one
b) the articular cartilage receives most of its nutrition from surrounding vessels
c) the articular cartilage is readily visible radiologically
d) the surrounding capsule has a very good nerve supply
e) the primary cartilaginous joints tend to lie in the midline

6 With regard to joints
a) in the limbs, there are many examples of the fibrous type
b) the secondary centers of ossification normally fuse at puberty
c) intra-articular discs may reduce the range of movements
d) there is a poor periarticular vascular anastomosis
e) the nerve that supplies a joint, supplies the muscles moving the joint and the skin overlying the joint

7 The skin
a) cannot absorb medication applied to its surface
b) plays a small part in the control of body temperature
c) appreciates and localizes pain poorly
d) is a hard-wearing organ
e) has a rich autonomic parasympathetic supply

8 With regard to the skin
a) that of the penis is liberally supplied with hairs
b) the arrectores pilorum muscles help to grease its surface
c) the arrectores pilorum muscles are under voluntary control
d) elastic fibers are abundant
e) the sweat glands are confined to the epidermis

9 With regard to muscle
a) striated muscle is mostly under voluntary control
b) cardiac muscle is completely under voluntary control
c) most muscles act individually
d) smooth muscle contracts rapidly in response to a stimulus
e) all the above are true

10 With regard to muscle

a) a unipennate muscle is usually more powerful than one which is parallel fibered

b) a substantial proportion of a nerve supplying a muscle relays back sensory information

c) section of its nerve supply leads to its degeneration into mostly fibrous or fatty tissue

d) any nerve traversing a muscle bound for other destinations always supplies that muscle

e) all the above are true

11 The spinal cord

a) may reach to the level of the second lumbar vertebra in the adult

b) reaches to the level of the third lumbar vertebra in the child

c) is surrounded by the same three membranes as surround the brain

d) is stabilized by ligamenta denticulata

e) all the above are true

12 With regard to the primary rami

a) the limbs are sometimes supplied by posterior primary rami

b) posterior primary rami have a far more extensive supply than anterior rami

c) the posterior primary rami supply sensation from the vault of the skull to the upper buttock, behind the posterior axillary line

d) they include preganglionic parasympathetic fibers

e) all the posterior cervical rami reach the skin

13 With regard to the autonomic nervous system

a) it is mostly a voluntary system

b) while most of it appears to be outside the central system, it is controlled to a variable degree by higher centers in the central system

c) the hypothalamus is concerned mainly with the control of respiration and circulation

d) it rarely supplies cardiac muscle

e) it always has ganglia near the organ of supply

14 With respect to the sympathetic system

a) this system has its outflow from the posterior horn cells of T1–L2

b) all the preganglionic fibers synapse in the sympathetic chain (trunk)

c) the sympathetic chain is a single trunk lying in the midline anterior to the vertebral column

d) it has myelinated postganglionic fibers passing to the spinal nerves

e) it sends preganglionic fibers to the medulla of the suprarenal gland

15 The efferent sympathetic supply to

a) the skin is mostly by blood vessels

b) the body viscera is mostly by the spinal nerves

c) the pupil originates in T1

d) the heart originates in T6–8

e) the pelvis comes from S2, 3, 4 segments

16 The efferent sympathetic supply to

a) the upper limb is distributed by the brachial plexus (C5–T1)

b) the lower limb is distributed by the lumbosacral plexus (L4–S2)

c) the abdominal viscera is via the splanchnic nerves which pierce the diaphragm

d) the pelvis mainly originates in L1 and 2 segments and may travel down the sympathetic trunk

e) all the above are true

17 The parasympathetic outflow

a) originates from a cranial and a lumbar outflow

b) is mostly antagonistic to the sympathetic system

c) involves cranial nerve nuclei of III, V, IX and X only

d) innervates the suprarenal medulla

e) travels with the sympathetic fibers in the sympathetic chain (trunk)

18 With regard to the circulation

a) the pulmonary arteries carry oxygenated blood

b) the walls of arterioles allow the free flow of the substances essential for cellular metabolism

c) the cornea of the eye is devoid of a capillary network

d) many arteries and veins are valved

e) the smaller the artery, the more elastic it is

19 In the circulatory system

a) cavernous tissue is found in the liver

b) sinusoidal systems are found in the clitoris and penis

c) in some areas, arteries can empty directly into veins and bypass capillaries

d) the diameter of an arteriole can be increased by the injection of epinephrine (adrenaline)

e) the heart can be slowed by sympathetic stimulation

20 With regard to the lymphatic system
a) the nodes manufacture the lymphatic fluid
b) all the body's lymph eventually reaches the thoracic duct
c) lymph eventually returns to the heart through large veins at the roots of the limbs
d) the vessels are often valved
e) there are numerous channels in the brain

Part II   The Upper Limb

3 Pectoral and scapular regions and axilla

Pectoral region · Scapular region · Axilla

Pectoral region

Bones

The bones of the pectoral region are the clavicle, scapula and humerus (see Fig. 36.5).

Clavicle

DEVELOPMENT

If you are an early bird, you will love the clavicle, as it is the first long bone in the body to ossify. It is also one of the bones that develops in membrane. The bone develops from the fusion of two ossification centers. Abnormal

Fig. 3.1   The clavicle with its two ossification centers. Bone formation then proceeds toward each end.

development of this bone may result in the absence of most of each clavicle, which enables the person to bring the shoulders together in the midline – a doubtful asset (see Fig. 2.4). Sometimes fusion fails, resulting in a defect that separates the outer third and inner two-thirds of the bone. A physician may see such a patient in the emergency room with an injury of the upper limb and confuse this congenital defect with a fracture of the

clavicle. Avoid this embarrassment by x-raying the opposite shoulder, since congenital defects are often bilateral.

FUNCTION

What is the function of the clavicle? The human clavicle is bigger than an elephant's but smaller than a gibbon's, an animal one-third man's size. This should suggest its function: it acts as a strut or pit-prop to hold the upper limb away from the body and allows the arm, and therefore the hand, to pass into extraordinary positions. Gibbons swing from tree to tree, elephants can barely abduct their shoulders, using their trunks as effective upper limbs, while humans can dunk a basketball.

The clavicle also transmits part of the weight of the limb to the skeleton. One would otherwise have tired muscles from supporting the upper extremity.

SHAPE

The clavicle is shaped in a double curve, the medial two-thirds being convex forward and the lateral one-third concave forward. Since the junction of two curves is the weakest point of a structure, this is the site of clavicular fractures.

MUSCULAR ATTACHMENTS (Fig. 3.14)

Two powerful muscles originate from the anterior surface of the clavicle: the large pectoralis major, the climbing muscle of the ape, arises medially; and the powerful abductor, the deltoid, attaches laterally. The seemingly insignificant subclavius muscle occupies a small groove on the undersurface of the clavicle as the latter crosses the first rib. It is difficult to believe that this muscle has any function, but it very likely has a purpose.

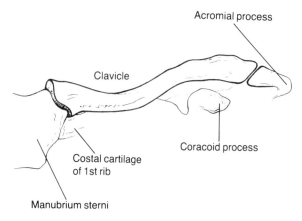

Fig. 3.2   Articulations of the clavicle. The clavicle, convex forward in its medial two-thirds and concave forward in its lateral one-third, articulates medially with the manubrium sterni and laterally with the acromion.

Running beneath the clavicle are the subclavian vessels and the large nerves for the upper limb. In a fracture of the clavicle (the most commonly occurring long-bone fracture), one might assume that these underlying structures are commonly injured. Yet this is rarely so, partly because of the protection by the subclavius muscle. Gaining attachment to the posterosuperior border are the trapezius laterally and the sternomastoid medially.

POSITION

The clavicle is subcutaneous in its whole length. The bone can be grasped between thumb and forefinger, and, if so desired, waggled around. However, it cannot be moved much, since it is firmly fixed medially and less firmly fixed laterally. The medial broad triangular expanded surface articulates

with a big notch on the sternum and with the first costal cartilage (see p. 67), itself attached to the sternum. When a person falls on the outstretched hand or receives a shoulder check in hockey, the weight is transferred along the clavicle to the body where it can be easily absorbed. Without this force transfer, the clavicle would fracture even more frequently, since it cannot sustain forces withstood by the thoracic cage. The lateral end of the clavicle is attached to the very mobile scapula (discussed later). This acromioclavicular articulation is, therefore, relatively weak, and subluxations (partial separations of the joint surfaces) or dislocations (complete separations of the joint surfaces) are common.

LIGAMENTS

Because of the extreme mobility of the upper limb, the clavicle may appear to be a mobile bone. The opposite is true, however. Very strong ligaments bind it medially to the first rib (costoclavicular) and laterally to the coracoid process of the scapula (coracoclavicular).

Scapula

Today's students are fortunate: students of yore had to memorize a wealth of descriptive anatomy concerning the scapula, with little profit. Currently, the scapula is of little interest, since it is merely a mobile plate of bone, completely encased in muscles. Fractures are therefore not only infrequent, but also require minimal treatment because of the prevention of displacement by the enclosing muscles.

The triangular-shaped bone covers the thoracic cage from the second to the seventh rib posteriorly – useful information when you needle the chest cavity. Exposure of the chest wall for this purpose can be improved by elevating the upper extremity, thereby displacing the bone anteriorly around the thorax. Projecting posteriorly from the body of the scapula is a large, easily palpable spine

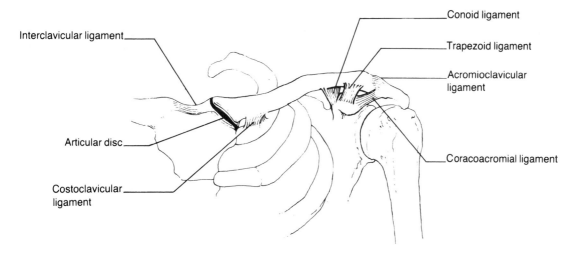

Interclavicular ligament

Conoid ligament

Trapezoid ligament

Acromioclavicular ligament

Articular disc

Costoclavicular ligament

Coracoacromial ligament

Fig. 3.3   Ligaments of the clavicle. The sternal end of the clavicle is held secure by an interclavicular ligament, an intra-articular disc and a costoclavicular ligament. The lateral end is held firmly to the coracoid process by the two components of the coracoclavicular ligament, the conoid and trapezoid, and less firmly to the acromion by the acromioclavicular ligament.

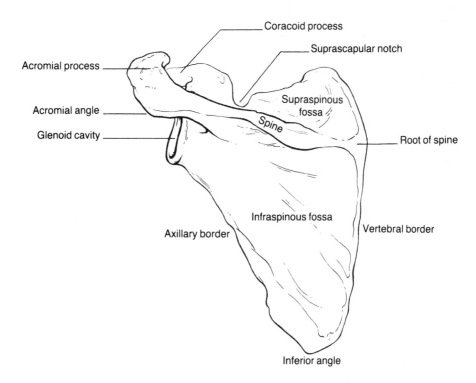

Coracoid process

Suprascapular notch

Acromial process

Supraspinous fossa

Acromial angle

Spine

Glenoid cavity

Root of spine

Infraspinous fossa

Vertebral border

Axillary border

Inferior angle

Fig 3.4   Dorsal surface of scapula. The spine bends forward at an angle to become the acromion.

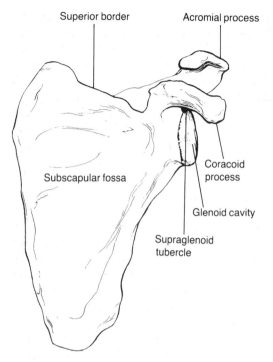

Superior border   Acromial process

Subscapular fossa

Coracoid process

Glenoid cavity

Supraglenoid tubercle

Fig. 3.5   Ventral surface of scapula. The shallow glenoid cavity is surmounted by the coracoid process.

that bends sharply at its lateral extremity to become the acromion. The upper limb is measured from this process, even by the tailor, because of its prominent subcutaneous position. The medial end of the spine of the scapula lies opposite the spine of the third thoracic vertebra, a useful landmark.

The glenoid fossa, a smooth pear-shaped area where the head of the humerus articulates, faces forward and laterally. The humerus lies in a corresponding position, placing the upper limb in a position of apparent medial rotation. Adjacent to the glenoid fossa is the finger-like coracoid process, palpable 2.5 cm (1 in) below the clavicle at the junction of its medial three-quarters and lateral quarter. When the coracoid process is first palpated by the eager young medical student, it is sometimes mistaken for an ominous tumor. Muscles arising from the surface and borders hold the scapula to the body wall but also allow free movement of the upper limb.

Humerus

The humerus is a long bone with a proximal head whose contour is a little less than half a sphere and which articulates poorly with the glenoid fossa. Beyond the head is a constriction, known as the anatomical neck. (This is the site of the epiphyseal plate.) Following this constriction are two large tuberosities for muscular attachments – the greater tuberosity located laterally, and the lesser tuberosity located medially. The greater tuberosity – not the deltoid muscle – is responsible for the smooth shapely shoulders of the sexy screen actress. Loss of this con-

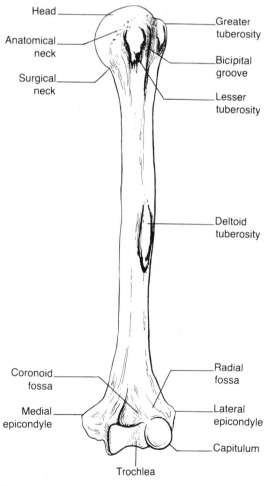

Head

Anatomical neck

Surgical neck

Greater tuberosity

Bicipital groove

Lesser tuberosity

Deltoid tuberosity

Coronoid fossa

Medial epicondyle

Radial fossa

Lateral epicondyle

Capitulum

Trochlea

Fig. 3.6   Anterior surface of the humerus. The bone above the deltoid tuberosity is circular, that below triangular in cross-section.

tour and its replacement by an ugly dent below the acromion are diagnostic of a dislocated shoulder joint, which is the commonest large joint to dislocate. The deformity is due to displacement of the humeral head and therefore the greater tuberosity, the attachments of the deltoid being unchanged. The bicipital groove for the tendon of the long head of biceps muscle separates the tuberosities. This tendon moves in the groove inside a well-developed sheath; wear and tear of this mechanism is a common cause of a painful shoulder. The tendon can become so worn in older people that it may even rupture.

Below the tuberosities the bone narrows into a neck, the surgical neck of the humerus, so called because of the frequency of fractures at this level. Below this the shaft is cylindrical until the deltoid tuberosity, halfway down the lateral surface. Excrescences on bones are partly or wholly produced by the traction of muscles: the greater tuberosity receives the insertion of muscles from the posterior aspect of the scapula, the lesser tuberosity the insertion of a muscle from the anterior scapula, the subscapularis, and the deltoid tuberosity the attachment of the multipennate deltoid.

Joints

A word about orientation before articulating the clavicle, scapula and upper end of the humerus. The costal surface of the scapula lies on the posterolateral part of the chest wall; when moving anteriorly, it passes forward around the chest wall along the circumference of an imperfect circle. The clavicle runs laterally and backward, crossing the first rib obliquely to articulate the acromion.

Sternoclavicular joint

The sternoclavicular joint transmits nearly all the stresses from the clavicle to the axial skeleton. The articulation has a thick cartilaginous intra-articular disc, which acts as a shock absorber. Powerful surrounding ligaments help to produce a stable articulation that dislocates only with difficulty (see Fig. 3.3). However, in some wrestling arenas, the clavicle may be pushed out of the joint. Should it be displaced posteriorly, it impinges on the trachea, producing shortness of breath. Forward dislocations sometimes occur but, apart from the cosmetic disability, are of little significance.

Acromioclavicular joint

The lateral articulation with the scapula is weak, with sloping surfaces that encourage subluxation or dislocation. This injury commonly affects football gladiators and is easily diagnosed by seeing and feeling the clavicle overriding the acromion process. Despite its unattractive appearance and temporary discomfort, it produces little disability. Only in a manual worker or a professional athlete should active treatment be considered.

The sternoclavicular and the acromioclavicular joints participate – the former to a much greater degree than the latter – in the composite movements of the scapula during movements of the shoulder joint. The clavicle moves in all directions as the scapula moves upward, downward, forward and backward, and during elevation the clavicle undergoes rotation. These movements can be easily confirmed clinically.

Shoulder or glenohumeral joint
(Figs. 3.8–3.13)

The shoulder joint between the glenoid cavity of the scapula and the head of the humerus is a compromise between mobility and stability. The bony articulation is pitiful, and, were it not for the support from other structures, dislocations would follow even a sneeze. Muscles cover nearly all aspects of the joint. The shallow socket of the glenoid cavity is somewhat deepened by a narrow rim of fibrocartilage, the labrum. At times a big price is paid for this modest assistance,

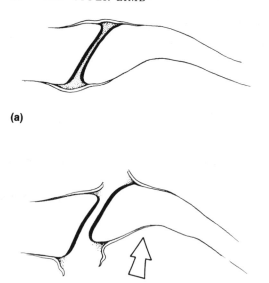

(a)

(b)

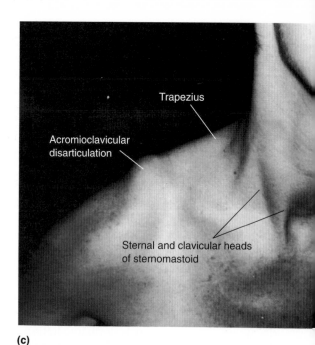

Trapezius

Acromioclavicular
disarticulation

Sternal and clavicular heads
of sternomastoid

(c)

Fig. 3.7 (a) The sloping weak acromioclavicular articulation. (b) Acromioclavicular subluxation (partial dislocation). (c) Acromioclavicular dislocation (complete separation). (d) X-ray photograph of the same patient.

(d)

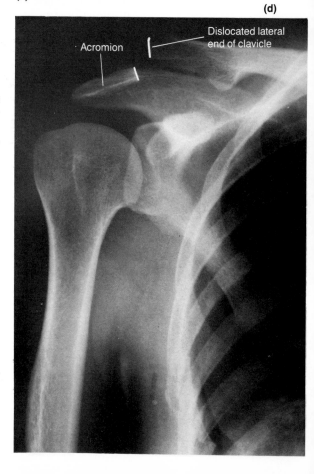

Dislocated lateral
end of clavicle

Acromion

for tears or separations of the labrum are fairly common and troublesome. The many ligaments around the shoulder joint are mostly thickenings of the capsule, but of some consequence is a ligament joining the coracoid process with the acromion, the coracoacromial ligament (see Fig. 3.3). It prevents upward displacement of the humeral head and acts with the subjacent subacromial bursa as a secondary beam across the roof of the shoulder joint. An important support of the shoulder joint is the long head of the biceps tendon that arches over the head of the humerus from a small supraglenoid tuberosity.

Factors preventing dislocation have been mentioned. Those encouraging it are the poor socket for the humeral head and the loose inferior attachment of the articular capsule. It is not surprising that the shoulder joint is the commonest large joint to suffer recurrent dislocation. It also explains the so-called 'double jointedness' of some contortionists.

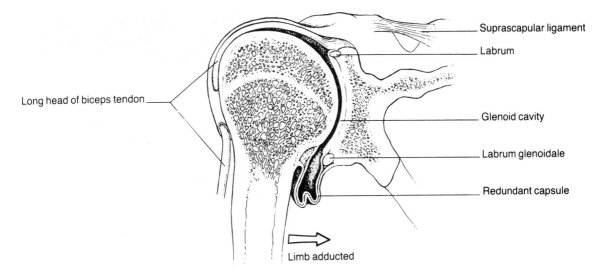

Fig. 3.8   The shoulder joint. The head of humerus makes an insecure articulation with the glenoid cavity. The capsule hangs loosely in the adducted position. The long head of biceps tendon passes through the upper part of the joint cavity.

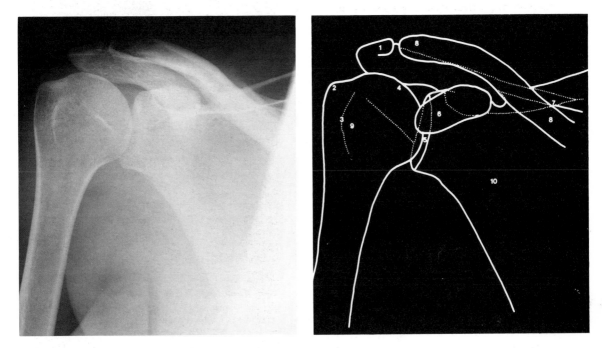

Fig. 3.9   Anteroposterior view of the shoulder. (From Weir and Abrahams 1978)

1 Acromion.
2 Greater tuberosity.
3 Intertubercular sulcus (bicipital groove).
4 Head of humerus.
5 Glenoid fossa.

6 Coracoid process.
7 Spine of scapula.
8 Clavicle.
9 Lesser tuberosity.
10 Subscapular fossa.

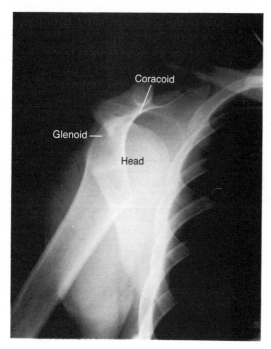

Fig. 3.10   Radiograph of dislocated right shoulder joint. The humeral head is lying below the coracoid process (subcoracoid position).

CAPSULE

The capsule is attached to bone immediately distal to the articular cartilages of both the scapula and the humerus, except inferiorly on the latter bone where the attachment is 1.25 cm (0.5 in) down the medial aspect of the shaft at the level of the surgical neck.

LIGAMENTS

The ligaments in the shoulder joint are weak to allow free movement. Strength is provided by the surrounding muscles, especially the 'rotator cuff', and by the long head of the biceps tendon.

JUXTA-ARTICULAR MUSCLES AND TENDONS

The deltoid muscle covers the tuberosities of the humerus into which are inserted three muscles from the posterior scapula, the supraspinatus, the infraspinatus and the

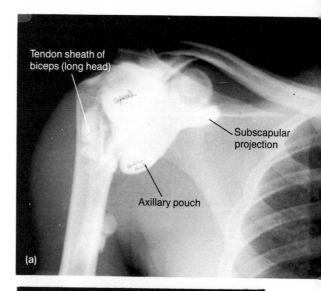

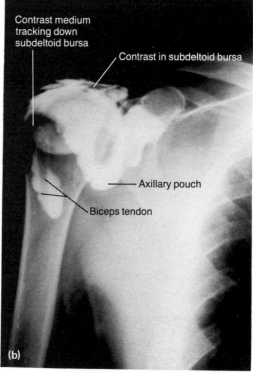

Fig. 3.11   (a) Arthrogram of shoulder joint. Contrast injected into the capsule of the shoulder joint outlines the extent of the cavity. The medial prolongation passes deep to the subscapular muscle/tendon, while the axillary pouch is seen with the arm in the adducted position. Contrast also occupies the tendon sheath of the long head of biceps. (b) Bad news – the contrast has leaked into the subdeltoid bursa through a capsular tear, caused by a ruptured supraspinatus tendon.

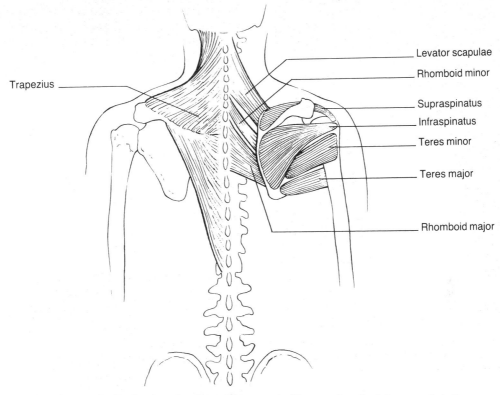

Fig. 3.12 Muscles attached to the dorsal surface of the scapula. The muscles attach the scapula to the spine, skull and upper arm.

teres minor, and one from the anterior scapula, the subscapularis. These muscles are inserted into the joint capsule and the tuberosities by a musculotendinous insertion, often called the rotator cuff. Their function is to hold the head of the humerus firmly in the shallow pear-shaped glenoid fossa, thus allowing other larger muscles to put the joint through extensive movements;

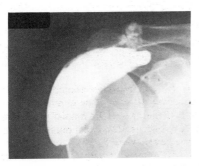

Fig. 3.13 Subdeltoid bursa. Radiograph of contrast medium injected into, and outlining, the subdeltoid (subacromial) bursa.

for example, the deltoid muscle would just pull the humerus up against the acromion during abduction if the head were not so well fixed.

The tendon of the supraspinatus muscle passes across the superior aspect of the shoulder joint, and, therefore, like the deltoid muscle, it is a true abductor. The supraspinatus tendon is generally considered to initiate the movement of abduction, which is then continued by the much more powerful deltoid. The two muscles together can produce about 120 degrees of abduction. The supraspinatus tendon is separated from the deltoid muscle by a large bursa, the subdeltoid bursa, which continues medially between the supraspinatus tendon and the acromion. The subdeltoid bursa is often called the subacromial bursa, but it is one continuous bursa. The supraspinatus tendon is so closely attached to the floor of the bursa that tendinitis, usually the result of wear and tear, often affects the bursa. Supraspinatus tendinitis or subdeltoid bursitis is responsible for much unhappiness in sportsmen and older people.

MOVEMENTS

Abduction is a composite movement of the glenohumeral joint (100–120 degrees) and the scapulothoracic (shoulder girdle) joint (55–65 degrees), producing a total movement of 180 degrees. The movement of abduction requires external rotation of the humerus for the following reason. The head of the humerus passes through the movement of abduction at the glenohumeral joint; when the humeral articular surface is 'used up', the angle is approximately 80 degrees. Further abduction in this position is impossible since there is no available articular surface on which the humerus can rotate. However, if one externally rotates the humerus, an additional humeral articular surface will be brought into play to allow about 120 degrees of abduction. Loss of external rotation means loss of some abduction. This occurs with fractures that are immobilized with the arm bandaged to the side in the internally rotated position, especially in elderly patients. The shoulder should therefore be immobilized in the abducted, externally rotated position (airplane splint).

In adduction, huge muscles (pectoralis major and latissimus dorsi) are available for bringing the arm to the side. This does not seem to be such a crucial movement since gravity will accomplish the same motion by merely allowing the arm to drop to the side. These huge muscles are needed for climbing. The arms are placed in the abducted position, and the body is brought up to the arms rather than the reverse. The huge adductor muscles of the apes exemplify this process. The same movement is used in swimming in which the body is again brought to the outstretched arm. It is normal to see an active swimmer with extremely well-developed pectoralis major and latissimus dorsi muscles.

Flexion and extension are simple movements. It is interesting to note that, because of the position of the glenoid fossa and humeral head, flexion brings the upper limb across the body and extension does the opposite. Bringing the arm across the body will bring the hand into an optimal position for popping some candy into the mouth, while extension brings the hand into a position where it is still visible, rather than behind the back where it would be out of view.

The purpose of medial or lateral rotation is to change the position of the hand. Without the infraspinatus and teres minor (the lateral rotators), one would have trouble writing a prescription or playing a back-hand stroke in tennis.

Scapular region

Attaching the scapula to the vertebral column are the posteromedial scapular muscles, consisting of a large superficial trapezius covering three deeper muscles (see Fig. 3.12).

Trapezius muscle

Deriving its name from the trapezoid shape of the combined muscles on each side, each trapezius arises from the medial third of the superior nuchal line of the skull (p. 183), the ligamentum nuchae, a ligament that joins the cervical spinous processes and the spinous processes of all the thoracic vertebrae. From this long origin the fibers pass in different directions: the upper obliquely downward, the lower obliquely upward, and the middle transversely. They insert continuously into the crest of the spine of the scapula, the medial border of the acromion and the lateral third of the clavicle. The upper fibers elevate the acromion, and the lower fibers depress the medial end of the scapular spine. This effectively rotates the scapula like a winged nut so that the glenoid fossa faces upward – abduction. This rotation is greatly assisted by the lowest four digitations of the serratus anterior which inserts into the inferior angle. The middle fibers assist the rhomboid in retracting the scapula and bracing the shoulders. In general the trapezius assists in the support of the upper extremity. Beneath the trapezius lie the rhomboids and the levator scapulae.

Rhomboids and levator scapulae

The rhomboids and levator scapulae draw the medial border of the scapula medially and upward and square the shoulders (not a particularly useful movement, but necessary for good posture). The levator scapulae arises from the upper cervical vertebrae, while the two rhomboids arise from the lower cervical and upper thoracic vertebrae. From above down, the levator inserts into the medial border of the scapula above the spine; the rhomboid minor inserts into the triangular area opposite the spine; and the rhomboid major attaches to the medial border between the base of the spine and the inferior angle.

NERVE SUPPLY

The trapezius is supplied by the spinal portion of the accessory nerve, while the levator scapulae and rhomboids are supplied by branches of the anterior rami of cervical nerves 3, 4, and 5.

Axilla

The axilla is a pyramid-shaped space between the upper limb and the chest wall. It contains the neurovascular trunks as they run from the neck to the upper arm and large numbers of lymph nodes draining lymphatics from the upper limb and anterior and posterior chest, including the breast. All these structures are encased in varying amounts of fat, and the base of the pyramid is supported by the axillary fascia (see Fig. 3.17). This fascia is important clinically. The

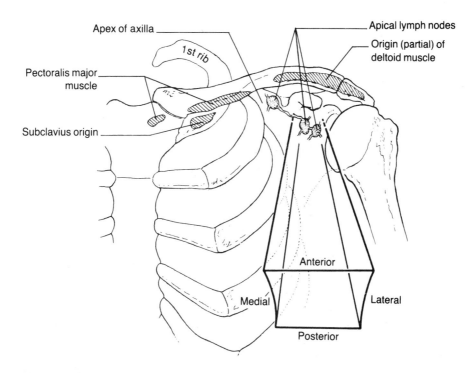

Fig. 3.14 The pyramid-shaped axilla. The anterior, posterior, medial and lateral walls narrow toward the apex, through which the neurovascular structures pass to and from the neck to the upper extremity. The apex is bounded by the upper border of the scapula, first rib and clavicle. (See also Fig. 36.6.)

lymph nodes lie deep to the fascia and can be palpated only when the fascia is relaxed, as when the arm is by the side. Elevation of the arm tenses the fascia and makes adequate examination impossible (their counterpart in the groin, the inguinal nodes, lie superficial to the fascia and normally are palpable in most people irrespective of the position of the limb).

The apex of the axilla is bounded by the clavicle, first rib and upper border of scapula, and their covering muscles. This narrow space is further narrowed by bracing of the shoulders. Compression of the neurovascular structures on their way from the neck to the arm is common and is referred to as the costo-clavicular (or thoracic outlet) compression syndrome. Malingerers can often mimic it by bracing the shoulders, and it is one way to dodge the draft (read on for other ways!).

The walls of the axilla are all muscular. The large anterior wall consists mainly of the pectoral muscles, while the equally large pos-terior wall consists of the latissimus dorsi, the subscapularis and the teres major muscles. The medial wall is the upper ribs covered by the digitations of serratus anterior, and the lateral wall is the upper medial surface of the humerus. The main neurovascular bundle runs from the upper medial wall across the axilla to reach the upper arm, veering post-eriorly to lie on the posterior wall. The pec-toralis minor, running from the upper ribs to the coracoid process, crosses this bundle and descriptively divides the structures into three portions.

Anterior wall

Pectoralis major muscle

A large muscle arising from the manubrium and body of the sternum, the upper six costal cartilages and the medial half of the clavicle, the pectoralis major runs laterally, twisting

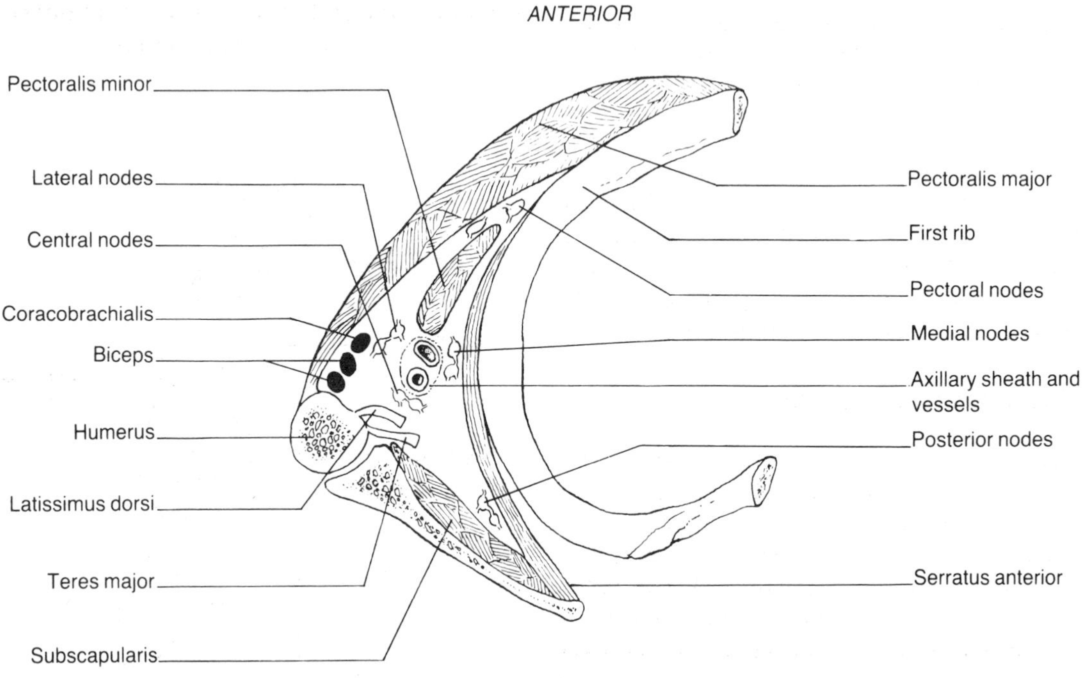

ANTERIOR

Fig. 3.15   The muscular walls of the axilla. The axilla is packed with fat, which encases the neurovascular bundles and lymph nodes. These last are grouped anteriorly (behind and between pectoral muscles), posteriorly (on subscapularis), medially (on serratus anterior), laterally (along upper lateral arm) and centrally in amongst the fat. They all drain eventually to the apical nodes.

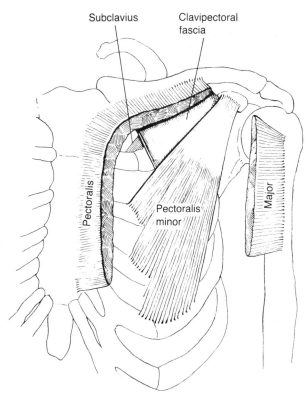

Fig. 3.16 The anterior wall of the axilla. A portion of the pectoralis major has been removed to show the underlying pectoralis minor, the clavipectoral fascia and the origin of the subclavius.

on itself to produce a rounded lower border, and inserts into the lateral lip of the bicipital groove of the humerus. It is a very powerful adductor of the arm and, in combination with the latissimus dorsi, is responsible for the flattened thorax of the adult primate.

Pectoralis minor muscle

Lying deep to the pectoralis major muscle, the pectoralis minor's attachments are ribs 3, 4 and 5 below, and the coracoid process above. Is it a useful muscle or does it merely facilitate anatomic descriptions? The latter, especially in older texts, would appear to be its more important function, but it does have some useful actions. First, as the scapula rotates around the chest wall, the pectoralis minor helps to keep it in place (although the serratus anterior is much more effective in

performing this function). Second, if the coracoid attachment of this muscle is fixed by holding firmly onto a table, it can be made to act at its costal attachment, elevating the ribs and acting as an accessory muscle of respiration. Although one does not wish to belittle the muscle, there are bigger and better muscles for both purposes. The pectoral muscles are supplied by pectoral nerves from the medial and lateral cords of the brachial plexus.

The anterior wall is completed by the fascia above and below the pectoralis minor. The fascia above this muscle, the clavipectoral fascia, splits at its clavicular attachment to enclose the subclavius. It is pierced by several structures, such as the acromiothoracic artery, lateral pectoral nerve, cephalic vein and important lymphatics from the breast. These lymphatics may carry cancer cells in breast cancer, and cannot be clinically detected because they lie deep to the thick pectoralis major. For this reason, tumors are sometimes treated by x-ray irradiation; in the past the muscles were removed with the breast (radical mastectomy).

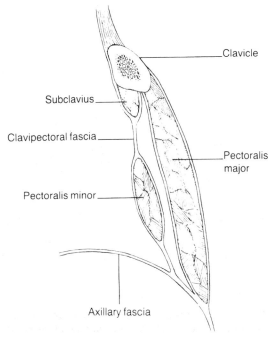

Fig. 3.17 Structures comprising the anterior axillary wall in coronal section.

Posterior wall

The posterior wall consists of three muscles that were embryologically a single muscle mass – the subscapularis, the teres major and the latissimus dorsi. Not surprisingly, these three muscles all derive their nerve supply from the posterior cord of the brachial plexus.

Subscapularis muscle

Occupying the subscapular fossa of the scapula, the subscapularis inserts into the lesser tuberosity of the humerus; this can be avulsed (pulled off) during a violent contraction.

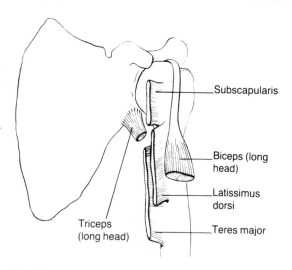

Fig. 3.19   Insertion of the muscles of the posterior axillary wall.

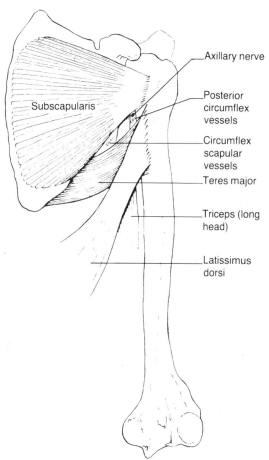

Fig. 3.18   The posterior wall of the axilla. The long head of triceps can be seen splitting the space between the subscapularis and teres major muscles into a medial triangular space for the circumflex scapular vessels, and a lateral quadrilateral space for the axillary nerve and posterior circumflex vessels.

Teres major muscle

In reality a separated portion of the muscle above or below, the teres major arises from the lower dorsal surface of the scapula and inserts into the medial lip of the bicipital groove of the humerus.

Latissimus dorsi muscle

The latissimus dorsi (the widest muscle of the back has an origin from the lower six thoracic spines, the posterior portion of the iliac crest and the lumbar fascia, through which it gains attachment to the lumbar and sacral spinous processes. On its way to its insertion by a surprisingly narrow tendon into the bicipital groove of the humerus, the latissimus dorsi picks up a few slips of origin from the lower ribs and inferior angle of the scapula. These last two origins seem unimportant – are they? Read on. If the arms, and therefore the humeral attachments of the muscle, are fixed, the costal slips elevate the ribs (as with the pectoralis minor), making the latissimus dorsi one of the accessory muscles of inspiration. By running across the inferior angle of the scapula, the muscle holds the scapula in position, allowing the bone to glide around and stay close to the chest wall.

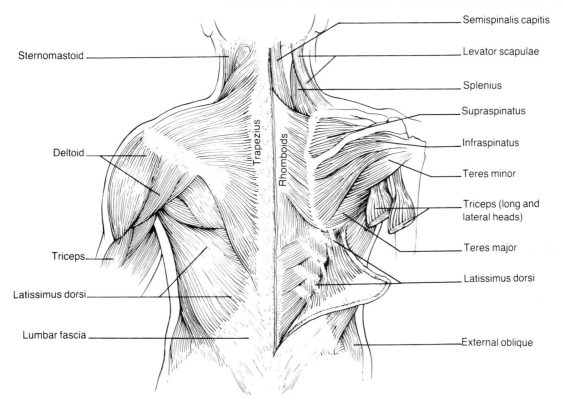

Fig. 3.20 Muscles of the posterior neck, thorax and abdomen. On the right, the latissimus dorsi can be seen picking up small costal slips and one from the inferior angle of the scapula.

This is evident clinically when the nerve supply is damaged. The scapula comes away from the chest wall during its forward movement, well described as winging, which is also seen with a denervated serratus anterior (Fig. 3.21). Interestingly, the latissimus dorsi acts as an expiratory muscle; contraction can be palpated over the lower chest during coughing, a useful clinical test for the presence of a functioning muscle. Severe paroxysms of coughing often produce pain in the muscle. Lastly, the muscle is a powerful adductor (see p. 66).

Medial wall

The medial wall of the axilla is formed by the *serratus anterior muscle*, on which lies the outer third of the breast. Arising by eight digitations from the outer surface of the upper eight ribs, the muscle is inserted into the medial border of the ventral surface of the

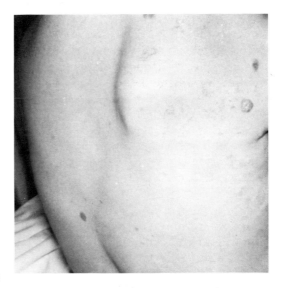

Fig. 3.21 This patient underwent a mastectomy, where the long thoracic nerve to serratus anterior was severed. The right winged scapula is easily seen. The dilated blood vessels result from postoperative radiotherapy.

scapula including, and especially, the inferior angle (see Fig. 3.15).

The serratus anterior muscle has two important actions in the movements of the shoulder joint: the scapula is pulled forward around the chest wall (as in punching), and, by virtue of its major insertion into the inferior angle of the bone, the glenoid fossa is rotated face upward, thereby producing abduction. The nerve supply of the muscle (the long thoracic nerve of Bell) runs inferiorly from the roots of C5, 6 and 7 in the neck, down the medial wall of the axilla to supply each digitation. It is at risk during axillary dissections.

Lateral wall

This is formed by two muscles covering the upper medial surface of the humerus, the coracobrachialis and the short head of the biceps.

Coracobrachialis muscle

This is discussed in Chapter 4.

Biceps muscle

The biceps muscle consists of two heads. The long head arises inside the shoulder joint from a supraglenoid tubercle and runs through the joint carrying a sheath of synovial membrane beyond the joint (see Fig. 3.11a). The short head arises from the coracoid process in common with the coracobrachialis, with which it passes inferiorly.

Axillary artery and vein

After crossing the first rib, the subclavian artery becomes the axillary artery. It runs across the axilla mediolaterally, ending on the posterior wall, where it becomes the brachial artery at the lower border of the teres major. The axillary artery is surrounded by numerous structures: the axillary vein lies on its medial side and the cords of the brachial

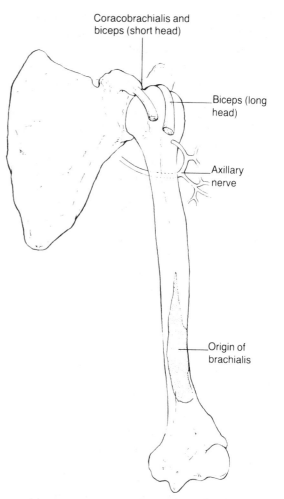

Coracobrachialis and biceps (short head)

Biceps (long head)

Axillary nerve

Origin of brachialis

Fig. 3.22   Anterior muscles of the upper arm, and the axillary nerve. The long head of the biceps tendon emerges from the shoulder joint. The common origin of the short head of biceps and coracobrachialis muscles arise from the tip of the coracoid process. The axillary nerve passes posterior to the surgical neck of the humerus.

plexus are arranged as indicated by their names. The branches of the cords retain this relationship except for one of the two branches comprising the median nerve. This branch crosses the artery and is said to occasionally compress the vessel while so doing – an unlikely story, since to our knowledge, no one has ever transposed the responsible nerve or rerouted the artery. The branches of the artery supply the axillary contents, the breast and the surrounding muscles. From above downward, these branches are the inconsequential superior thoracic, the consequential acromiothoracic which pierces the clavipectoral fascia to radiate into several branches, the lateral thoracic running along the inferolateral border of the pectoralis minor and assisting in the supply of the breast, the small anterior and the large posterior circumflex humeral arteries encircling the humerus. The largest and most distal branch, the subscapular artery, forms an anastomosis with several branches of the subclavian artery and comes into play in distal occlusions of the latter vessel.

The axillary vein is a large vessel that becomes the subclavian vein at the outer border of the first rib. Like many large veins, it is surrounded by numerous lymph nodes, enlargements of which may compress it.

Brachial plexus

The upper limb begins as a bud opposite the lower neck and upper chest of the fetus. This bud therefore receives its nerve supply from the lower cervical segments, C5 – 8, and the T1 segment of the thoracic part of the cord, with occasional help from C4 and T2. The center of the upper limb is approximately opposite the seventh segment. Although the limb elongates and descends somewhat, it never changes its nerve supply. This is unlike the vascular supply, which changes significantly during development with some vessels disappearing and other vessels being added. A nerve supply remains constant and indicates the embryologic history of any part. The anterior primary rami (roots) C5 – T1 emerge from their foramina. C5, C6, C8 and T1 join together to form the upper and lower trunks respectively, and the middle nerve (C7) goes it alone as the middle trunk. The trunks then split into anterior and posterior divisions; the three posterior divisions join to form the posterior cord, the upper anterior two divisions form the lateral cord, and the lower anterior division continues on its own as the medial cord (see Fig. 14.4). The roots and trunks are in the neck, the divisions behind the clavicle, and the cords are arranged around the axillary artery as their names indicate. At the lower border of the pectoralis minor they break up into their various branches, which are the named peripheral nerves. Like the axillary vessels, the plexus carries a sheath, derived in the neck from the prevertebral fascia, into the upper extremity (see p. 214).

The posterior cord supplies posterior structures, the sensation of the back of the arm, forearm and proximal hand, and the extensor muscles of these parts, mostly via its largest branch, the radial nerve, and the deltoid muscle together with the skin overlying its lower half by the axillary nerve.

The lateral cord supplies sensation to the lateral surface of the arm and forearm. To get there, its main branch, the musculocutaneous nerve, passes among the anterior muscles of the arm (the biceps, brachialis and coracobrachialis). Since it is a divine rule that nerves running among muscles supply these muscles, the muscular and cutaneous distribution of the lateral cord should be clear.

The medial cord supplies sensation to the medial side of the arm, forearm and hand, and through its main branch, the ulnar nerve, the muscles of the medial side of the forearm and the medial side of the hand at least (see Chapter 6 for details).

Finally, the median nerve, made up of contributions from the medial and lateral cords, supplies the remaining structures, which tend to lie in the midline of the forearm (median), and the part of the hand lying between the medial (ulnar) and posterior (radial) territories.

Dermatomes and myotomes

The dermatomes and myotomes of the upper limb have important clinical significance. Remembering the brachial plexus coming from C5–T1, and a little bit from C4–T2 on each side, you can easily work out the dermatomes. The skin has been stretched symmetrically over the developing limb, and C7,

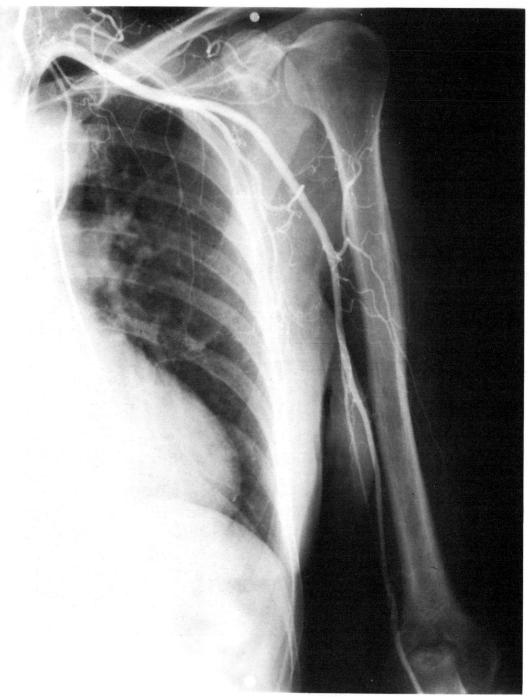

Fig. 3.23 Subclavian–axillary arteriogram. (From Weir and Abrahams 1978)

1 Catheter in left subclavian artery.
2 Aortic knuckle or knob.
3 Vertebral artery.
4 Inferior thyroid artery.
5 Suprascapular artery.

6 Internal thoracic artery.
7 Dorsal scapular artery.
8 Transverse cervical artery (superficial cervical).
9 Axillary artery.
10 Superior thoracic artery.

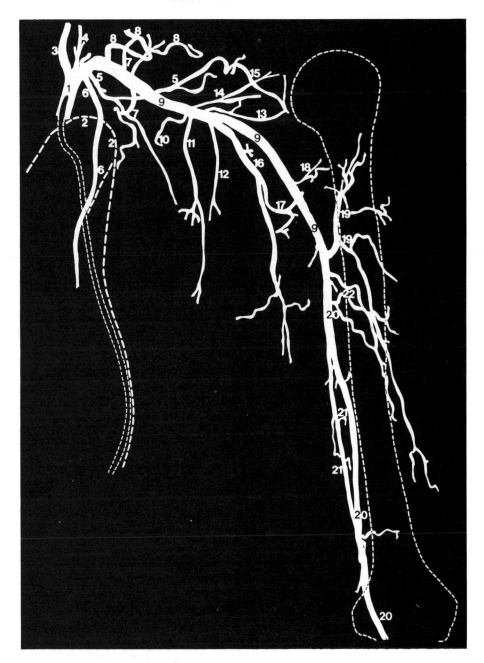

11 Lateral thoracic artery.
12 Pectoral branch, thoracoacromial artery.
13 Acromial branch, thoracoacromial artery.
14 Deltoid branch, thoracoacromial artery.
15 Anastomoses between suprascapular and acromial arteries.
16 Subscapular artery.

17 Circumflex scapular artery.
18 Anterior circumflex humeral artery.
19 Posterior circumflex humeral artery.
20 Brachial artery.
21 Profunda brachii artery.
22 Deltoid branch, profunda brachii artery.

the middle of the plexus, lies in the middle of the limb. The other dermatomes, therefore, must lie on either side of it, divided by the ventral and dorsal lines across which the overlap is minimal.

The muscles are not so simple because their movements are very precise. Nevertheless, they do progress from C5 for shoulder movements to T1 for intrinsic hand movements, with the elbow being served by C5 and 6 segments, the forearm by C6, the wrist by C6 and 7, and the fingers and thumb by C7 and 8. A little further detail is in order. C5 abducts and externally rotates the shoulder while C6, 7 and 8 adduct and internally rotate it (up with C5 and down with C6, 7 and 8, as in flapping your wings). Knowledge of the segmental muscle group innervation is important in testing nerve root compression syndromes.

Brachial plexus injuries are not uncommon. If the arm is stretched violently above the head, the lower part of the plexus (C8–T1) may suffer a traction injury. The symptoms are numbness along the inner aspect of the hand and loss of intrinsic hand muscles (Klumpke's paralysis). This occurs as a birth injury in which the baby is born bottom first (breech) and the arms are delivered in an extended position with the aftercoming head. More adventurous ways of sustaining this injury are grabbing hold of a convenient tree while falling down a mountainside, or skidding on a motorcycle and trying to catch a 'passing' lamp post.

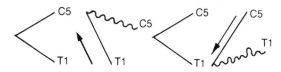

Fig. 3.25   Traction injuries of the brachial plexus. Since the brachial plexus is somewhat triangular in shape, traction of the lower part of the plexus stretches it and relaxes the upper, and vice versa.

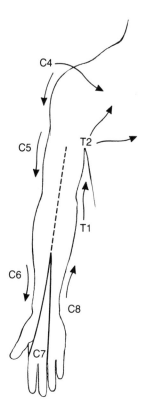

Fig. 3.24   The dermatomes of the upper limb follow each other from C4 to T2 in arithmetical progression. Single lines cannot truly represent the distribution because there is a great deal of overlap.

Sometimes the site of the plexus injury can be diagnosed. For example, if T1 is injured near the cord, the sympathetic outflow to the head and neck may be involved, producing Horner's syndrome, as evidenced by a small pupil (unopposed third cranial nerve), drooping upper eyelid (paralysis of the smooth muscle part of levator palpebrae superioris) and absence of sweating of half of the head and neck.

The upper part of the plexus (C5–6) is injured when the head and neck are forcibly separated, such as in a birth injury in which traction is applied to the shoulders to free an impacted aftercoming head in a breech delivery (Erb's palsy), or in falling off a cycle and landing on the point of the shoulder.

Erb's palsy involving C5 and 6 will result in an adducted and internally rotated shoulder (loss of abductors and external rotators) and an extended pronated elbow (loss of flexion and supination by the biceps) – the porter's tip position. In Britain it was considered poor taste to hand the porter a tip for a service; therefore he held the open palm of his hand facing posteriorly into which the tip was dropped without the exchange of a glance or a word.

4   Upper arm

Muscles · Vascular supply · Nerves · Lower half of the humerus

Muscles

Deltoid

The deltoid muscle is a large triangular muscle arising at its base from the lateral portions of the clavicle and the acromion, and narrowing to its insertion into the deltoid tuberosity on the lateral side of the humerus. It lies edge to edge with the pectoralis major. In this deltopectoral groove lies the large cephalic vein, a vein much used for cannulation. (See also Figs. 3.14, 3.21 and 4.3.)

    This powerful abductor of the shoulder is

supplied by the axillary branch of the posterior cord, which runs around the surgical neck of the humerus just below the capsule of the shoulder joint to reach the muscle and thereafter supplies the skin over the lower part of the muscle (see also Fig. 3.22). Fractures or dislocations of the humeral neck may damage the nerve, the function of which must be tested before treatment for those conditions is attempted. Failure to do so may lead to the suggestion that the nerve damage was the result of the treatment.

    The structures of the upper arm are simply arranged: the flexor muscles lie anteriorly, the extensor muscles posteriorly and the

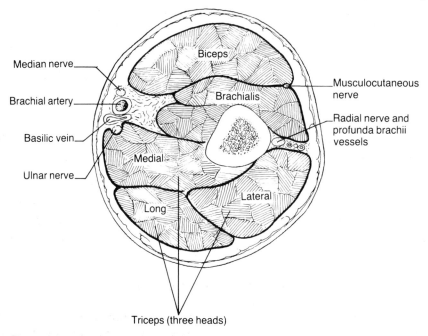

Fig. 4.1   Cross-section of the upper arm.

medially placed vascular bundle is protected by the body wall.

Flexor muscles

Figure 4.2 presents the arrangement of the flexors in the arm.

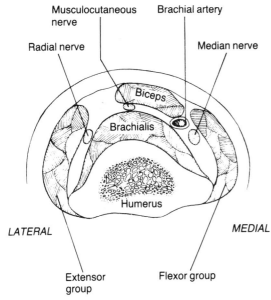

Fig. 4.2   The flexors of the upper arm. The biceps lie on the brachialis; both are flanked by the commencement of the forearm extensors laterally and the flexors medially

Biceps brachii

The long head of the biceps, having left the shoulder joint, continues between the tuberosities where it is secured by a ligament joining these prominences. After losing its synovial sheath, the tendon gives rise to a muscle belly, which is joined by the muscle belly of the short head. The two bellies end in a tendon which twists in front of the elbow joint (see Fig. 5.7) to insert into the posterior aspect of the radial tuberosity. The distal tendon gives rise to a bicipital aponeurosis on its medial aspect which passes anterior to the brachial vessels and median nerve to merge with the deep fascia and insert into the ulna.

When the tendon sheath, sliding to and fro in its bicipital groove, becomes inflamed, it gives rise to the second commonest cause of painful shoulder (bicipital tenosynovitis) – inflammation of the tendon sheath. Like an old belt, constant use may also cause fraying of the tendon and eventual rupture. Is this serious? Not unless the person is a weight lifter or the like, for other muscles will enlarge and compensate.

The action of the biceps is twofold. This muscle is not only a powerful flexor of the elbow joint, but also an equally powerful supinator of the forearm. The latter function explains why its tendon is inserted into the posterior aspect of the radial tuberosity; this permits maximal rotation of the bone. Supination can be carried out effectively only with the elbow flexed. With an extended elbow, the movement has little strength; prove this by driving in a difficult screw with the elbow extended.

Brachialis

This muscle arises from the anterior lower humerus and inserts into the coronoid process of the ulna. A sudden severe contraction can avulse the latter bony prominence. The brachialis is a pure flexor and can compensate for a ruptured biceps tendon, especially if helped by the brachioradialis.

Coracobrachialis

A small muscle, the coracobrachialis is of intriguing interest embryologically, but of little functional importance. It runs from the coracoid process to the medial side of humerus. It helps the woman keep her clutch-bag tucked under her arm as she runs for the bus, or the sergeant-major his baton in place!

Brachioradialis

The brachioradialis lies midway between the extensor and flexor groups and is responsible for the soft prominence of the lateral aspect of the elbow joint. A flexor in the midposition of the forearm (midway between pronation and supination), the brachioradialis arises from the upper part of the lateral supracondylar ridge of the humerus and inserts into the styloid process of the radius (see Fig. 5.5). It is known as the 'beer-drinker's muscle', a common syndrome affecting British first-year medical students.

It is worth stressing that few muscles act individually, and that movement is a synergistic affair. For example, when forcibly putting in a screw, each act of supination by the biceps to advance the screw is accompanied by contraction of the triceps (an extensor). This counteracts the flexion that would accompany the contraction of the biceps during supination and avoids putting the screwdriver through one's teeth.

Extensors

Triceps

The three heads of the triceps join to form a tendon which is inserted into the olecranon process of the ulna. Dropping the arm to the side straightens the elbow. Why, then, the need for this extensor muscle? For punching with a straight elbow or putting the basketball in the hoop, this muscle is a must.

Vascular supply

Brachial artery

The axillary artery becomes the brachial artery at the lower border of the teres major and ends by bifurcating opposite the neck of the radius into the ulnar and radial arteries. Although covered only by skin and fascia, it is injured surprisingly rarely because of its sheltered position on the medial aspect of the arm. Its branches supply surrounding muscles and anastomose with branches of the radial and ulnar arteries in front of and behind the elbow joint. The profunda, near the origin of the artery, is a large branch that

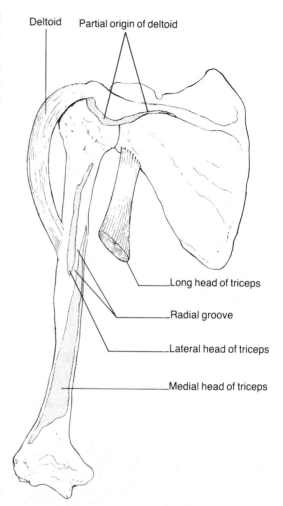

Deltoid   Partial origin of deltoid

Long head of triceps

Radial groove

Lateral head of triceps

Medial head of triceps

Fig. 4.3   The origin of the triceps muscle. The long head arises from the infraglenoid tuberosity, the lateral head from a strip of humerus above and lateral to the radial (musculospiral) groove, and the medial head from a large area below the groove but reaching up almost to the insertion of the teres major.

accompanies the radial nerve to reach the elbow anastomosis. It is a useful vessel in bypassing a brachial artery obstruction since it anastomoses with the large posterior circumflex humeral above. (See also Figs. 3.23 and 5.7.)

Veins

Veins of the upper limb are important practically because they are cannulated with increasing frequency by each succeeding

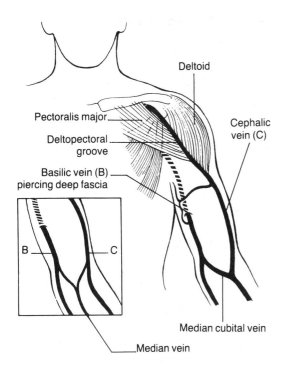

Fig. 4.4 **The cephalic and basilic veins. In the main diagram, they are joined by an 'H' arrangement; the inset shows an 'M' arrangement.**

generation of physicians. From an anastomosing network of veins of the back of the hand (notice the absence of veins in the palm of the hand), two large vessels arise at the radial and ulnar borders of the hand, the cephalic vein and basilic vein respectively.

Cephalic vein

The cephalic vein runs over the anatomical snuffbox (see p. 505 and Fig. 36.10), a remarkably constant relationship. While the position of most veins is variable, the cephalic vein and its counterpart in the leg, the saphenous vein, are the two most constantly placed veins in the extremities. One day when you desperately need to cannulate a vein, you may have cause to remember this gratefully. Running up on the radial side of the anterior forearm to the lateral side of the palpable biceps tendon, it continues on the same side of the muscle to reach the deltopectoral groove in which it lies before piercing the clavipectoral fascia to join the axillary vein.

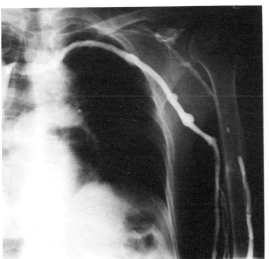

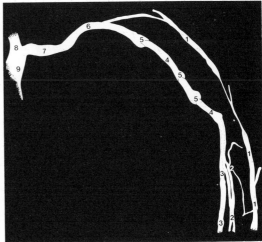

Fig. 4.5 **Axillary and subclavian venogram. (From Weir and Abrahams 1978)**

1 Cephalic vein.
2 Brachial veins (venae comitantes of brachial artery).
3 Basilic vein.
4 Axillary vein.
5 Site of valves (three in axillary vein).
6 Left subclavian vein.
7 Left brachiocephalic vein.
8 Right brachiocephalic vein.
9 Superior vena cava.

Basilic vein

Crossing the ulnar border to reach the anteromedial aspect of the forearm, the basilic vein ascends on the medial side of the biceps tendon, pierces the deep fascia in the upper arm, and joins the venae comitantes of the brachial artery to form the axillary vein at the lower border of the teres major.

Arrangement of veins at the elbow

There are two common venous arrangements at the elbow: the median vein of the forearm splits to join the cephalic and basilic veins, leading to an 'M' formation; or another vein, the median cubital, joins the veins, constituting the 'H' arrangement. These veins are all large, and it is tempting to use them for

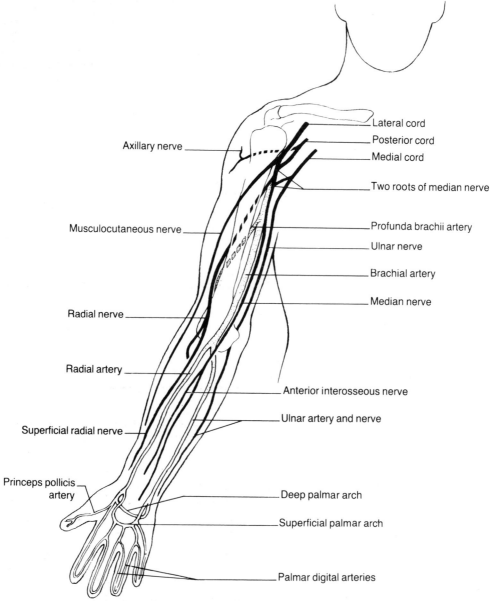

Axillary nerve

Musculocutaneous nerve

Radial nerve

Radial artery

Superficial radial nerve

Princeps pollicis artery

Lateral cord
Posterior cord
Medial cord

Two roots of median nerve

Profunda brachii artery
Ulnar nerve
Brachial artery
Median nerve

Anterior interosseous nerve

Ulnar artery and nerve

Deep palmar arch
Superficial palmar arch

Palmar digital arteries

Fig. 4.6 **The larger branches of the cords of the brachial plexus and the major arterial channels of the upper limb.**

injections or for the placement of continuous intravenous infusions. Yet wary anesthesiologists or physicians strain their eyes to find a vein on the back of the hand or forearm. Is this bravado? No! Deep to these inviting veins and separated only by the bicipital aponeurosis is the brachial artery. Injections into the artery by the uninitiated have produced disastrous results, such as gangrenous fingers.

Nerves

The three large branches of the cords of the brachial plexus pass in different directions.

Radial nerve

The large radial nerve leaves the axilla to pass posteriorly between the heads of the triceps muscles (which it supplies), and enters the radial (spiral) groove of the humerus to reach the lateral arm. In the lateral arm, at the level of the lateral epicondyle, it splits into the important motor posterior interosseus nerve (deep radial nerve) and the relatively unimportant sensory division (superficial radial nerve). The nerve is in danger as it lies in the radial groove: it can be compressed when one falls asleep with the arm over the back of the chair after celebrating the passing of the anatomy exam (Saturday-night palsy) or, less happily, it may be severed by a fracture of the middle third of the humerus.

Median nerve

Having been formed by branches of the medial and lateral cords, which enclose the axillary artery, the median nerve runs distally, crossing the brachial artery lateromedially, to reach the cubital fossa. It has no branches in the arm.

Ulnar nerve

Arising from the medial cord on the medial

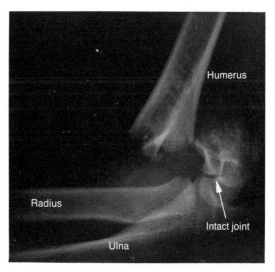

Fig. 4.7 Lateral view of a supracondylar fracture. Note the anterior displacement of the humerus, endangering the brachial artery and median nerve.

side of the brachial artery, the ulnar nerve maintains this relationship until the middle of the arm. At this point, it passes into the extensor compartment to lie in a groove on the posterior aspect of the medial epicondyle ('funny bone'), where it can be playfully twanged. It also has no branches in the upper arm.

Lower half of the humerus

The humerus flattens anteroposteriorly to end inferiorly in a pulley-shaped trochlea that articulates with the ulna, and a convex-shaped capitulum that articulates with the head of the radius (see Fig. 3.6). Above these articular surfaces are fossae in which the radius and ulna lodge in extremes of flexion and extension, and where the bone is so thin that it is sometimes deficient. No wonder that this is the site for the common supracondylar fracture of the humerus (Fig. 4.7). Two projections protrude from the nonarticular surface: the medial and lateral epicondyles.

5 Elbow, forearm and wrist

Elbow · Forearm · Lymph drainage of the upper limb

Elbow

The radius has a nicely polished head, the upper concavity of which fits into the convexity of the capitulum of the humerus. The circumference of the head fits into a circle formed by the radial notch of the ulna and the annular ligament, inside which it rotates. Below the head is a narrow neck leading to a bicipital tuberosity for the biceps tendon. The shaft of the bone has a gentle lateral curve, the purpose of which will be seen later. (See also Fig. 36.8.)

The ulna has a large trochlear notch for the trochlea of the humerus, surmounted by a prominent posterior olecranon process, on which you lean, and a smaller anterior coronoid process for the brachialis muscle. Below this, the shaft is a little sexier than the radius, with two gentle curves.

Elbow joint

The elbow joint is reliable and well fitting, an example of a pure hinge. It can do no tricks; it merely flexes and extends. The lower articular surface of the humerus fits snugly into the deep semicircular notch of the ulna, with the radius being less secure. The strongest flexor of this joint is the biceps muscle. However, should the biceps tendon rupture, as happens in older patients, one can manage very well on the brachialis and the brachioradialis. Extension is performed by the triceps.

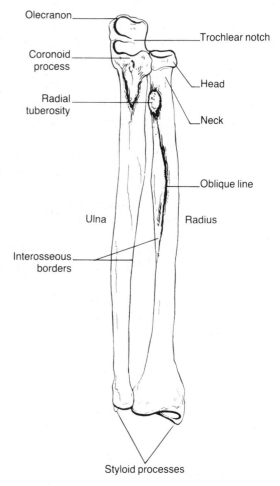

Fig. 5.1   Anterior aspect of the radius and the ulna.

Cubital fossa

In the cubital fossa, the superficial veins lie in either an 'H' or an 'M' pattern anterior to the pulsating brachial artery, from which

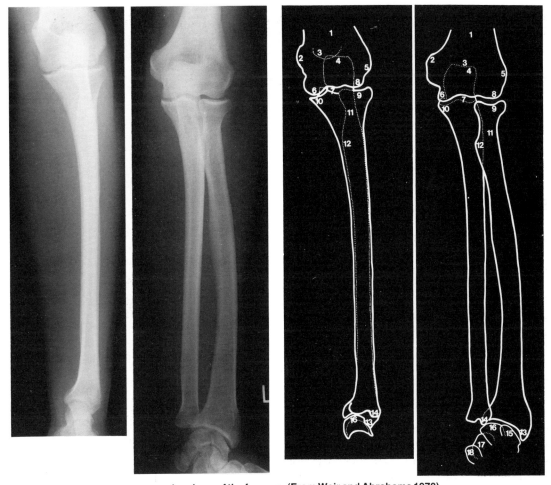

Fig. 5.2 Lateral and anteroposterior views of the forearm. (From Weir and Abrahams 1978)

1 Humerus.	6 Trochlea.	11 Neck of radius.	16 Lunate.
2 Medial epicondyle.	7 Trochlear notch.	12 Radial tuberosity.	17 Triquetral.
3 Olecranon fossa.	8 Capitulum.	13 Styloid process of radius.	18 Pisiform.
4 Olecranon process.	9 Head of radius.	14 Styloid process of ulna.	
5 Lateral epicondyle.	10 Coronoid process of ulna.	15 Scaphoid.	

they are separated by the bicipital aponeurosis. The artery has the median nerve on its medial side and the biceps tendon on its lateral side. These three structures lie on the brachialis muscle, which covers the lower end of the humerus. The two sides of the triangular cubital fossa are formed by important muscles: the pronator teres medially and the brachioradialis laterally. Two important nerves lie lateral to the biceps – the division of the radial nerve into its two terminal branches and the musculocutaneous nerve.

Forearm

Figures 5.1 and 5.2 present different views of the two bones in the forearm, the radius and the ulna.

The radius and the ulna have their bulk at different ends. The broad lower end of the radius articulates with the carpus and absorbs any impact from the hand. The force is then transmitted through the downward and medially directed fibers of the interosseus membrane to the ulna. The ulna passes the force through its large upper end to the broad lower end of the humerus. Both the radius and ulna are gently curved, but in the supine position (palm up), they lie parallel to each other.

Ulna

The ulna lies deeply anteriorly, covered by the muscles of the forearm, but one can palpate the posterior border subcutaneously all the way from its prominent upper end, the olecranon process, to the little projection at the distal end, the styloid process.

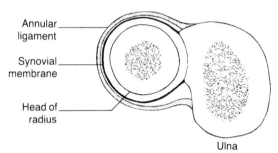

Fig. 5.3   Head of the radius inside the annular ligament.

Olecranon bursa

When you lean on your elbow, the skin is compressed between the bony subcutaneous surface and the table. To relieve this pressure, a bursa is interposed, the olecranon bursa. However, if the bursa is subjected to repeated pressure, its wall thickens to form a lump below the insertion of the triceps. This lump is present in all hard-working students! (student's elbow or olecranon bursitis).

Radius

The radius has a gentle, convexly curved shaft that ends in an expanded distal end. This end has a prominent, laterally directed styloid process that is much larger and reaches considerably lower than the ulnar styloid. In a common fracture of the distal radius (Colles' fracture), both processes are on the same level, indicating proximal displacement of the radial styloid, an important diagnostic point.

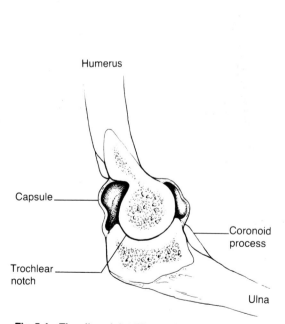

Fig. 5.4   The elbow joint. The trochlea of the humerus articulates with the trochlear notch of the ulna and is enclosed by the capsule.

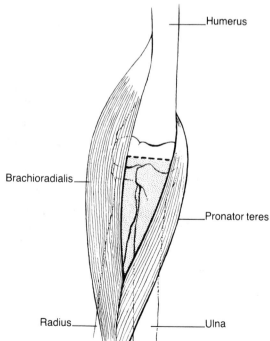

Fig. 5.5   The cubital fossa (stippled area), formed by the lower end of the humerus (dotted line), the brachioradialis laterally and pronator teres medially.

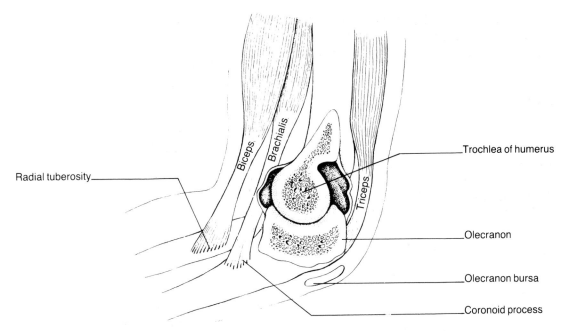

Fig. 5.6   The elbow joint in sagittal section. Anteriorly lie the biceps tendon inserting into the radial tuberosity and the brachialis attaching to the coronoid process. Posteriorly the triceps tendon inserts into the olecranon.

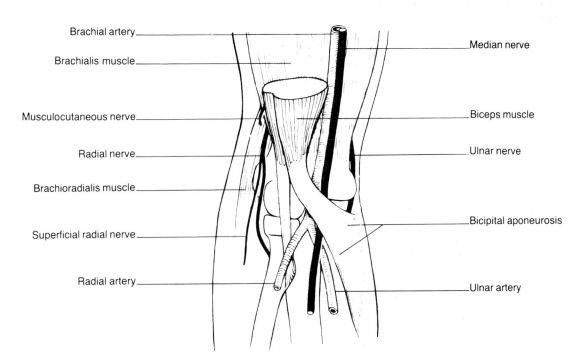

Fig. 5.7   Contents of the cubital fossa. These include the brachial artery and its bifurcation, the median nerve and the biceps tendon, with the aponeurotic expansion of the last covering and protecting the first two structures. Three nerves – the musculocutaneous, radial and ulnar nerves – appear at the outer limits of the fossa.

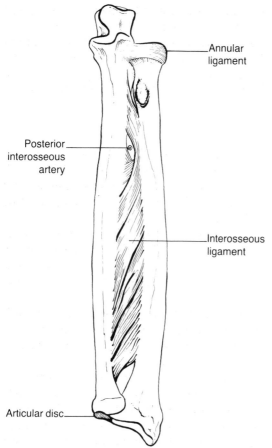

Annular ligament

Posterior interosseous artery

Interosseous ligament

Articular disc

Fig. 5.8   Anterior aspect of the interosseous membrane and annular ligament. Three structures join the forearm bones: the annular ligament, interosseous ligament and articular disc; they constitute part of the superior, middle and inferior radioulnar joints.

The upper part of the radius is buried in muscle, but the head is palpable, especially if rotated by supinating and pronating the forearm. Tenderness of the radial head is suggestive of a fracture of the head or neck of the radius, a common injury which occurs when the radius strikes the capitulum of the humerus during a fall on the outstretched hand.

The radius and ulna are joined by an interosseus membrane, a fibrous joint, whose fibers run downward and medially from the radius to the ulna. This membrane is one of several structures uniting the bones. The lower end of the radius lies superficially and is easily palpable anteriorly and posteriorly.

Posteriorly, since the muscles have become tendinous, both styloid processes are visible.

Muscles

The forearm is composed of what appears to be a bewildering number of muscles on both anterior and posterior surfaces. Attempts to memorize them are frustrating, and they should rather be considered functionally. Let's start with the movements of the forearm – pronation and supination.

Supination

There are two supinators: a superficial powerful biceps, and a deeper and less powerful supinator muscle. The supinator is composed of two heads, a deeper ulnar head whose transverse fibers encircle the neck of the radius, and a superficial humeral head whose ventral fibers insert into the radius near the anterior oblique line.

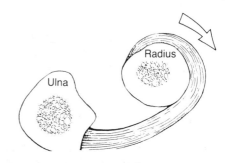

Radius

Ulna

Fig. 5.9   The supinator muscle, arising from the posterior part of the supinator fossa, surrounds and attaches to the neck of the radius. It is admirably arranged to roll the radius outward (supinate).

Pronation

Again we have a superficial and more powerful pronator, the pronator teres, and a much weaker deep pronator, the pronator quadratus. The pronator teres runs obliquely from the humerus to the area of maximal convexity of the radius and swings the radius across the ulna like a bucket handle. The pro-

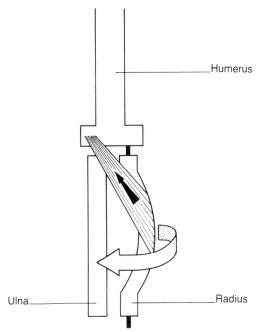

Fig. 5.10 The pronator teres muscle runs from the medial epicondyle to the apex of the curve of the radius. This enables it to rotate the radius across the ulna (pronate).

nator quadratus joins the distal ends of the radius and ulna and is seen deep to the tendons of the forearm as a small postage-stamp-shaped muscle.

Movements at the wrist joint

Muscles on both radial and ulnar sides, anteriorly and posteriorly, flex and extend the wrist joint. Without this arrangement the joint would cock excessively to one or other side (radial or ulnar deviation). These muscles are situated peripherally, and, because they act only on the wrist joint, insert no further than the carpus or proximal meta-Carpus. The flexor carpi radialis and flexor carpi ulnaris lie anteriorly and the extensors carpi radialis (2) and extensor carpi ulnaris posteriorly. Two radial muscles, extensor carpi radialis longus and extensor carpi radialis brevis, are required posteriorly to balance the insertion of the single flexor carpi radialis into the second and third metacarpal bones (the same bones into which the two

extensors are inserted). The grip of the hand is strongest with the wrist in radial deviation, which explains the more extensive radial insertions of the flexor and extensors of the wrist joint. Between these peripheral muscles lie those tendons destined for the digits.

FLEXOR ASPECT

On the flexor aspect there are tendons moving the metacarpophalangeal joints (MCP joints), the proximal interphalangeal joints (PIP joints) and the distal interphalangeal joints (DIP joints).

To understand this simplification it must be remembered that the palmaris longus, situated between the two reins, the radial and ulnar carpal flexors, is a relic of a muscle which once ran through the palm to insert into the first phalanx and so act on the MCP joint. It is a truism that any muscle that has a short belly and a long tendon is on sale; it is going out of style, and its action is being taken over by another muscle. This is the fate of the palmaris longus. This muscle acts weakly on the MCP joint because it is being replaced by sophisticated muscles, the lumbricals, which have developed over the last few million years. For our purposes, remember that the palmaris longus was originally a flexor of the MCP joints.

The palmaris longus (ye olde flexor of the MCP joints, but now relegated to the role of protector of palmar nerves and vessels by virtue of the palmar aponeurosis), is destined for the proximal phalanges, the flexor digitorum superficialis for the middle phalanges, and the flexor digitorum profundus and flexor pollicis longus for the terminal (distal) phalanges. It is interesting that the tendon running most distally, that of the profundus, as its name indicates, has to split the overlying palmaris longus and flexor digitorum superficialis tendons to reach the terminal phalanx. Presumably this is a protective mechanism, for the profundus is capable of flexing the metacarpophalangeal and both interphalangeal joints, should the sublimis and the palmaris longus be severed.

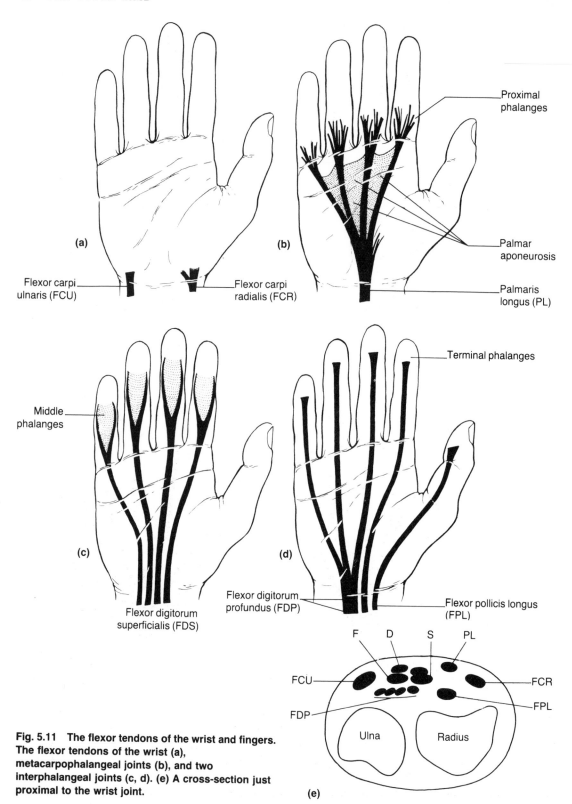

Flexor carpi ulnaris (FCU)

Flexor carpi radialis (FCR)

(a)

Proximal phalanges

Palmar aponeurosis

Palmaris longus (PL)

(b)

Middle phalanges

Flexor digitorum superficialis (FDS)

(c)

Terminal phalanges

Flexor digitorum profundus (FDP)

Flexor pollicis longus (FPL)

(d)

F   D   S   PL

FCU

FCR

FDP

FPL

Ulna   Radius

(e)

Fig. 5.11 The flexor tendons of the wrist and fingers.
The flexor tendons of the wrist (a),
metacarpophalangeal joints (b), and two
interphalangeal joints (c, d). (e) A cross-section just
proximal to the wrist joint.

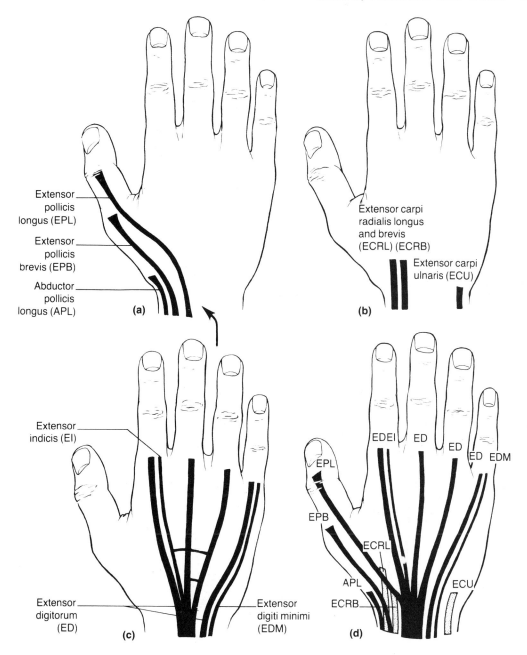

Fig. 5.12 The extensor tendons of the wrist and fingers. The extensor tendons of the thumb (a), wrist (b) and metacarpophalangeal joints (c) are depicted. In (d) are shown all the tendons crossing the posterior aspect of the wrist joint.

EXTENSOR ASPECT

With regard to the extensors, the two radial and the ulnar extensors lie peripherally and terminate in the carpus. Because of the large first web space, the muscles that move the

thumb end in tendons that must run obliquely. Mediolaterally, they are the extensor pollicis longus, extensor pollicis brevis and abductor pollicis longus, the last two keeping close company. During their oblique course, they cross the insertions of the radial

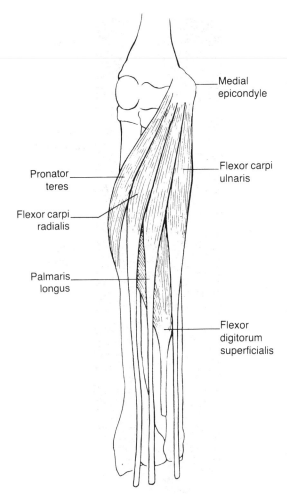

Medial
epicondyle

Pronator
teres

Flexor carpi
ulnaris

Flexor carpi
radialis

Palmaris
longus

Flexor
digitorum
superficialis

Fig. 5.13 Superficial flexor muscles of the forearm.

extensors. The hollow between the extensor
pollicis longus and the extensor pollicis
brevis, best seen on extending the thumb, is
called the anatomical snuffbox (see Fig. 36.10).
It is a convenient receptacle for sniffing snuff
(a tobacco derivative), a habit that persists
even today in parts of Europe.

The remaining muscles are destined for the
four digits, and therefore lie between the
peripheral carpal extensors. The second and
fifth fingers are, next to the thumb, the most
mobile digits and have therefore developed
extra tendons. The tendons passing down the
central aspect of the forearm therefore are the
extensor digitorum for all four fingers, the extensor indicis for the second finger and the

extensor digiti minimi for the fifth finger.
This arrangement explains the freer movement of second and fifth fingers and illustrates why the third and fourth fingers are not
good for pulling that ace out of your sock.
Limiting the third and fourth fingers further
are dorsal intertendinous bands that tether
the tendons to each other.

If we flank these digital muscles with the
radial and ulnar carpal extensors and watch
the tendons to the thumb disappearing on
their oblique way, we should have no difficulty in placing the structures where they
belong.

The muscles on the flexor aspect that have
been mentioned – the pronator teres, flexor
carpi radialis, flexor carpi ulnaris, palmaris
longus, flexor digitorum superficialis and
flexor digitorum profundus – all arise from
the anterior aspect of the medial epicondyle.
Except for the palmaris longus and the flexor
carpi radialis, all have additional origins,
some substantial, from the radius and/or
ulna. The flexor pollicis longus and pronator
quadratus arise from the anterior surfaces of
the shafts of the radius and ulna only, lying
far too low down to participate in the
common humeral origin. Those muscles
which cross the elbow joint are flexors of
this joint, although modestly so, for this is
not their prime function. Nevertheless, a
violent flexion contraction can avulse the
medial epicondyle, which may be displaced
into the joint. The muscles on the extensor
aspect – the extensor carpi radialis longus
and the extensor carpi radialis brevis, the extensor carpi ulnaris, the extensor digitorum
and the extensor digiti minimi – all arise
from the anterior aspect of the lateral condyle. This apparent paradoxical origin (extensors from anterior or flexor aspect of the
humerus) makes sense when it is realized
that the functional position of the forearm is
in the midposition. The other extensor
muscles arise from the humeral shaft (the extensor carpi radialis longus) or the radius and
ulna (the extensor pollicis longus and the extensor pollicis brevis, the abductor pollicis
longus and the extensor indicis). The

anconeus, a short muscle running from the posterior aspect of the lateral epicondyle to the posterior surface of the ulna, appears to be a piece of misplaced triceps; it is an extensor, innervated by the nerve to the medial head of triceps.

Radial and ulnar vessels

From their origin, the radial and ulnar arteries run to join the radial and ulnar nerves. The neuro-vascular bundles lie deep to the muscles forming the ulnar and radial borders of the forearm, the flexor carpi ulnaris and the brachioradialis respectively. The radial artery surfaces to lie so superficially on the lower end of the radius that it is the favorite place for palpating and counting the arterial pulse. It then passes posteriorly into the 'anatomical snuffbox'. The ulnar artery also surfaces near the wrist where it is palpable lateral to the tendon of the flexor carpi ulnaris. If you cannot feel it, it is not too late to switch professions! The central part of the forearm is supplied by the common interosseous branch of the ulnar artery, which divides into branches that run on the anterior and posterior surfaces of the interosseous membrane, the anterior and posterior interosseous vessels. The veins are venae comitantes that accompany the arteries.

Nerves

The three large nerves enter the forearm, each lying between two portions of the muscle (see Fig. 5.4). At times the nerves may become compressed; these entrapment neuropathies are well recognized.

Radial nerve

The radial nerve splits at the lateral epicondyle into the superficial radial nerve, which joins the artery under the brachioradialis, and the posterior interosseous (deep radial) nerve, which passes posteriorly between the two heads of the supinator muscle, which it supplies to reach the extensor aspect of the

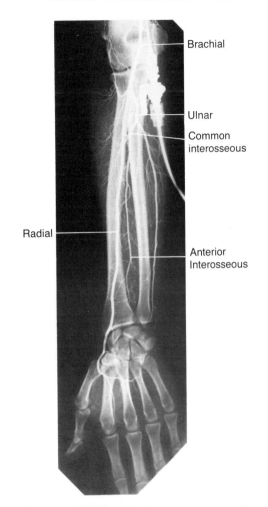

Fig. 5.14 Arteriogram illustrates the bifurcation of the brachial artery opposite the neck of the radius and the radial, ulnar and interosseous vessels. The radial artery is proceeding distally to form the deep carpal arch.

forearm. The posterior interosseus nerve supplies all the muscles of the extensor aspect of the forearm, namely the extensors and abductor of the thumb and the extensors of the fingers. The superficial radial is a bit of a washout. Reaching the radial side of the dorsum of the hand, it theoretically supplies the skin of the dorsum of the hand and the back of the radial three and a half fingers up to about the base of the second phalanges. In actual fact, if this nerve is severed, the deficit is confined to a small area around the base of

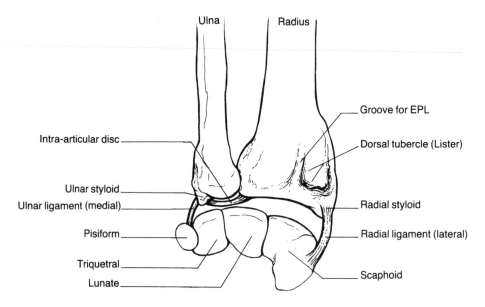

Fig. 5.15 The wrist joint. The ulna is excluded by the triangular intra-articular disc. The groove for the extensor pollicis longus (EPL) tendon is seen surmounted by the dorsal tubercle laterally.

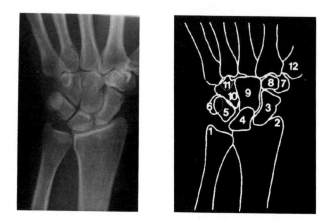

Fig. 5.16 Posteroanterior view of the scaphoid. (From Weir and Abrahams 1978).

1 Ulna, styloid process.	7 Trapezium.
2 Radius, styloid process.	8 Trapezoid.
3 Scaphoid.	9 Capitate.
4 Lunate.	10 Hamate.
5 Triquetral.	11 Hook of hamate.
6 Pisiform.	12 First metacarpal.

the thumb and index finger, which indicates its minor sensory role. Remember that the vital sensation in the fingers is in the pulp, in which the sensory system is so developed that one can differentiate between coins by feel alone.

Ulnar nerve

The ulnar nerve passes behind the medial epicondyle and between the two heads of the flexor carpi ulnaris. Running inferiorly deep to this muscle, the nerve is joined by the ulnar artery, following the example of its radial counterpart. Wedged between the flexor carpi ulnaris and the flexor digitorum profundus, the nerve supplies the former, and that half of the latter destined for the little and the ring finger. The nerve reaches and crosses the flexor retinaculum of the wrist joint.

Median nerve

As the name indicates, the median nerve runs in the midline. It enters the forearm between the two heads of the pronator teres muscle, to lie deeply between the flexor digitorum superficialis and the flexor digitorum profundus. At the wrist, the nerve escapes from the former to lie between the palmaris longus and the flexor carpi radialis tendons. This last relationship is important since the nerve must be identified in the common injury of cut wrist. In about 10 per cent of people the palmaris longus is missing, which confirms an earlier statement indicating the potential demise of the muscle. It is therefore a good example of an atavistic muscle. The nerve is then located to the medial side of the flexor carpi radialis. The nerve can also be blocked at this point to anesthetize a large portion of the hand. The median nerve, with its interosseous branch clinging to the interosseous membrane, supplies all the muscles on the flexor aspect of the forearm not supplied by the ulnar nerve. The median nerve supplies the more superficial muscles, while its anterior interosseous branch supplies the

deeper muscles between which it lies, the flexor pollicis longus and the lateral half of flexor digitorum profundus, before reaching and supplying the pronator quadratus.

Wrist joint

The proximal articulation is the lower end of the radius and its articular disc, which runs from the lower part of the notch of the radius to the ulnar styloid and excludes the ulna from the joint. This disc acts as a shock absorber and protects the delicate head of ulna from impacts. The distal articulation is the proximal row of the carpal bones. The movements are those of a condyloid joint. The joint flexes, extends, abducts and adducts, and, combining these movements, circumducts. The surface marking of the wrist joint is the distal crease at the lower end of the forearm.

We are now ready to enter the hand, and do so via the carpal tunnel.

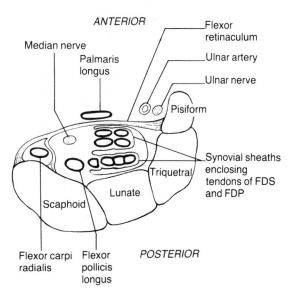

Fig. 5.17   Carpal tunnel at the level of the proximal row of carpal bones. The four tendons of the flexor digitorum superficialis (FDS) lie anterior to the flexor digitorum profundus (FDP).

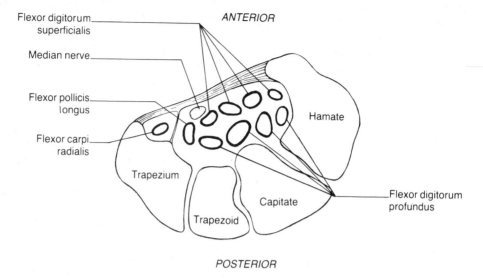

Fig. 5.18 Carpal tunnel at the level of the distal row of carpal bones. At this level the profundus tendons to the medial three fingers have separated off.

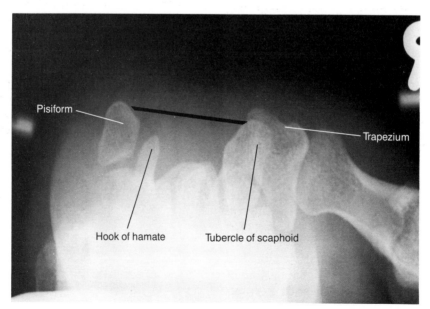

Fig. 5.19 X-ray photograph of carpal tunnel from above. The pillars on the left (medially) are the pisiform and hook of hamate, and laterally the tubercle of the scaphoid and the trapezium. The black line represents the flexor retinaculum.

Carpal tunnel

The carpal tunnel is a space not much bigger than one's thumb, through which pass most of the structures of the flexor aspect of the forearm. The tunnel is formed as a result of the concavity of the carpal bones: the pisiform and the hamate on the ulnar side, with the tubercle of the scaphoid and crest of the trapezium on the radial side. All are palpable bony points below the distal wrist crease. The tunnel is roofed in by a thick flexor retinaculum. Running over the retinaculum are the palmaris longus tendon, which is partially inserted into it, the ulnar blood vessels and nerve, and two small nerves (palmar branches of the ulnar and median nerves). Crowded together in the tunnel are nine tendons: four superficialis and four profundus heading for the digits, and the pollicis longus making for the thumb. Crunched in with these tendons, and lying most superficially, is the median nerve. A common condition is the carpal tunnel syndrome in which, for many reasons, the tight fit in this tunnel is exacerbated. The structure most liable to compression is the median nerve. Diagnosis is simple and so is the cure. The tunnel is decompressed by splitting the flexor retinaculum, and a grateful nerve emerges.

Lymph drainage of the upper limb

Although most lymph nodes of the upper extremity are congregated in the axilla, it is worth remarking on the supratrochlear node. This constant node is situated in the superficial fascia just above the medial epicondyle, lying alongside the basilic vein. The supratrochlear node receives lymph from the inner three fingers and inner half of the palm and forearm, which it passes to the axillary nodes.

Infection of the digits is a common event and so is the appearance of red streaks on the front of the forearm. These streaks are inflamed lymph vessels (lymphangitis). Note that these vessels are not seen on the back of the forearm, which suggests the route taken by these vessels. Like veins, the lymph vessels pass from the palm to the dorsum to avoid occlusion by the pressure of the grasping hand. From the dorsum, they make their way to the anterior forearm, from where they run vertically upward. Those on the medial side often stop at the supratrochlear node; those on the radial side make their way directly to the axillary nodes, with an occasional stop en route at a small node situated on the cephalic vein in the deltopectoral groove.

The axillary nodes eventually drain from the apical group via a subclavian trunk into the confluence of the jugular and subclavian veins in the neck on the right side (see Figs. 3.14 and 3.15). On the left side, this subclavian trunk drains in the same way, or it may join the thoracic duct.

6 Hand

Skeleton of the hand · Muscles and tendons · Blood vessels and nerves ·
Summary of innervation

The hand is a remarkable part of the body that separates humans from all their competitors. Its powerful grasping mechanism is matched by its ability to carry out finely controlled movements – it can crush a beer can or crack a combination safe. Were this not enough, it is the chief tactile organ of the body. It can act as an eye for the blind person, enabling her to read (Braille).

This versatile instrument, as strong as it is sensitive, is covered by skin which differs on the palmar and dorsal surfaces. On the palmar surface it is thick, ridged, immobile and hairless, and possesses a profusion of sweat glands. The thickness enables it to withstand wear and tear. The ridges increase the surface area and improve the grip; they also provide fun for the fingerprint department of the police. The skin is bound to the underlying palmar aponeurosis, again to improve the grip. This rigidity permits very little swelling, and inflammatory edema manifests on the looser dorsum on the hand, even though the infection is in the palm. Do not be fooled by the large dorsal flipper! The lack of hair implies the absence of sebaceous glands and a great grip. The sweaty palms moisten the skin to allay friction. This worked well for one's club-wielding ancestors, but less well when the clubs turned to tennis rackets.

The sensation of the skin of the hand is extraordinary and unmatched by skin anywhere else. Replacement by skin grafts from other parts of the body is therefore inferior. The dorsal skin is hairy, thinner and so loose that it can be picked up off the under-lying structures. However, this laxity does not mean it is superfluous: making a fist, one can see that there is no skin to spare.

Skeleton of the hand

Eight carpal bones are distributed in two rows. In the proximal row, lateromedially, are the boat-shaped scaphoid, the moon-shaped lunate and the triquetral supporting the pea-shaped pisiform. The second row has the trapezium with its distal saddle-shaped surface, a small trapezoid, a large capitate bone with a prominent head (caput) and a hamate with a hook. Distally lie the five metacarpals, slender bones except for the first, which is very stout and set at a right angle to the palm of the hand. The four meta-carpal bones are commonly fractured; displacement is minimal because of their surrounding muscles.

Each metacarpal bone ends in a big condyle (knuckle), which articulates with the proximal phalanx. This joint, like the wrist, moves in five directions: flexion, extension, adduction, abduction and circumduction. With regard to the thumb, the movements occur in relation to its position as it lies at right angles to the palm. Flexion and extension occur across the palm, while abduction and adduction are at right angles to the palm (Fig. 6.1).

In the medial four fingers, metacarpals give way to three phalanges, which become smaller as they proceed distally. The thumb, however, has two stout phalanges. The phal-

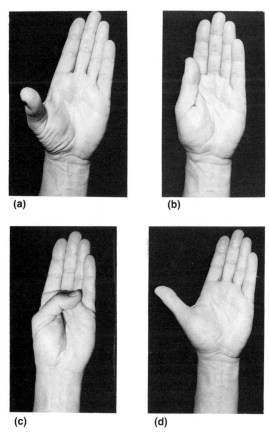

(a)   **(b)**

(c)   **(d)**

Fig. 6.1   Movements of the thumb at the metacarpophalangeal joint: (a) abduction, (b) adduction, (c) flexion and (d) extension.

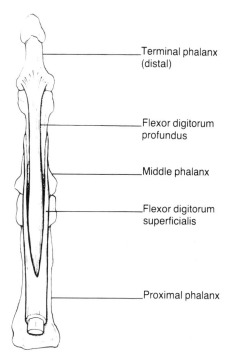

Terminal phalanx (distal)

Flexor digitorum profundus

Middle phalanx

Flexor digitorum superficialis

Proximal phalanx

Fig. 6.2   The flexor digitorum superficialis and profundus tendons overlying the three phalanges. The flexor digitorum profundus tendon passes through the split in the flexor digitorum superficialis tendon to insert into the terminal phalanx.

anges, except for the most distal, also end in condyles, which produce smaller knuckles. The interphalangeal joints are pure hinges, flexing and extending only; the interossei do not insert sufficiently distally to move them from side to side.

Muscles and tendons

Palmar aponeurosis

The palmar aponeurosis is the triangular-shaped termination of the palmaris longus tendon after it has crossed the flexor retinaculum (see Fig. 5.11). The aponeurosis is thick and splits near the base of the fingers into four divisions. It would not dare send a slip to the thumb (as its counterpart does to the big toe) because nothing is allowed to hamper the movement of this vital digit. Opposite the webs (the loose folds of skin joining the bases of the digits), each division fuses with the fibrous flexor sheath. The functions of the palmar aponeurosis are to tether the skin of the hand and to protect the underlying tendons, vessels and nerves.

The thenar and hypothenar eminences (the radial and ulnar prominences just beyond the wrist crease) are composed of the first intrinsic muscles we have encountered (intrinsic muscles are muscles which arise and end in the hand). There are three muscles in each group. The thenar muscles, which run from the flexor retinaculum to the metacarpal and the proximal phalanx of the thumb, are largely responsible for opposition, the most important movement of the thumb. The hypothenar muscles are medially attached in

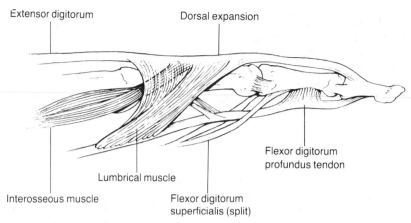

Extensor digitorum   Dorsal expansion

Flexor digitorum
profundus tendon

Lumbrical muscle

Interosseous muscle   Flexor digitorum
superficialis (split)

Fig. 6.3   The phalangeal attachments. The extensor digitorum, lumbrical and interosseous muscles join to form a dorsal expansion on the posterior aspect of the proximal phalanx. The flexor digitorum profundus passes through the split flexor digitorum superficialis to reach the distal phalanx.

a similar manner: they move the little finger comparatively minimally, but do assist in cupping the hand.

Deep to the palmar aponeurosis are the terminations of the arteries and nerves, and beneath these are the tendons, which, having originated in the forearm, are now on their way to the fingers. The tendons passing through the carpal tunnel are the flexor digitorum superficialis with its four tendons already separate, the flexor digitorum profundus with only the index finger tendon separated, and the flexor pollicis longus which has all along been its own man/ woman. These tendons enter the palm of the hand and run to their respective digits. The profundus splits the superficialis into two opposite the proximal phalanx (see Fig. 6.2). It then passes to the terminal phalanx, while the superficialis inserts into each side of the middle phalanx. In each finger both tendons lie within the fibrous flexor sheath, a thick structure which, like the flexor retinaculum, helps hold the tendons in position and prevents bowstringing during flexion. The sheath is thinner at the finger creases – the site of the joints – for maximal flexion. Inside the flexor sheath each pair of tendons is enclosed in a synovial sheath, a well-developed structure consisting of two continuous layers – a visceral layer adherent to the tendon and a parietal layer surround-

ing the visceral layer. This allows the tendons to move like a piston in a cylinder, the synovial fluid providing the lubrication.

Because muscles (lumbricals) are going to arise from the profundus, the finger sheaths are interrupted. The sheaths of the second, third and fourth fingers stop at the distal crease of the palm and start again at the proximal palmar crease, with the fifth finger sheath being partially interrupted. This sounds trite but is very important in the commonly occurring tendon injuries. The finger sheaths join in the proximal palm to form a common synovial sheath, which extends to the forearm just proximal to the wrist joint and encloses all the tendons in the carpal tunnel. If you squeeze the palm of your hand, synovial fluid in the common sheath is pushed proximally to produce a small bulge above the wrist joint.

The thumb has its own tendon sheath, and it runs without interruption to end above the wrist joint. In the region of the wrist joint there is a communication, in about 50 percent of persons, between the sheath of the thumb and the common sheath of the four fingers. This explains how a prick of the thumb involving the tendon sheath can give rise to an infection of the synovial sheath of the little finger.

The tendon sheaths on the dorsum of the hand (the paratenon), are much less highly developed than on the palmar surface because their movements are far less demanding. For this reason none reaches farther than the middle of the metacarpal bones. Severed extensor tendons are repaired much more easily than flexor tendons, for the same reason.

Lumbricals

The lumbricals are small muscles, named after the earthworm, which arise from the four profundus tendons and run on the radial side of the metacarpophalangeal joints to join the dorsal expansion. They have been developed to supersede the function of the palmaris longus.

Deep to the tendons lie the metacarpal bones with their interossei and the adductor pollicis. The latter muscle arises from the third metacarpal and part of the second metacarpal inserting into the proximal phalanx of the thumb; it adducts the thumb to the palm of the hand.

Interossei

The interossei appear impossible to understand. As their name indicates, they arise between the metacarpal bones. It is worth remembering that the movements of the fingers are related to a central axis that passes through the third finger (in the foot, the axis is through the second toe). Thus we adduct or abduct our fingers to or from the third finger; perhaps more descriptive words are oppose and separate. We must remember also that the thumb and little finger are highly developed. The thumb has more than its share of abductors and adductors and does not need any help from the interossei. The little finger has its own abductor, one of the hypothenar muscles, but it does not have an adductor.

Palmar interossei

The second, fourth and fifth fingers therefore require adductors; the third finger cannot be adducted to itself. A muscle passing from one

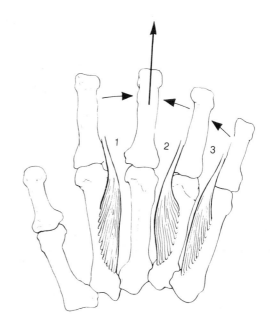

Fig. 6.4 The three palmar interosseous muscles adduct digits two, four and five to the midline.

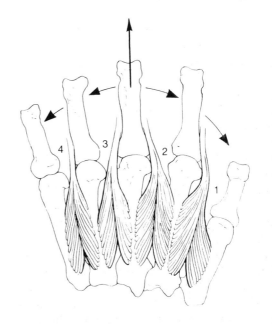

Fig. 6.5 The dorsal interosseous muscles abduct digits two, three and four, away from the midline. These muscles arise from both metacarpals between which they lie.

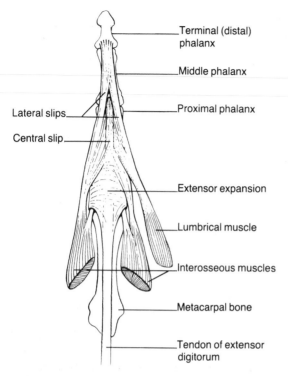

Terminal (distal) phalanx

Middle phalanx

Lateral slips

Proximal phalanx

Central slip

Extensor expansion

Lumbrical muscle

Interosseous muscles

Metacarpal bone

Tendon of extensor digitorum

Fig. 6.6 The extensor (dorsal) expansion, interosseus and lumbrical muscles. The dorsal expansion divides into a central slip for the middle phalanx and two lateral slips which join to insert into the terminal phalanx.

metacarpal bone to adduct the second, fourth and fifth fingers toward the middle finger must arise and insert on the middle finger side of the second, fourth and fifth metacarpal bones. PAD (**p**almar **ad**duct) is a useful reminder.

Dorsal interossei

The four dorsal interossei are simple to remember. The third finger can abduct away from itself to either side and so we have an abductor on either side of the third finger. We need abductors for the second and fourth fingers, as the first and fifth have their own abductors. These interossei, then, lie on either side of the middle finger and on the sides of the second and fourth fingers away from the middle finger. Remember DAB or

dorsal **ab**duct. Yes, somebody up there cleverly arranged room for all adductors and abductors to fit in between the four metacarpals, and somebody down here coined the terms DAB and PAD. All the interossei are supplied by the ulnar nerve.

Insertion of the extensor tendons

As the extensor digitorum communis, extensor indicis and extensor digiti minimi pass from the forearm to their insertion, they form an extension over, and insert mainly into, the proximal phalanx. This dorsal expansion is joined by the lumbrical and interosseous muscles. The dorsal expansion then divides distally into a central slip, for the middle phalanx, and two lateral slips, which pass on each side of the central slip to join and insert into the distal phalanx. By this means the lumbricals and interossei reach the terminal phalanx.

Actions of the extensor muscles

The following is a little bit of a fairy story, but it is so clinically correct that fine details, not entirely understood even by the experts, are unnecessary. Most of the extensor digitorum inserts into the first phalanx; its action, therefore, is mainly to extend the metacarpophalangeal (MCP) joint. The lumbricals and interossei cross the MCP joint from palmar to dorsal side and, via the extensor slips, insert into the middle and terminal phalanges. They have a unique action; they flex the MCP joint and extend the two interphalangeal (IP) joints, and are the only muscles capable of this. In summary, the MCP joint is extended by the extensor digitorum and flexed by the lumbricals and interossei, while the IP joints are extended by the lumbricals and interossei and flexed by the flexor digitorum sublimis and the flexor digitorum profundus (Table 6.1).

Table 6.1   Summary of action of the muscles in the hand

Joint(s)	Muscle(s)	Action	Nerve supply
MCP 1	Thenar group	Opposes the thumb—flexes, abducts and rotates it medially, acting at the carpometacarpal and metacarpophalangeal joints	Median
	Adductor pollicis	Adduction	Ulnar
	Extensors pollicis longus and brevis	Extension	Radial
MCP 2–5	Two lumbricals and interossei	Flexion	Ulnar
	Lateral two lumbricals	Flexion	Median
	Extensors digiti communis, indicis and digiti minimi	Extension	Radial
	Interossei	Abduction and adduction	Ulnar
IP 1	Flexor pollicis longus	Flexion	Median
	Extensor pollicis longus	Extension	Radial
PIP and DIP 2–5	Two lumbricals and interossei	Extension	Ulnar
	Two lumbricals	Extension	Median
	Flexor digitorum superficialis	Flexion	Median
	Flexor digitorum profundus	Flexion	Ulnar/median

Blood vessels and nerves

Radial and ulnar arteries

The arterial supply in the hand is a continuation of the radial and ulnar arteries. The ulnar artery is palpable as it passes across the flexor retinaculum next to the pisiform bone. It is the major contributor to the superficial palmar arch, which lies under the palmar aponeurosis and supplies digital branches to the medial three fingers. The arch is completed by a small branch of the radial artery before it passes to the dorsum of the hand. On the dorsum the radial artery lies in the snuffbox before it passes between the heads of the first dorsal interosseus into the palm. Here it lies deeply on the metacarpals while crossing the palm as the deep palmar arch. The arch sends arteries to the thumb and index (princeps pollicis and radialis indicis), and metacarpal branches to anastomose with the digital arteries. The blood supply of the hand is profuse and keeps one's fingers warm while skiing. The deep veins of the hand are mostly venae comitantes that accompany the arteries. The superficial veins of the dorsum drain mainly into the cephalic veins whilst those of the palm drain into the basilic.

Ulnar nerve

The ulnar nerve crosses the flexor retinaculum with the ulnar artery and splits into a superficial division, which supplies sensation to the medial one and a half fingers, and a deep division, which passes deeply in the hand to supply all the muscles not supplied by the median nerve – the hypothenar muscles, the interossei, the ulnar lumbricals arising from the ulnar part of the profundus, and the adductor pollicis.

Median nerve

The large median nerve passes under the retinaculum, crowded together with the tendons. As it escapes from the retinaculum, it immediately gives a branch to the thenar muscles. This branch lies very superficially just beyond the retinaculum, and a careless incision will sever the nerve with disastrous results: the thenar muscles will be paralyzed

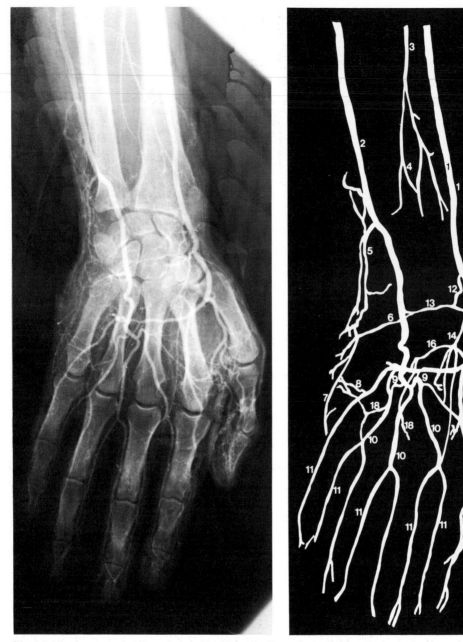

Fig. 6.7 Anteroposterior view of a hand arteriogram. (From Weir and Abrahams 1978)

1 Radial artery.
2 Ulnar artery.
3 Anterior interosseus artery.
4 Median artery.
5 Deep palmar branch of ulnar artery.
6 Palmar carpal branch of ulnar artery.
7 Dorsal carpal branch of ulnar artery.
8 Deep palmar branch of ulnar artery.
9 Superficial palmar arch.

10 Common palmar digital arteries.
11 Proper palmar digital arteries.
12 Palmar carpal branch of radial artery.
13 Palmar carpal arch.
14 Superficial palmar branch of radial artery.
15 Princeps pollicis artery.
16 Deep palmar arch.
17 Radialis indicis artery.
18 Palmar metacarpal artery.
19 Artery to radial aspect of thumb.

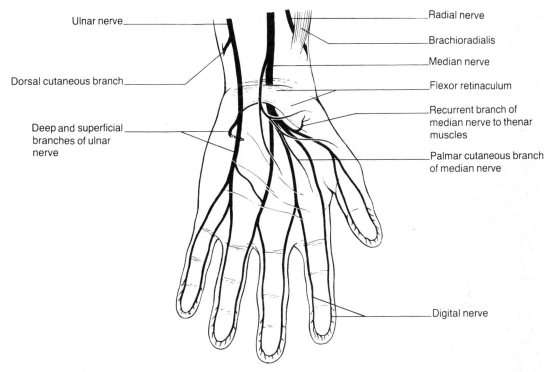

Ulnar nerve

Dorsal cutaneous branch

Deep and superficial
branches of ulnar
nerve

Radial nerve

Brachioradialis

Median nerve

Flexor retinaculum

Recurrent branch of
median nerve to thenar
muscles

Palmar cutaneous branch
of median nerve

Digital nerve

Fig. 6.8   The nerves of the palmar surface of the hand. They also supply the dorsal skin overlying the distal one and a half phalanges of the digits.

and opposition of the thumb will be lost. The median nerve then splits into two large divisions, both concerned with supplying sensation to the lateral three and a half fingers; two of these nerves supply the median two lumbricals.

Summary of innervation

Sensory

On the dorsum of the hand, the radial three and a half fingers are rather unconvincingly supplied by the radial nerve, and the ulnar one and a half fingers are supplied by a branch of the ulnar nerve that arises in the lower forearm and passes dorsally to reach the fingers. The palmar nerves, digital branches of median and ulnar nerves, supply the dorsal aspect of the fingers, including the nails, almost to the proximal interphalangeal

(PIP) joint, which means that the contribution to the nerve supply of the dorsal surface of the fingers by the dorsal nerves is minor.

Motor

The median nerve supplies all the anterior forearm muscles except the one and a half muscles supplied by the ulnar (flexor carpi ulnaris and one-half of flexor digitorum profundus). As if to get its own back, the ulnar nerve supplies all the muscles in the hand except two lumbricals and the thenar muscles. The posterior interosseous (deep radial) nerve expends itself in supplying all the muscles of the extensor aspect of the forearm and so fails to reach the hand.

Let's take a few clinical examples of nerve lesions. If the median nerve is cut at the wrist

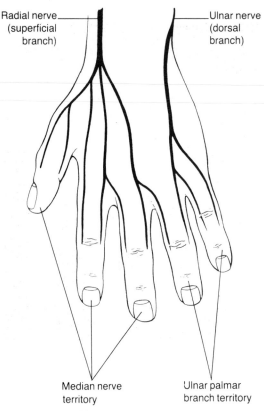

Radial nerve (superficial branch)

Ulnar nerve (dorsal branch)

Median nerve territory

Ulnar palmar branch territory

Fig. 6.9 **The nerves of the dorsal surface of the hand. The skin of the distal one and a half phalanges is supplied by palmar nerves; laterally three and a half digits (median), medial one and a half digits (ulnar).**

(as it often is), one loses sensation over all the palmar and distal part of the dorsal aspects of the lateral three and a half fingers. Opposition of the thumb is also lost, but the weakness of the two lumbricals will go unnoticed since the interossei will take over their function. If the ulnar nerve is severed at the wrist, a person will lose the hypothenar muscles and the two ulnar lumbricals; loss of the latter would not matter very much if the interossei could come to the rescue. However, the loss of all the interossei muscles with the inability to oppose and separate the fingers is very serious. Sensation to the medial one and a half fingers would be lost. If the superficial radial nerve is severed in the forearm, there will be a small area of anesthesia, as mentioned before, and no motor loss. Which is the most important nerve of the hand? Certainly the ulnar nerve with its wider muscular distribution would appear to be more important than the median nerve, but the loss of sensation to three and a half fingers is by far the more serious disability. Do not forget that the nail bed is supplied by palmar digital nerves, and when operating on the nail bed, the palmar nerves need anesthetizing. Just because the nails lie posteriorly, it does not

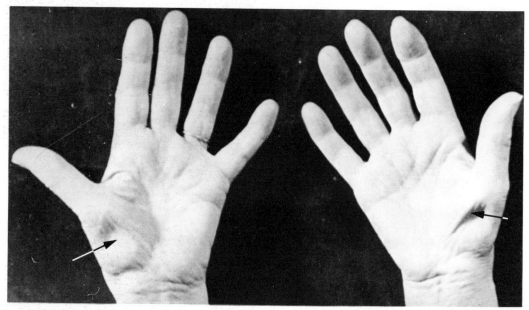

Fig. 6.10 **Bilateral median nerve compression. Note the loss of muscle bulk (arrows) in both thenar eminences.**

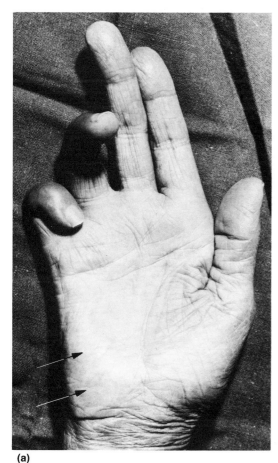

(a)

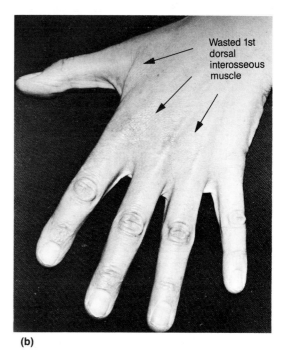

Wasted 1st
dorsal
interosseous
muscle

(b)

Fig. 6.11 Ulnar nerve palsy with clawing of the ring and little fingers (a). Note also the wasting of the hypothenar muscles (arrows, a) and the interossei (arrows, b), giving rise to 'guttering' in between the extensor tendons.

mean that the nerve supply originates posteriorly. Blocking a dorsal digital nerve will not affect the nail bed. When you incise the nail bed after blocking the wrong nerve, you will have scorn and many other things poured over you by the patient!

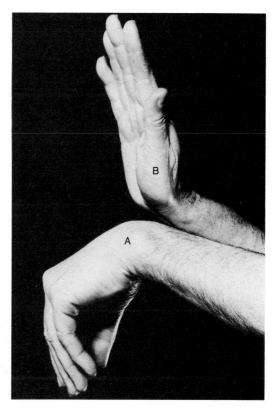

Fig. 6.12 Radial nerve lesion. The patient has been asked to extend both hands at the wrist and fingers. The right hand (B) responds normally but the left (A) has an obvious 'wrist drop'.

Table 6.2 Upper limb motor innervation – summary

	Flexor compartment	Extensor compartment
Arm	Musculocutaneous nerve	Radial nerve
Forearm	Median nerve	Radial nerve
	Except: 1. flexor carpi ulnaris 2. medial half of flexor digitorum profundus Ulnar nerve supplies the exceptions	
Hand	Ulnar nerve	
	Except: 1. three thenar muscles – flexor pollicis brevis abductor pollicis brevis opponens pollicis 2. lateral two lumbricals Median nerve supplies the exceptions	

Table 6.3 Segmental innervation of joint movements of the upper limb

Joint	Movements	Segments
Shoulder	Flexion	C5, 6
	Extension	C7, 8
	Abduction	C5, 6
	Lateral rotation	C5, 6
	Adduction	C6, 7, 8
	Medial rotation	C6, 7, 8
Elbow	Flexion	C5, 6
	Extension	C7, 8
	Pronation	C7, 8
	Supination	C6, 7
Wrist	Flexion	C6, 7
	Extension	C6, 7
Fingers (extrinsic muscle)	Flexion	C7, 8
	Extension	C7, 8
Fingers + thumb (intrinsic muscle)	All movements	T1 T1

Review questions on the upper limb

(answers p. 529)

21 Movements of the scapula play an important part in full usage of the upper extremity. In this respect, the following is/are true:
a) reaching far forward is aided by the rhomboids
b) abduction to 120 degrees necessarily involves the action of the trapezius
c) 'winging' of the scapula may result from paralysis of the lower serratus anterior
d) if spinal nerve C5 is torn out from the spinal cord, the scapula is totally immobilized
e) Scapular mobility can compensate to a large extent for a 'frozen shoulder' (immobility of glenohumeral joint)

22 The brachial plexus
a) enters the axilla anterior to the scalenus anterior muscle
b) has no branches from the trunks
c) has radial and axillary nerves that are formed from posterior primary rami
d) is enclosed in a fascial sheath derived from the prevertebral fascia
e) has five major terminal branches – the radial, ulnar, median, axillary and long thoracic nerves

23 A patient is seen with a cut over the palmar surface of the wrist. It is noticed that he cannot pick up a small coin between his thumb and index finger. His probable injury is severance of the
a) ulnar nerve
b) median nerve
c) anterior interosseous nerve
d) adductor pollicis
e) flexor carpi radialis

24 A deep cut on the back of the upper arm may lead to
a) claw hand
b) paralysis of thenar muscles
c) an inability to grasp a hammer strongly
d) anesthesia of the medial two fingers
e) loss of pronation of the forearm

25 A young man falls on his hand and complains of severe pain in his wrist in the 'anatomical snuffbox'. He probably has a fractured carpal bone. Which is the most lateral carpal bone articulating with the radius?
a) Lunate
b) Scaphoid
c) Hamate
d) Pisiform
e) Triquetral

26 Which of the following cannot flex the elbow joint?
a) Pronator teres
b) Coracobrachialis
c) Brachioradialis
d) Brachialis
e) Palmaris longus

27 Surgical relief for overstretching of the ulnar nerve may involve removal of the medial epicondyle of the humerus, allowing the nerve to be moved to the front of the elbow joint. Muscles attached to the medial epicondyle include the
a) flexor pollicis longus
b) pronator quadratus
c) flexor pollicis brevis
d) brachioradialis
e) flexor carpi radialis

28 The axillary nerve may be injured in an anterior dislocation of the shoulder. Such an injury may be diagnosed by demonstrating impaired skin sensation over the

a) acromion
b) upper half of the deltoid muscle
c) lower half of the deltoid muscle
d) spine of the scapula
e) medial wall of the axilla

29 A surgeon's hand is accidentally cut by his assistant. A few weeks later, flattening of the thenar eminence indicates that the nerve supply has been interrupted. The nerve in question arises from the

a) ulnar nerve at the distal wrist crease
b) radial nerve in the anatomical snuffbox
c) median nerve where it crosses the flexor retinaculum
d) median nerve at the distal wrist crease
e) median nerve immediately distal to the flexor retinaculum

30 The median nerve innervates the
a) adductor pollicis
b) skin of the entire palm
c) nail bed of the index finger
d) four lumbrical muscles
e) palmaris brevis

31 The cephalic vein
a) is found on the medial side of the arm and forearm
b) accompanies the brachial artery
c) joins the brachial veins halfway up the arm
d) drains the radial side of the dorsal venous arch of the hand
e) lies in the groove between the clavicular and sternal heads of the pectoralis major muscle

32 Nerves important in carrying out pronation of the forearm are the
a) median nerve alone
b) radial nerve with its posterior interosseous branch
c) musculocutaneous nerve alone
d) median nerve and its anterior interosseous branch
e) radial and ulnar nerves combined

33 In Erb's palsy, the upper trunk (or upper two nerve roots) of the brachial plexus is injured. Which of the following actions would be likely to be noticeably impaired?
a) Flexion of the elbow joint
b) Pronation of the hand
c) Internal rotation at the shoulder
d) Adduction at the glenohumeral joint
e) Flexion of the extrinsic hand muscles

34 External (lateral) rotation at the glenohumeral joint is essential for full abduction of the arm. A blacksmith therefore needs a strong external rotator muscle such as the
a) subscapularis
b) supraspinatus
c) infraspinatus
d) teres major
e) latissimus dorsi

35 The crowded carpal tunnel contains
a) eight tendons and a nerve
b) nine tendons and a nerve
c) the tendon of palmaris longus
d) the palmar cutaneous branch of the ulnar nerve
e) the terminal part of the radial nerve

36 The fingernail of the index finger requires removal because of an underlying abscess. The nerve that requires blocking by a local anesthetic agent is the
a) dorsal digital branch of the ulnar nerve
b) dorsal digital branch of the radial nerve
c) palmar digital branch of the ulnar nerve
d) palmar digital branch of the median nerve
e) terminal branches of the posterior interosseous

37 A 90-year-old boxer ruptures his biceps tendon during an argument. After the brawl it is noticed that he can still bring a flagon of beer to his mouth without difficulty. His flexion is now accomplished by the following muscles:
a) brachialis
b) brachioradialis
c) pronator teres
d) flexor carpi radialis
e) all of these

38 The lumbrical may not look like much of a muscle, but it

a) can extend the interphalangeal joints of the thumb

b) can flex the interphalangeal joints

c) flexes the metacarpophalangeal joint

d) abducts the metacarpophalangeal joint

e) adducts the interphalangeal joint

39 The interossei look much more impressive and

a) are essential in grabbing a rebounding basketball

b) the dorsal ones are innervated by the radial nerve

c) the palmar ones are innervated by the median nerve

d) each finger has two of them

e) their actions include extension of the metacarpophalangeal joints

40 A child sustains a fracture of the humerus just above the condyles (supracondylar fracture). She cannot extend the metacarpophalangeal joints of her fingers or oppose her thumb to the other fingers. She may have injured the

a) median nerve

b) ulnar nerve

c) median and ulnar nerves

d) radial nerve

e) radial and median nerves

41 In the upper arm the

a) coracoacromial ligament protects against upward dislocation of the shoulder joint

b) clavipectoral fascia is pierced by the basilic vein and lymphatics draining the breast

c) subacromial (subdeltoid) bursa is prone to inflammation from degenerative disease of the infraspinatus tendon

d) axillary nerve encircles the anatomical neck of the humerus

e) rounded contour of the shoulder is due to the rotator cuff muscles

42 Which of the following nerves has a cutaneous component?

a) The anterior interosseous

b) Long thoracic (nerve of Bell)

c) Suprascapular

d) Posterior interosseous

e) Axillary

43 A patient with a cervical rib suffers from symptoms referable to the C8–T1 component being stretched. This may cause

a) weakness in protraction of the shoulder

b) wasting of the biceps brachii

c) anesthesia over the thumb and index finger

d) wasting of the first palmar interosseous muscle

e) wasting of the forearm flexor muscles

44 The smooth rounded contour of the human shoulder is produced by the

a) acromion process

b) multipennate deltoid muscle

c) greater tuberosity of the humerus

d) acromial end of the clavicle

e) the deltoid tuberosity

45 You win a set of carpal bones in a medical school raffle. In assembling them correctly you must remember that the scaphoid articulates with all of the following except the

a) articular disc

b) trapezium

c) lunate

d) radius

e) capitate

Part III The Vertebral Column

7 Vertebral column

Development of spinal curvatures · A typical vertebra · Ossification of a vertebra and maldevelopment · Cervical vertebrae · Thoracic vertebrae · Lumbar vertebrae · Ligaments of the spinal column · Movements of the spinal column · Spinal cord localization · Intervertebral discs

The spinal column, also known as the vertebral column, consists of vertebrae superimposed on each other and separated by articular discs. There are 33 vertebrae, of which 24 are separate bones. The lower nine are fused, five forming the sacrum and four forming the coccyx.

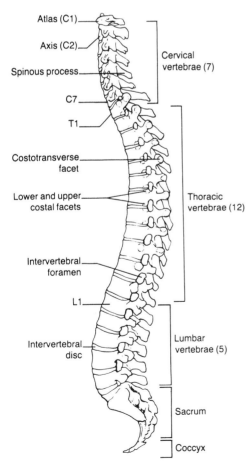

Fig. 7.1   Lateral view of the vertebral column.

- Atlas (C1)
- Axis (C2)
- Spinous process
- C7
- T1
- Costotransverse facet
- Lower and upper costal facets
- Intervertebral foramen
- L1
- Intervertebral disc
- Intervertebral disc
- Cervical vertebrae (7)
- Thoracic vertebrae (12)
- Lumbar vertebrae (5)
- Sacrum
- Coccyx

Development of spinal curvatures

The curves of the spinal column develop in the following way. The fetus, curled up in the uterus, remains that way in early infancy and shows two primary curves, thoracic and sacral. When the infant starts lifting his head at 3 months, he develops a cervical curve (B), and when he starts sitting up at 5 months, he develops a lumbar curve. Both of these secondary curves are convex forward (C). When, sadly, man passes into old age (D), he returns to his fetal position, a gentle C-shaped curve.

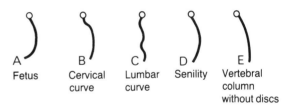

A	B	C	D	E
Fetus	Cervical curve	Lumbar curve	Senility	Vertebral column without discs

Fig. 7.2   The shape of the vertebral column through the ages.

The reason for this is that the shape of the vertebral column is determined largely by the intervertebral discs and to a much lesser degree by the bones. If the discs are removed from the vertebral column, the curves disappear (E). This has its counterpart in the degenerating discs of the elderly person. The bones do, however, play some part in shaping the vertebral column, especially in the thoracic region, where they are deeper behind than in front, resulting in the normal dorsal curve.

115

A typical vertebra   (Fig. 7.3)

Examine the construction of a vertebra. Note the body, which bears the weight of the vertebra above, and behind which is a hollow ring, called the vertebral foramen, surrounded by an arch. When you place the bodies on top of each other, the succession of vertebral foramina form a vertebral canal that contains the spinal cord with its surrounding structures. The vertebral arch consists of two pedicles, or pillars, attached to the body, giving way to sloping laminae which form the roof of the arch and join to form the spinous process (see Fig. 31.4).

At the junction of the pedicles and laminae are laterally projecting processes, the transverse processes, and two articular processes, the superior and inferior processes, pointing in various directions. The processes are levers for the attachment of muscles, which move the vertebra in different directions.

The transverse processes in the thoracic region also articulate with the ribs. The formation of the ribs in vertebrae other than the thoracic has been suppressed, the costal elements having been incorporated into the transverse processes in the cervical and lumbar regions and into the lateral masses of the sacrum.

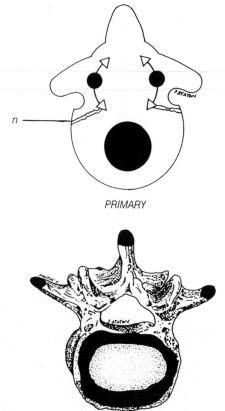

PRIMARY

SECONDARY

Fig. 7.4   Primary and secondary ossification centers of a typical vertebra (n denotes neurocentral synchondrosis).

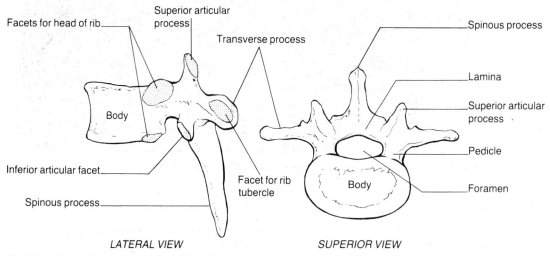

LATERAL VIEW   SUPERIOR VIEW

Labels: Facets for head of rib; Superior articular process; Transverse process; Body; Inferior articular facet; Facet for rib tubercle; Spinous process; Spinous process; Lamina; Superior articular process; Pedicle; Body; Foramen

Fig. 7.3   A typical thoracic vertebra. In the lateral view of this vertebra, extra facets are seen on the transverse processes and sides of the body for articulation with the ribs.

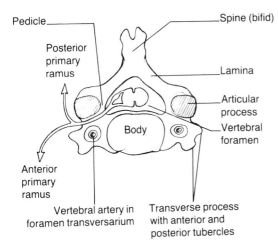

Fig. 7.5 A typical cervical vertebra.

Notice in Fig. 7.8 that the attachment of the pedicles to the body are such that gaps are left above and below. On assembling the vertebrae, these gaps form foramina for the passage of spinal nerves and vessels.

Ossification of a typical vertebra and maldevelopment (Fig. 7.4)

The body of the vertebra ossifies from two centers which fuse to form a single center. Each half of the neural (vertebral) arch has a separate center. The line of fusion between the arch and body centers lies across the posterolateral part of the body, which means that part of the body is actually ossified from the neural arch center. The line of fusion is called

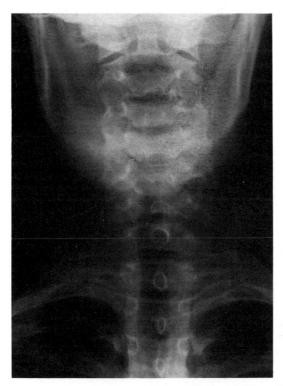

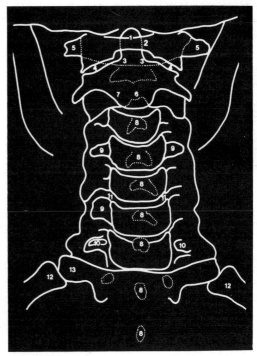

Fig. 7.6 Anteroposterior view of cervical spine. (From Weir and Abrahams 1978)

1 Odontoid process of axis (dens).
2 Anterior arch of atlas.
3 Posterior arch of atlas.
4 Atlantoaxial joint.
5 Transverse process of atlas.
6 Spinous process of axis (C2).
7 Lamina of axis.

8 Spinous processes (C3–T1).
9 Transverse processes.
10 Foramen transversarium of C7.
11 Synovial joint between arches of C5 and C6.
12 First rib.
13 Transverse process of T1.

the neurocentral synchondrosis, which becomes ossified in early life. Five secondary epiphyseal centers appear at puberty: one for the tip of the spine, one for the tip of each transverse process, and one for the cartilage on the upper and lower surfaces of each body. These epiphyseal centers fuse at maturity except for the central part of the cartilage that lies on the two surfaces of the body. These plates of cartilage lie on either side of the intervertebral disc and assist in their nutrition.

Some important clinical syndromes result from defects of this ossification process:

1 The two ossification centers for the body may not fuse. This is seen radiologically as a vertical gap. Much more serious is the development of only one center for half of a body, with a resultant very severe lateral bending of the spine (scoliosis).

2 The vertebral arches normally fuse posteriorly during the first year; failure to do so produces a classic defect known as spina bifida.

3 The neural arch sometimes ossifies from two ossification centers. The synchondrosis between these centers lies between the superior and inferior articular pro-

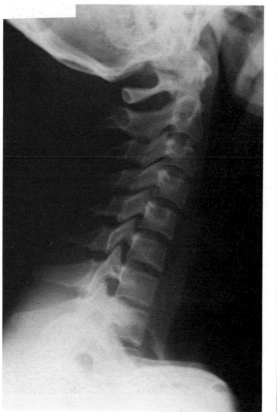
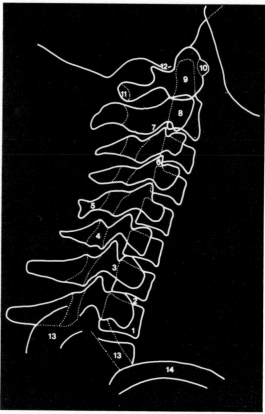

Fig. 7.7 Lateral view of cervical vertebrae. (From Weir and Abrahams 1978)

1 Body of T1.
2 Intervertebral space.
3 Pedicle of C7.
4 Lamina of C6.
5 Spinous process of C5.
6 Superior articular process of C4.
7 Inferior articular process of C2 (axis).
8 Transverse process of C2 (axis).
9 Odontoid process of axis (dens).
10 Arch of C1 (atlas).
11 Posterior tubercle of C1 (atlas).
12 Atlanto-occipital joint.
13 First rib.
14 Clavicle.

cesses; if ossification fails to occur, the body weight will produce a sliding forward of the anterior segment. A classic example of this condition is when the anterior part of the fifth lumbar vertebra slides forward toward the pelvis, which it may narrow sufficiently to interfere with childbirth (spondylolisthesis).

4 The secondary epiphysis of the transverse process may fail to fuse, and the shadow on x-ray examination may suggest either a fracture or even a calculus (stone) in the ureter.

Cervical vertebrae (Figs. 7.5–7.8)

The cervical vertebrae are relatively small because they bear less weight than the succeeding vertebrae. They sit on one another like a stack of easy chairs. The vertebral foramina are the largest in the spinal column because they house the large cervical enlargement of the spinal cord. The transverse processes are unique in that each possesses a large foramen (transversarium) to conduct the vertebral artery and

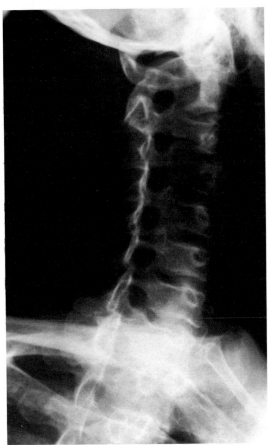 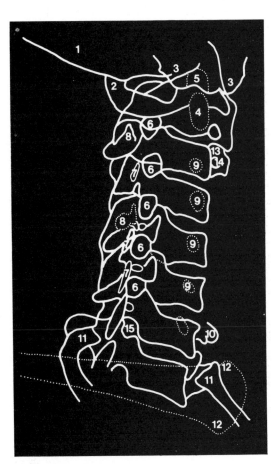

Fig. 7.8 Oblique view of cervical vertebrae. (From Weir and Abrahams 1978)

1 Occipital bone.
2 Posterior arch of atlas.
3 Mastoid air cells.
4 Transverse process of axis (C2).
5 Odontoid process of axis (dens).
6 Intervertebral foramina.
7 Laminae.
8 Bifid spinous processes.

9 Transverse processes.
10 Anterior tubercle of transverse process of C7.
11 First rib.
12 Sternal end of clavicle.
13 Transverse process of opposite side.
14 Foramen transversarium of C3.
15 Site of spinal nerve C8.

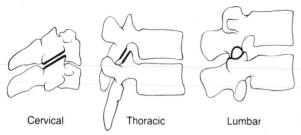

Cervical   Thoracic   Lumbar

Fig. 7.9 The articular surfaces of the three types of vertebrae: cervical sloping, thoracic anteroposterior, and lumbar medial and lateral (a portion of the superior articular process has been removed to show direction of the articulating facets). Only the cervical vertebrae can dislocate without fracturing.

accompanying structures to or from the cranial cavity. The articular processes slope – the superior mainly upward and the inferior mainly downward (Fig. 7.9). Place two cervical vertebrae together, and note that it is possible to slide one off the other. This is the danger in neck injuries. Unlike the other vertebrae, cervical vertebrae can be dislocated with less force than is required to fracture the bone. Furthermore, a cervical vertebra can dislocate and crush the large cervical cord, and the dislocation can then self-reduce. To one's astonishment, the routine x-ray examination of the neck in a patient with a crushed cord may show no abnormality. The true situation is seen when the x-ray is taken with the neck gently flexed.

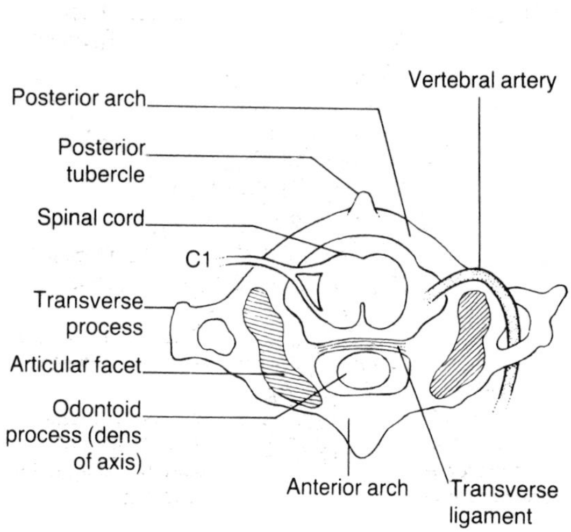

Two vertebra are completely atypical (the atlas and the axis); and two have slight variations, the sixth and seventh.

Atlas

The atlas is shaped like an irregular ring with anterior and posterior arches. It has lost its body, which is attached to the vertebra below, the axis. On the upper surface of the atlas are two curved articular facets, which articulate with the occipital condyles of the skull. On its lower surface are flat articular facets for the axis. Because of the shape of the atlanto-occipital articulation, only flexion and extension are possible at this joint, rotation taking place at the atlantoaxial joint. Nodding 'yes' takes place at the atlanto-occipital joint. Shaking one's head 'no' occurs mainly at the atlantoaxial joint and to a lesser degree at the joints below. This appears to be so difficult to understand that most politicians remain noncommittal and do not move their heads.

The transverse processes of the atlas are so large that they are palpable high up in the neck. Their foramina transmit the vertebral artery and accompanying structures. The artery then curves behind the facet, where it occupies a large groove before entering the foramen magnum. Because of the size of the transverse process, the vertebral artery has to

Posterior arch
Posterior tubercle
Spinal cord
C1
Transverse process
Articular facet
Odontoid process (dens of axis)
Vertebral artery
Anterior arch
Transverse ligament

Fig. 7.10 View of the atlas from above. The vertebral artery enters the foramen transversarium and passes medially into the foramen magnum.

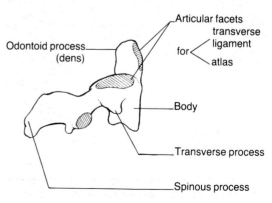

Odontoid process (dens)
Articular facets for transverse ligament / atlas
Body
Transverse process
Spinous process

Fig. 7.11 Lateral view of the axis.

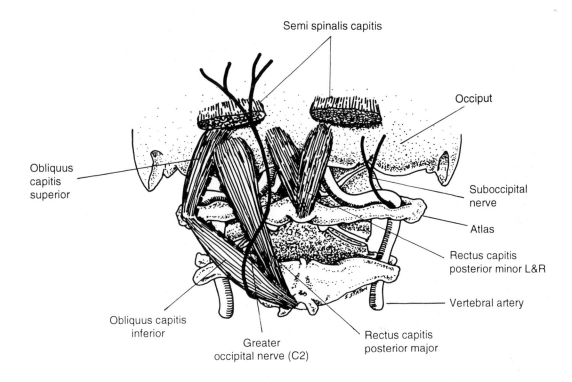

Fig. 7.12 The suboccipital triangle formed by the rectus capitis and the obliquus capitis superior and inferior. The vertebral artery courses through the triangle and the suboccipital nerve emerges from it.

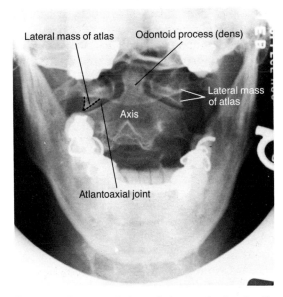

Fig. 7.13 Radiograph through the mouth, showing the axis with its odontoid process between the lateral masses of the atlas.

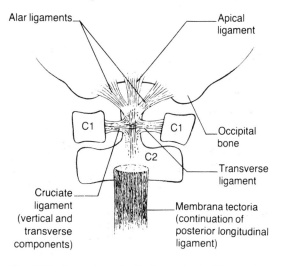

Fig. 7.14 Base of the skull, atlas and axis. Ligaments are seen joining the first two cervical vertebrae to each other and to the skull.

make a big lateral swing from axis to atlas, after which it curves back again (Figs. 7.10 and 7.12). This can give rise to vascular problems in the rigid arteries of the elderly. Crossing the anterior part of the large vertebral foramen is a transverse ligament, which divides the foramen into a smaller anterior area for the peg-like process of the axis and a large posterior space for the spinal cord.

Axis

The second atypical vertebra, the axis, is characterized by having the body of the atlas attached to it. It projects upward as a peg-like process, the odontoid process. This articulates posteriorly with the transverse ligament and anteriorly with the posterior surface of the anterior arch of the atlas. The odontoid process is further fixed by ligaments running from its apex and sides to the margins of the foramen magnum. It narrows inferiorly to a neck, through which it is united to the body of the axis. In young people the neck possesses a plate of cartilage that represents the intervertebral disc; this tends to disappear with time, although some remnants persist into old age.

A fracture of the odontoid process is a very dangerous affair because displacement compresses the cord and leads to the rapid onset of rigor mortis. This is the likely mode of death in execution by hanging. Sometimes inflammatory conditions in the area produce softening of the ligaments and dislocation of the odontoid. Under these conditions, a sudden movement, such as moving the patient from the bed to a chair, produces backward displacement of the process and a sudden demise.

The axis has a huge spine for the attachment of muscles acting on the head and neck. This should serve to indicate that it is the main site for rotation of the head.

Sixth and seventh cervical vertebrae

The two vertebrae with minor variations are

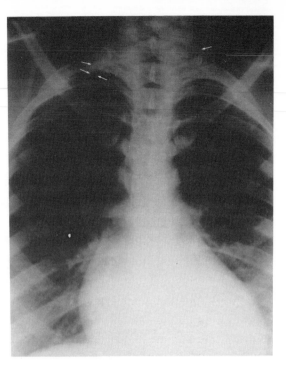

Fig. 7.15 Radiograph of the lower neck and chest demonstrating bilateral cervical ribs (indicated by white arrows).

the sixth and seventh cervical vertebrae. The anterior tubercle (p. 243) of the sixth cervical vertebra is called the carotid tubercle because it is against this vertebra that the carotid artery can be compressed to control unexpected hemorrhage. The seventh cervical vertebra, or vertebra prominens, has a long spine, second only to that of the first thoracic vertebra. The foramen transversarium is present as usual, but the vertebral artery does not know about this since it enters the foramen of the sixth vertebra. The vertebral vein, however, has heard about the foramen transversarium and some of its branches pass inferiorly through the seventh foramen.

The transverse process of the seventh cervical vertebra is important because its anterior part, the costal element, may get ideas of grandeur and enlarge sufficiently to form a rib. This cervical rib is capable of producing symptoms by making it more difficult for the neurovascular structures to pass from the neck to the axilla (Fig 7.15).

Thoracic vertebrae

The thoracic vertebrae are characterized by facets on the side of the bodies and on the transverse processes for articulation with the corresponding ribs (see Fig. 7.4). The vertebral foramina are small, and the laminae overlap like the tiles on a roof, preventing external penetration by sharp instruments. The spinous processes are long and slope severely downward to end in a knob. The articular processes face backward and forward, and dislocation without fracture of these processes is impossible.

Lumbar vertebrae

The large lumbar vertebrae enlarge as we descend the column; the fifth vertebra is huge and carries the weight of the whole vertebral column. The laminae do not overlap; the spines are massive and point directly backward, without overlap. This allows us to pass a needle with relative ease into the spinal canal, a procedure known as lumbar puncture. Articular processes face inward and outward, and dislocation without fracture is impossible. The intervertebral foramina in the lumbar region decrease in size, while the lumbar nerves increase in size, from above downward. This peculiar arrangement explains the commonness of sciatica, a condition in which the tight fit of L4 and 5 nerves increases for various reasons; for example, the new bone laid down in osteoarthritis narrows the foramina even further and causes shooting pains down the leg.

Ligaments of the vertebral column

The ligaments joining the vertebrae are the ligamenta flava, the supraspinous and interspinous ligaments, and the ligamentum nuchae. Also related to the vertebral bodies are the anterior and posterior longitudinal ligaments.

Ligamenta flava

The ligamenta flava connect the laminae. Composed of yellow elastic tissue, they prevent too-rapid flexion of the spine, thereby guarding against disc injury. By virtue of their elasticity, they assist restoration of the spine to the erect position.

Supraspinous ligaments

The supraspinous ligaments connect the tips of the spinous processes from C7 to the sacrum. They are often injured when sudden movements occur without the protective action of the surrounding long spinal muscles.

Interspinous ligaments

These connect the spines, and assist the supraspinous ligaments.

Ligamentum nuchae (see Fig. 2.11)

The ligamentum nuchae is really a continuation of the supraspinous and interspinous ligaments, but its attachments are modified by the forward convexity of the cervical spine. Animals with large heads have particularly strong nuchal ligaments; for example, the bull and the elephant. That of the giraffe, however, is the longest!

Movements of the vertebral column

The degree of movement between two successive vertebra is small, but the sum of these movements gives the column a fairly extensive range of movement. The column can flex, extend, laterally flex and rotate. The first three movements are freest in the cervical and lumbar regions, and rotation is most marked in the thoracic region.

The muscles responsible for these

movements are grouped anteriorly, laterally and posteriorly.

Muscles lying anteriorly are responsible for flexion, the sternomastoids and scalenes in the neck and, inferiorly, the psoas and abdominal wall musculature, especially the rectus abdominis (see later).

Laterally, the oblique muscles of the abdomen are significant rotators assisted by the oblique, deeply situated portions of the posterior musculature.

Posteriorly lies a large mass of powerful musculature that reaches from sacrum to spine (Fig. 7.16). This is not surprising, since they are responsible for one of man's greatest achievements, rising from the stooping to the erect position and earning the diploma of *Homo erectus*. The negative aspect to this is the dissection of earlier anatomists who produced detailed descriptions of the numerous individual muscles. These muscles are much more important clinically than anatomically for they, like the underlying ligaments, are frequently the site of injuries especially in the athletically inclined.

The reader is warned against memorizing the detailed attachments of the muscles. A simple schema based on and modified from the writings of others is offered (Table 7.1).

The actions of the above muscles are extension, side flexion and rotation of the spine. Extension follows action bilaterally, side flexion unilaterally and rotation as mentioned above. It should be mentioned that gravity is sufficient to initiate the spinal flexion that follows forward bending; simultaneously the trunk is supported by the spinal extensor muscles.

Spinal cord localization

It is important to relate the spinal cord to the body surface. The cord, until the third month of fetal development, extends the whole

Table 7.1 Deep muscle groups of the back

	Erector spinae spinalis *longissimus* *iliocostalis*	*Transversus spinalis semispinalis* *multifidus* *rotatores*
Width	Hand's breadth, from vertebral spines to rib angles	Thumb's breadth, from vertebral spines to tips of transverse processes
Length	From 4th sacral segment to mastoid process of temporal bone Three slender muscles reach the neck	From 4th sacral segment to occipital bone Three groups reach the neck, but one is very large, the semispinalis
Arrangement	Three muscle groups lying alongside each other Each muscle group is divided into three relays where the replacing muscle lies medial to the muscle it replaces	Three muscle groups lying on top of one another The largest, most superficial and most easily identifiable of the group, the semispinalis, replaces itself by relays which lies lateral to its predecessor
Nerve supply	All muscles are supplied by the posterior primary rami of the spinal nerves	

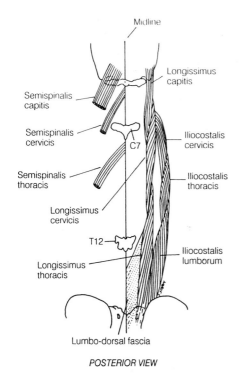

POSTERIOR VIEW

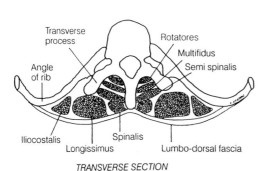

TRANSVERSE SECTION

Fig. 7.16 The deep muscles of the back seen from behind and on transverse section.

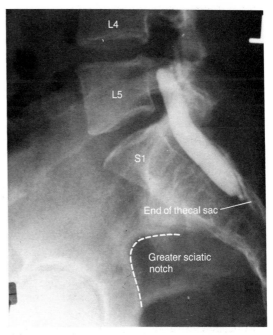

(a)

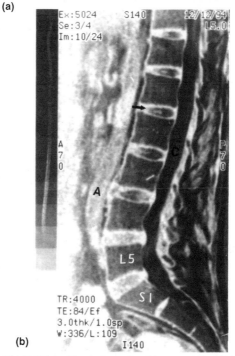

(b)

length of the canal, but after that it fails to keep pace with the more rapidly growing vertebral column. At birth the spinal cord reaches the level of the third lumbar vertebra. In the adult, it usually reaches only to the lower border of the first lumbar vertebra. This is a good reason for doing a lumbar puncture at a lower level; one avoids the danger of poking the needle into the cord, which may carry with it considerable anguish for the doctor and paralysis for the patient.

Fig. 7.17(a) The subarachnoid space extends to approximately S2, as indicated by contrast medium instilled into the space. (b) Magnetic resonance imaging study of the vertebral column T11 to S3. L5, fifth lumbar vertebra; S1, first sacral vertebra; A, abdominal aorta; C, spinal canal. Arrow demonstrates an intervertebral disc showing nucleus pulposus enclosed by annulus fibrosis.

The disparity of the lengths of the spinal column and spinal cord and the sloping of the spinous processes mean that the segments of the cord do not correspond to the spinous processes, the only palpable parts of the vertebral column. An easy rule to remember is that the spinous process of a cervical vertebra is one below the corresponding cord segment. The sloping upper thoracic spines lie two below the corresponding cord segment, whereas the more severely sloping lower spines lie three below; for example, the seventh thoracic spine will correspond to the tenth thoracic cord segment. It is important to note that the dural and arachnoid membranes extend down to the level of the second sacral vertebra, while the pia mater becomes the filum terminale and inserts into the coccyx (see Figs. 2.10 and 19.11).

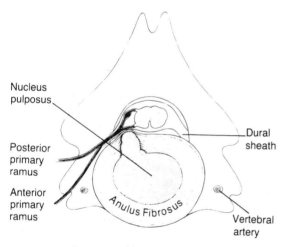

Nucleus pulposus

Posterior primary ramus

Anterior primary ramus

Anulus Fibrosus

Dural sheath

Vertebral artery

Fig. 7.18 Disc prolapse. The concentric fibers of the anulus fibrosus of the disc surrounds the nucleus pulposus. This last has prolapsed through the rupture of the anulus to press on both rami, causing sensory and motor disturbances (pain and paralysis). Note the vertebral artery in the foramen transversarium.

Intervertebral discs

Between the vertebrae are a series of fibrocartilaginous discs, the intervertebral discs, which give the vertebral column its shape and act as a series of shock absorbers. As one descends the spinal column, the discs become thicker.

Structure

The nucleus pulposus, the remains of the notochord, is a centrally lying jelly-like substance held under considerable tension by the concentric and radial fibers of the surrounding anulus fibrosus. These fibers attach firmly to the vertebrae above and below, and to the anterior and posterior ligaments uniting the vertebral bodies. The intervertebral disc receives its nutrition by diffusion through two cartilage plates on the underlying bone of each vertebra. The lumbar and cervical discs are the thickest, while the thoracic and sacral are considerably thinner.

Weakening of the anulus leads to escape of the nucleus. Although the nucleus can pass in any direction, it most commonly heads posteriorly, where it comes into contact with the spinal cord and/or its roots to produce characteristic symptoms of pain or weakness ('slipped disc' or acute lumbago). Much less commonly, the gelatinous nucleus passes anteriorly, or into the body of the vertebra above or below. Note that the disc does not 'slip', but prolapses its central portion like toothpaste being squeezed from its tube.

Review questions on the vertebral column

(answers p. 529)

46 With regard to the vertebral column, the
a) cervical articular processes slope anteriorly and posteriorly
b) lumbar articular processes lie anterior to the intervertebral foramina
c) nodding movement 'yes' takes place at the atlantoaxial joint
d) transverse processes of the axis are wider than those of the atlas
e) thoracic vertebrae easily dislocate without fracturing

47 The vertebral artery
a) usually passes through the foramen transverarium of all the cervical vertebrae
b) has to make a wide lateral curve to pass from axis to atlas
c) occupies a groove on the anterior arch of the atlas
d) passes into the cranial cavity through the vertebral foramen just lateral to the foramen magnum
e) is a branch from the third part of the subclavian artery

48 An anesthesiologist wishes to place a solution into the spinal canal to obtain anesthesia by paralyzing the spinal nerves. She
a) finds it easiest to reach the canal in the cervical region, as these laminae overlap less than in any other vertebrae from which site the solution can gravitate inferiorly
b) finds it easiest to reach the canal in the lower thoracic region, as these laminae overlap less than in any other vertebrae
c) will be able to place her needle below, and avoid possible damage to the spinal cord in an adult patient, if it is placed below the second lumbar vertebra, since the cord ends at the level of L1/2

d) may avoid penetrating the spinal meninges if the needle is placed below the fourth lumbar vertebra, since the meninges end at the level of the third lumbar vertebra
e) may inject through the sacral hiatus as the meningeal sac continues down to S5

49 The spinal column has
a) its shape largely determined by the intervertebral discs
b) its shape largely determined by the contour of the vertebrae
c) its widest vertebral foramen in the thoracic region
d) a forward convexity in the cervical region from birth
e) a lumbar curve from about 20 months onward

50 With respect to intervertebral discs,
a) those in the cervical region are particularly large
b) when the nucleus pulposus prolapses, it usually does so anteriorly
c) they receive their nutrition from the surrounding vessels
d) the nucleus pulposus is the remains of the embryonic notochord
e) the anulus fibrosus attaches only to the anterior longitudinal ligament

51 With regard to the development of a vertebra, the
a) body develops from two ossification centers
b) neural (vertebral) arch develops as an extension of ossification from the body
c) vertebral arches fuse at puberty
d) tips of the transverse processes have five separate ossification centers
e) secondary epiphyses appear at 5 years and fuse around puberty

Part IV   The Thorax

8 Thoracic wall, pleura and lungs

Breast · Thoracic wall · Mechanism of respiration · Pleura · Lungs · Bronchial tree · Lymph drainage of the thoracic wall and viscera

The thorax consists of a bony framework consisting of the 12 thoracic vertebrae with their attached rib/cartilage complexes, enclosing a cavity containing the major elements of the circulatory and respiratory systems, as well as other structures passing to destinations in the thorax, abdomen or neck. The bony framework is filled in by muscles and the chest wall so constituted is surmounted by a pair of breasts.

Breast

Location

The glandular tissue of the breast is based on the second to sixth ribs, and extends from the parasternal margin to the midaxillary line. An extension toward the axilla constitutes the axillary tail (of Spence). Since the superolateral breast contains the most glandular tissue, it is the commonest site for tumors. The breast tissue lies on three muscles – the pectoralis major, the serratus anterior and the aponeurosis and fibers of origins of the external oblique.

Development

Early in embryonic life an ectodermal thickening, called the milk line, runs from the axilla to the groin and upper thigh. In humans, this thickening develops in the pectoral region and becomes depressed beneath the surface as the nipple, from which ductules and lobes branch into the underlying mesoderm. In some animals, nipples form along most of the line. Sometimes humans are decorated with accessory nipples or even accessory breasts. Shortly before birth, the nipple evaginates as a protuberance. Not uncommonly, the invaginated nipple misbehaves and fails to evaginate – this must be differentiated from the very serious retracted nipple of the cancerous breast; the duration of the retraction is the critical differentiating point.

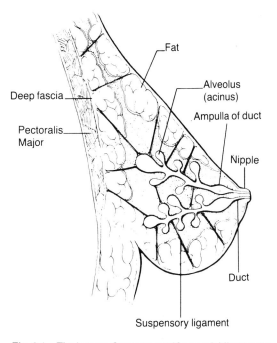

Fat

Deep fascia

Pectoralis Major

Alveolus (acinus)

Ampulla of duct

Nipple

Duct

Suspensory ligament

Fig. 8.1   The breast. Suspensory (Cooper's) ligaments support the secretory components and attach them to the skin and underlying deep fascia.

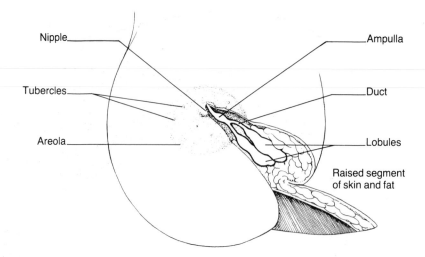

Nipple

Ampulla

Tubercles

Duct

Areola

Lobules

Raised segment
of skin and fat

Fig. 8.2   The areola surrounds the nipple and contains tubercles (Montgomery). A segment of skin and fat has been raised to demonstrate lobules that terminate in ducts.

Structure

The 15–20 ducts of the breast give rise to lobules, which are supported by fibrous tissue. The ducts dilate to form an ampulla immediately before they exit from the nipple. The nipple is surrounded by pigmented skin, the areola, which darkens during pregnancy. Smooth muscle causes erection of the nipple for suckling, and the areolar glands (Montgomery) secrete oils to assist in keeping the nipple supple during feeding. Fibrous bands (Cooper's or suspensory ligaments) connect the breast substance to the superficial subcutaneous tissue, skin and nipple, and are important structures aiding in the support of the secretory tissue. Their involvement in cancer, in which the tumor causes their contraction, results in a characteristic puckering of the skin. A variable amount of fat is interspersed with the fibrous tissue. The proportions of fatty to glandular tissue vary with age and with function.

Lymphatic drainage

A knowledge of the lymphatic drainage of the breast is crucial to the study of the spread of breast cancer. It is sad to say that cancer cells

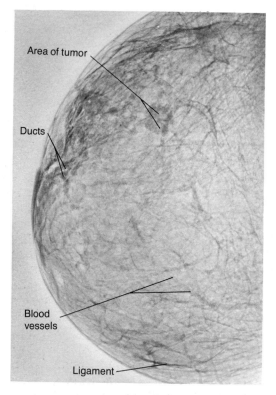

Area of tumor

Ducts

Blood
vessels

Ligament

Fig. 8.3   Xeroradiograph of the breast. Blood vessels, ducts and suspensory ligaments are outlined, as is a small tumor.

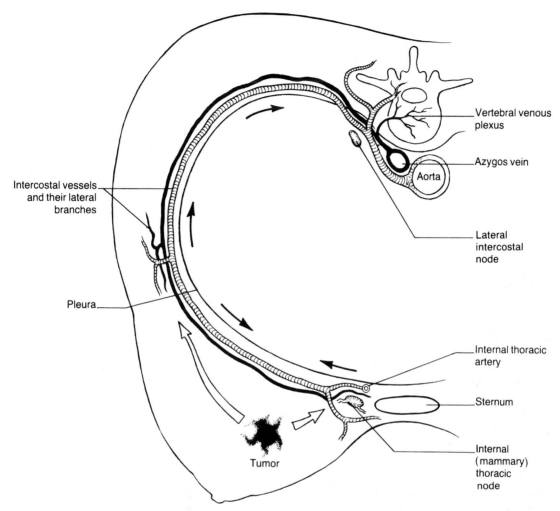

Fig. 8.4 Vascular supply of the breast. A tumor (carcinoma) is present in this breast. A tumor cell may reach the intrathoracic nodes by lymphatics accompanying the vessels, or by entering the veins themselves. As the veins participate in the valveless vertebral venous plexus (Batson), it is easy to understand the frequency of secondary vertebral deposits in cases of breast carcinoma.

do not obey rigid rules and may involve any nodes. This led surgeons, until recently, to perform massive operations in a desperate attempt to excise all involved lymphatics. The survival figures for patients failed to improve, and so the surgical approaches have become more conservative and limited. Nevertheless, it is relatively true to say that about 75 percent of the lymph draining from the breast tissue passes to the axillary nodes (see Fig. 3.15). Initially most lymphatics pass to the anterior group, but all eventually pass through the central to the apical nodes. From the deep part of the breast, vessels pass through the pectoral muscles to the internal thoracic and axillary nodes. While the internal thoracic nodes are more commonly involved in the spread of tumors from the medial half of the breast, lymph from the lateral half also reaches them. Lymph entering the thorax by vessels accompanying the branches of the intercostal and internal thoracic arteries usually passes to the internal thoracic nodes, but some reaches the lateral intercostal nodes.

Blood supply

The breast's arterial supply consists of branches of the internal thoracic, lateral pectoral and intercostal vessels. The venous drainage system is important because, in addition to the major drainage to the axillary vein, it provides a route for blood from the breast to reach the intercostal vessels, which themselves drain into the vertebral plexus (Batson's plexus). This plexus is a vast and valveless venous system running from the skull to sacrum and draining the vertebral column, skull and spinal cord. Tumors of the breast commonly spread to the skull and spine through this anastomosis.

Nerve supply

Although the activity of breast tissue is almost exclusively influenced by hormones, the somatic nerves supplying the overlying skin and nonsecretory tissue are branches of the second to sixth intercostal nerves.

Thoracic wall

Bony skeleton of the thorax

The bony skeleton comprises 12 thoracic vertebrae, 12 pairs of ribs (unless you are unduly blessed) and a sternum. The shape of the thorax is interesting. Circular in the child, it flattens anteroposteriorly in the adult. The circular chest of the newborn child results from the horizontal lie of the ribs. Since the diameter of a circle cannot be enlarged, the child breathes mainly with the diaphragm (abdominal breathing) rather than the ribs. As the child grows older, the ribs become oblique, and by the age of 7 years, breathing with the ribs is significant (thoracic breathing). The adult shape is caused partly by the angulation of the ribs as they project backward to articulate with the transverse processes. This makes it very comfortable for you to lie on your back. If you look at your dog, you will find that he cannot lie on

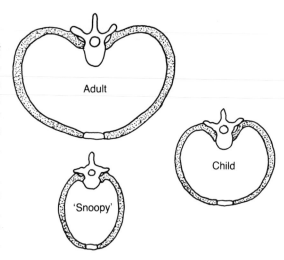

Fig. 8.5 The different shapes of the chest in the child, the adult and the canine.

his back and keeps rolling from side to side. The muscular forearms compress his thorax from side to side, and he lacks posteriorly projecting ribs. Can you remember the other reason for the flattening of the human thorax? (It is not too late to reread the discussion in Chapter 3 on the axilla.)

Sternum

The sternum consists of three parts, which articulate with each other. From above downward they are: the broad manubrium; the body, composed of four fused sternebrae; and the small pointed xiphoid process. The joint between the manubrium and the body does not fuse until a ripe old age; this is fortunate since fusion leads to marked respiratory problems. The manubrium articulates laterally with the clavicle and the first rib. Evidence of the fusion of the sternebrae of the body is easily discerned in the adult. Sometimes there is a hole in the lower part of the body, indicating that the double centers from which the bone developed failed to fuse. In New York City, however, there are often other causes for a hole in the sternum.

The xiphoid process is important only in that it is a variable structure; one will be consulted by patients who suddenly discover

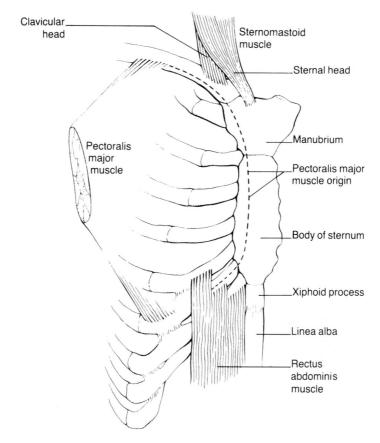

Fig. 8.6 The sternum and some of its attachments. The sternal origin of the sternomastoid, a considerable portion of the pectoralis major, the upper insertion of the linea alba and a small part of the diaphragm attaching to the xiphoid process (the last not shown).

that their bone differs from somebody elses and is therefore thought to be abnormal. It can be short, long or pointed in any direction, and patients need reassurance on this point.

A few definitions are in order. The angle of Louis is an easily palpable prominence, caused by the angulated joint between the manubrium and sternal body. It lies opposite the second costal cartilage. The suprasternal notch is the concave upper border of the manubrium. Above it is the suprasternal fossa, whose main content is the trachea. The subcostal angle is formed by the junction of the two costal margins. It is narrow in thin, slender people, and wide in those with broader builds. Projecting into the angle is the xiphoid process.

Ribs

There are 12 ribs. They all articulate with the 12 thoracic vertebrae. The first seven pass forward to articulate with the sternum; ribs 8, 9 and 10 articulate successively with each other; and the last two are free or floating ribs.

The head of each rib articulates posteriorly with the vertebra after which it is numbered and, to a slight degree, with the vertebra above it. After narrowing to form a neck, the rib shaft then bends at an angle and also twists, so that the anterior extremity of the rib is at a lower level than the posterior extremity. The rib bears a groove just above its lower margin, the subcostal groove, which houses the neurovascular bundle. At its

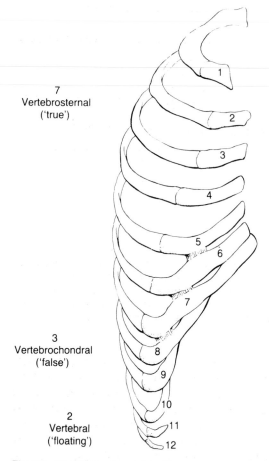

7
Vertebrosternal
('true')

3
Vertebrochondral
('false')

2
Vertebral
('floating')

Fig. 8.7   The ribs and costal cartilages.

anterior extremity the rib becomes cartilaginous, forming the costal cartilage, which runs in an upward direction to articulate with either the sternum or the cartilage. The cartilages of the last two ribs point forward but do not articulate.

In early fetal development the whole rib is cartilaginous; ossification, beginning at the angle, spreads anteriorly and posteriorly. Anteriorly the rib does not complete its ossification, the costal cartilage remaining as an unossified and relatively mobile part of the rib. However, as we get older, the cartilage starts calcifying and then ossifying, leading to a rigid chest wall with a diminished vital capacity, which is one of the reasons for the great tenor, Caruso, eventually sounding like Donald Duck.

Intercostal spaces

The intercostal spaces contain three muscles. The external intercostal muscle runs downward and forward, and the internal intercostal muscle runs upward and forward between adjacent ribs. The unimpressive innermost intercostal muscles, grouped anteriorly, laterally and posteriorly, run across more than one rib space. Perhaps the best defined of the innermost intercostal muscle group is the anterior muscle the sternocostalis.

The neurovascular bundle lies between the internal and innermost intercostal layers. (In the abdominal wall where the external, internal and transverse muscles are the equivalent

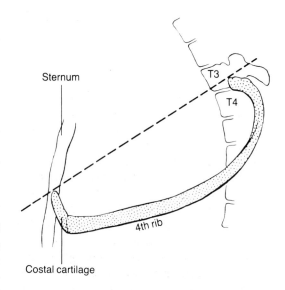

Sternum

T3

T4

4th rib

Costal cartilage

Fig. 8.8   Costosternal and costovertebral articulation. (Note: the head articulates with two vertebral bodies.)

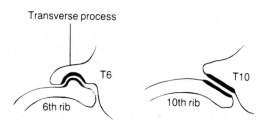

Transverse process

T6

6th rib

T10

10th rib

Fig. 8.9   Costotransverse articulations in upper and lower ribs.

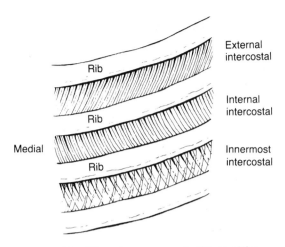

Fig. 8.10 The different arrangement of fibers of the three intercostal muscles.

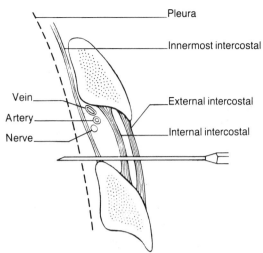

Fig. 8.11 The relation of the neurovascular bundle to the muscle layers and ribs. A needle passed immediately over the upper border of the rib will not damage the neurovascular bundle, which is lodged in and protected by the subcostal groove.

muscles, the vessels and nerves are similarly placed.)

Intercostal nerves and vessels

Having emerged from the spinal foramen, the intercostal nerve immediately divides into a posterior primary ramus to the paraspinal region, and an anterior primary ramus which runs around the chest wall in the muscular plane described above. The lower five intercostal spaces are shorter than the upper six, and nerves 7–11 therefore leave these spaces to enter and supply the abdominal wall (T12 passes below the twelfth rib as the subcostal nerve). Since each anterior ramus gives off a lateral and an anterior branch, both of which bifurcate, a complete segment of skin is supplied with sensation, While lying between the intercostal muscles the anterior rami naturally supply these muscles.

The neurovascular structures lie in the subcostal groove, well protected from injury. When aspirating the chest (e.g. to remove fluid) a needle is placed along the upper border of the rib to eliminate the danger of puncturing these vessels.

Posterior intercostal arteries arise from the thoracic aorta for the lower nine spaces, the upper two spaces being supplied by an artery from the neck running down into the thorax

(the significance of this seemingly unimportant superior intercostal artery will be discussed later). The posterior intercostal artery gives off a posterior branch, which runs with the posterior ramus to the paraspinal muscles, supplying on the way a very important spinal branch to the spinal cord (see p. 31). The artery continues around the chest wall with the anterior ramus, supplies lateral and anterior branches, and anastomoses with branches of the internal thoracic artery.

The veins of the intercostal spaces run posteriorly to the azygos system and anteriorly to the internal thoracic veins. Lymph nodes are located in the upper anterior and most of the posterior extremities of the spaces.

Azygos veins

The azygos system of veins is responsible for collecting the blood from the thoracic wall. These are vertical veins: one complete vein on the right, the azygos (vena azygos), joins the superior vena cava; and usually two incomplete veins on the left, the superior and the inferior hemiazygos, run behind the eso-

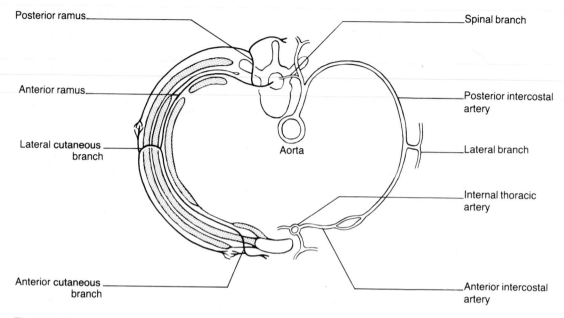

Fig. 8.12 **The distribution of a typical intercostal nerve and artery.**

phagus to join the azygos vein. The azygos collects the blood from the lower ten right intercostal veins and the two hemiazygos from the lower eight left intercostal veins. The upper left second and third intercostal veins drain by a special vein, the left superior intercostal; the first intercostal veins cross the first ribs to drain into their respective brachiocephalic veins.

Mechanism of respiration

If you remember four facts, it will be easy to understand the mechanism of respiration:

1 The anterior end of each rib is lower than the posterior end.
2 The rib at its circumference is lower than either the anterior or posterior end, exactly like a bucket handle lying over the side of the bucket.
3 The first seven ribs articulate with the transverse processes by a convexity fitting into a concavity, enabling the rib to rotate.
4 In ribs 8, 9 and 10, the articular facets are flat and slope backward, meeting similar facets on the transverse processes of the vertebrae. These facets cannot rotate; they slide in a backward direction.

Now let us increase the capacity of the thorax in three directions and produce a negative pressure:

1 Elevation of the rib pushes the sternum forward, thereby enlarging the anteroposterior diameter.
2 Elevation of the rib by the bucket-handle action increases the transverse diameter. The transverse diameter of the lower thorax is increased by ribs 8, 9 and 10 passing backward, such as occurs when a pincer is opened (because of their different articulations).
3 An arching musculotendinous partition, the diaphragm, separates the thorax and abdomen. Contraction straightens the arch with descent of the diaphragm and enlargement of the vertical extent of the chest, like the withdrawal of the plunger from a syringe.

The descent of the diaphragm necessitates some redistribution of the abdominal viscera. To do this, we relax our abdominal wall, and this increases the capacity of the abdomen. If we did not do this, we would produce such a rise in intra-abdominal pressure that we would have to try to find a way of reducing the pressure by getting rid of some of the abdominal contents. This is, in fact, partly the way we get rid of contents from the alimentary canal, the bladder and the pregnant uterus.

With the enlargment of the thorax, air is sucked in to prevent the formation of a vacuum, and the lungs expand. We cannot really elevate the ribs any more than by the mechanisms just detailed, but we can increase the rate of breathing by using muscles that will elevate the ribs more rapidly. We therefore bring into play accessory muscles of respiration, which include any muscles attached to the ribs that pass to the upper extremity or the neck. If we fix either the neck or the upper extremity, we will be able to elevate the ribs by this means. The latissimus dorsi and the two pectorales will elevate the ribs if we fix the arms; cardiac patients who are short of breath soon learn this trick. If we fix the head in extension, the scalene muscles of the neck, which insert mainly into the first rib, will elevate the whole rib cage. Note how Olympic athletes stand at the end of a race, with their arms fixed on their thighs or some nearby support.

Pleura

Each pleural sac is a complete sac of thin serous membrane invaginated by the lung, which is thereby enclosed by a visceral layer and surrounded by a parietal layer. The parietal layer lines the parietes and is subdivided according to the part of the pleura under discussion. For example, the costal pleura lines the chest wall; the diaphragmatic pleura, the diaphragm; the mediastinal pleura, the mediastinum; and finally, the cervical pleura is the portion of the pleural sac that extends into the neck above the first rib. The parietal and visceral

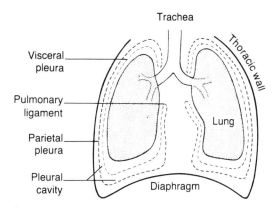

Fig. 8.13 Pleural sacs enclosing the lungs.

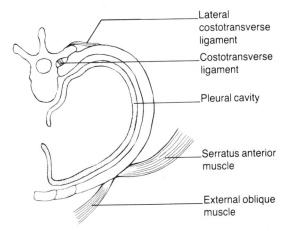

Fig. 8.14 Transverse section through the upper thorax. The ligaments joining the rib and the transverse process, and the origin of the external oblique and serratus anterior muscles, are seen.

layers are completely continuous and normally separated by a small serous-lubricated space, the pleural cavity. When adherent to one another, as in pleurisy (inflammation of the pleura), a leathery friction rub is audible.

Nerve supply

The visceral pleura has an autonomic nerve supply, but the parietal pleura has a somatic supply by the adjacent intercostal nerves as they run around the chest wall. It will be repeated later that the abdominal wall, lower thoracic wall (including costal pleura) and the diaphragm share their nerve supply, a fact

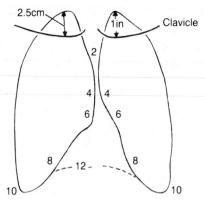

Fig. 8.15   Anterolateral surface relationship of the pleural sacs. The pleura ascend posteriorly to cut across the middle of the downward sloping 12th ribs. (See also Fig. 30.4.)

of great clinical importance. It means that diseases affecting the chest wall and pleura (e.g. pneumonia) may present as abdominal pain.

Surface marking

The first rib slopes downward and forward with its anterior end at a lower level than its posterior. The pleura does not reach beyond the neck of the first rib but, because of the slope, it projects above the anterior extremity of the rib by about 2.5 cm (1 in). Since the medial end of the rib is covered by the medial third of the clavicle, the practical marking would be 2.5 cm above the medial third of the clavicle. Here it is covered by a sheet of fascia, the suprapleural membrane (Sibson's fascia), which stretches from the inner border of the first rib to the transverse process of the seventh cervical vertebra. This fascia tends to prevent the pleura from projecting up into the neck upon deep inspiration. In some people it is thin, and the pleura can be seen bulging above the clavicle after severe exertion.

The medial border of each pleural sac descends behind the sternoclavicular joints to the level of the sternal angle (angle of Louis). Here the sacs meet, after which the two anterior borders run vertically downward in different ways. The right border runs straight down in the midline of the

sternum to reach the sixth costal cartilage, after which the lower margin passes laterally, crossing the nipple line at the eighth rib, and the midaxillary line at the tenth rib (its lowest level). After reaching the midaxillary line, the anterior border ascends to cross the twelfth rib at about its middle before reaching the spinous process of T12.

The posterior border ascends along the paravertebral region, but, because of the thick overlying muscles, this surface marking cannot be profitability used. On the left side, having reached the fourth costal cartilage, the pleura is separated from the chest wall by the protrusion of the pericardium. The pleura arches out and descends lateral to the border of the sternum, halfway to the apex of the heart, leaving the medial ends of the fourth and fifth spaces uncovered by pleura. This means that a needle passed through these spaces immediately next to the sternum (to avoid the internal thoracic artery) will enter the pericardial sac without traversing the pleura. This, therefore, is one of the sites where the pericardial sac may be safely aspirated. If the pleura is not avoided, air will rush in, create a positive pressure and collapse the lung; this is called a pneumothorax, which will add to the patient's woes. Having reached the sixth rib, the lower border of the pleura then passes laterally as on the right side. The even numbers (2, 4, 6, 8, 10, 12) will assist you in remembering where the pleura lies in relationship to the ribs and costal cartilages.

Lungs

The lungs are two spongy organs composed of small alveolar sacs. They conform to the shape of the pleural cavities, but do not occupy the whole cavity because they would be unable to expand. Each lung has an apex that passes up in the same manner as the cervical pleura, a base that is related to the diaphragm, a parietal surface that is related to the ribs, and a mediastinal surface that abuts

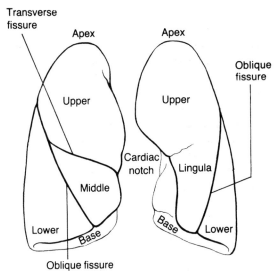

Transverse fissure

Apex

Apex

Oblique fissure

Upper

Upper

Cardiac notch

Lingula

Middle

Lower

Base

Base

Lower

Oblique fissure

Fig. 8.16 The lungs and their lobes.

on the vertebral column posteriorly and the mediastinum anteriorly.

Surface marking and disposition

The surface marking of the lung is the same as the pleura except for the lower margin and the cardiac notch. The lower margin of the lung passes around the chest wall more horizontally. The pleura and lung's surface markings are the same up to the sixth cartilage, after which the lower margin of the lung passes behind the sixth rib in the midclavicular or nipple line, to the eighth rib in the midaxillary line, to the tenth rib in the scapular line and, finally, to the tenth spinous process. The lower border of the lung is two ribs higher than the lower border of the pleura. The cardiac notch of the left lung, much bigger than in the case of the left pleural sac, extends from the fourth to the sixth costal cartilage and reaches the midclavicular line laterally. This leaves an area where the heart abuts against the chest wall and where percussion gives rise to a dull note, the cardiac dullness, indicating the absence of an air-filled lung and the presence of a fluid-filled chamber. The numbers to

remember for the surface relationship of the lung are 2, 4, 6, 6, 8 and 10.

Each lung is subdivided by an oblique fissure into upper and lower lobes. This important fissure runs posteriorly from the second or third spinous processes around to the sixth costal cartilage anteriorly. This is not a very useful description because the marked slope of the spinous processes makes exact identification of these processes difficult. It is easier to remember that the fissure runs roughly in the same line as the vertebral border of the scapula when the arm is elevated, the usual position for examining the posterolateral chest. The line of the sixth rib is also a reasonably accurate alternative surface marking for the fissure.

Another fissure on the right side passes horizontally and medially from the oblique fissure at the level of the fourth costal cartilage and demarcates a middle lobe. The equivalent of the middle lobe in the left lung is the lingula, a small area lying between the cardiac notch and the oblique fissure. These fissures may be complete or incomplete, and there may even be additional fissures.

The disposition of these fissures means the following. Clinical examination of the anterior chest would explore only the upper lobe of the left lung and the upper and middle lobes of the right lung. The middle lobe does not reach the periphery beyond the midaxillary line. Examination of the posterior chest would investigate mainly the lower lobe, the upper lobe being confined to a small area above the level of the second spinous process. The middle lobe does not reach beyond the midaxillary line and cannot be examined posteriorly.

Bronchial tree

This is a system of tubes that carries air from the trachea to the alveoli of the lung. The walls of the tubes are kept open by incomplete rings of cartilage in the larger tubes, which give way to small plates of cartilage as the diameter of the tubes decreases. The smallest tubes (the bronchioles) lack cartilage

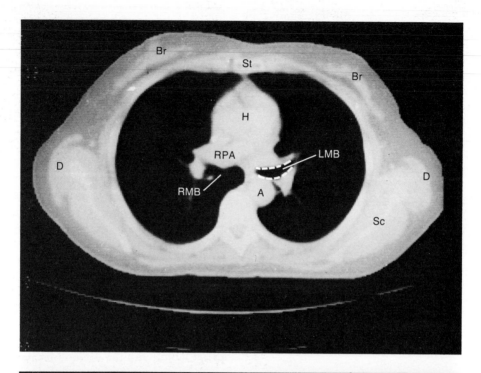

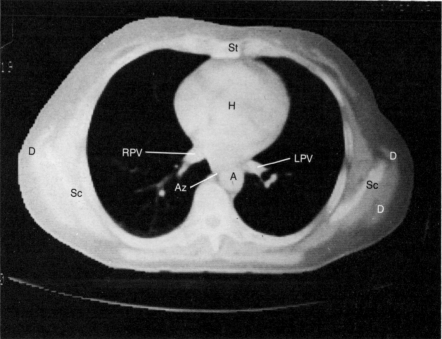

Fig 8.17   Computerized tomograms of a female chest at different levels, showing breast, lungs and pulmonary vessels. A, aorta; Az, azygos vein; Br, breast; D, deltoid; H, heart; LMB, RMB, left and right main bronchi; LPV, RPV, left and right pulmonary veins; RPA, right pulmonary artery; Sc, scapula; St, sternum.

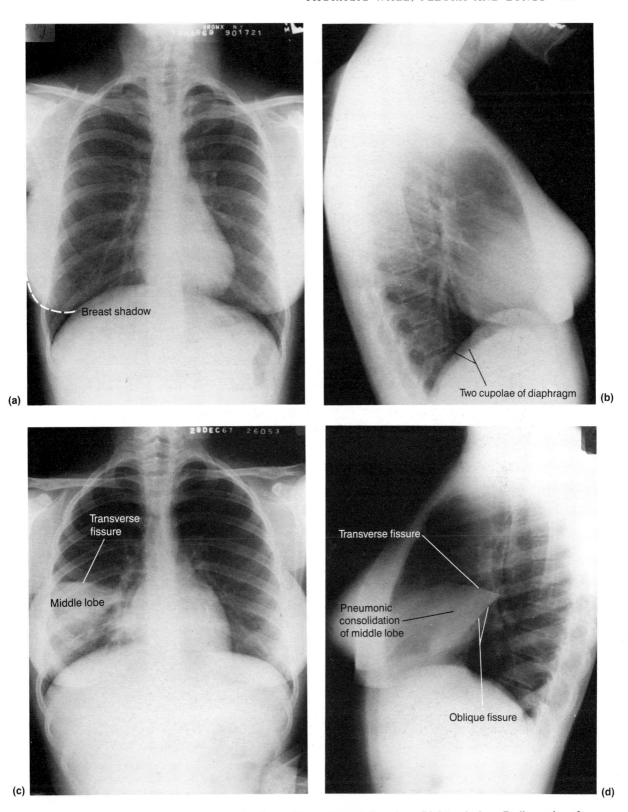

Fig. 8.18   Radiographs of normal female chest: (a) posteroanterior view, (b) lateral view. Radiographs of female patient's chest whose right middle lobe is outlined by right middle lobar pneumonia: (c) posteroanterior view, (d) lateral view.

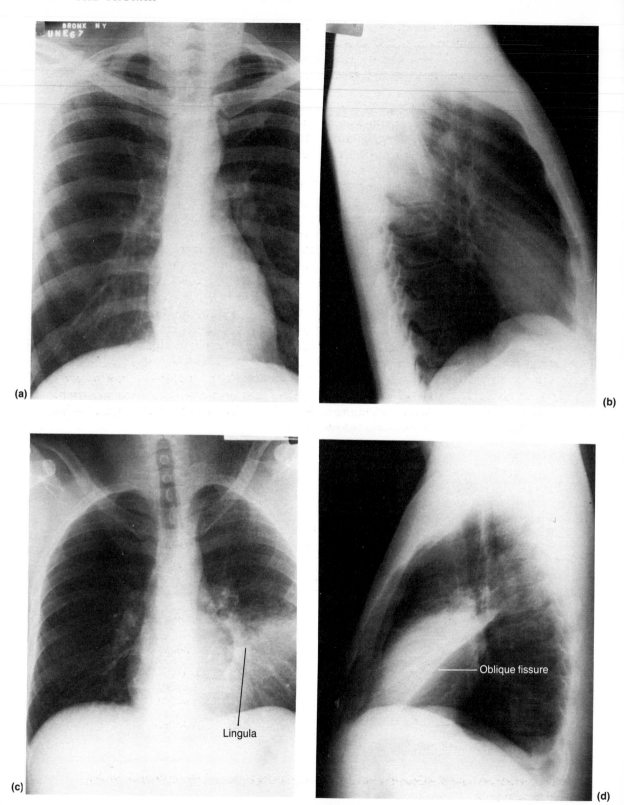

Fig. 8.19   Radiographs of normal male chest: (a) posteroanterior view, (b) lateral view. Radiographs of male patient with pneumonia, outlining the left middle lobe (lingula): (c) posteroanterior view, (d) lateral view. Note the lesser definition of the lingula as compared to the right middle lobe – there is no transverse fissure.

completely and are surrounded by smooth muscle. Excessive contraction of this muscle (e.g. an asthmatic attack) can almost close these tubes, as there is no cartilage to prevent the constriction.

The trachea bifurcates at the angle of Louis into the right and left main bronchus. The right bronchus is 2.5 cm (1 in) long, coming off at an angle of about 25 degrees and the left bronchus is about 5 cm (2 in) long and comes off at an angle of 45 degrees. The inclination of the bronchus is always in a posterior direction. (Remember that nearly all grouped structures lie vein–artery–duct (VAD) anteroposteriorly.) The bronchus therefore lies posterior to the large venous and arterial structures.

With the right bronchus lying more in line with the trachea, it is easy to understand why the peanut accidentally inhaled by a child passes down the right main bronchus, and if she is lying on her back in the cot, it will enter a posterior branch of the right bronchus, often the first one given off, the apical of the lower lobe. The right main bronchus is shorter than the left because it gives off a secondary bron-chus for the upper lobe, the eparterial bronchus (since it lies above the pulmonary artery), before reaching the lung. This is one of the two main differences between the distribution of the bronchi to each lung, the other being the crowding of the left lung by the impingement of the heart. Apart from these modifications, the two bronchial arrangements are almost symmetrical.

The right bronchus, after it enters the lung, gives off secondary bronchi to the right middle lobe and then continues to supply the lower lobe. On the left side there is no eparterial bronchus, the upper lobe bronchus supplying the upper lobe of the lung, including the part of the upper lobe corresponding to the middle lobe, the lingula. The bronchus then goes on to supply the lower lobe by roughly the same arrangement as on the right side, except for the anterior and medial basal bronchi arising from a common stem. On both sides, the first posterior bronchus is the dorsal (superior) branch to the apex of the lower lobe. This is the area where secretions will pool in a patient lying on his back.

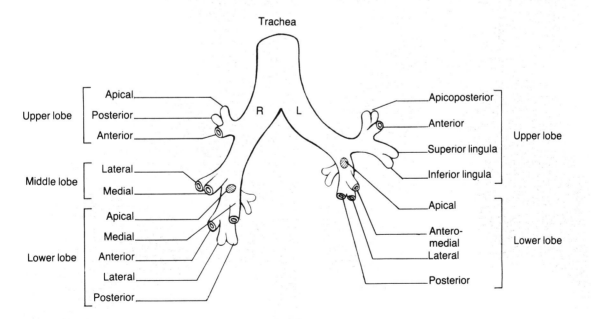

Fig. 8.20 The bronchial tree. Where the smaller branches appear closed, this implies that they are curving posteriorly (as they do).

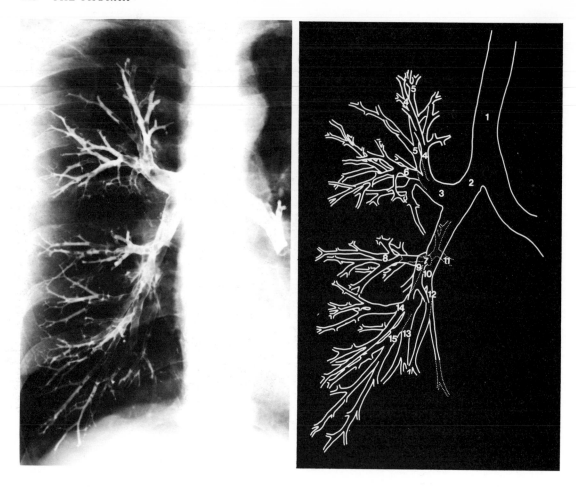

Fig. 8.21   Posteroanterior bronchogram of the right lung. (From Weir and Abrahams 1978)

1 Trachea.
2 Right main bronchus.
3 Right upper lobe bronchus.
4 Apical segmental bronchus.
5 Posterior segmental bronchus.
6 Anterior segmental bronchus.
7 Middle lobe bronchus.
8 Lateral segmental bronchus.

9 Medial segmental bronchus.
10 Right lower lobe bronchus.
11 Superior segmental bronchus (apical basal).
12 Medial basal segmental bronchus.
13 Anterior basal segmental bronchus.
14 Lateral basal segmental bronchus.
15 Posterior basal segmental bronchus.

Lymph drainage of the thoracic wall and viscera

Lymph nodes of the thorax

The lymph nodes of the thorax are distributed around the circumference of the thoracic wall anteriorly, posteriorly, laterally and inferiorly (see Fig. 8.4).

Anterior nodes

These are the internal thoracic nodes surrounding the vessels of the same name and occupying the upper two or three intercostal spaces. They drain deep structures of the upper abdominal wall and, very importantly, the breast, especially its medial half. Unfortunately, a tumor from the breast

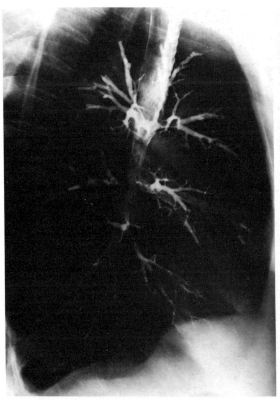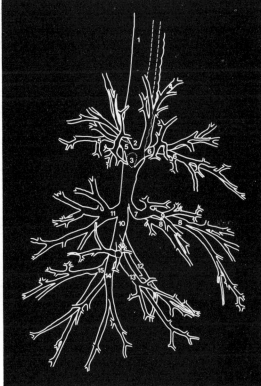

Fig. 8.22   Right lateral bronchogram of the right lung. (From Weir and Abrahams 1978)

1 Trachea.
2 Right main bronchus.
3 Right upper lobe bronchus.
4 Apical segmental bronchus.
5 Posterior segmental bronchus.
6 Anterior segmental bronchus.
7 Middle lobe bronchus.
8 Lateral segmental bronchus.

9 Medial segmental bronchus.
10 Right lower lobe bronchus.
11 Superior segmental bronchus (apical basal).
12 Medial basal segmental bronchus.
13 Anterior basal segmental bronchus.
14 Lateral basal segmental bronchus.
15 Posterior basal segmental bronchus.

reaching the thorax by this route makes treatment, at the present time, very difficult and markedly reduces the cure rate.

Posterior nodes

These lie behind the pericardium in the posterior mediastinum, and drain the esophagus and diaphragm.

Lateral nodes

The lateral nodes, lying in relationship to the

necks of the ribs, drain the deep structures of the side and back of the chest. The lymph from the lower spaces drains into the cisterna chyli and, from the upper spaces, into the respective ducts of each side in the supraclavicular region.

Inferior nodes

These nodes lie on the diaphragm, and drain the upper surface of the liver and the diaphragm.

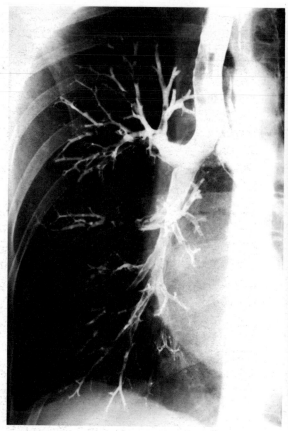
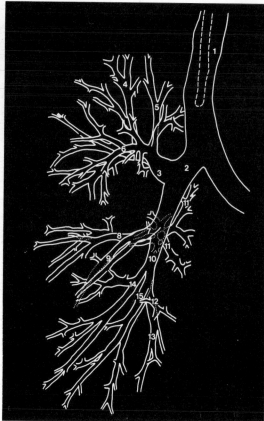

Fig. 8.23 Left anterior oblique bronchogram of the right lung. (From Weir and Abrahams 1978)

1 Trachea.
2 Right main bronchus.
3 Right upper lobe bronchus.
4 Apical segmental bronchus.
5 Posterior segmental bronchus.
6 Anterior segmental bronchus.
7 Middle lobe bronchus.
8 Lateral segmental bronchus.

9 Medial segmental bronchus.
10 Right lower lobe bronchus.
11 Superior segmental bronchus (apical basal).
12 Medial basal segmental bronchus.
13 Anterior basal segmental bronchus.
14 Lateral basal segmental bronchus.
15 Posterior basal segmental bronchus.

Visceral nodes

These consist of two main groups: the superior mediastinal and tracheobronchial nodes.

SUPERIOR MEDIASTINAL NODES

Lying in relationship to the trachea and aortic arch, these nodes drain the trachea, esophagus and heart.

TRACHEOBRONCHIAL NODES

These numerous nodes are among the largest in the body, and, at autopsy, they are nearly always darkly stained with the products of city living (such as soot). On occasion, cross-sections of a node will even show a cigarette butt! These glands are grouped around the trachea and bronchi, including the divisions of the latter, and drain both the lungs and the bronchi. Enlargement of these nodes at the

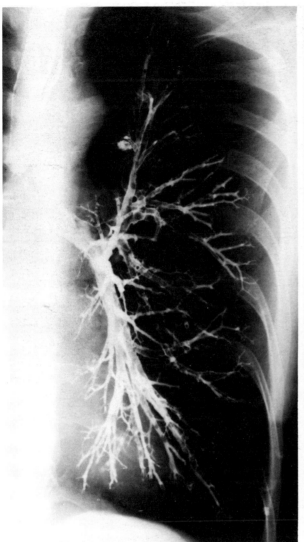
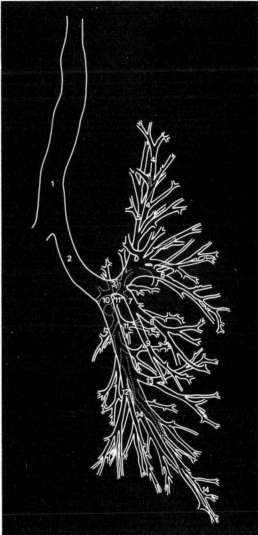

Fig. 8.24 Posteroanterior bronchogram of the left lung. (From Weir and Abrahams 1978)

1 Trachea.
2 Left main bronchus.
3 Left upper lobe bronchus.
4 Apical segmental bronchus.
5 Posterior segmental bronchus.
6 Anterior segmental bronchus.
7 Lingular lobe bronchus.

8 Superior lingular segmental bronchus.
9 Inferior lingular segmental bronchus.
10 Left lower lobe bronchus.
11 Superior segmental bronchus (apical basal).
12 Anteromedial basal segmental bronchus.
13 Lateral basal segmental bronchus.
14 Posterior basal segmental bronchus.

bifurcation of the trachea (carina) may be an ominous sign seen at bronchoscopy.

Lymph drainage of the lung

The lymph drainage of the lung is interesting and important. The lymph drainage from the whole of the right lung and from the lower half of the left lung (the lower portion of the upper and the whole lower lobe) is via the tracheobronchial nodes to the right side. The upper half of the left lung drains to the left

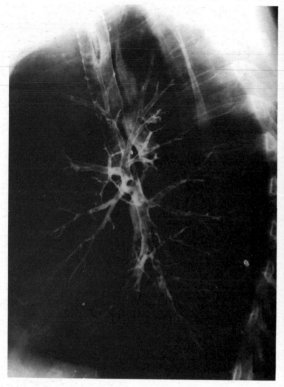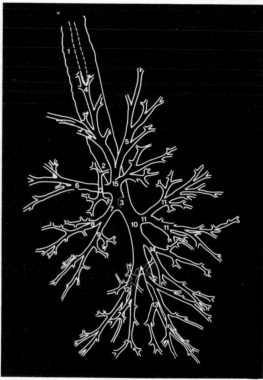

Fig. 8.25 Left lateral bronchogram of the left lung. (From Weir and Abrahams 1978)

1 Trachea.
2 Left main bronchus.
3 Left upper lobe bronchus.
4 Apical segmental bronchus.
5 Posterior segmental bronchus.
6 Anterior segmental bronchus.
7 Lingular lobe bronchus.
8 Superior lingular segmental bronchus.

9 Inferior lingular segmental bronchus.
10 Left lower lobe bronchus.
11 Superior segmental bronchus (apical basal).
12 Anteromedial basal segmental bronchus.
13 Lateral basal segmental bronchus.
14 Posterior basal segmental bronchus.
15 Apicoposterior segmental bronchus.

side. The implication of this arrangement is that a lesion above the middle of the left lung would eventually drain to the lymph nodes of the left side of the neck, whereas one in the lower half of the left lung would drain to those of the right side (Fig. 16.3). Tumors of the lung are notorious for spreading from the tracheobronchial nodes to the cervical lymph nodes. It is vital to know whether lymph nodes are involved: if they are, surgery is usually ineffective. It is difficult to obtain meaningful lymph nodes from the tracheobronchial area for examination, although some rather intrepid surgeons pass a mediastinoscope down the anterior aspect of the trachea from the neck. Occasional biopsy reports reading 'portion of aortic wall' have reduced the popularity of this method. It is far safer to explore the base of the neck to locate positive nodes. In a lesion of the right lung, the right lower neck should be explored, whereas for a lesion of the upper half of the left lung, the left side of the neck requires exploration. However, for a lesion of the lower part of the left lung, the right side should be investigated (see p. 265), and for a lesion at the junction of the upper and lower halves of the left lung, both sides require attention to cover all possibilities. Current radiographic techniques are able to demonstrate lymph nodes, and will probably replace the need for obtaining biopsy.

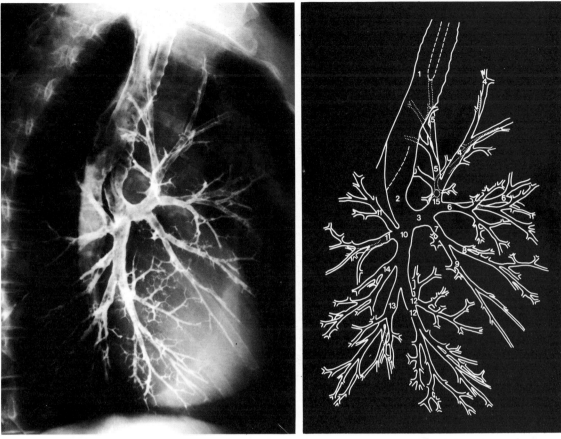

Fig. 8.26 Right anterior oblique bronchogram of the left lung. (From Weir and Abrahams 1978)

1 Trachea.
2 Left main bronchus.
3 Left upper lobe bronchus.
4 Apical segmental bronchus.
5 Posterior segmental bronchus.
6 Anterior segmental bronchus.
7 Lingular lobe bronchus.
8 Superior lingular segmental bronchus.
9 Inferior lingular segmental bronchus.
10 Left lower lobe bronchus.
11 Superior segmental bronchus (apical basal).
12 Anteromedial basal segmental bronchus.
13 Lateral basal segmental bronchus.
14 Posterior basal segmental bronchus.
15 Apicoposterior segmental bronchus.

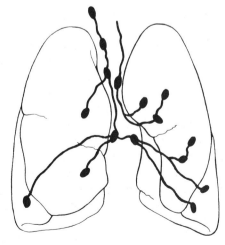

Fig. 8.27 Lymphatic drainage of the lungs. Note that the lymphatics of the lower half of the left lung cross over to the right side.

9 Mediastinal contents

Heart · Superior and posterior mediastinum · Nerves of the thorax

The mediastinum is the area that lies between the two pleural sacs and extends from the inlet of the thorax to the diaphragm. The inclination of the first rib is responsible for the downward-and-forward-sloping thoracic inlet. The outlets of the thorax are the apertures of the diaphragm through which structures pass from thorax to abdomen and vice versa. Lines projected posteriorly from the upper border of the manubrium passing through the first thoracic vertebrae and from the lower border through the disc between the fourth and fifth vertebrae will enclose the superior mediastinum. Below this lies the inferior mediastinum, a term seldom used. Rather, one refers to the middle

mediastinum, a small anterior interval, the anterior mediastinum, and a much larger posterior space, the posterior mediastinum. Clinically one will hear the terms superior, anterior, middle and posterior mediastinum, but these labels are purely artificial and descriptive.

The middle mediastinum contains the heart, which lies within its sac, the pericardium. The contents of the superior and posterior mediastinum include the trachea, esophagus, thoracic duct, part of the thymus, and nerves, vessels and lymphatics. The anterior mediastinum has no significant contents except for the thymus gland.

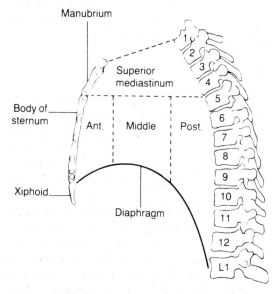

Fig. 9.1   Sternum, vertebral column and subdivisions of the mediastinum.

Heart

The heart is a muscular organ about the size of a clenched fist and packs as much power. It occupies the middle mediastinum and is enclosed by the pericardium, which consists of two different portions. Externally is the fibrous pericardium, which stabilizes the heart, prevents over-dilatation, and is perforated by six large veins. Within the fibrous pericardium is the serous pericardium, consisting of two layers: the visceral layer, which is attached to the heart itself, and the parietal layer, which lines the fibrous pericardium. Between the two layers is the usual serous-lubricated interval (this same arrangement exists for the pericardial, pleural, peritoneal and scrotal cavities). Posteriorly, the two serous layers are so separated as to form a space, the oblique sinus. The serous pericardium encloses not only the heart, but also the commencement of the two great vessels leading from it, the aorta and the pulmonary artery. If a finger is passed behind these two vessels, and in front of the intrapericardial superior vena cava, it will lie in a serous-lined space, the transverse sinus.

The heart deviates considerably from the simple four chambers of early fetal life (two atria and two ventricles). Functionally the right side of the heart, consisting of the right atrium and right ventricle, is completely separate from the left atrium and left ventricle. During fetal development, the atria come to lie posterior to the ventricles. Furthermore, the heart rotates so as to bring the right-sided structures, the right atrium and the right ventricle, anterior to the left-sided chambers.

The shape of the heart may be confusing on first description. Returning to the analogy of the closed right fist, the sternocostal surface (the surface facing the chest wall) is the dorsum of the hand, the flattened diaphragmatic surface is the proximal phalanges, and the base of the heart is the distal phalanges and the palm. Confusion arises because, when people talk about the base of the heart, it may be thought that this is the flattened diaphragmatic surface, but the base of the heart refers to the posterior aspect described.

Because of the rotation of the heart, the sternocostal surface is made up mostly of the right atrium and the right ventricle. A small portion of the left ventricle comprises the left border, with the left auricle of the atrium peeping out above it. The interventricular and interatrial septa, as seen in a transverse section, run across the heart in an almost coronal plane, separating the anterior right side from the posterior left side (Fig. 9.3).

Remembering that the blood of the right side of the heart is destined for the lungs, right-sided heart failure will result in the blood being dammed up in the peripheral venous system. The head and neck, abdomen

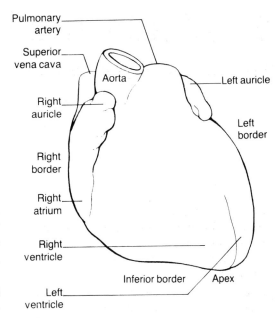

Fig. 9.2 Sternocostal surface of the heart.

and lower extremities show these effects by the swelling of the abdomen, particularly the liver, and the legs. The head and neck suffer less because drainage is assisted by gravity. In left-sided heart failure, blood is dammed up in the lungs and the left atrium fails to distribute it to the left ventricle in sufficient amounts. This congestion of the lungs causes shortness of breath, or dyspnea.

Position

Looking at the sternocostal surface of the heart, one notes that most of the heart extends to the left, 8.75 cm (3.5 in) from the midline, with a lesser amount, 3.75 cm (1.5 in), to the right. Vertically the heart lies behind the body of the sternum from the second to the sixth costal cartilages. It is easy to outline the heart if one joins the points opposite the left second, right third and right sixth sternocostal junctions, and the apex beat, 8.75 cm (3.5 in) from the midline in the fifth intercostal space. Joining points one and three will outline the atrioventricular groove, which separates the atria and ventricles. This is the

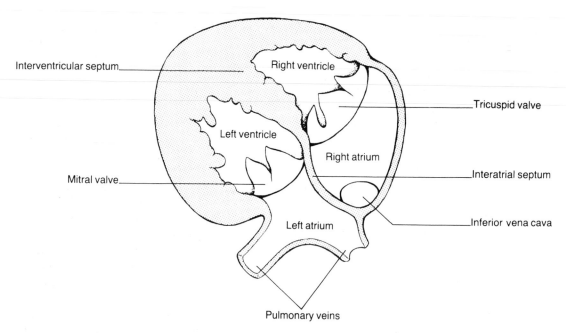

Fig. 9.3 Transverse section of the heart. Note the atria lying posterior to the ventricles, the two-cusped mitral valve and the three-cusped tricuspid valve.

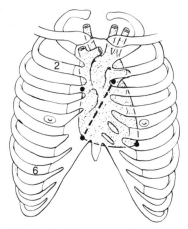

Fig. 9.4 The heart and great vessels in relationship to the bony skeleton. The heart is outlined by four points, and the broken line represents the atrioventricular groove.

disposition of the cadaver's heart in the department of anatomy; we trust that your patients will have a heart that varies considerably from this.

The position in life depends on the following factors:

1 Height of the diaphragm: this varies as to whether one is standing or lying down, the degree of elevation by the abdominal contents and the state of respiration. The diaphragm descends on inspiration, pulling the pericardium and the heart with it.

2 Position of the body: the position of the heart changes when one leans forward or turns to the side.

3 State of contraction: ventricular contraction changes the heart's position. The atria move minimally due to their fixation by the large entering veins. The apex is the most mobile portion, giving you a friendly slap on the chest wall during contraction.

It is worth knowing which chambers are represented on the surfaces of the heart. On the sternocostal surface, parts of all the chambers are visible. On the diaphragmatic surface, both ventricles and a small portion of the right atrium appear. On the posterior surface, both atria are obvious, the left predominating. The apex is represented only by the left ventricle.

Right atrium

The right atrium of the heart contains a smooth area, which is part of the absorbed sinus venosus, and an auricle, the original atrial chamber, projecting to the left like an ear (it really resembles a pugnacious dog's ear in that it looks all chewed up). The auricle lies on the anterior surface and partly covers the root of the aorta. The two parts of the atrium are demarcated by a groove, the sulcus terminalis, which projects into the cavity as a vertical ridge, the crista terminalis. The original auricle can be recognized by the presence of the internally projecting musculi pectinati (teeth of a comb). Opening into the chamber above and below are two large veins, the superior and inferior vena cavae. The inferior vena cava possesses a vestigal valve that directs the blood up to the now closed foramen ovale. This is represented by a shallow depression, the fossa ovalis, the site of fusion of the primary and secondary septa, surmounted by a prominent margin, the anulus ovalis. The opening of the coronary sinus lies between the inferior cava opening and the tricuspid valve, which leads to the ventricle. There are also other small openings where cardiac veins (venae cordis minimae) open directly into the right atrium.

Right ventricle

The right ventricle projects to the left of the right atrium (see Figs. 9.2, 9.3 and 9.5). The inside of this chamber is thrown into a series of ridges, the trabeculae carneae. One of these ridges becomes partly disconnected and runs between the septum and the wall of the ventricle to form the moderator band. Previously the band was thought to prevent over-distention of the ventricle, but it is now known that its main function is to carry the important right branch of the atrioventricular (AV) bundle.

Some of the trabeculae carneae are conical shaped and are called the papillary muscles. From their apices, thin cords, the chordae tendineae, pass to the marginal and ventricular surfaces of the three cusps of the tricuspid valve. When the right ventricle contracts the blood is forced up the pulmonary artery. The force of the blood pushes the cusps of the tricuspid valve toward the

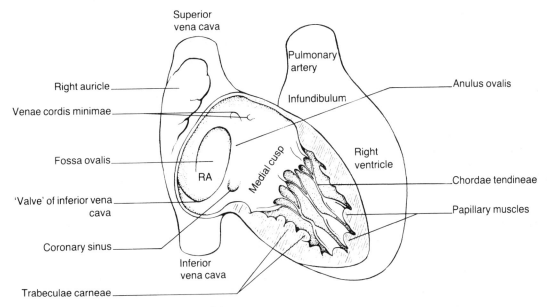

Fig. 9.5 Right atrium (RA) and right ventricle.

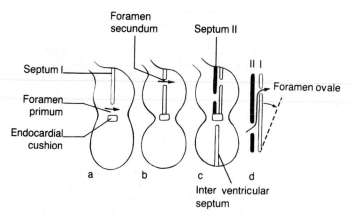

Fig. 9.6 The primary and secondary septa in the developing heart. The primary septum (septum primum) grows inferiorly (a), the blood flowing from right to left between its lower border and the endocardial cushion. When the septum fuses with the cushion, it breaks down superiorly (b). To the right of it a second septum (septum secundum) forms, leaving a gap for the passage of blood; concurrently a septum has been growing superiorly to separate the ventricles (c). The blood flows as indicated in (d), the septum primum opening like a door. When the pressure in the left atrium rises after birth, the door is shut and the two septa fuse to form the adult interatrial septum. Sometimes this fusion fails, a patent interatrial septal defect (foramen ovale) resulting.

right atrium, and would evert them into the atrium were it not for the action of the tendinous cords. The upper part of the ventricle narrows into the infundibulum, which contains more fibrous tissue than muscle and which leads to the pulmonary artery. Guarded by a three-cusped pulmonary valve, the pulmonary trunk passes upward and backward to divide into the right and left pulmonary arteries.

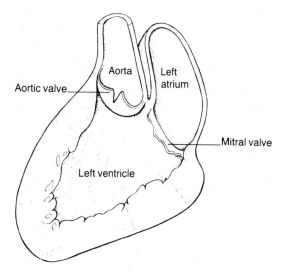

Fig. 9.7 The left atrium and left ventricle. Note the thick wall of the left ventricle.

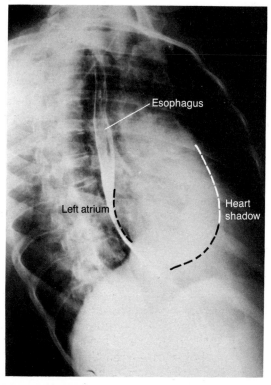

Fig. 9.8 An x-ray study of an oblique view of the left chest. This view demonstrates the close relationship of the left atrium to the barium-filled esophagus.

Left atrium

The left atrium lies behind the interatrial septum and makes up most of the base of the heart (see Fig. 9.3). Like the right atrium, it possesses a small auricular appendage, which creeps around the left border of the heart onto its anterior surface. The left atrium receives the four pulmonary veins (carrying arterial oxygenated blood), from which the atrium partly develops. The atrium of fetal days was the anteriorly projecting auricle. An important relation of the posterior surface of the atrium is the esophagus; an enlarged left atrium can be demonstrated by an indentation of the barium-filled esophagus (Fig. 9.8). The bicuspid mitral valve guards the atrioventricular orifice.

Left ventricle

The left ventricle is three times as thick as the right ventricle because it has to pump the blood around the whole body (see Figs. 9.2, 9.3 and 9.7). It has the same trabeculae carneae and papillary muscles, but no moderator band. Blood passes over both surfaces of the anterior cusp of the mitral valve, over the inner surface from atrium to ventricle, and then over the outer surface from the ventricle to the aorta.

Heart valves

The atrioventricular and arterial valves differ both structurally and functionally.

Atrioventricular valves

The valves guarding the atrioventricular orifices are closed by the propulsive force of blood produced by contracting ventricles. They depend on the chordae tendineae to prevent their ejection from the ventricles into the atria. They therefore have these strands attached to both the edges and the ventricular surfaces of the valve cusps. This also prevents the chordae tendineae from obstructing

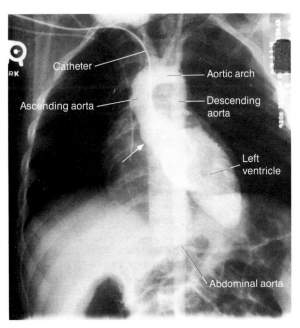

Fig. 9.9 An angiogram showing the left ventricle and the thoracic and abdominal aorta. The arrow points to the aortic sinuses.

the flow of blood from the atria to the ventricles.

Arterial valves

Arterial valves close by a different principle. The edges of the three cusps comprising these valves have fibrous nodules in their center; the three cusps meet, and the nodules come into apposition. It is thought that this may prevent the cusps from slipping. The cup-shaped valves are closed by the downward regurgitation of blood in the aorta and pulmonary arteries; the filled cusps force their edges together. They fill and close during diastole (relaxation of the ventricle).

Surface relationship of the valves

The valves lie behind the sternum in a nearly vertical plane. The highest valve opposite the third costal cartilage is the pulmonary valve, with the aortic valve just below. At the level of the fourth costal cartilage is the mitral valve, and behind the middle of the sternum,

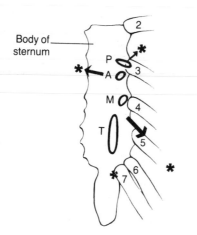

Body of sternum

Fig. 9.10 The surface markings of the cardiac valves and the direction of the blood flow (arrows). The black stars indicate the optimal sites for valvular auscultation. P, pulmonary; A, aortic; M, mitral; T, tricuspid.

reaching almost to the sixth costal cartilage in the vertical midline, is the tricuspid valve. These, however, are not the optimal sites to auscultate the valvular closures. The turbulence caused by the blood carries the sound in the direction of its flow as indicated in Fig. 9.10.

Ascending aorta

Passing upward, forward and to the right from the aortic orifice is the large ascending aorta (see Figs. 9.9 and 9.16). It is elliptical rather than circular in shape because the blood tends to be ejected against its right border. Immediately above the aortic orifice are three bulges, known as the aortic sinuses or sinuses of Valsalva, one anterior and two posterior. These sinuses are important; the open cusps might stick to the aortic wall were it not for these bulges. Furthermore, when the blood rushes retrogradely down the aorta after the cessation of the ventricular contraction, it fills the sinuses and pushes the cusps together. From the sinuses arise the two coronary arteries, the right from the anterior and the left from the left posterior. The ascending aorta is about 5 cm (2 in) long and

is enclosed with the pulmonary artery in a common serous sheath.

Blood supply

The heart is supplied by the coronary arteries. The right coronary artery arises from the anterior aortic sinus, and passes to the right to run in the atrioventricular groove around the circumference of the heart to the posterior surface, where it anastomoses with the left coronary artery. Its branches are the marginal artery along the inferior margin of the heart and the inferior (posterior) interventricular branch, which runs inferiorly in relationship to the interventricular septum to anastomose with the anterior interventricular branch of the left coronary artery.

The left coronary artery passes from the sinus to the left to enter the atrioventricular groove in which it runs to meet the right artery. The left coronary artery gives off a large anterior interventricular branch, which runs in relationship to the septum to meet the

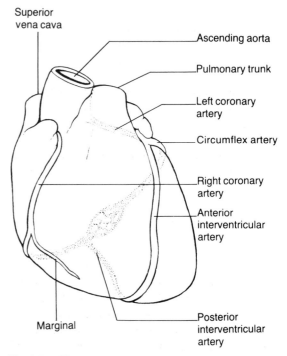

Superior vena cava

Ascending aorta

Pulmonary trunk

Left coronary artery

Circumflex artery

Right coronary artery

Anterior interventricular artery

Marginal

Posterior interventricular artery

Fig. 9.11 The coronary arteries and main branches.

interventricular artery of the right side. After this branch, the left coronary is often named the circumflex artery.

The right coronary artery supplies the right atrium, the right ventricle and, by its inferior interventricular branch, the left ventricle. The left coronary artery is larger, and it supplies most of the left ventricle and part of the left atrium. Occlusions of the left artery are therefore more dangerous.

All the arteries lie superficially and are likely to be injured by stab wounds. Occlusion of the origin of a coronary artery produces a dangerous situation. There are numerous anastomoses inside the heart muscle, but this does not help much if the main stems of the vessels are narrowed or occluded. Currently, these occlusions can be successfully bypassed by grafts from the aorta to the coronary arteries beyond the site of the radiologically demonstrated block.

The main vein of the heart is the coronary sinus, which runs from left to right in the posterior part of the atrioventricular groove and empties into the posterior wall of the right atrium. The sinus receives three large veins: the great cardiac vein, which runs with the interventricular branch of the left coronary artery; the small cardiac vein, which runs with the marginal branch of the right coronary artery; and, while it lies in the posterior part of the atrioventricular groove, the middle cardiac vein. Some inferior ventricular veins from the diaphragmatic surface may end in the coronary sinus. Several small veins enter the right atrium directly from its walls (venae cordis minimae).

Conducting system

The conduction of an impulse by cardiac muscle satisfies the metabolic needs of cold-blooded vertebrates, but it is too slow to meet the requirements of warm-blooded birds and mammals. Therefore, a specialized conduction system has evolved. The pacemaker, or initiator of the heart beat, is the sinoatrial (SA) node, which is situated in the upper part

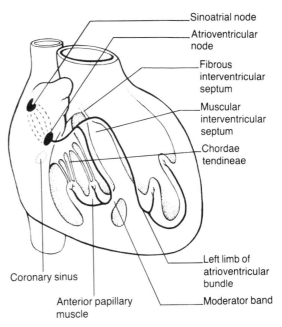

Fig. 9.12 **The conducting system of the heart.**

of the sulcus terminalis of the right atrium. It is supplied by the right coronary artery and the right vagus nerve.

The atrioventricular (AV) node is situated in the interatrial septum close to the coronary sinus opening. It is supplied by the right coronary artery, but by the left vagus nerve. The atrioventricular bundle arises from the AV node and runs to the fibrous part of the interventricular septum. Passing below the interventricular septum, the AV bundle meets the muscular part of the septum, where it divides into right and left branches. These pass on each side of the septum to the bases of the papillary muscles of each ventricle, the right one doing so via the moderator band. The finer branches then ramify through the myocardium.

Because the right coronary artery supplies the septal system, it therefore supplies the AV bundle and the bundle's right branch. The left coronary artery supplies the left branch only. It is very common *pari passu* for arterial occlusions to destroy part of this conducting system (either the bundle or its branches), with a resultant bundle branch block. This block results in a slower heart rate, since the impulse initiated in the atrium fails to reach the ventricles.

Superior and posterior mediastinum

We will discuss the superior mediastinum and follow the structures down toward either the middle or the posterior mediastinum. A few important facts make the disposition of the contents of the superior mediastinum easy to remember:

1 With the venous part of the heart on the right side, all venous structures are directed to that side.
2 The left portion of the heart is the arterial side, and most arterial structures begin and remain on the left side.
3 The disposition in the superior mediastinum follows the pattern of 'VAD': the large veins lie anteriorly with the arteries behind them, and the ducts lie posteriorly (the trachea, esophagus and thoracic duct).
4 The phrenic nerves go as far laterally as they possibly can; the right nerve hugs the venous structures, and the left nerve the arterial structures. The vagus nerves do the opposite. They try to get to the midline and run as posteriorly as possible, and they succeed.

With these principles in mind, we will describe the various structures.

Veins

The brachiocephalic veins begin behind the medial ends of each clavicle at the confluence of the internal jugular and the subclavian veins. In the fetus there are two veins, a right vein and a left vein, both running inferiorly and called the anterior cardinals. As the right side becomes the dominant venous side, a cross-branch develops and diverts the blood from the left to the right. The left anterior cardinal beyond this communication disappears. The cross-branch remains as the left brachiocephalic vein, which runs obliquely across the superior mediastinum to meet the right brachiocephalic vein to form the superior vena cava. This runs vertically downward to enter the right atrium. The brachiocephalic vein has many branches, which mostly correspond to the branches of the first part of the subclavian artery.

Masses in the superior mediastinum may cause pressure effects (superior mediastinal syndrome). The most likely signs result from compression of venous structures as they offer less resistance than the arteries or the ducts (Fig. 9.13).

Arteries

Aorta

The aorta consists of ascending, arch and descending divisions (see also Fig. 9.9). The ascending aorta runs up from the left ventricle to reach the angle of Louis, where it arches directly backward and to the left behind the lower half of the manubrium, to reach the fourth thoracic vertebra. As the arch passes backward, it lies above the left bronchus and crosses the left side of the trachea and esophagus. Lesions of the aorta, trachea, bronchus and esophagus tend to involve one another. It is an impressive sight when the aorta is seen pulsating against the bronchus or the esophagus during esophagobronchoscopy. One of the more dramatic ways of departing this life occurs when a lesion of the aorta ruptures into one of these two tubes.

Three very large branches arise from the arch of the aorta, slightly to the left of the midline and behind the lower half of the manubrium: the brachiocephalic, left carotid and left subclavian arteries. The brachiocephalic artery runs across the trachea to the right and divides into the right subclavian and the carotid branches behind the right sternoclavicular joint. The left common carotid and subclavian vessels pass cranially behind the left sternoclavicular joint. None of these arteries has any branches, except the brachiocephalic artery, which has an occasional small branch (the thyroidea ima) to the thyroid gland.

The walls of the ascending aorta and arch contain clever baroreceptors. These baroreceptors initiate reflex lowering of the blood pressure and heart rate when the aortic pressure rises excessively.

The descending aorta (also known as the thoracic or dorsal aorta) passes downward through the posterior mediastinum to enter the abdomen through the aortic hiatus. The aorta lies on the left side of the vertebral column from which it is partly separated by the two hemiazygos veins, while anteriorly the leftward-inclining esophagus veers across it. The aorta curves to the right to pass between the crura of the diaphragm anterior to the vertebral column.

The aorta gives off the lower nine posterior intercostal arteries to the intercostal spaces, the twelfth being called the subcostal, since it runs below the last rib. Coarctation of the aorta is a congenital defect resulting in a nar-

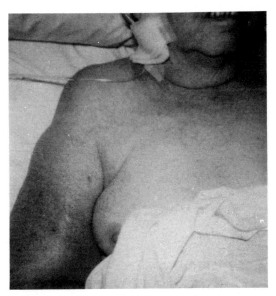

Fig. 9.13 A patient exhibiting the signs of venous obstruction; edema of the upper extremity and chest wall, as well as dilated superficial veins.

rowing of the aortic arch immediately beyond the origin of the left subclavian branch, in the region of the ductus arteriosus. The blood reaches the aorta beyond the constriction by the branches of the subclavian

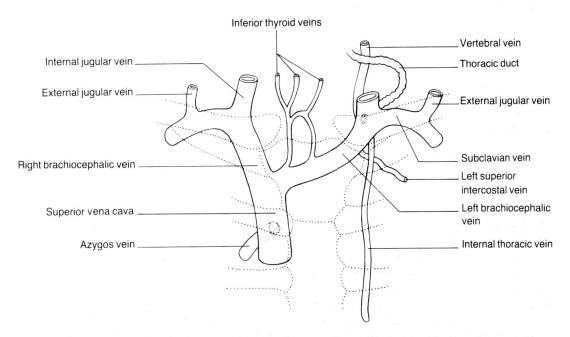

Fig. 9.14 The tributaries of the large venous channels of the superior mediastinum and the root of the neck. The left brachiocephalic also receives thymic veins. NB. Surface markings of the bony structures.

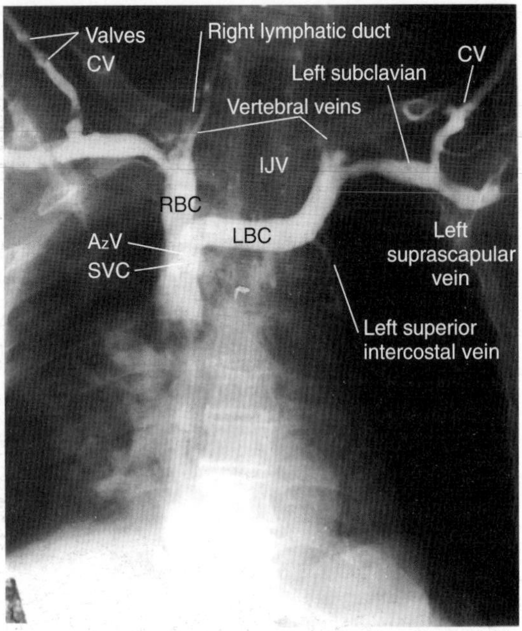

Fig. 9.15 **A venogram outlining the veins of superior mediastinum and root of neck. AzV, azygos vein; CV, cephalic vein; IJV, internal jugular vein; LBC, RBC, left and right brachiocephalic veins; SVC, superior vena cava.**

artery which arise above the aortic constriction. Thus the anterior intercostals of the internal thoracic pass the blood by a reversed flow to the lower posterior intercostals, with which they anastomose. The superior intercostal artery also helps by its anastomosis with the upper posterior intercostals. Bronchial branches, usually on the left side, supply oxygenated blood to the non-respiratory tissues of the lungs.

Radiological cardiac shadow (Fig. 9.20)

Now we can read an x-ray film of the cardiac silhouette. The latter contains fluid and therefore appears white, in contrast to the air-containing lungs which appear black. The right margin of the cardiac shadow is formed above downward by the right brachiocephalic vein, giving way to the superior vena cava, the right atrium (which forms the bulk of the

right border of the heart) and finally a small portion of the inferior vena cava as it enters the right atrium. The inferior border of the heart consists almost entirely of the right ventricle, with a little bit of the left ventricle represented at the apex. The left margin of the heart is the left ventricle for most of its border. Proceeding superiorly, a small portion of the left auricle peeps forward from behind, with the pulmonary artery above it, and, finally, the arch of the aorta is seen running anteroposteriorly (seen in profile and called the aortic knuckle). More cranially the branches of the aortic arch on the left, the common carotid and subclavian arteries, are seen by radiologists but seldom by other physicians. In summary, the right border is all venous, the left is all arterial and the inferior border of the heart is mixed.

Thymus

The thymus is an irregularly shaped, bilobed gland lying mostly in the anterior mediastinum but extending into the superior mediastinum and slightly into the lower neck. Small at birth, it grows to reach maximal size at puberty, after which it regresses and almost disappears. It lies behind the manubrium sterni anterior to the large veins, into one of which it drains (the left brachiocephalic vein). A producer of lymphocytes, the thymus is intimately involved with the immune mechanism of the body.

Ducts

Trachea

The trachea, the continuation of the larynx, passes inferiorly in the lower neck to enter the thorax in the midline behind the manubrium. It bifurcates at the carina, opposite the angle of Louis, into the right and left bronchi. It is about 10 cm (4 in) long and 1.9–2.5 cm (0.75–1.0 in) in diameter, being wider in the male. On the right side it is closely related to the pleura and lung, while on the left side the left common carotid and subclavian arteries

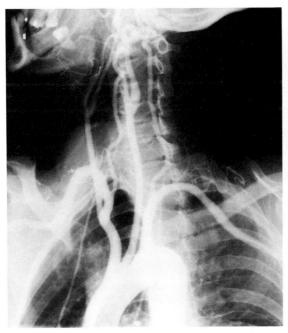

Fig. 9.16 Aortic arch arteriogram, left anterior oblique view. (From Weir and Abrahams 1978)

1 Ascending aorta.
2 Brachiocephalic artery.
3 Left common carotid artery.
4 Left subclavian artery.
5 Right common carotid artery.
6 Right subclavian artery.
7 Internal thoracic artery.
8 Vertebral artery.
9 Inferior thyroid artery.

10 Ascending cervical artery.
11 Costocervical trunk.
12 Internal carotid artery.
13 External carotid artery.
14 Lingual artery.
15 Facial artery.
16 Suprascapular artery.
17 Axillary artery.

separate it from the pleura and lung. Consequently, when listening to the right and left apices of the upper lobes of the lung, louder and more bronchial-type breath sounds will be heard on the right side, which must not be mistaken for an abnormality. The trachea is supported by a series of incomplete cartilage rings, and the completion of these rings by muscle permits alterations in its diameter.

Esophagus

The esophagus enters the thorax in the midline and runs for a short distance in that position before gradually veering to the left (see Figs. 9.17 and 9.18). This muscular tube lies posteriorly against the vertebrae with the

trachea lying anteriorly. From above downward, the esophagus is related to three important structures: the aorta running antero-posteriorly; the left bronchus below this; and the fibrous pericardium with the left atrium lying anteriorly. All these relationships are of great significance.

In the lowermost part of the posterior mediastinum, the esophagus curves markedly to the left, crossing the dorsal aorta as the latter makes for the midline. The tube is about 25 cm (10 in) long from its origin at the lower end of the cricoid cartilage in the neck. Because the distance from the teeth to the cricoid is 15 cm (6 in), it means that a tube must be 40 cm (16 in) long to enter the stomach.

Three areas of normal narrowing may be

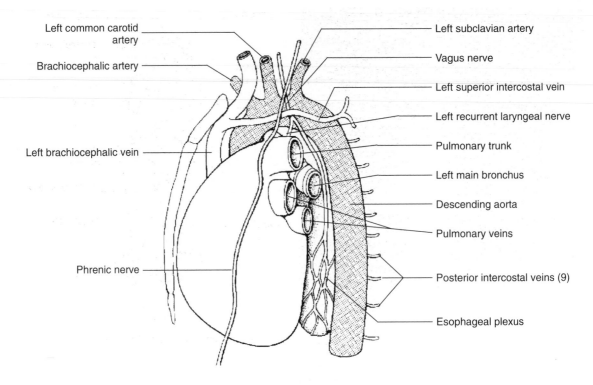

Left common carotid artery

Brachiocephalic artery

Left brachiocephalic vein

Phrenic nerve

Left subclavian artery

Vagus nerve

Left superior intercostal vein

Left recurrent laryngeal nerve

Pulmonary trunk

Left main bronchus

Descending aorta

Pulmonary veins

Posterior intercostal veins (9)

Esophageal plexus

Fig. 9.17 Oblique view of the left mediastinum. The aortic arch is seen crossing the left bronchus and esophagus and being crossed by the left phrenic and vagus nerves. The pulmonary artery is bifurcating in the concavity of the aortic arch.

seen on x-ray examination: at the origin of the esophagus by a circular muscle (cricopharyngeus); at the arch of the aorta; and at its lower end, where it enters the stomach.

The esophagus has striated muscle in its upper third, which produces rapid contraction and swallowing, and smooth muscle in its lower two-thirds, which exhibits the more leisurely type of peristalsis characteristic of the rest of the alimentary tract. The mucous membrane is thick and covered by squamous epithelium, although its lower part sometimes contains gastric mucosa – this may cause a peptic ulcer of the lower esophagus.

Thoracic duct

The thoracic duct begins at the upper end of a lymphatic sac in the abdominal cavity, the cisterna chyli. The duct, about 45 cm (18 in)

long passes upward to the right of the thoracic esophagus. At about the level of T4, it crosses behind the esophagus to reach its left side, and finally enters the neck to empty into the junction of the left subclavian and jugular veins (Fig. 9.14, see Fig. 2.24). The duct drains the lymph of the whole body except that from the right side of the head and neck, right arm, right thorax and the lower half of the left lung (see Fig. 2.25). The latter areas drain by a right lymph duct, which enters the large veins on the right side (Fig. 9.15).

Nerves of the thorax

Phrenic nerves

The right nerve crosses the subclavian artery to reach the thoracic cavity and immediately makes for the venous side of the heart. It runs

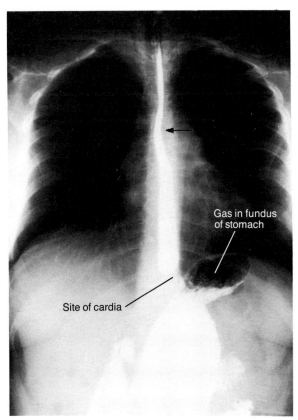

Gas in fundus
of stomach

Site of cardia

Fig. 9.18   Radiograph of a barium swallow outlining the esophagus. Arrow indicates an esophageal indentation by a normal aorta.

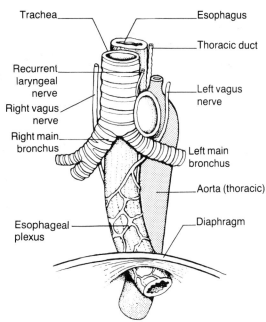

Trachea

Esophagus

Thoracic duct

Recurrent laryngeal nerve

Right vagus nerve

Left vagus nerve

Right main bronchus

Left main bronchus

Aorta (thoracic)

Esophageal plexus

Diaphragm

Fig. 9.19   The three ducts and the aorta.

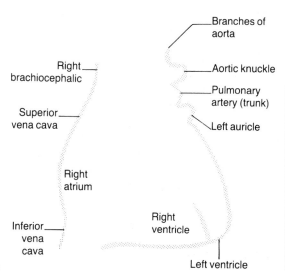

Branches of aorta

Aortic knuckle

Pulmonary artery (trunk)

Left auricle

Right brachiocephalic

Superior vena cava

Right atrium

Right ventricle

Inferior vena cava

Left ventricle

Fig. 9.20   An outline of the cardiac silhouette as seen on a posteroanterior chest x-ray. (To see all of this, you need to have descended from a long line of hawks.)

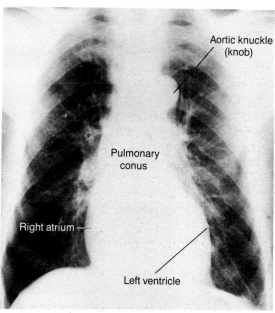

Aortic knuckle (knob)

Pulmonary conus

Right atrium

Left ventricle

Fig. 9.21   Radiograph of a posteroanterior study of the chest.

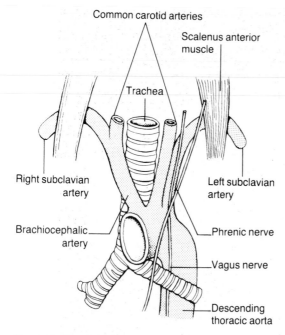

Common carotid arteries

Scalenus anterior muscle

Trachea

Right subclavian artery

Left subclavian artery

Brachiocephalic artery

Phrenic nerve

Vagus nerve

Descending thoracic aorta

Fig. 9.22   The trachea and aortic arch.

Vagus nerves

The right vagus nerve enters the thorax by crossing the subclavian artery, and makes its way posteriorly and to the midline to reach the posterior mediastinum. It therefore passes inferiorly on the side of the trachea and behind the root of the lung, where it gives off extensive branches to the pulmonary plexus and reaches the esophagus to become part of the esophageal plexus. The left vagus nerve runs in a similar direction but has to cross the aorta, which keeps it away from the trachea, to reach the posterior aspect of the left bronchus and the root of lung. It finally meets the right vagus in the esophageal plexus. The vagal branches have important functions.

Recurrent laryngeal nerves

As the right vagus crosses the subclavian artery and the left crosses the aorta, they each give off a recurrent laryngeal branch. These branches pass superiorly to reach the groove between the esophagus and the trachea, in which each runs to reach the larynx. The left recurrent nerve is clinically affected much more often than the right, since it has a close relationship to the aortic arch and the structures that the latter crosses, the left bronchus and the esophagus. In this area lie also numerous groups of lymph nodes.

Bronchial, esophageal or aortic lesions or lymph node enlargements due to diseases of the first two conditions may involve the left recurrent laryngeal nerve. An elderly man, his pockets bulging with packs of cigarettes, who comes to the physician complaining of hoarseness, which is found to be due to a recurrent laryngeal nerve paralysis, is the victim of a neoplasm of the lung where the secondary lymph node enlargement has involved the left recurrent laryngeal nerve as it passes under the aortic arch. In the same patient, one may also find that the phrenic nerve (lying adjacent as it crosses the arch of the aorta) is paralyzed, and an x-ray examin-

inferiorly on the right brachiocephalic vein, the superior vena cava, the pericardium covering the right atrium and the inferior vena cava, finally passing through the caval opening of the diaphragm to be distributed to its undersurface. The left nerve reaches the thoracic cavity as on the right side, and latches onto the arterial structures. It passes inferiorly close to the left subclavian artery, crosses the aortic arch and, incidentally, the vagus nerve, to reach the pericardium-covered left ventricle, and finally pierces the diaphragm.

The phrenic nerve is a mixed nerve, conducting motor impulses to the diaphragm and sensory impulses from all those structures beginning with 'P' – the pleura (mediastinal and diaphragmatic), pericardium and peritoneum of the abdominal cavity. The surgeon easily locates this nerve on the pericardium, where it may be crushed to temporarily paralyze the diaphragm, (e.g. to rest the diaphragm after it has been repaired).

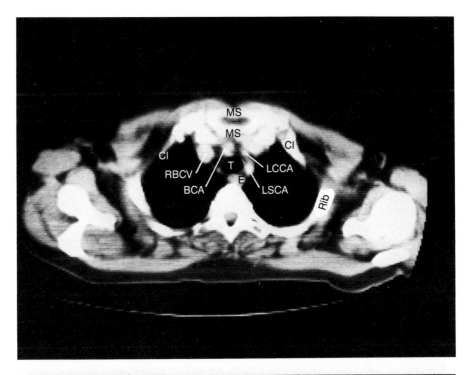

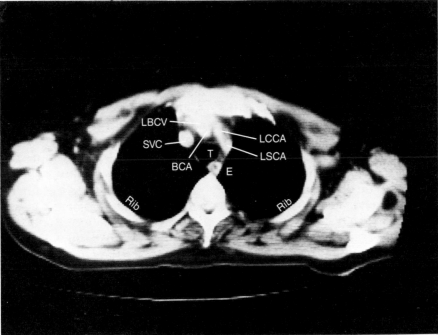

Fig. 9.23   Computerized tomograms of the great vessels of the neck at different levels. BCA, brachiocephalic artery; Cl, clavicle; E, esophagus; LBCV, RBCV, left and right brachiocephalic veins; LCCA, left common carotid artery; LSCA, left subclavian artery; MS, manubrium sterni; SVC, superior vena cava; T, trachea.

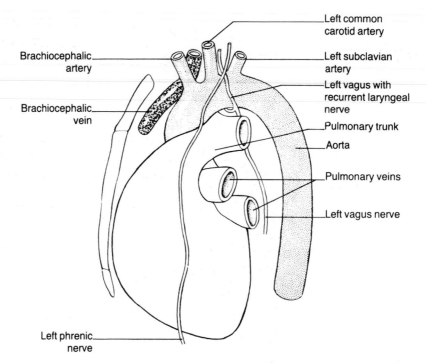

Brachiocephalic artery

Brachiocephalic vein

Left common carotid artery

Left subclavian artery

Left vagus with recurrent laryngeal nerve

Pulmonary trunk

Aorta

Pulmonary veins

Left vagus nerve

Left phrenic nerve

Fig. 9.24 Left oblique view of the mediastinum. Note the left phrenic and left vagus nerves crossing the aortic arch and the pulmonary artery bifurcating in the concavity of the aortic arch.

ation may confirm this by demonstrating an immobile diaphragm.

It is important to note that the recurrent nerve hooks around the ligamentum arteriosum, which is the fibrous remnant of the ductus arteriosus that joins the left pulmonary artery and the aortic arch. Surgeons who ligature a persistent ductus arteriosus know this relationship well: a child whose patent ductus has been surgically closed may be hoarse from accidental injury to the nerve but her lawyer may not be!

Cardiopulmonary plexuses

Both the vagi and the recurrent laryngeal nerves give off branches to the cardiac plexuses which lie anterior to the tracheal bifurcation. From here, offshoots pass via the pulmonary arteries as the pulmonary plexuses. The cardiac branches of the vagi arise not only in the thorax but also in the neck (cervical branches). It is worth noting that the vagus acts as the brake of the heart, for

which reason it is sometimes called the depressor nerve. The vagi also carry afferent impulses from the aorta and aortic bodies that act as baroreceptors and chemoreceptors.

Esophageal plexus

In the posterior mediastinum, the nerves form an esophageal plexus surrounding the esophagus. This plexus usually gives rise to two nerves, the anterior and posterior gastric nerves (sometimes there are three or four nerves), which pass through the esophageal hiatus to reach the stomach. Both nerves have branches from both vagi. At the present time, section of the vagi to reduce the amount of acid secretion of the stomach is the preferred form of surgery for peptic ulcer. Originally, the vagi were severed in the thorax, but this was discontinued in favor of an abdominal approach, since it is more difficult to locate the many and varied branches of the thoracic plexuses than the relatively well-collected fibers of the abdominal gastric nerves.

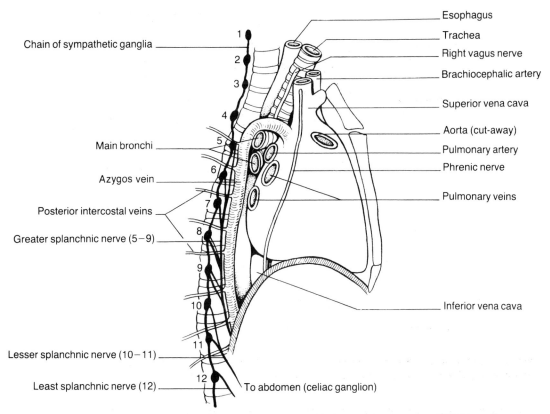

Chain of sympathetic ganglia

Main bronchi

Azygos vein

Posterior intercostal veins

Greater splanchnic nerve (5-9)

Lesser splanchnic nerve (10-11)

Least splanchnic nerve (12)

1
2
3
4
5
6
7
8
9
10
11
12

Esophagus

Trachea

Right vagus nerve

Brachiocephalic artery

Superior vena cava

Aorta (cut-away)

Pulmonary artery

Phrenic nerve

Pulmonary veins

Inferior vena cava

To abdomen (celiac ganglion)

Fig. 9.25 **A view of the right mediastinum. The root of the lung (with the structures grouped VAD) is well seen.**

Sympathetic trunk

Entering the thorax over the neck of the first rib, each sympathetic trunk proceeds distally anterior to the intercostal vessels. The trunk gradually edges medially from in front of the necks of the ribs, to the sides of the lower vertebral bodies, until it passes under the origin of the diaphragm from the medial arcuate ligament. There are usually 12 thoracic ganglia, but one or more may become fused. The first thoracic ganglion of the chain fuses with the inferior cervical ganglion to form the stellate ganglion in 50 percent of individuals.

Branches of the sympathetic trunk

The major outflow of the sympathetic system occurs in the thorax. Each ganglion receives from each intercostal nerve a white or myelinated ramus, which usually relays in this or a more distal ganglion. After relay, the ganglion gives off a gray or nonmyelinated ramus back to the intercostal nerve for distribution to the body wall. The major supply to the abdominal viscera is given off from the thoracic sympathetic chain. Branches from the fifth to ninth ganglia join to form the greater splanchnic nerve, branches from the tenth and eleventh ganglia join to form the lesser splanchnic nerve, and from the twelfth ganglion comes the least splanchnic nerve. These splanchnic branches pierce the crural origin of the diaphragm to join the abdominal plexuses for distribution to the abdominal structures. Branches are also given off to the thoracic viscera, as described below.

Nerve supply of the heart

It has been mentioned that the brake of the heart is the vagus nerve. The accelerator of the heart is the sympathetic system, which increases the heart rate (see Fig. 2.19). Cervical branches arise in the neck from three cervical ganglia and pass inferiorly into the thorax to join the cardiac plexuses. In addition, sympathetic branches from the upper three to five thoracic anterior nerve roots run to the thoracic ganglia and, via the thoracic cardiac nerves, to the cardiac plexuses. These branches are distributed to the coronary arteries and the conduction system. Sensation from the heart and its vessels is relayed back to the upper five thoracic posterior nerve roots. The pain from ischemic heart disease (angina) was once treated by sectioning the posterior nerve roots. Now, however, the cause is treated and the deficient blood supply is increased.

Nerve supply of the lungs

The afferent nerve supply is via the vagus nerve, which carries fibers subserving the cough reflex as well as pain. Efferent vagal fibers constrict the bronchi and stimulate secretion (premedication for general anesthesia includes drugs to paralyze the vagus and 'dry up' the patient so that she does not drown in her bronchial secretions). The sympathetic supply to the lungs constricts the blood vessels and dilates the bronchial tubes (epinephrine (adrenaline) and its derivatives are used in the treatment of the common bronchospastic condition, asthma).

Nerve supply of the esophagus

The nerve supply to the esophagus is both vagal and sympathetic. The afferent paths are not clear, and pain is poorly localized, being referred anywhere from the base of the neck to the xiphoid process.

Review questions on the thorax

(answers on p. 529)

52 In labored breathing during the inspiratory phase, the following muscles may be noticeably active except for
a) pectoralis major
b) serratus anterior
c) sternocleidomastoid
d) trapezius
e) external intercostals

53 The sinoatrial node
a) receives nerve fibers from the right vagus nerve
b) is located in the wall of the right atrium
c) lies near the crista terminalis
d) receives blood from the right coronary artery
e) all of the above are true

54 Bronchial arteries
a) carry oxygen-poor blood to alveoli in the lung
b) carry oxygen-poor blood to conducting and supporting tissues in the lung
c) are branches of pulmonary arteries
d) are usually branches of the aorta
e) are normally four in number

55 A patient is stabbed in the chest, the knife passing straight posteriorly through the anterior extremity of the fifth left intercostal space. The injured structures may be the
a) left pleural sac
b) right pleural sac
c) pericardial sac
d) superior epigastric artery
e) inferior vena cava

56 A barium swallow shows the patient's midesophagus to be compressed by a structure anterior to it. This could be the
a) left ventricle
b) right ventricle
c) right bronchus
d) left atrium
e) parathyroid gland

57 Lying in the posterior mediastinum is the
a) pericardial sac
b) tracheal bifurcation
c) superior vena cava
d) thoracic duct
e) phrenic nerve

58 An aneurysm (abnormal dilatation) of the aortic arch may affect the
a) left bronchus
b) right bronchus
c) vena azygos
d) oblique sinus
e) left ventricle

59 The ductus arteriosus joins the aorta and the
a) pulmonary vein
b) left pulmonary artery
c) bronchial vein
d) coronary sinus
e) pulmonary trunk (right branch)

60 Instead of closing after birth, the ductus arteriosus may remain patent and cause symptoms. In ligating it the surgeon must be careful not to injure a nerve lying very close to it. This is the
a) phrenic nerve
b) sympathetic trunk
c) recurrent laryngeal
d) fourth intercostal nerve
e) sixth intercostal nerve

61 When the diaphragm descends in inspiration, which of the following is enlarged?
a) Esophageal hiatus
b) Aperture for the inferior vena cava
c) Aperture for the internal thoracic artery
d) Aortic hiatus
e) Medial lumbocostal arch

62   In severe coarctation of the aorta, reversal of blood flow will be observed in the

a)   suprascapular artery
b)   internal thoracic artery
c)   superior mesenteric artery
d)   lower posterior intercostal arteries
e)   inferior thyroid artery

63   Cancer of the apex of the lung may affect certain vital structures, which cross the neck of the first rib. This could result in

a)   paralysis of one hemidiaphragm
b)   loss of sweating from half the face
c)   impaired pronation of the forearm
d)   inability to close the eyelid completely
e)   hiccups

64   With respect to the pleural sacs, they

a)   are normally in contact behind the xiphoid
b)   normally communicate behind the lower manubrium
c)   extend two rib spaces below the lungs posteriorly during quiet breathing
d)   are in positive pressure which collapses the lung when the diaphragm is relaxed
e)   reach the upper border of the medial half of the clavicle at their highest points

65   With respect to the trachea and bronchi,

a)   the carina is at the level of the suprasternal notch
b)   inhaled safety pins are more apt to enter the right main bronchus
c)   the left bronchus is longer and branches outside the lung
d)   the trachea is prevented from collapse by a series of bony rings
e)   there are more segmented bronchi in the left lung

66   With respect to the heart,

a)   the function of the papillary muscles is to open the atrioventricular valves
b)   the circumflex branch of the left coronary artery supplies the greater part of the left ventricular muscle
c)   contraction of the atria is essential in completely filling the ventricles
d)   the base of the heart rests on the diaphragm
e)   it ascends during inspiration

67   With respect to the innervation of the heart,

a)   all sympathetic innervation is by branches from the superior and middle cervical ganglia
b)   the sinoatrial node is supplied by the right vagus nerve
c)   the left vagus nerve supplies the left lung, but not the heart
d)   part of the sympathetic innervation of the heart is via the greater splanchnic nerve
e)   the atrioventricular node is supplied by the right vagus nerve

68   The surgeon dissecting the root of the right lung may locate the bronchus by appreciating that it

a)   is the most anterior structure
b)   lies between the pulmonary veins and arteries
c)   is posterior to the pulmonary veins and arteries
d)   is the most inferior structure
e)   none of the above

69   The sinoatrial node

a)   lies outside the pericardium
b)   is supplied by the right coronary artery
c)   is innervated by the left vagus nerve
d)   gives rise to the bundle of His
e)   relays the beat to the ventricular muscles

70   Which of the following is true of the left phrenic nerve?

a)   It is a pure motor nerve
b)   It supplies sensory innervation to the apex of the left ventricle
c)   It is derived from spinal cord segments C6, 7 and 8
d)   Pain impulses carried by it may be referred to the left shoulder
e)   Pain impulses carried by it may be referred to the left ear

71   The internal thoracic artery, which is sometimes used to bypass a blocked coronary artery,

a)   is a branch of the axillary artery
b)   has no terminal branches
c)   descends anterior to the sternum
d)   Is a branch of the brachiocephalic artery
e)   often supplies branches to the thymus gland

72 **The pericardial sac may become filled with blood and interfere with the heart's action (cardiac tamponade). To remove the blood, a syringe needle can safely be passed into the pericardial sac without entering the pleural cavity**
a) immediately to the right of the sternum in the sixth interspace
b) immediately to the left of the sternum in the fifth interspace
c) 2.5 cm (1 in) to either side of the sternum in the third interspace
d) between the xiphoid and the eighth costal cartilage
e) between the fourth and fifth right interspace in the midclavicular line

73 **The parietal (costal) pleura is innervated by the**
a) intercostal nerves
b) vagus nerve
c) phrenic nerve
d) spinal accessory nerve
e) greater splanchnic nerve

74 **The superior vena cava**
a) receives the thoracic duct
b) receives the azygos vein
c) forms part of the left cardiac border in an x-ray photograph
d) has a large valve at its entrance into the atrium
e) lies completely within the pericardial sac

75 **The mediastinum is conventionally divided into four areas in anatomical and clinical descriptions. Which of the following is correct?**
a) The phrenic nerve is found in the anterior mediastinum
b) The right recurrent laryngeal nerve appears in the superior mediastinum
c) A plane passing through the sternal angle (of Louis) and vertebral body T2 separates the superior and inferior mediastina
d) The cupola of the pleura forms the upper limit of the superior mediastinum
e) The arch of the aorta lies entirely within the superior mediastinum

76 **A stab wound that nicks the anterior surface of the heart could injure the**
a) great cardiac vein
b) coronary sinus
c) atrioventricular node
d) left atrium
e) mitral valve

77 **The left border of the 'cardiac shadow' in an x-ray photograph includes the**
a) left pulmonary artery
b) left atrium
c) left ventricle
d) the aortic arch ('knuckle' or 'knob')
e) all of the above

78 **The sternal angle (of Louis) is a surface landmark for all of the following *except***
a) the origin of the main bronchi (carina)
b) counting ribs – the second costosternal joint being at this site
c) the apex of the pericardial sac
d) the convergence of the pleural sacs
e) the thoracic duct crossing from the right to the left side anterior to the esophagus

79 **With respect to the fibrous pericardium,**
a) it is highly elastic
b) it should always be incised vertically (that is, sagitally) so as to avoid cutting the vagus nerve
c) six large veins pass through it to reach the heart
d) it occupies the anterior mediastinum
e) its only blood supply is from the internal thoracic arteries

80 **The surgeon may have to remove the thymus gland to alleviate myasthenia gravis (easy tiredness of the muscles). In doing so he has to take particular care to secure the thymic veins, which are large and drain into the**
a) superior vena cava
b) inferior vena cava
c) right brachiocephalic vein
d) left brachiocephalic vein
e) coronary sinus

81 **A patient has an enlarged lymph node above the inner one-third of the right clavicle. The primary disease may be present in the**
a) upper lobe of the left lung
b) upper lobe of the right lung
c) left main bronchus
d) left ventricle
e) thoracic duct

82   On x-ray photograph, one sees an abnormal shadow in the left posterior mediastinum. It could arise from the

a)   thymus
b)   left bronchus
c)   aortic arch
d)   azygos vein
e)   esophagus

83   A patient has had a severe cough and hoarseness for 1 month. Examination reveals that the left side of his larynx is paralyzed. He might have

a)   enlarged lymph nodes of the posterior mediastinum
b)   a tumour of the basal segments of the right lung
c)   an aneurysm of the aortic arch
d)   an inflamed pericardium
e)   a phrenic nerve injury

84   A tumor of the breast may involve the subjacent muscles, which are

a)   pectoralis major
b)   pectoralis major and serratus anterior
c)   pectoralis major, serratus anterior and internal oblique
d)   pectoralis major and pectoralis minor
e)   pectoralis minor and serratus anterior

85   Disturbances of the conduction system of the heart are common. Which of the following statements are true?

a)   Purkinje conduction fibers may cross the cavity of the left ventricle by means of the moderator band
b)   The atrioventricular bundle divides into two branches at the upper end of the muscular part of the interventricular septum
c)   The sinoatrial node receives its blood supply from the right pulmonary artery
d)   Impulses are conducted from the sinoatrial and atrioventricular nodes by means of Purkinje fibers
e)   A stab wound through the fossa ovalis would cut the main conduction bundle

Part V   The Head and Neck

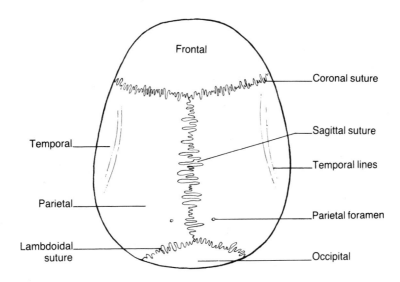

Fig. 10.4 Superior view of the vault of the skull.

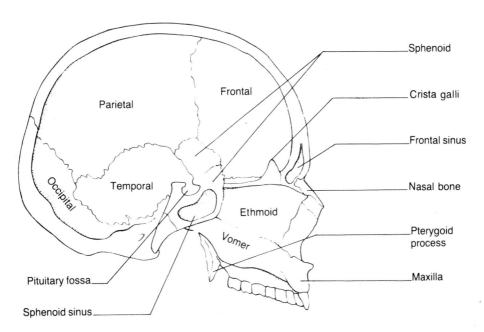

Fig. 10.5 Sagittal section of the skull bones.

followed by the most posterior bone, the occipital, articulating with it at the lambdoidal suture. The side of the skull is formed mainly by the temporal bone, and to a lesser degree by the part of the sphenoid bone (the greater wing) lying anterior to it.

The frontal and parietal bones, paired bones at birth, are separated by the sagittal suture. The sagittal suture begins at the junction of the two frontal bones at an area called the glabella and runs anteroposteriorly to end at the occipital bone. In adult life the

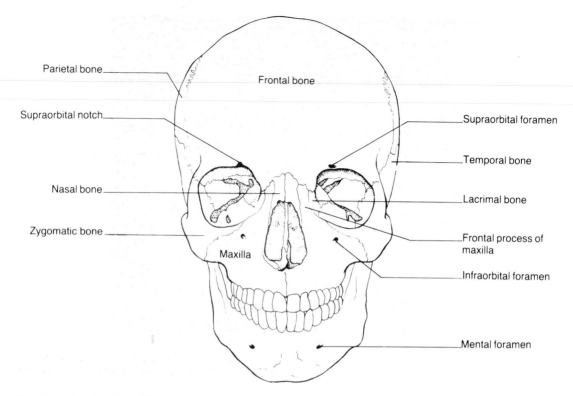

Fig. 10.6 Anterior view of the skull.

suture between the frontal bones disappears in 90 percent of skulls; it remains as the metopic suture in 10 percent.

Frontal bone

If you run your finger along the superior rim of the orbital cavity, you will feel, at the junction of the medial and middle thirds, a notch (which is sometimes a foramen) through which passes the large supraorbital nerve and its accompanying vessels. Running posteriorly, this nerve supplies a large area of the skull and scalp. A local anesthetic agent, deposited at the notch, will anesthetize a substantial area of scalp. Following the rim laterally, you will find the margin turning downward, forming an important zygomatic process that articulates with the zygomatic, or cheek, bone. Under cover of and slightly above this articulation lies the lacrimal gland, which produces tears to moisten the eye. Moving to a correspond-

ing site on the medial side of the orbit, where the frontal bone runs down to articulate with the nasal bones, the same finger might be able to feel a little spine, or trochlea. To this spine is attached the pulley for one of the muscles of the eye (superior oblique). Running above and below the trochlea are the supratrochlear and infratrochlear nerves and vessels, respectively, which are destined for a small area of the skin of the medial forehead and upper eyelid.

Just above the sharp rims of the orbit are two prominences, the superciliary ridges, which become larger in elderly persons and which meet in the midline at the glabella. Deep to the superciliary arches and between the inner and outer tables of each frontal bone lie a pair of cavities that extend up between the tables as the frontal air sinuses. These two sinuses vary in size and are separated by a septum, which seldom lies in the midline.

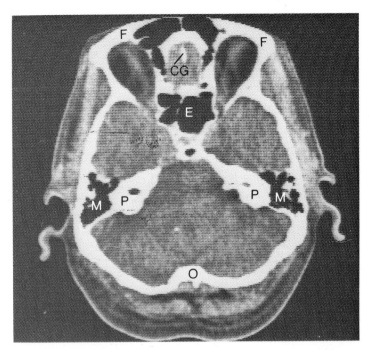

Fig. 10.7 CT scan. Base of skull. CG, Crista galli; E, Ethmoid sinus; F, Frontal bone; M, Mastoid air cells; O, Occiput; P, Petrous temporal bone. (Reproduced from Weir and Abrahams: An Atlas of Radiological Anatomy, 2nd Edition, Churchill Livingstone, 1986).

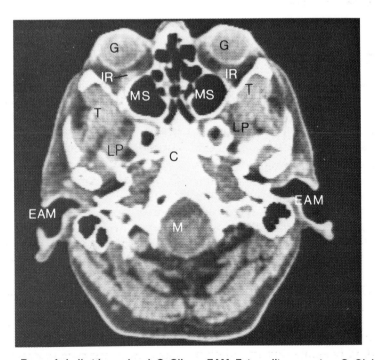

Fig. 10.8 CT scan. Base of skull at lower level. C, Clivus; EAM, Ext. auditory meatus; G, Globe; IR, Inf. rectus m.; LP, Lat. pterygoid m.; M, Medulla; MS, Maxillary sinus; T, Temporalis m. (Reproduced from Weir and Abrahams: An Atlas of Radiological Anatomy, 2nd Edition, Churchill Livingstone, 1986).

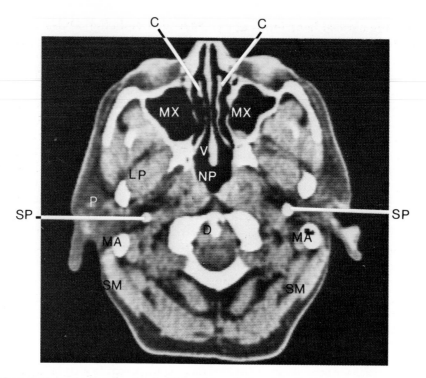

Fig. 10.9 CT scan. Upper neck region, level of C1 vertebra. C, Conchae; D, Dens of axis; LP, Lat. pterygoid m.; MA, Mastoid process; MX, Maxillary air sinus; NP Nasopharynx; P, Parotid gland; SM, Sternomastoid m.; SP, Styloid process; V, Vomer. (Reproduced from Weir and Abrahams: An Atlas of Radiological Anatomy, 2nd Edition, Churchill Livingstone, 1986).

The surgeon has to know the extent of these sinuses before he opens the skull; one would like to avoid passing through them, for they contain all sorts of yucky material. Furthermore, when a mugger delivers a smart bash to his victim, he may fracture the frontal bone. By itself this is bad enough, but if the fracture passes into the sinuses, serious complications occur since the fracture now becomes infected. Should the fracture penetrate the whole thickness of the bone and enter the cranial cavity, the implications are even more serious.

10% of adult skulls show a suture between the two frontal bones. Although this metopic suture disappears in most of the adult population, its persistence should not be mistaken for a fracture line.

When you slide your fingers over the frontal bone toward the coronal suture, you will note an area that is somewhat convex; this is the frontal tuberosity. This prominence is usually not very convincing in the adult. But wake up your baby brother, sister or infant progeny, and palpate the frontal area. Children have very pronounced frontal (and parietal) tuberosities.

Parietal bone

The parietal bone is curved, somewhat quadrilateral in shape, and the most mobile of the fetal skull bones. The sagittal suture separates the two parietal bones in the midline; the coronal suture separates the parietal bone from the frontal bone anteriorly; the lambdoidal suture separates the parietal from the occipital bone posteriorly; and a long suture line laterally separates it from the temporal and sphenoid bones, which form the side of the skull.

It is worth mentioning the names given to

the angles that the parietal bone forms at its articulations (otherwise you will be tormented during your stay in obstetrics). The coronal suture meets the sagittal suture at the bregma. The angle at the sagittal and occipital sutures is shaped like the Greek letter lambda (λ). The two inferior angles are, posteriorly, the asterion (named because of its stellar configuration), and anteriorly, the pterion (Fig. 10.3).

Just in front and slightly to each side of the sagittal–lambdoidal junction are the fairly large parietal foramina for veins (emissary) running between the scalp and sagittal sinus. The sagittal sinus is a large venous structure lying in the sagittal sulcus of the inner table. Emissary veins are numerous channels that join the intracranial and extracranial venous circulations. They can be both friend and foe. (For an explanation of this apparent contradiction, read the saga of the scalp in Chapter 11.)

Occipital bone

This diamond-shaped bone is perforated by the foramen magnum, which descriptively divides it into four parts: the squama(e) behind, the basilar part in front and a lateral part on each side.

If you run your finger over the posterior aspect of the squama, you will come across a big knob, known as the inion or external occipital protuberance. This prominence lies in the center of the squama or flat part of the occipital bone. Its function is not to support your hat, but to act as the upper attachment of the ligamentum nuchae (see Chapter 7). The inion demarcates an area above, covered by the scalp, and an area below, the nuchal area, for the attachment of several neck muscles. Running laterally from the inion is the superior nuchal line for the attachment of the important trapezius and sternomastoid muscles. The other muscles in the area are not of great practical or clinical significance. The superior nuchal line leads one laterally to a big boss, the mastoid process, which is described in the following section.

The two lateral parts bear the kidney-shaped occipital condyles which articulate with the atlas. The basilar part articulates anteriorly with the body of the sphenoid, at first by cartilage, but at 25 years by bone. The lateral and basilar parts share in the development of the condyles, the basilar part forming the anterior one-third of each condyle. The line of fusion of the component bones is often visible on the condyle. The four parts of the bone, separate at birth, soon fuse into a single structure. The foramen magnum transmits the lower end of the medulla with its covering meninges, the spinal roots of the accessory nerves, the vertebral vessels with their sympathetic plexuses, as well as the spinal vessels and the apical ligament.

Lateral aspect of the cranium

Temporal bone

The temporal bone, a bone composed of five developmentally different subdivisions, constitutes a large portion of the lateral or temporal area. The subdivisions are the squama(e), tympanic plate, petrous, styloid process and mastoid.

The squamous part is a flat portion, articulating superiorly with the parietal bone and anteriorly with the sphenoid.

Viewing the lateral aspect of the skull as a whole, one notes two closely apposed lines which run from the zygomatic process of the frontal bone upward and posteriorly across the frontal and parietal bones, and then curve anteriorly to cross the squama and reach the zygomatic arch. The superior temporal line is for the attachment of the temporalis muscle. These lines run into the zygomatic process of the temporal bone, which projects slightly laterally before turning forward to meet the temporal process of the zygomatic, or cheek bone. Together, these two processes constitute an important and easily palpable landmark, the zygomatic arch, beneath which the temporalis muscle passes on its way to insert into the mandible.

Below and behind the zygomatic process of the temporal bone is a large hole, the external

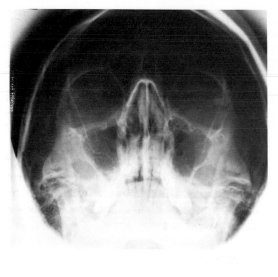
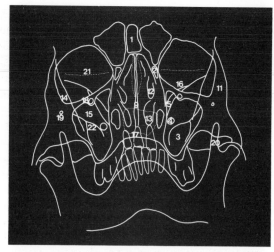
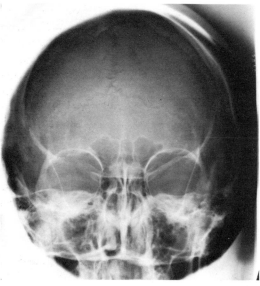
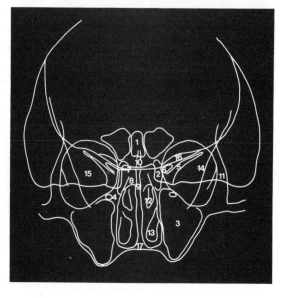

Fig. 10.10   Occipitomental and occipitofrontal views of the skull. (From Weir and Abrahams 1978)

1 Frontal sinus.
2 Ethmoid sinus.
3 Maxillary sinus.
4 Foramen rotundum.
5 Supraorbital fissure.
6 Anterior clinoid process.
7 Posterior clinoid process.
8 Floor of hypophyseal fossa.
9 Nasal septum.
10 Crista galli.
11 Frontal process of zygomatic bone.

12 Middle concha (turbinate).
13 Inferior concha (turbinate).
14 Lateral border of greater wing of sphenoid.
15 Greater wing of sphenoid.
16 Lesser wing of sphenoid.
17 Hard palate.
18 Infraorbital foramen.
19 Zygomaticofacial foramen.
20 Coronoid process of mandible.
21 Soft tissue lower lid.
22 Pterygoid plates of sphenoid.

acoustic or auditory meatus. The meatus consists mostly of the tympanic plate of the temporal bone. Its free outer border affords attachment for the cartilage of the external ear, while medially it fuses with another part of the temporal bone, the petrous temporal (Fig. 10.7).

If you look above and posterior to the external auditory meatus, you will see a little projection, the suprameatal spine. This is an important landmark. Between the spine, the backward prolongation of the zygomatic process, the supramastoid crest and a line drawn vertically down behind the meatus, one can make out a small triangle (MacEwen's triangle). Beneath the triangle lies the mastoid antrum, a backward prolongation of the cavity of the middle ear, which is commonly involved in ear infections. The antrum leads posteriorly and inferiorly to the mastoid process, a large palpable and visible landmark, which points inferiorly and to which are attached several muscles. The process is not present at birth but starts developing when the baby begins to lift its head. When the child is about 2 years of age, a respectable mastoid process will have been formed (see Fig. 10.23). The cavity of the antrum is prolonged into the process by means of intercommunicating air cells. Middle ear disease

spreads by this route to cause mastoiditis, a serious infection.

The last division making up the temporal bone is the styloid process, which pokes down from its socket in the tympanic plate. It is a long process that is usually broken off by successions of students while they are studying the skull. It is, however, formed from the second pharyngeal arch cartilage.

In summary, the temporal bone consists of five parts. Except for the mastoid process, which develops from the petrous, they are separate pieces of bone that fuse during the first year after birth.

Sphenoid bone

The disarticulated sphenoid bone resembles an owl and one would think that the word sphenoid means 'like an owl'. However, it actually means 'wedge', because the bone is wedged between the frontal and occipital bones. Being shaped like an owl, the sphenoid bone has a body, but, unlike owls that we have known, it has four wings, two lesser and two greater. The legs on which it stands are called the pterygoid processes. If you look at the bone in a sagittal section, you will see that it is a hungry owl in that the body is hollow and occupied by two air sinuses,

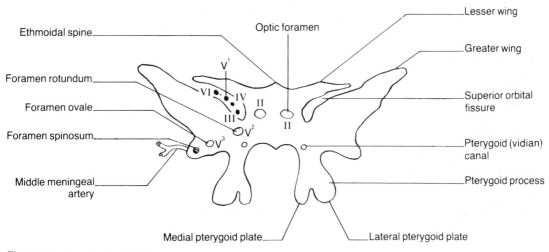

Fig. 10.11 The sphenoid bone.

separated (like the frontal sinuses) by a deviated septum. It has, more recently, been likened to an 'asteroid monster' from a Space Invader game.

The upper surface of the body is hollowed out to form a fossa for the pituitary gland. In front of this fossa is a small bump, the tuberculum sellae, the site through which the anterior lobe of the pituitary gland passes from its origin in the oral cavity to the interior of the skull (the craniopharyngeal canal). Tumors may arise from some of the cells that become lost on this trip (craniopharyngiomas).

Anterior to the tuberculum sellae is the optic chiasma, a slightly grooved area (chiasmatic sulcus) where the optic nerves meet. The pituitary fossa is overhung posteriorly by a large dorsum sellae, which ends in two posterior clinoid processes. This area is sometimes called the sella turcica, since it is said to resemble a Turkish saddle. The sella is easily seen on lateral x-ray films of the skull, when its size is crucial in the diagnosis of suspected tumors of the pituitary gland (see Fig. 11.22).

The posterior surface of the body is united to the occipital bone by a cartilaginous joint, which becomes ossified in adult life (a fact sometimes used in detective novels in determining the age of a skull unexpectedly turning up during a game of bridge). The lesser wing of the sphenoid bone is attached to the body by two roots that surround a large foramen for the passage of the optic nerve and ophthalmic vessels. The wing ends medially in a fairly sharp spine, the anterior clinoid process, to which the tentorial process of the dura mater is attached.

The greater wing of the sphenoid has a horizontal part, which is seen on the base of the skull. The horizontal part turns upward as a vertical part that articulates on the side of the skull with the frontal and temporal bones. The horizontal part has three foramina through which pass important structures. The most medial foramen is the foramen rotundum, which conducts the large maxillary nerve out of the skull. Posterolateral to

this is the foramen ovale, which transmits the mandibular nerve. Behind this is the small foramen spinosum for the passage of the important middle meningeal vessels. Between the greater and lesser wings of the sphenoid is a large gap, the superior orbital fissure, which admits the cranial nerves supplying the eye and its muscles (III, IV, ophthalmic division of V, VI), together with various veins and arteries. The pterygoid processes project inferiorly and serve as attachments for the muscles of the palate and upper pharynx. The medial pterygoid plate ends in the hamulus, around which hooks the tensor palati muscle.

Ethmoid bone

The ethmoid bone takes part in both the cranial and facial skeletons. This bone has a perpendicular plate in the midline. On each side

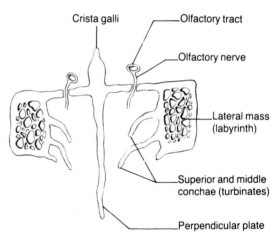

Fig. 10.12   **The ethmoid bone.**

of this plate lies a horizontal plate of bone with two inferiorly projecting lateral masses. Each part can be seen in different areas of the skull. The perpendicular plate projects sharply up into the inner aspect of the base of the skull immediately behind the glabella, forming the crista galli. To the crista galli is attached the falx cerebri, a part of the dura mater. The lower part of the perpendicular plate forms a large part of the septum of the

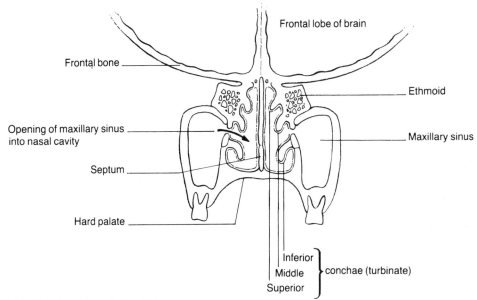

Fig. 10.13 Relationships of ethmoid bone.

nose, which is completed by other bones. The horizontal or cribriform plate, seen best from the inside of the skull, has many little holes for the passage of the olfactory nerve rootlets. This plate also closes the gap between the frontal bones on each side and the body of the sphenoid behind.

The two lateral masses are honeycombed by many air cells, which are called anterior, middle and posterior sinuses, according to their location. They are best seen on the medial wall of the orbit, where the lateral mass articulates with the lacrimal bone anteriorly and the lesser wing of the sphenoid posteriorly. Infection of these air sinuses is common and is known as ethmoiditis. The site of the pain and tenderness will be in the medial wall of the orbit

Facial skeleton

Zygomatic bone

If you run your finger along the zygomatic arch, you will reach an expanded area lying beneath the cheek – the zygomatic bone. This star-shaped bone has several processes, two of which have been mentioned: the posterior temporal process articulates with the zygomatic process of the temporal bone, and the ascending frontal process articulates with the zygomatic process of the frontal bone. The zygomatic bone forms the prominence of the cheek and also part of the inferior boundary of the orbit.

Nasal, lacrimal and maxillary bones

The nose has fixed and mobile portions, the former consisting of two nasal bones and the latter consisting of nasal cartilages, which permit changes in the caliber of the nasal passages. The two nasal bones form most of the bridge and bony portion of the nose, meeting laterally with the frontal processes of the maxillae to form the sides of the nose, or alae.

Pass your finger beyond the frontal process of the maxilla into the orbit and note a hollow fossa, where the maxillary process articulates with the lacrimal bone. This fossa contains the lacrimal sac, which absorbs the tears produced by the lacrimal gland after they have flowed across the eye.

Now note the lower margin of the orbit. It is formed by the zygomatic bone and the maxilla. Each maxilla is a large bone that bears the teeth of the upper jaw. It articulates with the nasal bone medially, the lacrimal bone posteriorly, the zygomatic bone laterally, and with its partner in the midline. To the free margins of the maxillae and nasal bones are attached the nasal cartilages. Most of the maxilla can be palpated above the upper teeth; multiple ridges are often felt, which are caused by the roots of these teeth. The maxilla contains a large air cavity, the maxillary air sinus. In the upper part of the bone, immediately below the margin of the orbit, lies the large downward-directed infraorbital foramen, where the terminal part of the maxillary nerve exits onto the face.

Palatine bone

This relatively inaccessible bone is L-shaped, with a horizontal portion forming the posterior part of the hard palate and a vertical portion wedged between the pterygoid process of the sphenoid and the maxilla.

Inferior nasal concha (turbinate)

Each bone is an elongated, scroll-like lamina projecting from the lateral wall of the nose and articulating with the bones comprising that wall.

Vomer

This bone is a major component of the bony nasal septum. Shaped like a plowshare, it articulates with the bones forming the roof and floor of the nose.

Mandible

The mandible, which bears all the teeth of the lower jaw, is the only movable part of the skeleton of the skull. The two original halves have fused to produce a bone that consists of a central body and an ascending ramus on each side. The body and ramus meet at an important landmark, the angle of the jaw. The upper end of the ramus ends anteriorly in a coronoid process and posteriorly in a condyle. Both of these are palpable, the former with a finger in the mouth and the latter (after wiping the finger) on the face anterior

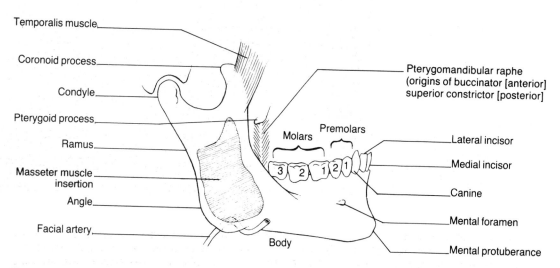

Fig. 10.14 Lateral aspect of mandible.

to the ear. The condyle is felt more easily by opening and closing the mouth. The coronoid process provides attachment for the temporalis muscle, and the condyle, which has its long axis running transversely, articulates with the mandibular fossa of the temporal bone. Between these two processes is a notch over which passes a muscular (masseteric) branch of the mandibular nerve.

The body of the mandible has a prominent mental protuberance in the midline. About halfway between the upper and lower border, opposite a point between the two premolar teeth is a large upward-directed mental foramen for the terminal branches of the mandibular nerve. It is useful at this point to mention that the supraorbital and mental foramina lie in the same vertical line, which can be located either in relationship to the two premolar teeth just mentioned or in relationship to the supraorbital foramen at the junction of the inner and middle thirds of the supraorbital margin (one thumb's breadth from the midline). These foramina are great places in which to deposit a local anesthetic solution, thus rendering large areas of skin painless for surgery.

On the inner surface of the mandible is a large mandibular foramen, overlaid by a plate

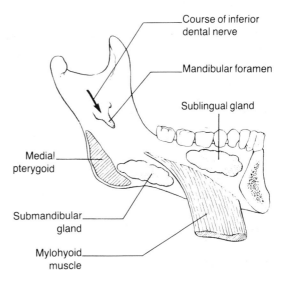

Fig. 10.15 Medial aspect of mandible.

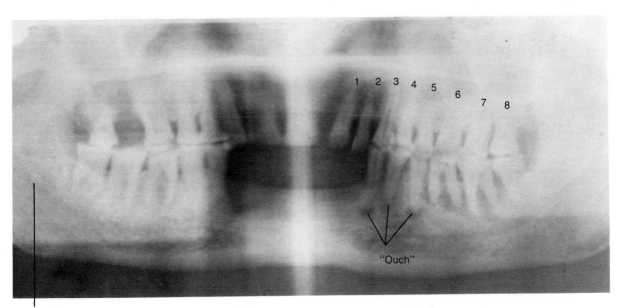

Mandibular canal

Fig. 10.16 Radiograph of mandible (Panorex). The mandible could well belong to one of the Golden Gloves finalists – which teeth are missing? The only complete half jaw is the left maxilla, which has all eight teeth. The roots of the left lower two premolars and canine are surrounded by small dark spaces, areas of resorbed bone: these are caused by root abscesses.

of bone, the lingula. The dentist searches for this landmark, and, if he is a good friend, he will find it and easily anesthetize the large inferior dental nerve, which enters the foramen and supplies all the teeth of the lower jaw. Running downward from the region of the third molar (the most posterior tooth) to the lower border at the midline is the mylohyoid line for the attachment of the mylohyoid muscle, the diaphragm of the mouth that separates the mouth from the neck.

Base of the skull

Outer surface

The outer surface of the base of the skull will be examined only in relationship to its clinical relevance. Notice first the row of teeth, behind which is the hard palate formed by the palatal processes of the two maxillae and horizontal portions of the two palatine bones. The two maxillae form the anterior two-thirds of the hard palate, and the horizontal portions of the two palatine bones complete the posterior portion of the palate. The horizontal suture separating the fused maxillae and palatine bones is easily seen on each side. A midline suture separates all four bones. In a young skull, one may also see the remains of a suture line demarcating the premaxilla, which has an interesting developmental history. The premaxilla, triangular in shape and containing the four incisor teeth, develops separately from the rest of the maxilla, with which it ultimately fuses.

The two halves of the palate fuse from front to back. The condition of the cleft palate involves various degrees of failure of fusion of the palatine and/or maxillary bones. When the maxillary bones do not fuse, the cleft runs up to the premaxilla, and, when the premaxilla also fails to fuse, the cleft passes along one or both sides of the premaxilla. The

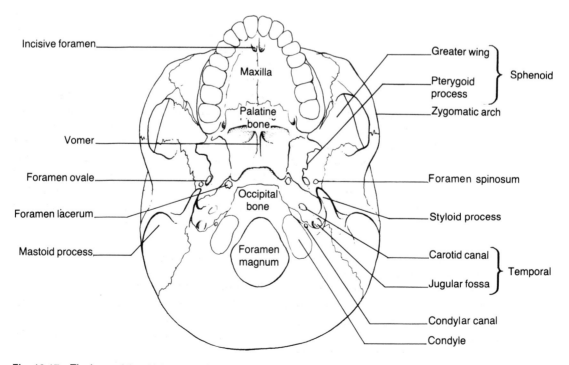

Fig. 10.17   The base of the skull (external view).

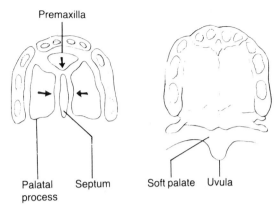

Premaxilla

Palatal process

Septum

Soft palate   Uvula

Fig. 10.18   **The development of the palate.**

latter condition may be associated with failure of the elements of the lip to fuse, or harelip.

There is a lot of action taking place at the junction of the palatal processes of the maxillary and palatine bones. Patients may consult you about a large lump at this junction, which lies toward the posterior part of the hard palate. You may reassure them that this is not some dastardly tumor, but a common abnormality called the torus palatinus. Since it is present in about 22 percent of the population, it is worth knowing about.

Passing posteriorly beyond the palate, note the undersurface of the greater wing of the sphenoid with the important foramina transmitting the large mandibular branch of the fifth cranial nerve and the middle meningeal artery. Continuing posteriorly, one sees the large styloid process of the temporal bone pointing downward from its socket in the tympanic plate, which forms the bony part of the external auditory meatus.

In front of the external auditory meatus is a smooth concavity, the mandibular fossa, which receives the condyle of the mandible at the temporomandibular joint. Passing further on medially, one notes large foramina, visible radioiogically, which indicate the position of important structures. At the most medial part of the undersurface of the petrous temporal is a large irregular gap, the foramen lacerum, which is closed by cartilage in life.

On the intracranial aspect of the foramen lacerum lies the internal carotid artery. Posterolateral to the foramen lacerum lies the large carotid foramen, admitting the internal carotid artery and its accompanying sympathetic plexus. From these two statements one can deduce that the internal carotid artery, after traversing the carotid foramen, runs medially to enter the cranial cavity above the upper surface of the foramen lacerum. An arteriogram shows this as the first of several curves made by this sinuous artery. It is said that these curves exist so as to reduce the impact of a pulsatile inflow into a closed rigid cavity.

Behind the carotid foramen is the large jugular fossa, which accommodates the jugular bulb, a dilatation of the internal jugular vein. Passing still laterally and looking at the undersurface of the projecting mastoid process, one sees a sharp notch, the digastric fossa, for the attachment of the digastric muscle. Medial to the digastric fossa is a small groove for the occipital artery. The latter artery was used to fashion intracranial and extracranial anastomoses in cases of cerebral ischemia. Between the styloid process and the mastoid process is the very important stylomastoid foramen transmitting the seventh cranial or facial nerve.

Continuing posteriorly, one reaches the smooth condyles of the occipital bone, which articulate with the atlas. Immediately in front of each condyle is the anterior condylar canal, which lies between the lateral and basilar elements of the occipital bone and through which the twelfth (hypoglossal) nerve emerges. Anterior and posterior to the condyles are the basilar and squamous parts of the occipital bone respectively.

Interior of base of skull

The interior of the base of the skull is more important clinically than its exterior surface. It is divided into the anterior, middle and posterior cranial fossae. In the anterior cranial fossa is the upward-projecting crista galli, with the cribriform plate of the ethmoid

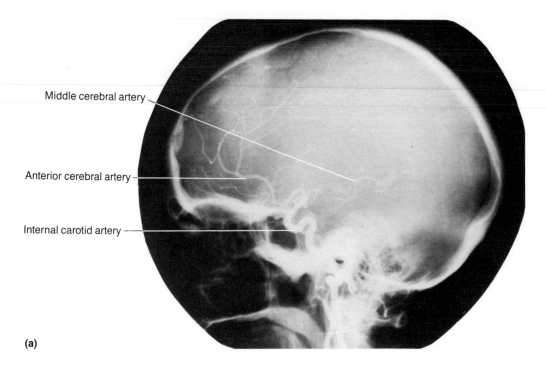

Middle cerebral artery

Anterior cerebral artery

Internal carotid artery

(a)

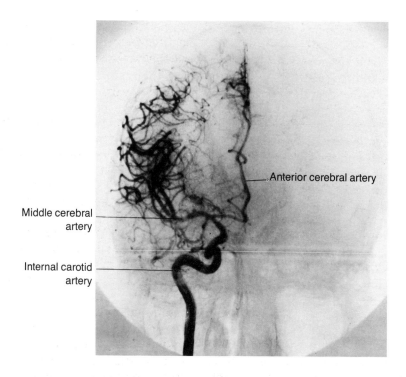

Anterior cerebral artery

Middle cerebral
artery

Internal carotid
artery

(b)

Fig. 10.19   Internal carotid artery angiogram: (a) lateral view; (b) anteroposterior view with subtraction.

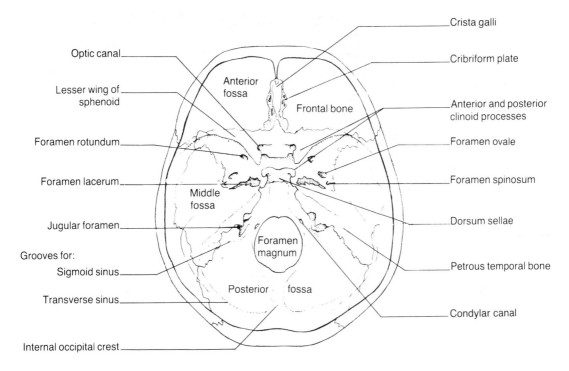

Optic canal

Lesser wing of sphenoid

Foramen rotundum

Foramen lacerum

Jugular foramen

Grooves for:

Sigmoid sinus

Transverse sinus

Internal occipital crest

Crista galli

Cribriform plate

Anterior and posterior clinoid processes

Foramen ovale

Foramen spinosum

Dorsum sellae

Petrous temporal bone

Condylar canal

Anterior fossa

Frontal bone

Middle fossa

Foramen magnum

Posterior   fossa

Fig. 10.20   Superior view at base of skull.

bone on either side. Lateral to the ethmoid bone are the horizontal portions of the frontal bone, the orbital plates. Passing posteriorly, one notes the prominent posterior limit of the anterior cranial fossa, the sharp edge of the lesser wing of the sphenoid.

Entering the middle cranial fossa, the pituitary fossa is seen in the midline, surmounted in front and behind by the anterior and posterior clinoid processes. On each side the greater wing of the sphenoid joins the squama of the temporal bone to form the floor of the middle cranial fossa.

Passing over the ridge of the superior border of the petrous temporal bone, one enters the posterior cranial fossa, the largest fossa. It also has the largest foramen, the foramen magnum.

Clinical aspects

Fractures of the skull are common injuries. While fractures of the vault are seen relatively easily radiologically, fractures of the base, with its numerous bony irregularities and foramina, are difficult to visualize and the diagnosis is often made clinically.

Anterior fossa

Since the nose is roofed over by the cribriform plate of the ethmoid and the orbit by the orbital plate of the frontal bone, a fracture will cause bleeding into the nose and/or orbit. When the fracture of the roof of the nose also involves the meninges, cerebrospinal fluid leaks into the nose, causing cerebrospinal rhinorrhoea. This unfortunately allows for easy infection from the nose into the meninges, with subsequent meningitis. The olfactory nerves are often damaged as they run through the cribriform plate, leading to loss of sense of smell (anosmia), which may be an advantage in some occupations (e.g. a colonoscopist). In fracture of the orbital plate a characteristic type of subconjunctival hemorrhage occurs; as the bleeding fills the orbit and passes forward beneath the conjunctiva, its posterior limit cannot be seen, unlike a blow to the eye, where the posterior edge of the subconjunctival hemorrhage can be delineated. The bleeding may produce bulging of the eyeball itself or even paralysis of the eye muscles.

Middle fossa

Fractures in the middle fossa are even commoner because the bone is weakened by the numerous foramina and canals. A fracture of the roof of the middle ear, the tegmen tympani, part of the petrous temporal bone, results in bleeding into the middle ear, visible as a bluish bulging of the tympanic membrane (eardrum). Excessive bleeding ruptures the drum, and blood is discharged from the ear. The fracture may involve meninges, which causes the leakage of cerebrospinal fluid from the ear (otorrhea). Because the seventh and eighth cranial nerves run in the petrous bone, they may be injured, especially the seventh nerve.

Posterior fossa

Fractures of the posterior fossa may involve the basilar part of the occipital bone anteriorly where it forms the roof of the pharynx. Bleeding occurs into the pharynx and usually regurgitates through the mouth. (Remember the scene in the movie when the heroine's pappy has received a bad head injury; blood trickles out the corner of the mouth, and death occurs before he can tell her where the gold is hidden.) Fractures of the squama of the occipital bone may lead to bleeding deep in the back of the neck, which usually surfaces under the skin a few days later.

Development of the skull

The vault, sides of the skull and the facial skeleton are developed in membrane, whereas the bones of the base of the skull first pass through a cartilaginous stage. It is almost as if the vault of the skull was required to grow so quickly so as to accommodate the rapidly enlarging brain that the bones did not have time to pass through the cartilage stage. The newborn child is all head and very little face in the proportion of eight to one. The components of the cranium are joined by membrane from the chin to the occiput. The two halves of the mandible fuse early in the second year, the two frontal bones later except in those individuals with metopic sutures, and the sagittal suture between the parietal bones usually obliterates in adult life. The two maxillae and nasal bones, on the other hand, seldom fuse.

The process of molding, made possible by the membranous articulations of the parietal bone, has been discussed. Ossification of both the parietal and the frontal bones begins at the prominent tuberosities of the youngster's skull and spreads throughout the rest of the bones. As the parietal bone ossifies, for a time there are fairly large membranous areas, called fontanelles, where the bone articulates at each of its angles. Of the four fontanelles – anterior, posterior and two laterals – the anterior is the largest and the most important.

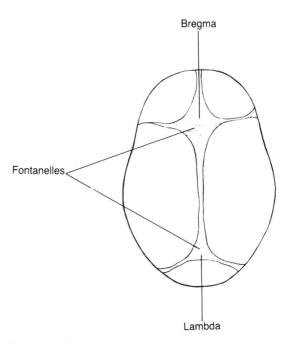

Fig. 10.21 The fetal skull.

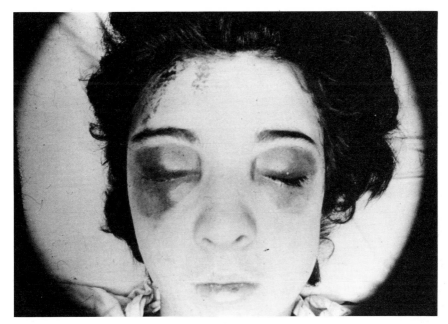

Fig. 11.3   An injury to the right forehead, producing two 'black eyes'.

Superficial nerves and vessels of the head

Arteries

The blood supply of the head comes from both the external and the internal carotid arteries. The internal carotid artery is usually thought of as supplying only internal structures, which was the reason for its name. However, the terminal distribution of its ophthalmic branch reaches the scalp as the supratrochlear and supraorbital arteries. The branches of the external carotid, which supply the rest of the head, are named according to their distribution: the facial artery supplies the face; the superficial temporal artery, the temple; and the posterior auricular and occipital arteries, the area behind the ear.

The blood supply to the scalp by the branches of the internal carotid artery is important. At the inner angle of the eye, an anastomosis occurs between the facial branch of the external carotid and the cutaneous branches of the internal carotid artery. Narrowing of the internal carotid artery, a common occurrence as one ages, often results

in the intracranial structures being supplied by the facial artery by way of the ophthalmic artery. This can be demonstrated by listening over the inner angle of the eye with an amplified stethoscope (Doppler). An increased pulse amplitude unilaterally indicates blocking of the internal carotid artery on that side. This is a much safer way of diagnosing a narrowing of the internal carotid artery on that side than injecting radiopaque material directly into the internal carotid artery.

The superficial temporal artery is the vessel palpable just in front of the tragus of the ear. It is a convenient place for feeling the pulse, especially when surgeons have hidden the rest of the body from view. This vessel is affected by some diseases, and biopsies may be necessary. It is important to remember that the auriculotemporal nerve runs closely with the artery and should not be included in the biopsy.

The occipital artery supplying the occipital area of the scalp is a large vessel, and, at one time, techniques were developed to anastomose this artery to intracranial arteries in cases of intracranial ischemia.

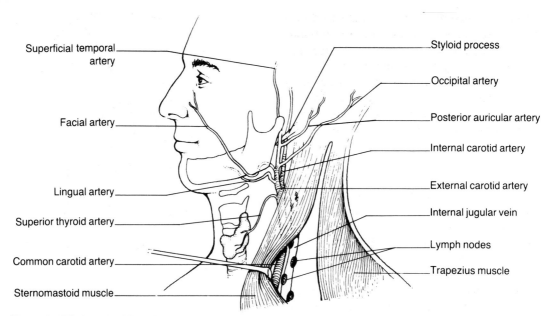

Fig. 11.4 The major arteries of the neck. Note that the three large anterior branches of the external carotid artery all have loops, to allow them to adjust to the movements accompanying swallowing. The retractor is pulling the sternomastoid forward to reveal the common carotid and internal jugular vein, which lie deep to the muscle.

The facial artery crosses the body of the mandible just in front of the masseter muscle. It then passes up in a corkscrew manner to the inner angle of the eye, where its anastomosis has been described.

Veins

The veins of the head accompany the arteries just mentioned, but two important points require emphasis. The supraorbital and supratrochlear veins are connected to both the anterior facial vein and the intracranial cavernous sinus. The facial vein runs inferiorly more superficially and with a straighter course than the facial artery. As it passes over the buccinator muscle, the vein connects via a deep facial vein with the pterygoid plexus of veins, which itself connects with the cavernous sinus. This means that two venous connections exist between the face and the cavernous sinus, either of which may allow an infected facial lesion to spread to the cavernous sinus. It is for this reason that the area around the side of the

nose and the upper lip is dubbed the dangerous area of the face. Infections in this area should receive vigorous treatment to prevent this serious complication.

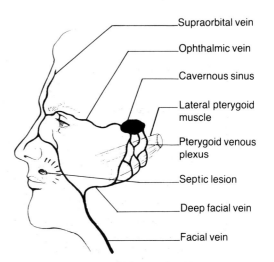

Fig. 11.5 The tributaries of the anterior facial vein. These connections explain how septic lesions in the area of face drained by the anterior facial vein may lead to intracranial infection.

Another constant vein is the mastoid emissary vein passing from the intracranial sigmoid sinus to the posterior auricular vein (another route by which extracranial infections may spread intracranially). Infections of the head and neck are usually treated successfully, but the intra- and extracranial anastomoses may render them dangerous.

Nerves

The skin of the face is largely supplied by the three divisions of the fifth cranial nerve. The ophthalmic division supplies the side of the nose, the upper eyelid, and the scalp as far back as the vertex; the maxillary division supplies the area of the cheek; and the mandibular division supplies the area of the lower jaw. These three areas, well demarcated from each other, meet at the outer angles of the eye and mouth.

The skull has grown so big that the rest of the skin of the head and a small portion of the face is supplied by cervical nerves. The posterior scalp is supplied by the greater and lesser occipital nerves, branches of the posterior and anterior rami of C2 respectively. A small area of skin over the angle of the mandible is innervated by the greater auricular, derived from the anterior primary ramus of C2.

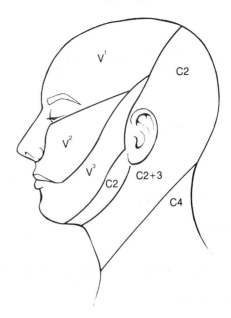

Fig. 11.6 Segmental cutaneous distribution to head and neck.

Face

The face possesses a subcutaneous sheet of striated muscle, broken up into components that dilate or constrict the orifices they surround (the eyes, nostrils and mouth). At the same time, the muscles influence facial expression, although this is probably not

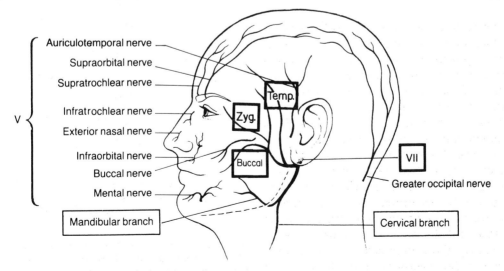

Fig. 11.7 Branches of fifth and [seventh] nerves to head and neck.

their primary action. Some of these muscles arise from bone, and most insert into skin; this correctly implies that the face does not possess deep fascia. For this reason, injury or infection is accompanied by marked swelling. All the muscles are supplied by terminal branches of the seventh cranial nerve (the patient with a facial nerve paralysis, who cannot close the eyes or mouth, is well known to clinicians).

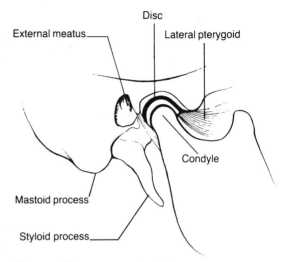

Fig. 11.8 The temporomandibular joint and surrounding bony relations. Note the insertion of the lateral pterygoid muscle into the joint capsule.

Temporomandibular joint

The temporomandibular joint is located between the transversely elongated articular surface of the condyle of the mandible and the anterior part of the mandibular fossa. These two parts fit so badly that they require the aid of an articular disc to render the surfaces congruous. The articular disc is large and attached to the capsule of the joint at its circumference. The mandibular fossa leads forward to a prominence, the articular eminence, and backward to the postglenoid tubercle, behind which is the tympanic plate, itself separated from the joint by the glenoid lobule of the parotid gland.

The socket of the joint is composed of rather thin bone, and a blow on the mandible from one of the heavyweight boxing contenders can fracture the middle fossa of the skull. Reinforcing the joint capsule on each side are collateral ligaments: they run to the mandible from the lower border of the zygoma laterally and from the sphenoid process medially.

One lonely night, you may be sitting at a bar and, from the noise, you may think that the man next to you is throwing dice onto the bar counter. However, he is actually chewing, with the condyle of his mandible rattling around. This should remind you that the disc is sometimes loose and clicking of the jaw is commonplace. This condition is acceptable as long as it is painless. When it is painful, removal of the disc may then become necessary. If you have had mumps (inflammation of the parotid gland), you will painfully remember the relationship of the glenoid lobule of the parotid gland as it passes between the tympanic plate and the condyle of the mandible. The glenoid lobule is squeezed every time you open your mouth, thus turning you into a strong silent type, and also a very hungry one.

Dislocation of the temporomandibular joint is common. With the jaw widely open, the condyle is precariously perched on the articular eminence. At this point, a sudden contraction of the opening muscles will send the condyle forward over the eminence into the infratemporal fossa. This usually happens when one yawns during an anatomy lecture or laughs uproariously. It does not help when someone bangs you on the back in a friendly gesture during either of these states because this finally sends the condyle over the hill. As with all dislocations, the neighboring muscles go into spasm to prevent painful movement. The condyle is pulled upward by the masseter and medial pterygoid and held there like a vise. The patient comes into the emergency room with the jaw half open if unilaterally dislocated, or open and stuck in an odd position if bilaterally dislocated. Reduction is effected by pulling the mandible downward with the

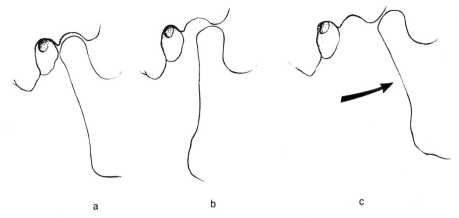

Fig. 11.9 The temporomandibular joint: (a) closed, (b) open and (c) dislocated.

thumbs to overcome the spasm of the masseter and clear the articular eminence. Now a backward push will click it into position. It is strongly recommended that the two thumbs be heavily padded with gauze. If they are not, the snap of the reducing mandible will result in yet another member in the club of eight-fingered doctors. The temporomandibular joint is also a notorious place for recurrent dislocations, as the capsule and disc attachment become looser with each succeeding dislocation.

Movements of the joint

On opening the mouth, the condyle of the mandible is tilted and passes forward onto the articular eminence, which prevents further movement. Closing the jaw is the opposite movement, with the condyle passing backward and being limited by the postglenoid process and tympanic plate. In protrusion the mandible moves forward, and in retraction the reverse movement occurs. In chewing movements, the protrusion of the

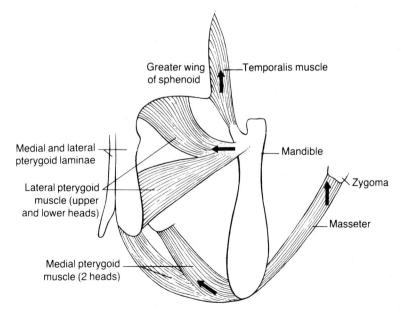

Fig. 11.10 The muscles of mastication and the directions of their pull.

one condyle turns the chin toward the opposite side.

Muscles that move the joint

The temporalis muscle, running from the temporal fossa beneath the zygomatic arch to the coronoid process, and the masseter muscle, running from the zygomatic arch to the angle of the mandible, elevate the mandible and close the mouth. Depression of the mandible, or opening the mouth, is performed by all muscles running between the mandible and the hyoid bone, the digastric, mylohyoid and geniohyoid. The two pterygoid muscles are the bane of students' lives, but they can be simplified. The lateral pterygoid muscle arises from the outer surface of the pterygoid plate and surrounding bone, and runs straight backward to be inserted into the capsule of the joint and articular disc. The medial pterygoid muscle arises mostly from the medial surface of the pterygoid plate and runs downward and outward to the angle of the mandible. The lateral pterygoid muscle pulls the mandible forward, thus protruding it. The medial pterygoid muscle is very large in the cow because the medial pull makes it the chewing muscle. These muscles of **m**astication are supplied by the **m**otor root of the fifth cranial nerve via its **m**andibular division.

One of the main retractors of the protruded mandible is the temporalis muscle, the posterior fibers of which run horizontally forward from the temporalis fossa. It is for this reason that this fossa extends so far posteriorly. The temporalis is the muscle of the canine teeth, and is particularly strong in carnivores such as the tiger.

For completeness, one should include the buccinator, the fibers of which arise from the maxilla and mandible close to the three molar teeth and run forward to decussate at the angle of the mouth before passing into the upper and lower lips. This muscle is responsible for passing the food from the space between the teeth and the cheek (vestibule) into the mouth for proper mastication. The buccinator, however, is supplied by the seventh cranial nerve. In patients with seventh nerve paralysis, food tends to lodge in the vestibule of the mouth, and, in combination with the malfunction of other facial muscles, drooling is a common complaint.

Salivary glands

Salivary tissue is mostly represented by three paired glands: the parotid (the largest), the submandibular and the sublingual. Areas of salivary tissue are also scattered throughout the palate, lips, cheeks and tongue. It may come as a surprise to learn that the commonest tumor of the palate is a salivary gland tumor.

Parotid gland

The parotid gland is wedge shaped, with the base facing outward and the apex thrust deeply inward between the mandible and the mastoid process. Since both bones have two muscles attached to them, one superficial and one deep, the gland lies in contact with these muscles: the mandible has the masseter externally and the medial pterygoid internally, the mastoid process has the sternomastoid laterally and the posterior belly of digastric medially. The apex of the gland almost

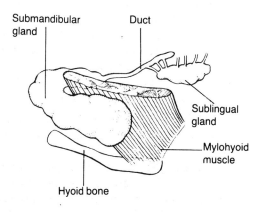

Fig. 11.11 The submandibular and sublingual salivary glands.

reaches the styloid process and its muscular attachments.

The base of the wedge of the gland extends forward onto the face. From its anterior border emerges the parotid duct, which is palpable as it crosses the contracted masseter (clench the teeth). The duct pierces the buccinator muscle opposite the third molar tooth to enter the mouth opposite the second upper molar tooth. This oblique passage is necessary to prevent reflux into the duct from the mouth; the buccinator closes the duct when the intraoral pressure is raised. The protective mechanism is very effective, except in trumpeters such as Louis Armstrong and the glass blowers of Venice, who all possessed big puffy parotids and in whom the excessive intraoral pressure overcame the sphincter mechanism. Reference has been made to the upper extremity of the gland, the glenoid lobule (p. 202).

Some important structures lie in the parotid gland.

1 The facial nerve splits into its two terminal divisions from which branches emerge at the gland's anterior border to supply the facial muscles. A swelling of the gland with paralysis of the nerve is an ominous sign.

2 The retromandibular vein leaves the gland at its lower border, splits into an anterior division, which helps to form the common facial vein, and a posterior division, which forms part of the external jugular vein (Fig. 12.6).

3 The external carotid artery enters the gland and bifurcates into two terminal divisions, the internal maxillary and superficial temporal vessels, both of which emerge from the gland.

The facial nerve is the most superficial structure and the external carotid artery the deepest.

Finally, the gland contains important lymphatic nodes. A parotid mass may, in fact, represent a primary tumor in the pharynx that has spread to the parotid lymph node to simulate a primary parotid tumor.

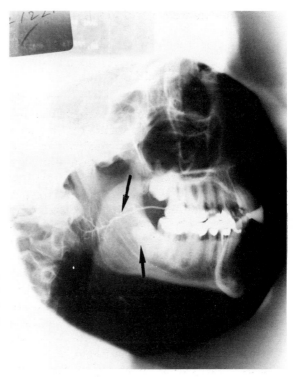

Fig. 11.12 Parotid gland and duct sialogram (upper arrow). The contrast is injected through the buccal opening of the duct, opposite the crown of the second upper molar tooth. Note the menacing unerupted upper and lower third molars (lower arrow) – a hint of trouble to come, often manifested by earache!

Lying in close relationship to the gland, but not in it, is the auriculotemporal nerve, which passes cranially to join the superficial temporal artery on the side of the face. The parotid gland secretes mostly thin serous material, which makes it an uncommon site for the formation of calculi (stones).

Sublingual gland

The sublingual gland, situated under the anterior tongue, has between 8 and 20 ducts that open into the mouth (see Fig. 11.11). Its secretion is mostly mucous, but problems with this gland are rare. It seems that the large number of ducts prevents difficulties; should one duct be blocked, the secretion merely passes up the others.

Submandibular gland

The submandibular gland is situated in the digastric triangle, the triangle formed between the two bellies of digastric muscle and the lower border of the mandible (see Fig. 11.14). It can be felt tucked beneath the mandible, but it is best palpated bimanually between an internal finger in the mouth and an external one pushing the gland internally.

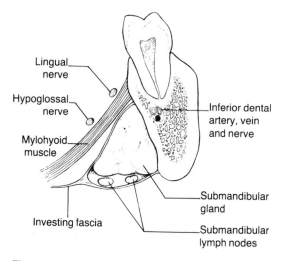

Fig. 11.13 **The submandibular gland. Note the lymph nodes enclosed in the fascial sheath surrounding the gland.**

A good way to study the relations of the submandibular gland is to practice its operative removal. The incision is made well below the lower border of the mandible to avoid severing the mandibular branch of the facial nerve as it turns up over the mandible. The incision goes through skin, platysma and an investing layer of fascia. The digastric bellies are identified, with the submandibular gland lying over them. The facial vessels are seen crossing the lower border of the mandible just in front of the masseter, being crossed themselves by the mandibular branch of the facial nerve. The submandibular gland is grasped and freed from the muscles. It lies on both the mylohyoid and the hyoglossus muscles, which run from the hyoid bone to the mandible and the tongue

respectively. The mylohyoid has a free posterior border, around which hooks part of the gland, and the quadrilateral hyoglossus lies on a slightly deeper plane.

On elevating the gland, we notice that we cannot lift it very far without pulling on the facial artery. This artery comes off the external carotid artery, passes up under the posterior belly of the digastric muscle, and curves down again to bury itself in the posterior portion of the gland before turning up between the gland and the lower border of the mandible, which it crosses. We divide the facial artery and are pleased to note the gland becomes instantly more mobile. In a moment of glory, we start waving the gland around, but notice that it is still stuck to two structures lying deep to it on the hyoglossus muscle. From above downward, we notice three structures running across the hyoglossus: (1) the very large lingual nerve, with parasympathetic secretory branches for the gland; (2) the submandibular duct sneaking behind the posterior border of the mylohyoid to get entangled with the lingual nerve; and (3) the hypoglossal nerve with its accompanying vein. To remove the gland we need to divide the branches from the lingual nerve to the gland, and the submandibular duct. Dividing these structures and the facial artery will reward us with the gland in the palm of

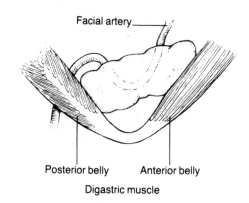

Fig. 11.14 **The intimate relationship of the submandibular gland and facial artery is well seen.**

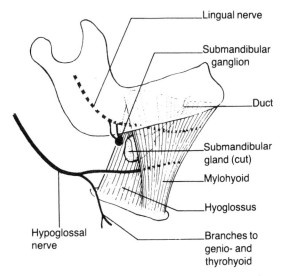

Fig. 11.15 The mylohyoid and hyoglossus muscles. The deep (cut) part of the gland is seen passing behind the free posterior border of the mylohyoid, and giving rise to the duct. The three structures lying on the hyoglossus, from above down, are the lingual nerve with its submandibular ganglion, the deep part of the gland and its duct, and the hypoglossal nerve.

our hand. It should be noted that in the palm lie not only the submandibular gland but also lymph nodes adherent to the gland. In fact, this is the most common reason for removing the submandibular gland; the diseased lymph nodes cannot be removed without removal of the salivary gland because of their close relationship. The second commonest reason for removing the gland is a stone which blocks the submandibular duct (Wharton). The orifice of the duct can easily be seen lateral to the frenulum of the tongue.

Teeth

Each tooth consists of a crown projecting above the gum and shaped according to its function, a neck that is grasped by mucous membrane, and a root (or roots) embedded in the maxilla or mandible (see Fig. 10.14). The tooth is composed of dentin, which is a calcified material; the crown is capped with enamel, as hard as any diamond; and the root is surrounded by cement, which is modified bone. All three parts contain a pulp cavity, which has a rich neurovascular supply. Between the cement and the mandible is the periodontal membrane, which is really modified periosteum and which shows up as a very important black line on x-ray films. Incisors are like sharp chisels, with a cutting edge; the canines have long roots with a single cusp and are the teeth which hold the

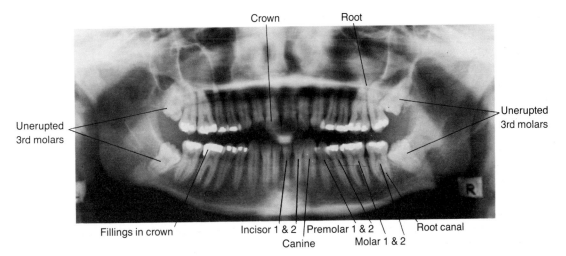

Fig. 11.16 An orthopantogram showing a full set of teeth (a full house). This triumph is sobered by four knaves – four unerupted third molars, a promise of trouble to come. The root canals are well seen. The price of these dazzling teeth is multiple fillings.

flesh (well developed in carnivores); and the molars are for grinding, the premolars having two or three cusps and the molars three to five cusps. There are two sets of teeth: primary (deciduous, temporary, milk) and permanent teeth.

Development of the teeth

Both the temporary and the permanent teeth develop from a shelf of epithelium, which arises on each gum and from which ingrowths pass deeply to form the primitive tooth buds. The permanent teeth arise as buds on the inner, or lingual, aspects of the temporary buds. Displaced portions of this epithelium give rise to cysts and tumors associated with teeth.

Primary teeth

It is common to use a formula for the disposition of teeth, as each half of each jaw has the same number of teeth. There are five primary teeth in each half of the jaw, and the mandibular teeth appear before the maxillary. The first teeth to appear are the lower central incisors, erupting at about 6 months of age – a time when the child, who has been guzzling milk cocktails for the previous five months, moves on to something more substantial. Between 6 and 24 months, all 20 teeth have erupted, and each half jaw has two incisors, a canine and two molars.

Permanent teeth

The first tooth to erupt lies posterior to the five temporary teeth, and it appears before any of these teeth have been shed. It is the 'sixth year' first molar, and its eruption can cause earache in a 6-year-old child (more on the subject later). The second permanent molar does not erupt until 12 years of age; it used to be called the 'factory molar' because, in nineteenth century Britain, the age for employment was gauged by its presence. Between 6 and 12 years of age the five de-

ciduous teeth are replaced by permanent teeth. From 18 years of age onward, the last of the three molars, the wisdom tooth, appears. Sometimes the wisdom tooth fails to erupt and remains impacted in the mandible, where it may cause trouble (see Figs. 11.12 and 11.16).

The formula for the permanent teeth is two incisors (medial and lateral), a canine, two premolars (bicuspids) and three molars. The first five teeth replace the five deciduous teeth, and the next three develop posterior to these teeth. This gives the formula 2, 1, 2, 3, with a total number of eight teeth to each half jaw. Dentists talk about teeth like exits on an expressway; for example, upper left seven is UL7 and lower right four is LR4. It is great gamesmanship to call up your dentist and complain of pain in RU6 (the receptionist will see that you don't get a parking spot).

Roots of the teeth

Embedded in the upper and lower jaws, the roots of the teeth have important relationships. The roots of the upper premolars and molars, especially the second bicuspid and first molar, project into the floor of the maxillary air sinus. Sometimes the relationship is such that they are separated from the sinus by its mucous membrane only. At times it is difficult to know whether the patient has sinusitis or abscesses in the roots of the teeth. The source of pain must be clearly differentiated, since pulling teeth will not cure sinusitis.

In the lower jaw, the roots of the second and third molars lie below the mylohyoid line, the attachment of the muscular diaphragm separating the mouth from the neck. Root abscesses affecting these teeth present in the neck; those of the other teeth present in the mouth.

Nerve supply to the teeth

The maxillary teeth are supplied by the terminal branches of the maxillary nerve, and the mandibular teeth are supplied by the terminal branches of the mandibular nerve. The mandibular nerve has a wider distribution than the maxillary nerve, supplying, among other areas, the ear. It is common for a patient whose real trouble lies in the teeth to present with earache. For example, little Johnny, 6 years old, complains of severe earache. Don't blame the otoscope when you cannot find an explanation in the ear. Just remember that the sixth-year molar (the first permanent tooth) erupts at that age. Ear drops may be fun to instill but will do nothing for the child. Because of the wide distribution of the maxillary and mandibular nerves, many cases of unexplained facial pain originate in an unsuspected bad tooth.

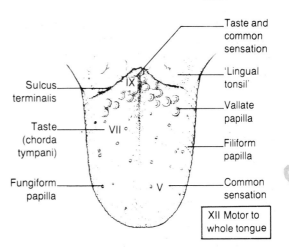

Fig. 11.17 Superior view of the tongue.

Tongue

The tongue is a highly mobile and articulate organ that develops its scaffolding from the first and third arches and its muscle from occipital myotomes, the second arch being buried in the substance of the tongue. The differences between the anterior two-thirds and the posterior one-third reflect this development. The anterior two-thirds is covered by thick mucous membrane with projecting filiform and fungiform papillae.

Filiform papillae give rise to the roughness of the tongue (well marked in the cat). Fungiform papillae are small pink spots lying near the tip and sides of the tongue that bear taste buds. The posterior third of the tongue is separated from the anterior two-thirds by a V-shaped furrow, the sulcus terminalis. Immediately anterior to the sulcus terminalis are the vallate papillae, a series of about a dozen papillae surrounded by a trough, arranged in the form of a 'V' with the apex pointing backward. Tastebuds abound in the walls of the vallate papillae. At the apex of

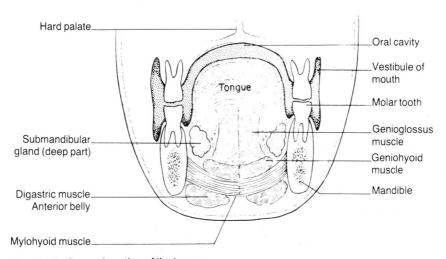

Fig. 11.18 Coronal section of the tongue.

the sulcus terminalis is a small pit called the foramen cecum, the origin of the thyroglossal duct, which passes down toward the neck to form the thyroid gland. The anterior two-thirds is flat, while the posterior one-third is lumpy from its contained aggregations of lymphoid tissue, the lingual tonsil.

Muscles of the tongue

The muscles of the tongue are intrinsic, which alter its shape, and extrinsic, which change its position. The intrinsic muscles are several groups of fibers running at angles to one another. The extrinsic muscles arise from surrounding bones: the genioglossus arises from the friendly genial tubercles of the inner surface of the body of the mandible; the hyoglossus, from the body and greater cornu of the hyoid bone; and the styloglossus, from the styloid process. The genioglossus is by far the largest muscle and makes up the bulk of the tongue. It runs anteroposteriorly and is the main protruder of the tongue. The hyoglossus, lying laterally, pulls the tongue downward and sideways, and the stylo-glossus, arising from the posteriorly placed styloid process, retracts the tongue. The whole organ itself is elevated by the contraction of the mylohyoid muscle. This muscle elevates the tongue, the edges and tip of which have been heaped up by the contraction of its intrinsic muscles, against the hard palate, thereby imprisoning a piece of food. Retraction of the tongue will now propel the bolus into the oropharynx to complete the voluntary stage of swallowing.

Neurovascular supply

The vascular supply to the tongue is from the large lingual artery. This vessel has a convenient loop that allows protrusion of the tongue at umpires who make bad decisions.

Lymph drainage

The tip drains to both submental nodes and to the main lymph node of the tongue. The side of the anterior two-thirds drains to the submandibular nodes, and the posterior third to the deep cervical glands. The posterior third and the area near the midline drain bilaterally; cancers in these areas therefore carry a worse prognosis.

The nerve supply of the tongue illustrates its development. The anterior two-thirds of the tongue is supplied with common sensation by the lingual branch of the mandibular

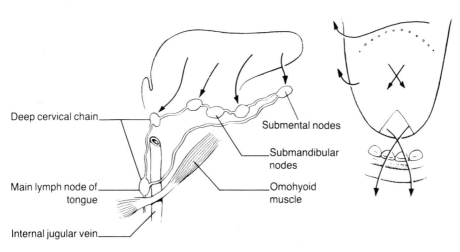

Deep cervical chain

Main lymph node of tongue

Internal jugular vein

Submental nodes

Submandibular nodes

Omohyoid muscle

Fig. 11.19 Lymph drainage of the tongue.

nerve, but the special sensation of taste is subserved by the chorda tympani of the seventh cranial nerve (the nerve of the second pharyngeal arch, which became buried in the tongue). The posterior third of the tongue, including the vallate papillae, is supplied by the glossopharyngeal nerve, which subserves both common sensation and taste. The motor supply is by the twelfth cranial nerve, which supplies every muscle that ends in 'glossus', except the palatoglossus, which is innervated by the tenth and eleventh cranial nerves via the pharyngeal plexus.

Can one gain anything by knowing the nerve supply of the tongue? You must have a feeling that the answer is yes. The fifth cranial nerve has a vast distribution in the head, and its lingual branch very often refers its sensation to the ear (like the tooth). A physician may see patients with a cotton pledget in a painful ear, but the examination of the ear might be annoyingly negative. The patient may have a lesion on the tongue, which is causing the earache. When patients develop a lesion of the facial nerve, they may lose taste from the anterior two-thirds of the tongue. In a patient who suffers a lesion of the hyoglossal nerve, that half of the tongue will be paralyzed; when the patient sticks her tongue out at the home plate umpire, the intact side will push it to the paralyzed side, toward first or third base. The paralyzed tongue points to the side of the lesion.

Pituitary gland (hypophysis cerebri)

Classically dubbed the 'leader of the endocrine orchestra', this gland, no bigger than $12 \times 8\,\mathrm{mm}$ ($0.5 \times 0.3\,\mathrm{in}$), controls the secretion of all the other endocrine organs. It has an interesting double development. The larger anterior lobe passes from the buccal cavity through the base of the skull to reach the intracranial cavity. Here it fuses with the posterior lobe, a downgrowth of the base of the brain (the floor of the third ventricle). Sandwiched between the two lobes is the pars intermedia, made up by a portion of each lobe at the site of fusion.

Sitting comfortably in the Turkish saddle, or sella turcica, of the sphenoid bone (see Figs. 10.2 and 10.5), the pituitary gland retains its connection with the base of the brain by a stalk, the infundibulum. This passes through a small aperture in the diaphragma sellae, the dura mater that roofs over the fossa.

Diseases of the pituitary gland come in two forms: the secretion of the gland becomes excessive or insufficient, or the gland enlarges and compresses the surrounding structures. It is therefore imperative to know these neighboring structures.

Laterally lie large venous lakes, the cavernous sinuses, which contain cranial nerves III, IV, two divisions of V, and VI, and the internal carotid artery. These structures are not

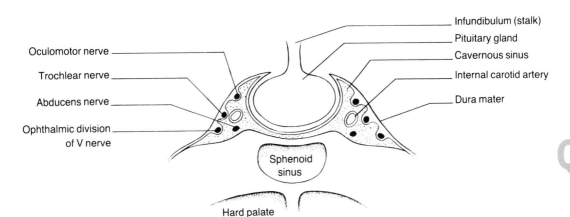

Oculomotor nerve

Trochlear nerve

Abducens nerve

Ophthalmic division of V nerve

Infundibulum (stalk)

Pituitary gland

Cavernous sinus

Internal carotid artery

Dura mater

Sphenoid sinus

Hard palate

Fig. 11.20   Coronal section of the pituitary gland and cavernous sinus.

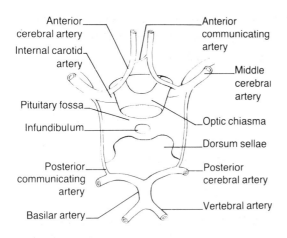

Fig. 11.21 Superior view of the pituitary fossa and the arterial circle of Willis.

often involved by a pituitary tumor.

Anteriorly lies the optic chiasma; when involved by a pituitary tumor, a visual field defect results, the defect depending on the particular nerve fibers affected. Superiorly a tumor may compress the base of the brain, usually near the site of origin of the posterior lobe (the floor of the third ventricle). Inferiorly the tumor excavates the sella turcica and encroaches on the sphenoid sinus, producing a characteristic radiographic appearance. Sometimes the clinoid processes are also eroded, and this can be recognized radiologically. The preferred approach for reaching the fossa to remove a tumor is through the nasal cavity and the sphenoid sinus. This sounds like a very unattractive septic approach, but the surgeon stays in a plane deep to the mucous membrane, thereby avoiding the nasal contents.

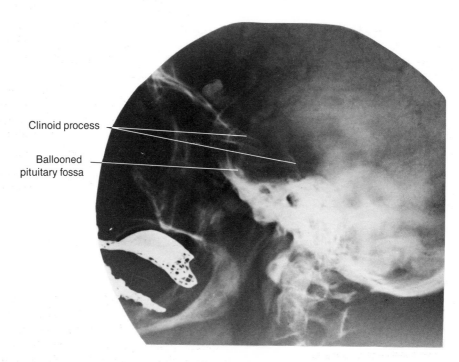

Fig. 11.22 Lateral radiograph of skull demonstrating a pituitary tumor. Note the enormous ballooning of the floor of the fossa by the tumor. This patient presented with loss of peripheral vision of both eyes.

12 Neck

Skin and superficial fascia · Deep cervical fascia · Superficial nerve supply · Superficial veins · Midline structures · Strap muscles · Sternomastoid muscle · Thyroid gland · Parathyroid glands · Great vessels

Skin and superficial fascia

The skin of the neck contains a subcutaneous muscle layer, the platysma. This sheet of muscle, arising in the region of the clavicle and deltoid muscles, sweeps up to the lower border of the mandible. It is able to wrinkle the skin, a rather unimportant action in the human. However, when suturing wounds of the neck, it is worth remembering that if the skin only is sutured and not the platysma as well, the wound will be distracted by the contracting severed muscle and will heal in a broad, ugly scar.

The collagen bundles in the neck, known as Langer's or crease lines, run transversely to form the creases of the neck. An incision in a vertical direction across the creases and/or failure to suture the platysma will reward you with a dissatisfied customer. On the other hand, if you remember these points, the patient will be eternally grateful. Within this layer lie the superficial veins and nerves.

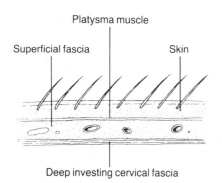

Fig. 12.1 **The subcutaneous platysma muscle.**

Deep cervical fascia

Investing layer

As with the limbs, the neck is enclosed in a continuous layer of fascia, like an unrolled polo-neck collar. Attached above and below

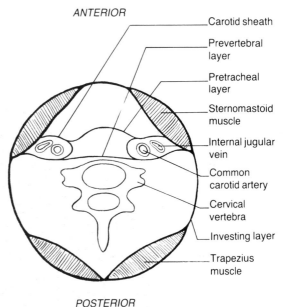

Fig. 12.2 **General arrangement of the cervical fascia.**

to bone, the investing layer splits to enclose two glands, the parotid and submandibular, two muscles, the sternomastoid and trapezius, and possesses two thickenings to bind down two intermediate tendons, those of the omohyoid and digastric muscles.

Pretracheal layer

The pretracheal layer, which passes medially from the investing layer, splits to enclose the thyroid gland. In the midline, it attaches superiorly to the laryngeal cartilages, while inferiorly it descends in front of the trachea to fuse with the fibrous pericardium. The attachment to both the thyroid gland and the larynx means that the gland will move on the laryngeal movement accompanying swallowing. Therefore, any swelling in or near the thyroid gland that moves with swallowing is a swelling of the gland, an important clinical point.

Prevertebral layer

The prevertebral layer passes medially from the investing layer anterior to the cervical vertebrae. It therefore covers those muscles arising from the vertebrae, such as the scalene muscles, and is prolonged into the axilla as a sheath for the brachial plexus.

Carotid sheath

The medial reflections from the pretracheal and prevertebral layers pass in front of and behind the carotid artery, respectively, to constitute the carotid sheath. This layer contains the common and internal carotid arteries (around which it is fairly thick), the internal jugular vein (around which it is very thin to allow for venous dilatation), and the vagus nerve, the last lying behind and between the artery and vein.

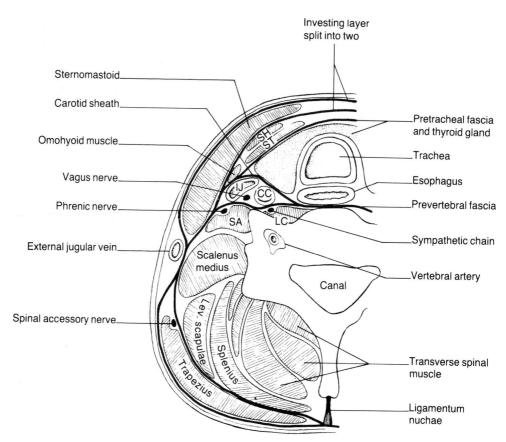

Fig. 12.3 The investing layer of the cervical fascia, as seen in a cross-section of the neck. CC, common carotid artery; IJ, internal jugular vein; LC, longus colli; SA, scalenus anterior; SH, sternohyoid, ST, sternothyroid.

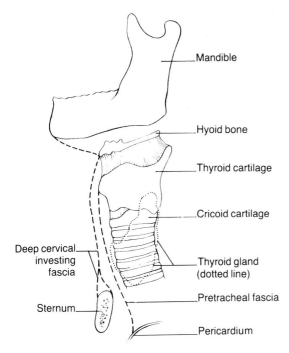

Mandible

Hyoid bone

Thyroid cartilage

Cricoid cartilage

Deep cervical
investing
fascia

Thyroid gland
(dotted line)

Sternum

Pretracheal fascia

Pericardium

Fig. 12.4   Vertical extent of the cervical fascia.

Superficial nerve supply

The superficial nerves of the neck arise from the anterior rami of second, third and fourth cervical nerves, and emerge at about the middle of the posterior border of the sternomastoid muscle (see Fig. 14.6). Blockage of these nerves at this point by an anesthetic agent will allow painless surgery on a large area of the neck. From here, the lesser occipital nerve runs along the posterior border of the sternomastoid muscle to supply the posterior aspect of the ear and scalp. The greater auricular nerve innervates both surfaces of the lower part of the ear; the transverse cervical nerve of the neck passes anteriorly across the sternomastoid muscle; and the three supraclavicular nerves run inferiorly to cross the medial, middle and lateral thirds of the clavicle.

The supraclavicular nerves arise from C3 and 4, and supply the skin over the first intercostal space and the shoulder. The phrenic nerve arises from the same rami (C3 and 4 and a little bit of C5) and supplies the diaphragm with both sensory and motor impulses. Since sensation from the skin is appreciated by the

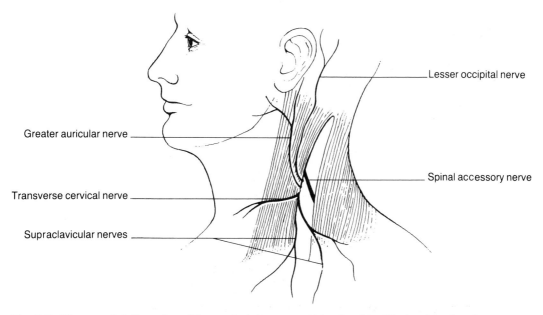

Lesser occipital nerve

Greater auricular nerve

Spinal accessory nerve

Transverse cervical nerve

Supraclavicular nerves

Fig. 12.5   The superficial branches of the cervical plexus, supplying the skin of the head, neck and upper thorax.

sensory cortex of the brain more than nearly any other structure (the card trickster relies on this with his marked cards), impulses passing up the phrenic nerve are interpreted by the cortex as originating in the skin rather than diaphragm. This results in the classic symptom of pain at the point of the shoulder from irritation of the phrenic nerve. Blood in the abdomen lapping against the diaphragm (phrenic nerve innervation) produces shoulder pain (supraclavicular nerve innervation), an important diagnostic point and a clear example of 'referred pain'.

Superficial veins

When the posterior facial vein emerges from the lower border of the parotid gland, it divides into anterior and posterior branches. The anterior branch joins the anterior facial vein to form the common facial vein, and the posterior branch joins the posterior auricular vein to form the external jugular vein.

External jugular vein

The large external jugular vein passes inferiorly across the sternomastoid muscle, pierces the investing fascia immediately above the clavicle and joins the subclavian vein. This has become an important vein because it can be used to thread a catheter through the subclavian vein to the superior vena cava and heart. It is not an ideal route since the angle of union with the subclavian vein makes the passage difficult (the preferred route is through the internal jugular vein, but sometimes beggars cannot be choosers). The surface marking, a line from the interval between the mastoid process and the angle of mandible to the middle of the clavicle, finds its way into most anatomy texts, perhaps unnecessarily: the vein can be easily seen by lowering the head of the bed and raising the intrathoracic pressure by expiring against a closed glottis.

Where the external jugular vein pierces the investing fascia, the latter and the vein wall

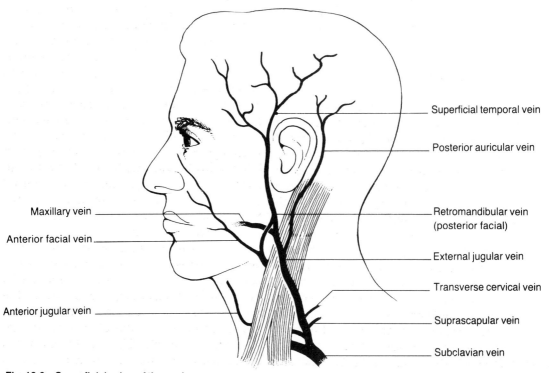

Maxillary vein

Anterior facial vein

Anterior jugular vein

Superficial temporal vein

Posterior auricular vein

Retromandibular vein (posterior facial)

External jugular vein

Transverse cervical vein

Suprascapular vein

Subclavian vein

Fig. 12.6   Superficial veins of the neck.

adhere to each other. If the vein is severed near this point at surgery or during unscheduled attacks on the neck (by Dracula, for example), the vein is held open by the fascia, and air is sucked into its lumen by the negative intrathoracic pressure. This constitutes the highly dangerous air embolism. If you hear a churning noise in the chest and the patient is becoming cyanotic (blue), put your finger on the opened vein until it can be closed by ligature.

Anterior jugular vein

The two anterior jugular veins, not always present or symmetrical, begin below the chin and run inferiorly on either side of the midline to the lower neck. Here they turn laterally deep to the sternomastoid muscle to join the external jugular veins. The anterior jugulars are the veins usually severed in a suicidal cutthroat, since the person extends the neck while performing the act. Usually, however, faintness ensues before the internal jugulars can be severed. The anterior jugular veins may be damaged when performing an emergency procedure such as laryngotomy.

Midline structures

Below the chin are the anterior bellies of the digastric muscles, as they run upward and medially to insert into the digastric fossae of the mandible on each side. The intermediate tendons of these muscles are held down to the hyoid bone. The two muscles and the body and greater cornua of the hyoid bone constitute a submental triangle, the floor of which is the mylohyoid muscle. This muscle runs down from the mylohyoid line on each side to insert into the body and greater cornua of the hyoid bone. The posterior borders of the mylohyoid muscle are free, and certain stuctures (lingual nerve, submandibular duct and hypoglossal nerve) pass behind them to enter the mouth. By separating the mouth from the neck, the mylohyoid acts as the diaphragm of the mouth.

The submental triangle contains several submental lymph nodes, the enlargement of which usually means disease in its drainage area (to be discussed later). Below the triangle is the easily palpable body of the hyoid bone, which moves on swallowing. This U-shaped bone consists of a central body and a pair of lesser and greater horns passing superolaterally. Being the only bone in the body not to articulate with another, it serves as the point of origin or insertion of a considerable number of muscles. Inferior to the hyoid is the cartilaginous skeleton of the larynx, whose component parts are connected by membranes. These cartilages are partly covered by the ribbon or strap muscles, by the thyroid gland, and, most inferiorly, by the large vessels entering or leaving the superior mediastinum.

Strap muscles

The relatively unimportant strap muscles are worthy of consideration only because they are in the surgeon's way when a tracheostomy is being performed. The sternohyoid muscle arises from the posterior aspect of the sternoclavicular joint and the bone on either side, while the sternothyroid muscle arises lower down from the posterior aspect of the manubrium and the first costal cartilage. These muscles run upward in contact with the muscles of the opposite side. The sternohyoid passes directly to its insertion into the hyoid bone, while the sternothyroid attaches to the oblique line of the thyroid cartilage, and then continues as the thyrohyoid muscle to reach the hyoid bone.

The two sternohyoids, in particular, almost meet in the midline, and this can be annoying when attempting to expose the trachea to perform a tracheostomy in a bit of a hurry. Rather than attempting to separate the muscles, it is much more practical to separate the fibers of one or the other side and go through the muscle. The attachment of the sternothyroid is important because it straps down the upper lobe of the thyroid gland,

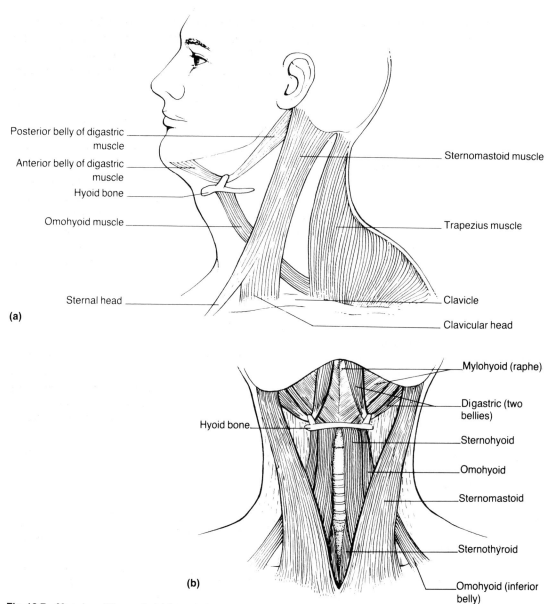

Posterior belly of digastric muscle

Anterior belly of digastric muscle

Hyoid bone

Omohyoid muscle

Sternal head

Sternomastoid muscle

Trapezius muscle

Clavicle

Clavicular head

(a)

Hyoid bone

Mylohyoid (raphe)

Digastric (two bellies)

Sternohyoid

Omohyoid

Sternomastoid

Sternothyroid

Omohyoid (inferior belly)

(b)

Fig. 12.7 Muscles of the neck: (a) lateral view, (b) anterior view.

which it covers, and prevents it from enlarging upward. The thyroid gland has to enlarge anteriorly or posteriorly or inferiorly.

Lying in the same plane is the third member of this muscular gang, the omohyoid muscle. This is an interesting little muscle, which, from its attachment to the hyoid bone, passes down alongside the sternohyoid, and, at the level of the cricoid cartilage, passes laterally as an intermediate tendon across the carotid sheath to its inferior belly. The inferior belly runs almost horizontally to the suprascapular notch on the upper border of the scapula. The intermediate tendon is tethered to the clavicle by a sheet of fascia. This muscle originally inserted into the clavicle but strayed laterally during its development. Its only importance is that it is often seen in surgery, where it acts as a good landmark. The action of these strap muscles (so called because of their shape) is to steady the hyoid bone, allowing the muscles passing between the hyoid and the mandible to open the jaw. They also help control the pitch of the voice.

Sternomastoid muscle

The sternomastoid muscle is a prominent and important muscle in the neck (see Fig. 14.6). It is attached inferiorly to the manubrium and clavicle; a depression between these attachments provides a useful landmark for the internal jugular vein. Sloping obliquely upward, it attaches superiorly to the superior nuchal line and the mastoid process.

In its lower half the sternomastoid muscle covers large neurovascular structures: the common carotid and subclavian arteries, including the branches of the latter; the internal jugular and subclavian veins and their branches; the brachial plexus; as well as the strap muscles. As the upper half slopes backward, the carotid and internal jugular vessels escape from under it, allowing easy palpation of the former vessel.

The sternomastoid muscle divides the neck into an anterior triangle – bounded by the midline, anterior border of the muscle and lower border of the mandible – and a posterior triangle – limited by the posterior border of the sternomastoid, anterior border of the trapezius and middle third of the clavicle. Geometric afficionados divide the anterior triangle into three further triangles, the demarcating muscles being the digastrics and the superior belly of the omohyoid, but this is not very helpful clinically. The accessory nerve passes through the muscle obliquely, entering at the junction of the upper and second quarters and leaving at the junction of the upper and second thirds, after which it makes for the trapezius.

The anterior and posterior fibers of the sternomastoid pass, respectively, in front of

and behind the atlantoaxial joint. Acting together, both muscles can therefore both flex and extend the head. As the muscles lie anterior to the cervical spine, they flex the neck. Finally, acting alone, the muscle rotates the face to the opposite side, while bending the head to the same side (feel the muscle contract when you turn your face to the opposite side).

In summary, the sternomastoid muscle enables you to flex your neck, moving it forward, and to extend your head, elevating it. This enables you to look over the fence and watch the match for free. It can also turn your face sideways when you hear the approaching steps of the security guard, and on escape it is a most useful accessory muscle of inspiration!

Thyroid gland

The thyroid is a large important endocrine gland. It develops from a duct, which begins at the junction of the anterior two-thirds and posterior third of the tongue (foramen cecum) and passes inferiorly between the muscles of the tongue to the anterior surface of the hyoid bone. The duct curves around the lower

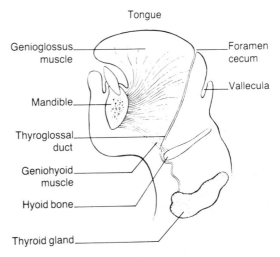

Fig. 12.8 Development of the thyroid gland: the thyroglossal duct is seen passing inferiorly from the foramen cecum. Note the close relation to the hyoid bone.

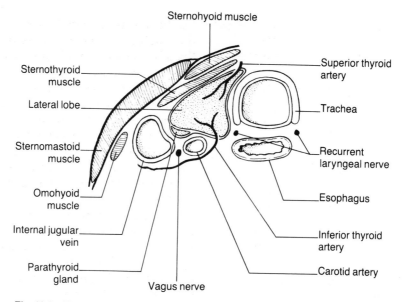

Fig. 12.9   The lateral lobe of the thyroid gland and its relations.

border of the bone to its posterior surface before it continues downward, deviating to one or other side (usually the left) because of the prominence of the thyroid cartilage. Upon reaching the trachea, the duct gives rise to an isthmus and two lateral lobes. The isthmus usually covers the second, third and fourth tracheal rings. The lateral lobes extend from the sixth tracheal ring to the oblique line of the thyroid cartilage, where the attachment of the sternothyroid muscle prevents upward pathologic enlargement. The gland has been likened to a bow-tie in both site and shape.

The pyramidal lobe is the lower portion of this track; it is represented by a small prominence on the superior border of the isthmus, usually on its left side. Sometimes the fibrous remains of the duct passes from the gland toward the hyoid bone. A little muscle may be present in this fibrous band, which then goes by the fancy name of levator glandulae thyroideae.

The lobes of the thyroid are covered by the strap muscles, with the sternomastoid muscle overlapping. Medially, the surface of each lobe is related to two tubes, the esophagus and the trachea, and to two nerves, the recurrent and external laryngeal nerves; all these structures may be affected by glandular enlargements. The posterior aspect of the gland usually bears the parathyroid glands.

Vascular supply

The superior thyroid artery, the first branch of the external carotid, runs downward with a bit of a loop to reach the apex of the lateral lobe of the thyroid gland. The inferior thyroid artery, an indirect branch of the subclavian artery, enters the gland from its deep aspect and supplies more of the glandular tissue. The inferior thyroid artery has a very close relationship to the recurrent laryngeal nerve, which may lie in front, behind or between the branches of the artery (see Fig. 13.9). The thyroid gland also receives a blood supply from vessels to the esophagus and trachea; the four large thyroid arteries can be tied off and yet the gland will function perfectly well.

The superior thyroid veins, unlike the other thyroid veins, accompany the artery of the same name. The middle thyroid vein (Fig. 12.11) leaves the thyroid gland at about its middle, and has a short course into the internal jugular vein. This vein is a pest to the surgeon and needs to be divided to mobilize the

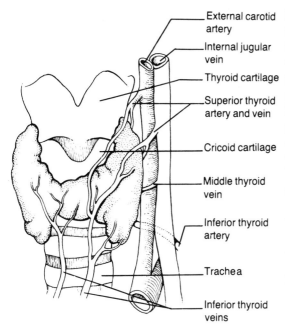

External carotid artery

Internal jugular vein

Thyroid cartilage

Superior thyroid artery and vein

Cricoid cartilage

Middle thyroid vein

Inferior thyroid artery

Trachea

Inferior thyroid veins

Fig. 12.10   The blood supply of the thyroid gland.

gland. The inferior thyroid veins leave the isthmus at its lower border and run down in front of the trachea to enter the left brachiocephalic vein as the latter crosses the trachea. The inferior thyroid veins are in danger if one does a tracheostomy below the isthmus of the thyroid gland instead of through it.

The thyroid gland often enlarges downward toward the mediastinum, as upward enlargment is difficult (see 'Strap muscles'). One of the signs of extension of the thyroid gland into the mediastinum is compression of the inferior thyroid veins and the brachiocephalic veins. This compression produces engorgement of the veins' tributaries, easily seen over the upper chest wall, a sign of mediastinal compression.

Parathyroid glands

The parathyroid glands are pea-sized glands, situated, as the name indicates, in close relationship to the thyroid gland. In spite of their small size, they are fundamental to the control of calcium metabolism, and vital to

health. When the parathyroids are diseased, the calcium metabolism goes awry. One of the advantages to modern living is the ease with which we can measure the calcium content of the blood. Whereas parathyroid disease used to be a diagnosis cleverly made by an astute clinician with bushy eyebrows, it is now made by the technician working the automatic analyzer. The result is that the parathyroid gland is frequently being attacked by the surgeon to correct an abnormal blood calcium.

The glands vary in number, but there usually are four, an upper pair and a lower pair. The development is interesting. The upper pair develops from the fourth pharyngeal pouch and comes into close relationship with the thyroid gland, whereas the lower pair develops from the third pouch together with the thymus gland. The thymus gland passes downward in front of the trachea to end up in the mediastinum, where the lobes from each side fuse. It sometimes happens that the thymus carries the third pouch parathyroid glands with it (these glands should perhaps be called parathymus rather than parathyroid).

Often, in examinations, one is asked questions concerning the origin of the two glands. The so-called superior parathyroid arises from the fourth pouch and the inferior parathyroid arises from the third pouch, this anomaly being explained by the interest that the third pouch parathyroid displays in following the path of the thymus. More important than the examination question, however, is that the upper glands may follow the thymus into the superior mediastinum. Although most tumors of the parathyroid gland are found in their usual position in relationship to the thyroid, the surgeon may have to search for the parathyroid tumor in an area extending from the thyroid gland above to the mediastinum below. In this situation, perhaps the saying 'rather a lucky surgeon than a good surgeon' is apt. The lucky surgeon finds the gland in the usual position in relationship to the thyroid gland, whereas the surgeon who has invoked the wrath of the

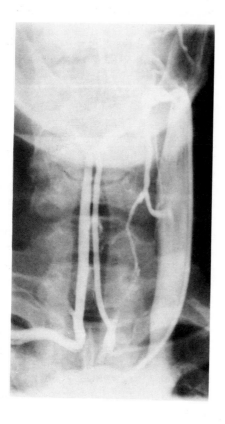

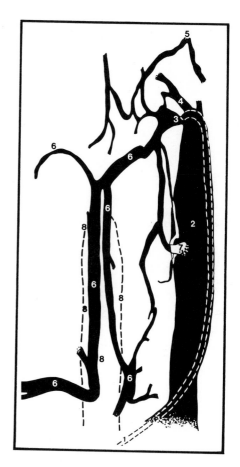

Fig. 12.11 Venogram of the neck. This x-ray was performed to study parathyroid hormone levels from tumors of the parathyroid glands. (From Weir and Abrahams 1978)

1 Catheter in left brachiocephalic vein.
2 Internal jugular vein.
3 Superior thyroid vein.
4 Lingual vein.

5 Facial vein.
6 Anterior jugular veins.
7 Middle thyroid vein.
8 Air within trachea.

gods has to hunt deep down in the thorax for the mischievous tumor.

The parathyroid glands, when normally situated, lie in close relationship to the posterior aspect of the thyroid gland inside its fascial sheath. Like all ductless glands, it pours its secretion into vessels. Being closely related to the thyroid, it is not surprising that the superior and inferior thyroid arteries are the vessels of supply.

Great vessels

Carotid system

Each common carotid artery enters the neck behind the sternoclavicular joint, where it is hidden by the sternomastoid and the strap muscles (see Fig. 11.4). The artery then makes for a point between the angle of the mandible and the mastoid process, but it does not reach this point. The artery bifurcates at the level of the upper border of the thyroid cartilage into the internal and external carotid arteries.

tragus of the ear with the auriculotemporal nerve and divides into anterior and posterior branches for the temporal area of the scalp. It is an artery that sometimes is afflicted with severe disease (temporal arteritis).

Internal carotid artery (Figs. 10.19 and 11.4)

The internal carotid artery, lying postero-lateral to the external carotid, runs upward with it, deep to the posterior belly of the digastric and stylohyoid muscles, and then deep to the styloid process to reach the base of the skull, where it enters the carotid foramen for its intracranial distribution. The internal carotid artery is one of the student's best friends, because it does not have any branches in the neck. However, it is not man's best friend because, if things go wrong with this artery, it has no collateral anastomoses in the neck and has to rely on its intracranial collateral circulation to bypass any deficiency, a risky state of affairs. It carries into the cranial cavity a rich plexus of sympathetic nerves from the superior cervical ganglion, which partly controls the caliber of the intracranial vessels. As well as supplying the cerebrum, it gives off the all-important ophthalmic artery.

Internal jugular vein (Figs. 11.4 and 12.11)

The internal jugular vein passes through the jugular foramen as the continuation of the intracranial sigmoid sinus. Since the carotid foramen is adjacent, the internal jugular vein and the internal carotid artery are closely related. Grouped around these vessels are the last four cranial nerves: IX, X and XI pass through the jugular foramen and XII exits just medially. The internal jugular vein runs deep to the posterior digastric and stylohoid muscles, following the internal and common carotid arteries and lying in the carotid sheath. This sheath is not a thick structure around the vein. A glance at any athlete finishing the 100-meter dash will reveal a huge internal jugular vein, an unlikely event were there a nondistensible thick sheath.

After meeting the common carotid artery, the internal jugular vein continues inferiorly, being crossed by the tendon of the omohyoid muscle and then overlapped by the sternomastoid. It makes for the medial quarter of the clavicle, where it has a very important surface marking; it lies in the notch between the clavicular and sternal heads of the sternomastoid muscle. The obvious disadvantage of puncturing the vein at this point is that the needle may penetrate the cervical pleura, with a resultant pneumothorax.

The internal jugular vein is joined just above the clavicle by the subclavian vein to form the brachiocephalic vein. The branches of the internal jugular are from the pharynx, face, tongue and thyroid. The common facial vein is worth mentioning. It is formed by the anterior facial vein and the anterior branch of the retromandibular vein. It crosses the submandibular gland to enter the internal jugular vein. The common facial vein is a much bigger and more constant vein than is usually thought, and it is a popular route in pediatric patients for passing a cannula into the internal jugular vein for feeding purposes.

The internal jugular vein is joined below the base of the skull by the inferior petrosal sinus, which also traverses the jugular foramen. This sinus is capable of carrying a considerable amount of intracranial blood when the larger intracranial veins are blocked. Lymph ducts enter the jugulo-subclavian venous junction on each side. There are two bulbs in the vein: one located superiorly, where it lies in the floor of the tympanic antrum and at which point uncommon tumors sometimes arise; and an inferior bulb, slightly above a bicuspid valve, 1.25 cm (0.5 in) above the clavicle. Large lymph nodes of the neck surround the internal jugular vein, and are known as the deep cervical chain.

13 Cranial nerves V, VII, IX–XII, and cervical sympathetic system

Trigeminal (V) nerve · Facial (VII) nerve · Position of nerves IX–XII at base of skull · Glossopharyngeal (IX) nerve · Vagus (X) nerve · Accessory (XI) nerve · Hypoglossal (XII) nerve · Cervical sympathetic trunk

Trigeminal (V) nerve

The trigeminal nerve is the largest cranial nerve. It supplies sensation to the face, the anterior scalp, teeth, mouth and nasal cavity, and it is the motor supply to all the muscles of mastication, except the buccinator. We will pick it up at its large trigeminal ganglion that lies in a recess of the dura mater in the posterior part of the cavernous sinus (Meckel's cave), occupying a small hollow near the apex of the petrous temporal bone, the trigeminal impression. Passing centrally to the pons, with its cell bodies in the trigeminal ganglion, is a large sensory root. Three large branches arise from its antero-lateral convexity and pass forward to their distribution. Passing peripherally deep to the ganglion is a small motor root on its way to join the mandibular division.

Ophthalmic (V¹)

The ophthalmic nerve is the smallest branch of the trigeminal nerve and is wholly sensory. It runs through the cavernous sinus, and, before passing through the superior orbital fissure, divides into its three branches, the lacrimal, frontal and nasociliary.

The lacrimal nerve, the smallest branch, enters the lateral part of the fissure, runs along the upper border of the lateral rectus muscle, and receives a twig from the zygomatic branch of the maxillary nerve which it carries to the lacrimal gland. It finally enters this gland, and, having supplied it, passes to the skin of the upper eyelid.

The frontal nerve, the largest branch, passes through the fissure above the ocular muscles. It runs forward above the levator palpebrae superioris, and divides halfway along the orbit into small supratrochlear and large supraorbital nerves, both of which pass through foramina or notches on the superior orbital rim. The supratrochlear supplies the skin of the lower part of the forehead near the midline, and the supraorbital supplies the scalp almost as far as the lambdoid suture; blocking of the latter nerve by anesthetic solution results in extensive anesthesia of the scalp.

The nasociliary nerve runs through the fissure within the tendinous ring (see Chapter 18) and passes medially across the optic nerve to reach the medial side of the orbit. After giving off the posterior ethmoidal and infratrochlear nerves, it enters the anterior cranial fossa through the anterior ethmoidal foramen as the anterior ethmoidal nerve, and crosses the cribriform plate to reach the nasal slit through which it descends inside the nasal cavity. After supplying the nasal mucous membrane, it emerges as the external nasal branch to the nose.

It receives a branch from the ciliary ganglion, which carries sensation from the eye via the short ciliary nerves; it gives off long ciliary nerves as it crosses the optic nerve, which usually contain the sympathetic fibers for the dilator pupillae muscle. You should be reminded that these fibers have gone to a lot of trouble getting there. Arising from the posterior hypothalamus of the brain, they run in the brain stem, through the midbrain, pons and medulla, to reach a center in the cervicothoracic region of the spinal cord. From there they pass via the anterior root, mixed nerve and anterior ramus of the first thoracic nerve to the sympathetic chain, and ascend to the superior cervical ganglion, where they synapse. The postganglionic fibers pass via the internal carotid plexus and the ophthalmic artery, and finally along the long ciliary nerves to the dilator pupillae muscle.

Maxillary nerve (V²)

Supplying the maxillary process, this division of the fifth nerve is purely sensory and intermediate in position and size between the ophthalmic and mandibular nerves. Arising at the middle of the trigeminal ganglion, it passes through the lateral wall of the cavernous sinus, and leaves the skull through the foramen rotundum to reach the upper part of the pterygopalatine fossa. It then enters the orbit through the inferior orbital fissure, changing its name to the infraorbital nerve, which passes through a groove, a canal and a large infraorbital foramen before appearing on the face; here it branches to supply the skin of the maxillary process, namely the lower eyelid, the side of the nose and the upper lip.

While still intracranial, the maxillary nerve gives off a meningeal branch to the meninges of the middle fossa.

In the pterygopalatine fossa, it gives two branches to the large pterygopalatine (parasympathetic) ganglion, which lies immediately below it. This ganglion receives the nerve of pterygoid canal, functionally the seventh nerve, which, after relaying in the ganglion, passes in the zygomatic branch of the maxillary nerve to reach the lacrimal gland and by other branches supplies the nasal and palatal glands. The zygomatic nerve, arising in the pterygopalatine fossa, enters the orbit through the inferior orbital fissure, runs along its lateral wall, gives off the secretory branch to the lacrimal nerve for the lacrimal gland, and divides into two branches which perforate the zygomatic bone to supply the cheek and temple. Before entering the inferior orbital fissure the maxillary nerve gives off posterior superior dental branches which run inferiorly in small canals in the maxilla; in this bone it joins the anterior and middle superior dental branches to form a superior dental plexus for the teeth of the upper jaw. The anterior and middle dental branches arise in the infraorbital canal and run in the walls of the maxillary air sinus toward the teeth; for this reason it is often difficult to differentiate facial pain as arising from sinusitis or from a diseased root of a tooth. Ganglionic branches appear to arise from the ganglion, but in fact most of the fibers leaving the ganglion pass through it without relay. The only nerves to synapse are the parasympathetic branches from the greater superficial petrosal nerve, which are distributed to lacrimal, nasal and palatal glands. The sensory nerves from the roof of the mouth, soft palate, tonsil and nose which pass through the ganglion are the palatine and sphenopalatine nerves. The greater and lesser palatine nerves pass inferiorly through the greater palatine canal, and emerge through the greater and lesser palatine foramina to supply the palate, including the uvula and tonsil. During its descent, the greater palatine gives off nasal branches which supply the lower part of the lateral wall of the nose. Long and short sphenopalatine nerves enter the nose through the sphenopalatine foramen. The former runs obliquely downward and forward along the nasal septum and enters the mouth through

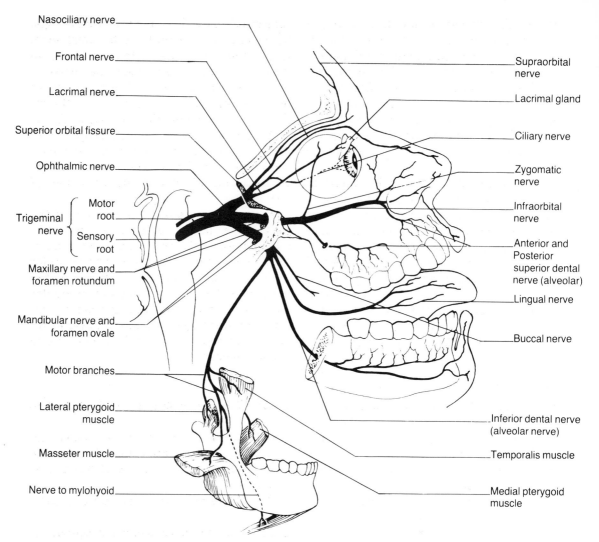

Nasociliary nerve

Frontal nerve

Lacrimal nerve

Superior orbital fissure

Ophthalmic nerve

Trigeminal nerve { Motor root / Sensory root }

Maxillary nerve and foramen rotundum

Mandibular nerve and foramen ovale

Motor branches

Lateral pterygoid muscle

Masseter muscle

Nerve to mylohyoid

Supraorbital nerve

Lacrimal gland

Ciliary nerve

Zygomatic nerve

Infraorbital nerve

Anterior and Posterior superior dental nerve (alveolar)

Lingual nerve

Buccal nerve

Inferior dental nerve (alveolar nerve)

Temporalis muscle

Medial pterygoid muscle

Fig. 13.1 Distribution of fifth cranial nerve.

the incisive canal at the anterior aspect of the hard palate, where it communicates with the nerve of the opposite side and with the greater palatine nerve. It mostly supplies the septum, whereas the short sphenopalatine is destined for the upper lateral wall of the nose.

Mandibular nerve (V³)

The third division of the trigeminal nerve exits from the foramen ovale together with the only motor component of the fifth nerve which soon joins it. The mandibular nerve is therefore a mixed nerve, supplying sensation to the derivatives of the mandibular process and motor power to the muscles of mastication. It splits into two divisions. The anterior division is mostly motor and supplies most of the muscles of mastication, including the temporalis, masseter, the two pterygoids and the tensor palati; the only sensory component is the long buccal nerve, which supplies the gums of the upper jaw, and the skin and mucous membrane in the lower cheek. The posterior division, mostly sensory, has three important branches.

The smallest branch, the auriculotemporal, passes posteriorly by two roots, which encircle the important middle meningeal artery, runs by the parotid gland to which it gives salivary fibers (originally ninth nerve), and then turns up in front of the ear to run with the superficial temporal vessels and to supply the upper portions of the auricle and the temple. It is responsible for the referral of pain from the temporomandibular joint (to which it is closely related) to the side of the scalp.

The second branch, the very large inferior dental (alveolar) nerve, passes inferiorly to enter the large mandibular foramen of the mandible. It runs forward in the mandibular canal, sending a branch of supply to each tooth, to finally emerge from the mental foramen and turn upward to supply the lower lip. We noted previously the mandibular foramen with an overlying plate of bone, the lingula, for which your dentist searches in order to anesthetize the nerve supply to the teeth of your lower jaw – a successful block will render your lower lip anesthetic as well. Just before it enters the foramen, this nerve gives off the only motor contribution of the posterior division, branches of supply to the anterior belly of the digastric and mylohyoid muscles, both muscles of mastication. Although small, this nerve leaves its impression on the bony mandible.

The third branch, the large lingual nerve, runs inferiorly and is joined at an acute angle by the chorda tympani from the facial nerve which hitches a ride to the anterior two-thirds of the tongue. The lingual nerve passes in close relationship to the third molar tooth of the lower jaw (a good spot to anesthetize the nerve), runs under the mucous membrane of the floor of the mouth which it supplies, to reach the tongue, where it mediates common sensation from the anterior two-thirds (the chorda tympani carries the sensation of taste).

In summary, the anterior division of the mandibular nerve supplies motor power to all the muscles of mastication, and sensation to the gum of the upper jaw. The posterior division supplies sensation to the side of the scalp, part of the ear, the temporomandibular joint, the floor of the mouth, the anterior two-thirds of the tongue and all the lower teeth, and is motor to the anterior belly of digastric and mylohyoid.

Facial (VII) nerve (Figs. 11.7, 13.2–13.6)

It is hoped that a course in neuroanatomy will cover the central connections of the facial nerve. If not, look at Figs 13.3 and 13.4, and read on.

The axons which innervate the facial nerve

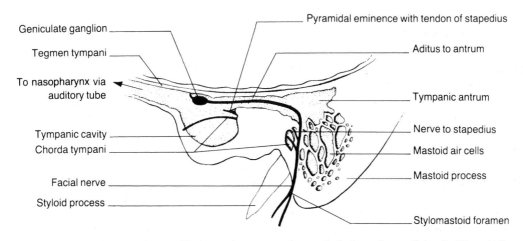

Fig. 13.2   Middle ear and facial nerve. The nerve is seen running posteriorly on the medial wall of the middle ear, before turning inferiorly to run vertically and exit at the stylomastoid foramen. Two branches pass anteriorly from the nerve during the vertical part of its course, the nerve to the stapedius muscle and the chorda tympani.

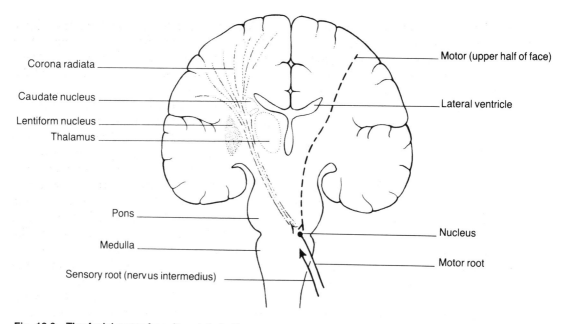

Fig. 13.3 The facial nerve from its origin in the cerebrum to its nucleus in the brain stem (pons).

nucleus arise in the facial area of the motor cortex, run in a spiral manner as the corona radiata, and pass through the internal capsule between the basal ganglia. Traversing the midbrain, the fibers then decussate to reach the facial nerve nucleus in the opposite side of the pons. This nucleus also receives fibers from the same side of the cortex which are going to subserve movements of the upper face, especially those of emotion; you will see the importance of this later. After synapsing at this nucleus, a new set of fibers emerge from the facial nucleus. They take a peculiar course, twisting backward around the nucleus of the sixth nerve to raise a bump in the floor of the fourth ventricle as they do so, the colliculus facialis. The nerve emerges from the lateral aspect of the pons in close relationship to the eighth nerve, both nerves passing in the angle between the cerebellum and the pons to enter the internal meatus. Here the facial nerve parts company with the eighth nerve, and runs over the vestibule of the inner ear (that part lying between the cochlea and the semicircular canals) to reach the medial or inner wall of the middle ear. Here the nerves make a severe bend poster-

iorly; at this genu is situated a large sensory ganglion, the geniculate ganglion. The nerve then passes in a backward direction along the inner or medial wall of the middle ear to reach the aditus to the antrum, which is a small passage between the middle ear and the tympanic antrum. Having reached the aditus,

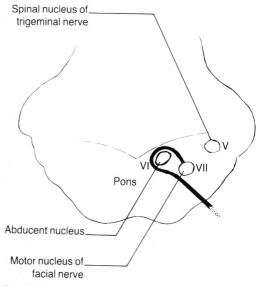

Fig. 13.4 The peculiar course of the seventh nerve around the nucleus of the sixth nerve.

it makes yet another curve, this time inferiorly, and passes down posterior to the middle ear to emerge through the stylomastoid foramen between the styloid and mastoid processes. Rather pleased with itself at having made this tortuous journey, the nerve then splits in two large divisions, and immediately passes forward to enter the parotid gland. In the gland, the nerve further subdivides, branches intermingling and re-grouping, and eventually emerges as five main divisions. The divisions that appear at the upper and lower borders of the parotid gland are relatively unimportant; the former, the temporal, passes to the forehead to innervate the frontalis, and the latter, the cervical, passes into the neck to innervate the platysma. Both muscles attach to the skin and can cause wrinkling – loss of the ability to wrinkle the skin of our neck or our forehead is a minor disability. The three divisions emerging from the anterior border of the gland, however, are vitally important in that they reach and control the orifices of the eye, nose and mouth. The upper divisions, the zygomatic branches, cross the zygomatic bone to reach the orbicularis oculi, and are thus responsible for closing the eye. For this reason, an incision should never be made along the zygomatic bone, as it would certainly sever the nerves. The middle group of nerves, the buccal, run with the parotid duct to control the nostrils and mouth. The buccal branch supplies the buccinator, which is the muscle responsible for transferring food from the vestibule into the oral cavity. The lowest division, the mandibular, takes an important course. It runs for a short distance into the neck, where it lies about 6 mm (0.25 in) behind the angle of the mandible, and then curves forward across the mandible and the facial vessels to reach the angle of the mouth. If the mandibular branch is severed, it will produce a very ugly depression of the angle of the mouth – for this reason incisions are never made along the lower border of the mandible. There are patients with leery smiles pursuing surgeons who made such incisions.

Thus far, we have described the motor fibers of the facial nerve, but it has other important components. Sensory fibers subserving taste pass from the anterior two-thirds of the tongue along the lingual nerve. The fibers leave the lingual nerve just below the base of the skull as the chorda tympani, and pass through a small fissure of the base into the middle ear. Here it passes across the lateral wall of the cavity (the tympanic membrane) to enter the trunk of the facial nerve. The sensory fibers then reach the geniculate ganglion where they have cell bodies and continue through the bony facial canal. The sensory nerve emerges from the internal auditory meatus, sandwiched between the seventh and eighth nerves (thereby its name, nervus intermedius), and enters the pons, from which fibers pass to the sensory cortex where they register taste.

The facial nerve also carries parasympathetic secretomotor fibers which are destined for the salivary glands, which they reach via the chorda tympani, and the lacrimal gland, which they reach via the greater superficial petrosal and the maxillary nerves.

Branches of the facial nerve

There are both branches of communication and branches of supply.

Branches of communication

These are many, but we will mention two which hold high honor in the court of anatomy. Where the facial nerve bends at the geniculate ganglion, it gives off the greater petrosal branch (so named because it grooves the petrous temporal bone) that passes forward to join the deep petrosal nerve that arises from the sympathetic carotid plexus to form the nerve of the pterygoid canal. This nerve passes through the canal of the pterygoid process to the spheno(pterygo)palatine ganglion where parasympathetic fibers, originating in a lacrimal nucleus in the brain stem, relay and reach the lacrimal gland

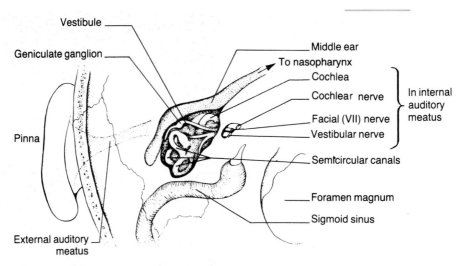

Fig. 13.5 Internal ear and facial nerve in the petrous temporal bone. The facial nerve, accompanied by the eighth nerve, enters the internal auditory meatus, and crosses the vestibule of the internal ear to reach the geniculate ganglion. Here it bends sharply posteriorly to run along the medial wall of the middle ear.

through the zygomatic branches of the maxillary nerve. Most of the fibers of the greater petrosal nerve, however, are taste fibers from the palate, which pass from the palatine branches of the fifth nerve over or through the sphenopalatine ganglion without relay to reach their cell bodies in the geniculate ganglion.

At the geniculate ganglion, filaments pass forward to exit the skull and join the lesser petrosal nerve which arises from the tympanic plexus of the ninth nerve. Relaying in the otic ganglion, fibers pass via the auriculotemporal nerve to stimulate the parotid gland. These fibers are said to come from the inferior salivary nucleus in the brain stem. Throughout its course, the seventh nerve has close connections with the fifth nerve.

Branches of distribution

These are very important if we wish to understand the symptoms and signs of the commonly occurring facial paralysis. Within the facial canal, two branches are given off: a branch to the stapedius muscle, the auditory muffler, which controls the movement of the stapes; and the chorda tympani. The latter arises about 6 mm (0.25 in) before the nerve

emerges from the stylomastoid foramen, and runs through the tympanic cavity, crossing the tympanic membrane at the level of the upper end of the manubrium of the malleus, one of the auditory ossicles. This is fortunate since the tympanic membrane is commonly incised in inflammations of the middle ear; incisions are made below the ossicles, thereby avoiding the nerve. The chorda tympani then leaves the skull to join the lingual nerve, which carries it to its ultimate distribution. The lingual nerve passes close to the submandibular and sublingual salivary glands on its way to the tongue. The chorda tympani does not miss this opportunity; the secretomotor fibers which it carries (originating in the superior salivary nucleus) to the salivary glands are given off as the lingual nerve passes by these glands. Also travelling in the chorda tympani, but in the other direction, are the taste fibers from the anterior two-thirds of the tongue (excluding the vallate papillae), which pass through the geniculate ganglion.

Exiting from the stylomastoid foramen, the facial nerve gives off three relatively inconsequential branches to muscles in that area – the posterior belly of digastric, the stylohyoid and the rudimentary muscles behind the ear. You will not be the object of derision if you lose the nerve supply to the first two. With regard to the auricular muscles, they may be important to the elephant, but in man they come into play only as a party trick.

Having given off these branches, the nerve immediately enters the parotid gland, in which it is the most superficial structure, a matter of clinical importance. It divides into two divisions, a temporofacial running sharply upwards, and a cervicofacial continuing the course of the parent trunk. These divisions subdivide to form what is called the pes anserinus. Next time you are at the zoo, examine a goose's foot, and you will immediately note the similarity to the branching of this nerve!

We should now be able to elucidate the effects of facial paralysis.

1   Intracranial. Lesions above the nucleus (supranuclear), which usually involve the cortex or corona radiata, will produce paralysis of the lower half of the opposite face; remember! – the upper half is bilaterally innervated. With regard to nuclear lesions, we mentioned the close proximity of the sixth nerve – therefore there is often a sixth nerve paralysis accompanying the paralysis of the whole of the same side of the face.

2   Cranial. The lesion is usually either a fracture of the base of the skull or an inflammation of, or operation on, the middle ear. Paralysis of the whole of the same side of the face occurs; it will be accompanied by loss of taste if the lesion is above the origin of the chorda tympani, and by alteration in hearing if the nerve to the stapedius muscle is involved. In the latter case, hyperacusis results – excessively loud sounds due to the loss of the damping effects of the stapedius muscle.

3   Extracranial. Here we come across the very common illness described by Sir Charles Bell, Bell's palsy. The cause of this disease is not yet known, but it is thought to be due to an inflammatory con-

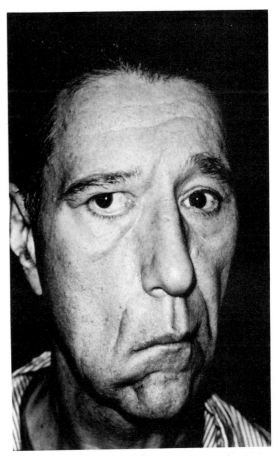

Fig. 13.6   Patient with paralysis of right facial nerve. Note the drooping of the angle of mouth, the flattening of the nasolabial fold, and the drooping of the lower lid on the right side.

dition of the nerve, causing it to swell at its tight exit from the stylomastoid foramen. This leads to complete paralysis of the whole of the same side of the face, but as this occurs below the chorda tympani branch and the nerve to the stapedius, these branches are unaffected. Other causes of malfunction of the nerve are operative injury to the nerve, and diseases, usually malignant, of the parotid gland. Nearly all patients with Bell's palsy recover, but, if you consult some otorhinolaryngologists, they will decompress your nerve in the stylomastoid foramen – your nerve will also recover but your wallet will be a little lighter.

Branches of the nerve, as they lie on the face, are mostly injured from flying broken glass – the paralysis depends on the nerve involved.

Position of nerves IX–XII at base of skull

Looking at the base of the skull, we notice the carotid canal situated in front of the jugular foramen. Through the latter foramen pass the ninth, tenth and eleventh nerves. The twelfth nerve exits through the condylar foramen, situated a little way medial to these nerves. Therefore, the last four nerves lie between the internal carotid artery and the internal jugular vein, and take different courses to reach their different destinations. The tenth nerve maintains its original position between the artery and the vein, and runs all the way down in the carotid sheath to enter the thorax. To reach the tongue and pharynx, the ninth nerve passes forward between the two carotid vessels. The twelfth reaches the tongue by also passing forward, but at a lower level, where it crosses both arteries as well as the loop of the lingual artery in so doing. The eleventh nerve passes backward across the internal jugular vein to reach the sternomastoid. Let us examine these nerves in more detail.

Glossopharyngeal (IX) nerve

This nerve is one of the structures separating the two carotid arteries. It runs forward deep to the hyoglossus muscle to supply (as the name indicates) the posterior one-third of the tongue, including the vallate papillae, with both taste and common sensation, and the surrounding oropharynx (tonsil, pharyngeal arches and part of the soft palate) with sensory fibers. Some of these fibers are important afferent contributions to the swallowing reflex. A small but important branch to the carotid sinus (sinus nerve) has the ability to alter the blood pressure when stimulated. A tympanic branch to the middle ear supplies the middle ear, mastoid air cells and part of the nasopharyngeal tube, and leaves to join a branch of the seventh nerve in the base of the skull to form the lesser petrosal nerve. This last sends a branch via the otic ganglion and the auriculotemporal nerve to the parotid gland; rather a complicated route to produce some spit!

Vagus (X) nerve

This nerve derives its name from its wide distribution (vagrant or wandering). In the neck it runs within the carotid sheath lying behind and between the internal jugular vein and common carotid artery as far as its entry into the thorax. The vagus nerve has two ganglia: the upper smaller one lies in the jugular foramen (jugular ganglion) and the lower large ganglion nodosum receives a large part of the accessory nerve. In fact, this is how the eleventh cranial nerve gets its alternative name; most of it is accessory to the vagus.

The branches of importance

While in the jugular foramen, the vagus nerve gives off two sensory branches, a meningeal branch to the posterior fossa, and an important auricular branch. The latter reaches the external auditory meatus, where it supplies the posterior half of the eardrum and meatus, and terminates by supplying a small strip of skin between the ear and the scalp.

Joining the ganglion nodosum is the cranial part of the eleventh nerve, which innervates the striated muscle supplied by the vagus – i.e. pharynx, larynx and soft palate. The vagus then gives off very important cardiac branches, both in the neck and in the thorax, after which it enters the abdomen through the esophageal hiatus to supply a large section of the gastrointestinal tract. So, in summary, the vagus is sensory to the meninges, the ear, the gastrointestinal and the respiratory tracts, and motor to the pharynx and larynx; cardiac branches inhibit the heart, and the nerve is visceromotor to the

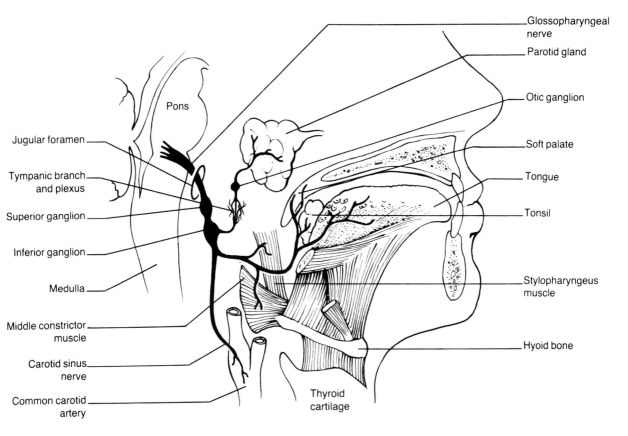

Glossopharyngeal nerve

Parotid gland

Otic ganglion

Soft palate

Tongue

Tonsil

Stylopharyngeus muscle

Hyoid bone

Jugular foramen

Tympanic branch and plexus

Superior ganglion

Inferior ganglion

Medulla

Middle constrictor muscle

Carotid sinus nerve

Common carotid artery

Pons

Thyroid cartilage

Fig. 13.7 Glossopharyngeal nerve.

respiratory and alimentary tracts. In the abdomen, generally, it stimulates the muscle and secretory digestive glands and relaxes the sphincters; the afferent supply from the gastrointestinal and respiratory tracts is mostly concerned with reflex actions. The vagus does not play a large part in conveying impressions of pain, the sympathetic system subserving this function. Let us now translate some of this knowledge into practice.

1   The meningeal branch may well play a part in producing a characteristic slowing of the pulse (bradycardia), when the meninges are stretched. Cases of raised intracranial pressure are commonly associated with bradycardia.

2   Auricular branch. If your ear is irrigated with water, it may produce (a) nausea and vomiting via the nerve supply of the vagus to the stomach, (b) a possible fainting attack from slowing of the heart, and (c) coughing via the nerve supply to the respiratory tract. To be even more practical, (a) some patients with gastric disturbances complain of persistent earache; for instance, sufferers from hiatus hernia will recognize the earache of an acute attack; (b) when syringing the wax out of the ear of an elderly patient, be sure to have him seated securely as he may faint due to the sudden slowing of the heart; and (c) a small boy may have a persistent cough which may be traced to a peanut which he is storing in his ear.

3   The pharyngeal nerves form a major part of the pharyngeal plexus for the supply to the pharyngeal constrictor muscles. A patient with a lesion of the brain stem involving the origin of these nerve fibers

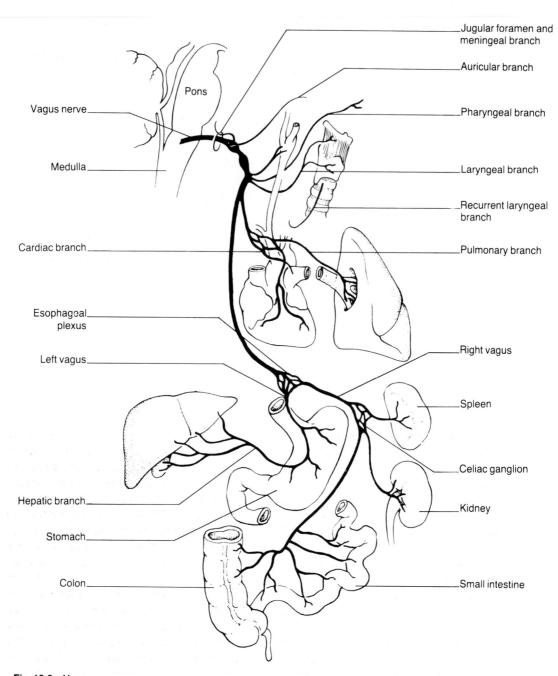

Jugular foramen and
meningeal branch

Auricular branch

Pharyngeal branch

Laryngeal branch

Recurrent laryngeal
branch

Pulmonary branch

Right vagus

Spleen

Celiac ganglion

Kidney

Small intestine

Pons

Vagus nerve

Medulla

Cardiac branch

Esophageal
plexus

Left vagus

Hepatic branch

Stomach

Colon

Fig. 13.8   Vagus nerve.

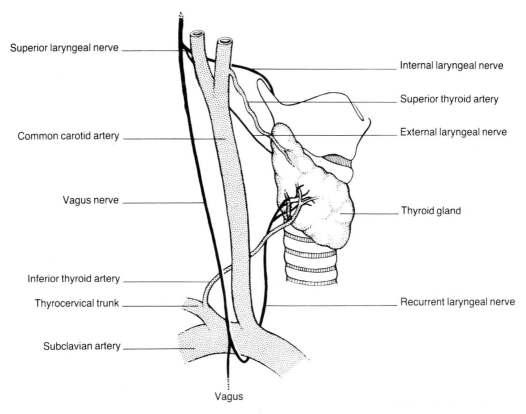

Superior laryngeal nerve

Common carotid artery

Vagus nerve

Inferior thyroid artery

Thyrocervical trunk

Subclavian artery

Internal laryngeal nerve

Superior thyroid artery

External laryngeal nerve

Thyroid gland

Recurrent laryngeal nerve

Vagus

Fig. 13.9 The laryngeal branches of the right vagus nerve.

has great difficulty in swallowing, a common and difficult clinical situation (bulbar palsy).

4 Recurrent laryngeal nerves. These are given off depending on the whims of the embryologic sixth aortic arch. On the right side the fifth and the anterior part of the sixth arches disappear, the recurrent nerve then coming into relationship with the fourth arch, which is going to form a portion of the adult subclavian artery; the right nerve, therefore, hooks around this artery. On the left side, the sixth aortic arch is more persistent and forms part of the pulmonary artery and the ligamentum arteriosum. The left nerve therefore passes below the ligamentum arteriosum before it turns cranially. Translated into adult anatomy, the right nerve hooks around the subclavian artery, and the left

nerve the ligamentum arteriosum (sixth arch). Both nerves make for the groove between the trachea and the esophagus and run cephalad in this position, passing in close relationship to the posterior aspect of the thyroid gland. At the level of the thyroid gland each nerve lies amongst the branches of the inferior thyroid artery, after which it enters the larynx deep to the lower border of the inferior constrictor. The nerves, fortunately, lie outside the sheath of the thyroid gland and, if the surgeon is wary and stays inside the thyroid capsule during operations on the thyroid gland, she will avoid damage to the nerve. The recurrent nerve is very important, supplying sensation to the larynx below the vocal fold, and motor power to all the muscles of the larynx except the cricothyroid.

5   Superior laryngeal nerve. This shares the nerve supply of the larynx with the recurrent branch. Arising high up in the neck, the superior laryngeal nerve passes deep to both carotids and divides into a large laryngeal branch, which pierces the thyrohyoid membrane to supply sensation to the larynx above the vocal cord, and a rather dainty external laryngeal branch, which runs down on the inferior constrictor to supply the cricothyroid muscle, the only laryngeal muscle not supplied by the recurrent nerve.

Accessory (XI) nerve

This nerve escapes from between the internal carotid artery and internal jugular vein by passing backward across the latter vessel to reach the junction of the upper and middle quarter of the sternomastoid muscle. Running obliquely through the muscle, it leaves it at its posterior border at the junction of its upper and middle thirds. The nerve then passes across the posterior triangle of the neck deep to the investing fascia, to enter the trapezius muscle in which it runs down

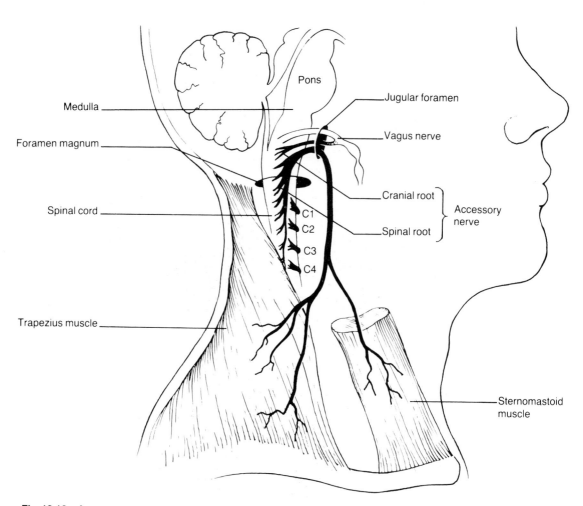

Fig. 13.10   Accessory nerve.

to as far as the 12th thoracic spine, supplying the muscle en route. The spinal part of the XIth nerve has a long journey from origin to destination. Roots from the upper five or six segments of the spinal cord join to form a spinal nerve, which runs cephalad posterior to the ligamentum denticulatum, and passes through the foramen magnum to meet, temporarily, the cranial division of the XIth nerve, the latter really a lower part of the vagus. The spinal part soon leaves the cranial part (which makes for the big ganglion nodosum of the vagus nerve) and passes through the jugular foramen in the direction described above. As it has a close relationship to the internal jugular vein, itself surrounded by large lymph nodes, and also because the nerve runs amongst lymph nodes which abound in the posterior triangle, the accessory nerve may be involved by pathologic lymph nodes; while removing these lymph nodes, it may be difficult to preserve the nerve. At one time, the nerve was sacrificed rather lightly, in the belief that the disability of the lost trapezius would not be too great. Older people, however, find it a disabling condition, and have great difficulty in reaching a shelf above head level. It might

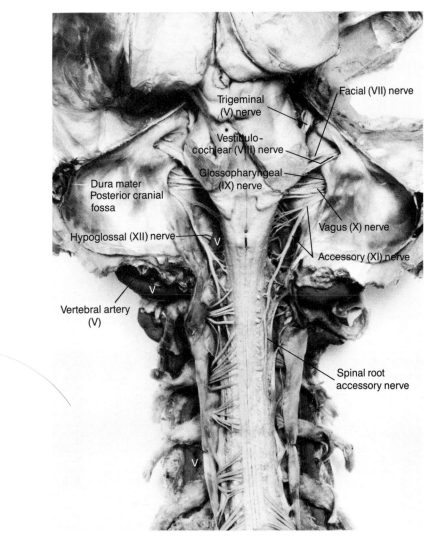

Fig. 13.11   Dissection of the cranial nerves in the posterior cranial fossa.

be pertinent to mention that the trapezius and sternomastoid have fairly large additional cervical nerves supplying them, the sternomastoid from C2 and the trapezius from C3 and 4. Some inquiring investigator often asks whether these branches are enough to supply motor power, should the accessory nerve be severed. No, they are purely proprioceptive sensory nerves; they help to tell you the position of your head in space. This is the reason for the origin of the accessory nerve from the upper cervical segments of the cord. A movement of the head relays its position via the sensory nerves; the motor fibers are then able to act reflexly and change that position if so desired.

Hypoglossal (XII) nerve

The XIIth nerve is a purely motor nerve which passes out of the cranial cavity through the anterior condylar canal, which lies medial to the jugular foramen. The nerve immediately comes into relationship with the internal carotid artery and jugular vein. Running caudally between the artery and vein in a slightly anterior position to the vagus nerve, it reaches the occipital branch of the external carotid. Here it hooks anteriorly around this branch to pass over the loop of the lingual artery and the hyoglossus muscle and disappears beneath the mylohyoid to supply the tongue. It supplies all muscles ending in glossus, except the palatoglossus. As it hooks around the occipital artery, it gives off a branch which intrigues students; it is called the superior root (descendens hypoglossi), which passes inferiorly on the carotid sheath to meet a branch that arises from the second and third cervical nerves, the inferior root (descendens cervicalis), to form a loop on the carotid sheath, the ansa cervicalis, which supplies

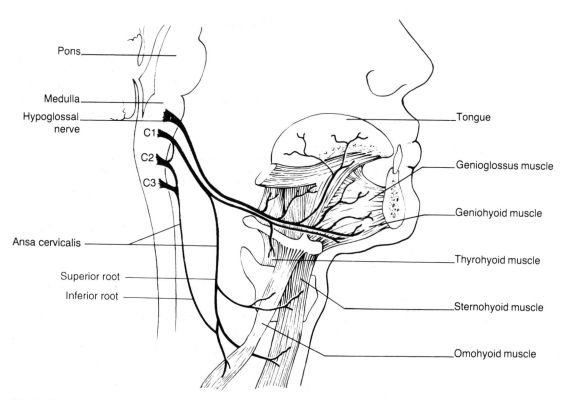

Fig. 13.12 Hypoglossal nerve.

the strap muscles. In actual fact, the descendens hypoglossi derives from a communication from the first cervical nerve to the hypoglossal; the strap muscles are therefore supplied by C1, 2 and 3 ventral rami.

Cervical sympathetic trunk

The cervical sympathetic trunk lies between the carotid sheath and the prevertebral fascia. It is continuous with the thoracic sympathetic trunk and, in fact, the fibers that it carries arise mostly from the first and second thoracic spinal segments. Unlike the fairly regular arrangement in the thorax and abdomen, the ganglia of the cervical trunk fuse as follows. The superior ganglion, the largest ganglion in the neck, represents the first four ganglia; the middle, the smallest of the three, represents ganglia five and six; and the inferior, intermediate in size, represents ganglia seven and eight. The inferior ganglion often joins the first thoracic ganglion to form the stellate ganglion.

Branches of ganglia

Each ganglion gives off similar branches to cervical nerves, the heart and arteries. The white rami, passing to the chain from the thoracic nerves, relay in the ganglia, and pass as gray rami to the eight cervical nerves, as they lie in close relationship to them; for example, the superior ganglion gives off gray rami to the first four nerves, the middle ganglion to nerves five and six, and the inferior ganglion to nerves seven and eight. The ganglia give off cardiac nerves, the accelerator nerves of the heart, and these branches are called superior, middle and inferior according to their origin. The ganglia give branches to the large arteries as they lie in close relationship to them: the superior to the carotids, the middle to the inferior thyroid, and the inferior to the subclavian and vertebral. The superior ganglion gives a pharyngeal branch, which joins the pharyngeal plexus. The chain ends above where the superior ganglion gives off a plexus of nerves which surround the internal carotid artery and pass into the skull as the internal carotid plexus.

Situation

The superior ganglion lies in front of the second and third cervical vertebrae, behind the carotid sheath, the middle lies in close relationship to the inferior thyroid artery on the sixth cervical vertebra, and the inferior

Table 13.1 Sympathetic ganglia of head and neck

	Superior	Middle	Inferior
Site of ganglion	C2/3, hyoid	C6, cricoid	C7, first rib
Nature	Fusiform, 3 cm long	Pinhead	50% separate; 50% fused = stellate
Preganglionic outflow	Upper thoracic segments but mainly from T1 outflow		
Gray rami	C1, 2, 3, 4	C5, 6	C7, 8
Vascular branches	Internal and external carotid	Inferior thyroid artery	Subclavian and vertebral artery
Cardiac branches	R to deep plexus; L to superficial plexus	All to deep cardiac plexus	All to deep cardiac plexus
Other branches	Pharyngeal plexus	Ansa subclavia to inferior ganglia	Ansa subclavia to middle ganglia
	Via carotid plexus to dilator pupillae	—	—

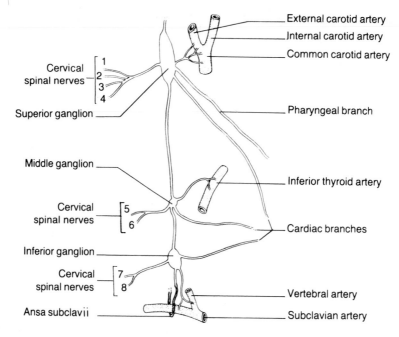

Fig. 13.13   The cervical sympathetic chain and its distribution.

ganglion lies on the neck of the first rib often as part of the stellate ganglion (found in 50 percent of normal individuals). This last ganglion may be involved in diseases which paralyse the cervical sympathetic system and cause Horner's syndrome.

Table 13.2   Parasympathetic ganglia of head and neck

Ganglion	Ciliary	Pterygopalatine	Submandibular	Otic
Nucleus of origin	Edinger–Westphal in midbrain	Superior salivatory in pons	Superior salivatory in pons	Inferior salivatory in medulla
Site of ganglion	Posterolateral to optic nerve in orbit	Pterygopalatine fossa inferior to maxillary nerve	On hyoglossus muscle inferior to lingual nerve	Inferior to foramen ovale outside skull
Preganglionic pathway	III nerve via branch to inferior oblique muscle	VII nerve via greater superficial petrosal and nerve of pterygoid canal	VII nerve via chorda tympani and lingual nerves	IX nerve via tympanic and lesser petrosal branches
Postganglionic pathway	Short ciliary nerves	Greater and lesser palatine nerves Sphenopalatine nerve Lacrimal nerve	Lingual nerve	Auriculotemporal nerve
Organ of destination	Ciliary body Circular muscles of pupil (sphincter)	Palate Nose Lacrimal gland	Sublingual submandibular glands	Parotid gland

14 Deep structures of the neck

Deep muscles of the neck · Root of the neck · Cervical nerves · Thoracic duct

Deep muscles of the neck

The details of many of the deep muscles of the neck are relatively unimportant, so let us concentrate only on those of clinical significance.

It is worth remembering that the transverse processes of all the cervical vertebrae, but especially 3–7, have deep grooves for the large anterior rami of the cervical nerves. These grooves separate the lateral extremities of the transverse processes into anterior and posterior tubercles. When you mention tubercles to muscles, they immediately make

a grab for them, as epitomized by the scalene muscles.

The scalene muscles are anterior, medius and posterior. The scalenus anterior arises from the large anterior tubercles of the third to sixth cervical vertebrae, and runs inferiorly to a tubercle on the inner border of the flat first rib, the scalene tubercle. This gives the muscle a triangular shape (scalene). It has many closely related important structures. The scalenus medius comes from most of the posterior tubercles, and runs to an impression of the outer aspect of the first rib, posterior to the scalene tubercle. It should be clear that,

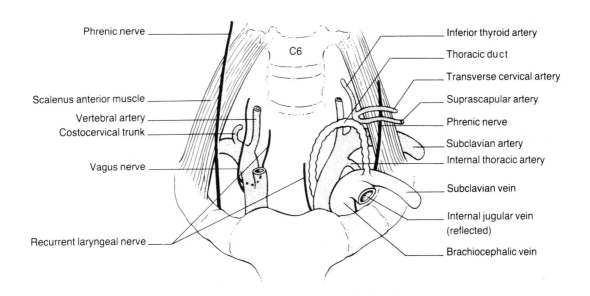

Fig. 14.1   The root of the neck.

Phrenic nerve

Scalenus anterior muscle

Vertebral artery

Costocervical trunk

Vagus nerve

Recurrent laryngeal nerve

C6

Inferior thyroid artery

Thoracic duct

Transverse cervical artery

Suprascapular artery

Phrenic nerve

Subclavian artery

Internal thoracic artery

Subclavian vein

Internal jugular vein (reflected)

Brachiocephalic vein

as the large anterior rami of the lower cervical nerves emerge from their grooves, they lie between the scalenus anterior and medius. The scalenus posterior comes from the posterior tubercles of the cervical vertebrae and runs inferiorly to insert into the second rib.

The levator scapulae arises from all the posterior tubercles and runs downward and backward to reach the medial border of the supraspinous portion of the scapula (see Fig. 3.20). The longus capitis muscle arises from the same cervical tubercles as the scalenus anterior, but runs cephalad. The longus cervicis lies in front of the vertebral bodies and runs from the lower to the upper cervical vertebrae. As the prevertebral fascia runs anterior to the cervical vertebrae, it will cover all the above muscles.

Root of the neck

Suprapleural membrane (Sibson's fascia)

Projecting up from the thorax into the root of the neck is the apex of the lung covered by its pleura. This last is further covered by a suprapleural membrane, which runs from the inner border of the first rib to the transverse process of the seventh cervical vertebra. This fascia varies in thickness; its function is to prevent the inflated lung with its pleura from rising high into the neck. You will see some people with thin fascia who, after exertion, have the apices of the lung protruding up and down in the root of their neck, a habit considered undignified in Victorian drawing rooms. Running across Sibson's fascia on their way to the upper limb are the subclavian vessels.

Subclavian artery

The subclavian artery passes into the neck behind the sternoclavicular joint, indents Sibson's fascia and the underlying pleura and lung, passes laterally between the scalenus anterior and the scalenus medius,

and finally crosses the first rib, at the outer border of which it changes its name to the axillary artery. Remembering VAD (veins are anterior to the artery), the subclavian vein passes across Sibson's fascia, in front of the scalenus anterior, and, after crossing the first rib, becomes the axillary vein.

Branches of the subclavian artery

Anatomists of bygone years were the most important people in the medical school and held as much power as the Internal Revenue Service of today. They spent a lot of time manufacturing triangles and quadrangles to add elegance to the subject. So it is that the subclavian artery is divided into three, the part before (I), behind (II) and beyond (III) the scalenus anterior. The branches arise mainly from the first two parts, and they are very significant clinically.

1 The first and largest branch is the *vertebral artery*, which runs cranially to enter the foramen transversarium of the sixth cervical vertebra. That makes you wonder about the purpose of the foramen transversarium of the seventh cervical vertebra. From the foramen transversarium of the sixth cervical vertebra, the artery runs cephalad through the foramina to reach the foramen in the axis. It then makes a wide curve medially to lie in the suboccipital triangle, which is formed by three small deep muscles (Fig. 7.12), to reach the foramen transversarium of the atlas. You need this wide curve, otherwise every time you turned your head sideways you would kink the artery and go gaga. Having left the foramen of the atlas, it then passes medially lying on a broad groove on the posterior arch of the atlas in close relationship to the branches of the first cervical nerve; finally it runs upward through the foramen magnum to reach the brain stem, where it meets the vertebral artery of the opposite side (see Fig. 7.10). A fascinating syndrome (the subclavian steal syndrome) has been described where the

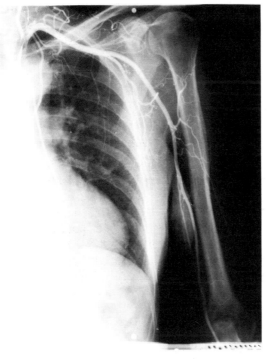

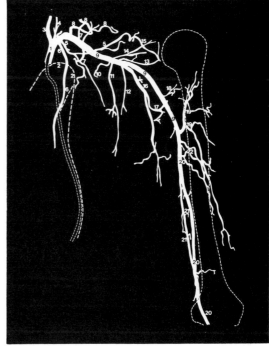

Fig. 14.2 Subclavian – axillary arteriogram. (From Weir and Abrahams 1978)

1 Catheter in left subclavian artery.
2 Aortic knuckle or knob.
3 Vertebral artery.
4 Inferior thyroid artery.
5 Suprascapular artery.
6 Internal thoracic artery.
7 Dorsal scapular artery.
8 Transverse cervical artery (superficial cervical).
9 Axillary artery.
10 Superior thoracic artery.
11 Lateral thoracic artery.
12 Pectoral branch, thoracoacromial artery.

13 Acromial branch, thoracoacromial artery.
14 Deltoid branch, thoracoacromial artery.
15 Anastomoses between suprascapular and acromial arteries.
16 Subscapular artery.
17 Circumflex scapular artery.
18 Anterior circumflex humeral artery.
19 Posterior circumflex humeral artery.
20 Brachial artery.
21 Profunda brachii artery.
22 Deltoid branch, profunda brachii artery.

subclavian artery becomes narrowed or occluded at its origin. To compensate for this, the opposite vertebral artery fills the poorly supplied subclavian artery by diverting the blood away from the basilar artery (which is formed by the junction of the two vertebral arteries) down the vertebral artery to fill the affected subclavian artery from above. This leads to a reduction in the amount of blood reaching the medullary region of the brain stem: the syndrome of ischemia of an upper limb, with symptoms of medullary insufficiency such as giddiness, makes

the diagnosis and leads to successful treatment. As aspiring surgeons may read these pages, it is important for them to know the inferior cervical or stellate ganglion is situated immediately behind the origin of the vertebral artery, a useful landmark to remember when searching for this elusive structure.

2 Running inferiorly is the large *internal thoracic artery*. Passing behind the clavicle, it lies posterior to the costal cartilages about 1.25 cm (0.5 in) from the lateral margin of the sternum. Reaching the sixth costal cartilage, it divides into

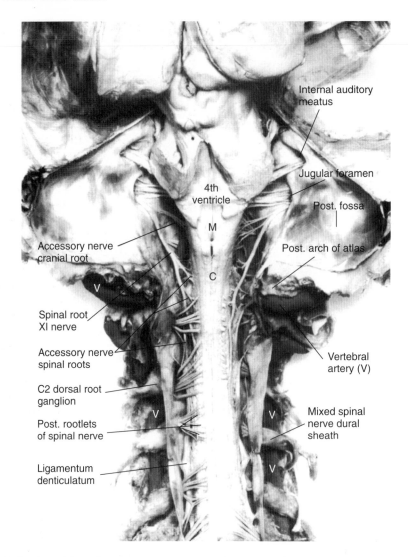

Fig. 14.3   Posterior view of deep dissection of head and neck. The vertebral artery is seen running lateromedially over the posterior arch of the atlas. The spinal accessory nerve is running superiorly between the anterior and posterior rootlets of the cervical nerves, into the posterior fossa. V, vertebral artery; M, medulla; C, spinal cord.

its two terminal branches – the musculo-phrenic, supplying the diaphragm and adjacent muscles, and the superior epigastric, which enters the rectus sheath to anastomose with the inferior epigastric artery. The internal thoracic artery gives off to each intercostal space two branches that anastomose with similar branches from the posterior aortic intercostals. It also gives off a quaint artery, the pericardiacophrenic, a small branch running with and supplying the phrenic nerve. Other branches supply the thymus gland, pleura and the pericardium. When the aorta is congenitally narrowed beyond the origin of the subclavian artery (coarctation), the blood reaches the aorta beyond this constriction partly via the anastomosis between the intercostal branches of the internal thoracic and the aorta. With this reversed flow, the internal thoracic artery and its intercostal branches become dilated; the intercostal branches may excavate the costal grooves, which is easily recognizable radiologically. The internal thoracic artery is often used in coronary artery occlusive disease when it is implanted into the vessel beyond the narrowed segment.

3 Near the medial border of the scalenus anterior is the origin of the thyrocervical trunk, with its three radiating branches. The inferior thyroid artery runs superiorly along the inner border of the scalenus anterior, curves medially at the level of C6 posterior to the carotid sheath, and, lying in close relationship to the recurrent laryngeal nerve, enters the thyroid gland to supply the bulk of its glandular structure. It also supplies the parathyroid glands, and surgeons have become rather wary of ligating this vessel lest they induce hypofunction of these glands. The two other branches of the thyrocervical trunk, the transverse cervical and the suprascapular arteries, run laterally amongst the branches of brachial plexus, to supply the muscles in the scapular region. The importance of these vessels is that they take part in the scapular anastomosis which may come into play in coarctation of the aorta.

4 Lastly, the costocervical branch is a small trunk which divides into two branches. The superior intercostal artery arches over the apex of the pleura to supply the first two intercostal spaces, where it anastomoses with the intercostal branches of the aorta. The deep cervical artery runs up to supply the deeper neck muscles. Both branches are of importance when anastomoses are needed because of vascular occlusion.

Subclavian vein

The subclavian vein receives most, but not all, of the analogous branches of the subclavian artery. Those veins not draining into the subclavian vein empty into its entering tributary, the external jugular, or the continuation of the subclavian vein, the brachiocephalic vein.

Cervical nerves

Brachial plexus

Nerves C1–4 form the cervical plexus, and nerves C5–8 form the brachial plexus with some help from C4 and a large contribution from T1. Reference to a diagram will show the configuration better than a description (see Fig. 14.4). In brief, C5 and 6 join to form the upper trunk, C7 goes it alone as the middle trunk, and C8 and T1 form the lower trunk. As mentioned above, the nerves lie sandwiched between the two scalenes, and, after further subdivisions into anterior and posterior divisions behind the clavicle, they regroup to form cords in the axilla. The plexus lies above and lateral to the subclavian vessels, approaching them as they pass inferiorly; in fact, the lower trunk lies between the artery and the rib. Purists would have it that it is the lower trunk that causes the subclavian groove

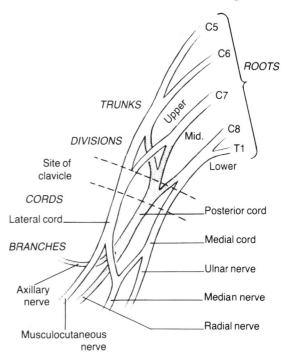

Fig. 14.4 The brachial plexus.

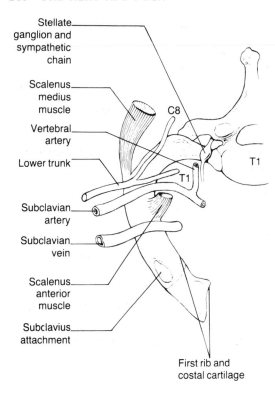

Stellate ganglion and sympathetic chain

Scalenus medius muscle

Vertebral artery

Lower trunk

C8

T1

T1

Subclavian artery

Subclavian vein

Scalenus anterior muscle

Subclavius attachment

First rib and costal cartilage

Fig. 14.5   The relations of the first rib.

on the first rib, and not the artery; it is certainly true that few arteries actually make impressions on bones.

The thoracic outlet syndrome (scalenus anterior syndrome, costoclavicular syndrome) results from the subclavian vein, artery and plexus passing through a narrow gap between the clavicle and the first rib on their way from the neck to the axilla (Fig. 14.6). Many abnormalities, bony, muscular or neurogenic may occur which lead to compression of any or all of these three components. It is likely that the function of the subclavius muscle is to act as a cushion to prevent this compression. By bracing back or pulling down on the arm, it is sometimes possible to obliterate the radial pulse, which indicates the tightness of this channel. Surgeons are always eager to enlarge this channel and often do so by removing either the scalenus anterior, or first rib, or any unwelcome intruders such as an extra cervical rib – a prolongation of the anterior tubercle of C7 (see Fig. 7.15).

Cervical branches of the brachial plexus

These are all destined for muscle, and all begin with 'S'. They supply the subclavius, the scapular muscles via the suprascapular nerve, and the serratus anterior. This last nerve, also referred to as the long thoracic nerve (Bell), arises from C5, but also receives additional branches from C6 and 7. The scapular muscles that have been mentioned were muscles related to the dorsal and medial borders of the scapula, namely the supraspinatus and infraspinatus. The rhomboids incidentally, receive their supply from the brachial plexus but by their own branch from C5.

Cervical plexus

This plexus is made up of the first four cervical nerves, the fourth one being a little undecided and splitting into two to participate in both the cervical and brachial plexuses. The nerves emerge from the intervertebral canals and join each other by looping branches. The plexus lies mostly on the scalenus medius. Its branches are to the muscles in the immediate surroundings, and to skin (Fig. 12.5). Its most important branch is the phrenic nerve which comes from C3, 4 and 5 but mainly from 4. This nerve runs a characteristic course, curving around the lateral edge of the scalenus anterior to reach its anterior surface on which it runs obliquely, covered by the prevertebral fascia. It passes over the lower part of the medial border of the muscle to run over the apex of the pleura into the chest. Remember that the phrenic nerve is a mixed nerve carrying sensory as well as motor fibers.

Thoracic duct

This duct arises in the abdomen, passes through the thorax and enters the neck on the left side of the esophagus, after which it makes an outward curve at the level of C7,

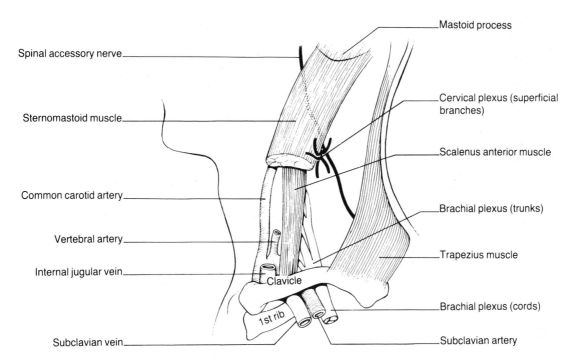

Fig. 14.6   The posterior triangle of the neck with its major contents.

between the carotid and vertebral vessels, to enter the junction between the subclavian and internal jugular veins (see Fig. 14.1). It carries the lymph from the whole body, except for the right chest, arm, right head and neck and lower half of the right lung. It is also the route along which malignant cells may pass from the rest of the body to the venous system; sometimes they pass from the cervical part of the duct to the neighboring supraclavicular nodes. Enlargement of the latter nodes is common; they were described a century ago, and are called the Virchow–Troisier nodes in honor of the physicians who first described them.

15 Upper respiratory and alimentary passages

Nose·Paranasal air sinuses·Pharynx·Palate·Larynx

Nose

This organ comes in many shapes and sizes and has many functions. If large enough, it will enable you to have one side of your face shaded from the sun. Its more important functions, however, are to warm, humidify and filter the inspired air. It is cleverly constructed for this purpose; it is lined by tall, columnar, ciliated cells, interspersed with mucus-secreting goblet cells, beneath which lies highly vascular tissue with large amounts of lymphoid aggregates and many mucous and serous glands. The air is heated by the vascular plexus, which raises the temperature of the air to that of the body by the time it reaches the nasopharynx; it is humidified by the secretion of the glands which produces 100 percent saturation; and it is filtered by the secreted mucus, which is

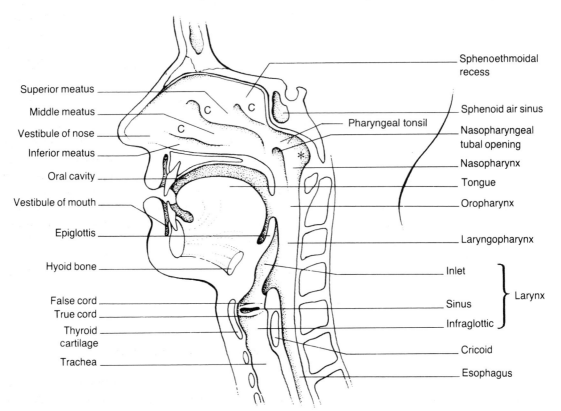

Fig. 15.1 **Sagittal section showing the nose, oral cavity, larynx and pharynx. C, conchae (turbinates); *, pharyngeal recess.**

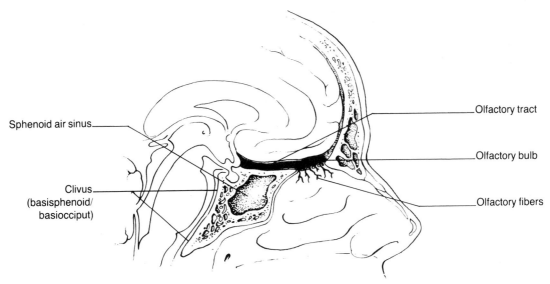

Fig. 15.2   The olfactory nerve.

swept toward the pharynx by the cilia and then swallowed. More objectionable material is removed by the sneeze reflex.

The cavity of the nose is divided into two by a septum, and each nasal cavity has an external and an internal opening. The nasal skeleton itself is composed of bone and cartilage, the cartilaginous portions enabling you to change the caliber and, less importantly, the direction of your nose – it is easier to move your head away from some objectionable smell than to swivel your nose sharply to one side.

The septum lies in the midline theoretically, but by adulthood only about a quarter maintain this position. Deviation is very common, probably the result of the traumas of one's childhood. The entrance to each nostril is slightly dilated to form a vestibule, which is lined by squamous epithelium and guarded by short, stiff hairs. The cavity of the nose is bounded by a medial wall, the nasal septum, a roof, a floor which also forms the roof of the mouth, and a lateral wall which is noteworthy for its three projections – the conchae or turbinate bones, with spaces or meati between them. The upper two conchae are processes of the ethmoid bone, while the inferior one is an independent bone. The nasal sinuses open into the meati, mostly the middle meatus, and the tear (nasolacrimal) duct empties into the anterior part of the inferior meatus. The columnar, ciliated, mucous membrane is very thick over the septum and conchae. This explains why slight swelling of the mucous membrane easily produces stuffiness and nasal obstruction, particularly if the septum is deviated. The superior part of the nasal cavity is lined by a special olfactory epithelium from which arise nerve fibers that carry the sensation of smell to the olfactory tracts via the cribriform plates. The nerve supply of the nasal cavity is by the first and second branches of the fifth nerve. The arterial blood supply is from branches of the maxillary artery posteriorly and the facial artery anteriorly. One site of anastomosis between these arteries in the region of the vestibule is a frequent site for nosebleeds (epistaxis). The rich venous submucosal plexus drains by veins that accompany the arteries.

Paranasal air sinuses

Bilateral outpouchings (diverticula) of the nasal mucous membrane project into the surrounding bones, the sphenoid, ethmoid, frontal and maxillae (see also Chapter 10).

The sphenoidal sinus occupies the body of the sphenoid bone. Each sinus drains into the sphenoethmoidal recess, which lies above the superior concha.

The ethmoidal sinus occupies the lateral mass of the ethmoid and consists of anterior, middle and posterior air cells according to position. Since the lateral mass forms part of the medial wall of the orbit, this is the site of pain and tenderness in ethmoidal sinusitis (inflammation of the sinus). Most of the air cells drain into the middle meatus.

The frontal sinus is situated deep to and forms the superciliary ridges of the frontal bone; it might be thought of as a very anterior ethmoidal air cell extending into the frontal bone. Its duct, the frontonasal, drains downward into the middle meatus. Frontal sinusitis is relatively easily diagnosed; a frontal headache with tenderness over the sinus and pus in the middle meatus is diagnostic.

A maxillary sinus occupies each maxilla. Its opening into the middle meatus is situated for some inexplicable reason in the upper part of the sinus near its roof, making it an antigravity drainage system; purulent collections usually require drainage by an opening in the lower part of the sinus—or by the patient standing on his head! The roots of some of the upper teeth, especially the bicuspids and molars, project into the floor of the antrum.

Pharynx

Structure

Air which has been inspired through the nose and mouth, and food from the mouth, now reach a common cavity, the pharynx. Air and food have to be directed along their respective passages. The pharynx should be thought of as a fairly capacious tube, which opens anteriorly, from above downward, into the nasal cavities (the nasopharynx), into the mouth (the oropharynx), and finally into the larynx (the laryngopharynx). The tube is

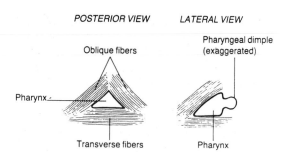

POSTERIOR VIEW   LATERAL VIEW

Oblique fibers

Pharyngeal dimple (exaggerated)

Pharynx

Transverse fibers   Pharynx

Fig. 15.3   The arrangement of the fibers of the inferior constrictor muscle.

composed mainly of a sheet of muscles enclosed by fibrous tissue externally and internally, and lined by squamous mucous membrane, except in the nasopharynx, where it is 'respiratory' – i.e. ciliated and liberally supplied with submucous glands. The muscles are called the constrictors and they run from the base of the skull to the level of C6, where the pharyngeal tube becomes the esophagus (Fig. 15.4). The three constrictors are tradionally likened to three

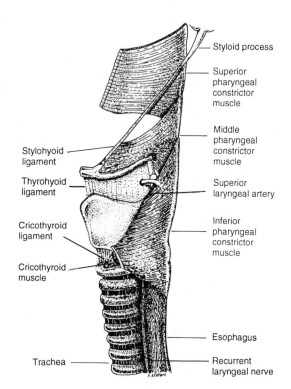

Styloid process

Superior pharyngeal constrictor muscle

Middle pharyngeal constrictor muscle

Superior laryngeal artery

Inferior pharyngeal constrictor muscle

Stylohyoid ligament

Thyrohyoid ligament

Cricothyroid ligament

Cricothyroid muscle

Esophagus

Trachea

Recurrent laryngeal nerve

Fig. 15.4   The pharyngeal constrictors (lateral view).

flower pots fitted into each other; perhaps three dixie cups or beer glasses would be more appropriate in the twentieth century. The superior arises from the skull and mandible, the middle from the hyoid bone, and the inferior from the upper two laryngeal cartilages. The fibers widen as they pass backward to insert into a common raphe, which runs from the pharyngeal tubercle of the occipital bone inferiorly to the origin of the esophagus. The fibers of the inferior constrictor, arising from the thyroid and cricoid cartilages, run in different directions, the former obliquely and the latter transversely. The latter circular fibers act as the sphincter and enclose a passage which is said to be the narrowest part of the gastrointestinal tract; anything that can pass through this canal should be able to reach the anal canal. Strong propulsive efforts tend to evaginate a dimple of mucous membrane through the weakest part of the pharyngeal wall, the area between the transverse and the oblique fibers of the inferior constrictor, often called the dehiscence of Killian. Should the transverse fibers of the muscle fail to relax in the face of an advancing peristaltic wave, the pharyngeal dimple may be forced through the dehiscence and enlarge to form a pouch or diverticulum (a good place to hide things from the customs officer). This pharyngeal pouch, however, can be so symptomatic as to require surgical removal. The constrictor muscles are supplied by the pharyngeal nerve plexus, which consists mainly of the fibers of the accessory nerve carried in the pharyngeal branch of the vagus, with some additional fibers from the superior cervical ganglion and the external and recurrent branches of the vagus; one way or another, it is the tenth nerve. Diseases of the brain stem involving the origin of the tenth and eleventh nerves lead to paresis of the constrictor muscles, resulting in severe swallowing difficulties (bulbar palsy).

Nasopharynx

This division of the pharynx lies between the base of the skull and the soft palate. The pharyngeal opening of the pharyngotympanic (eustachian) tube lies about 1 cm (0.4 in) behind and just below the posterior end of the inferior nasal concha. Posterosuperior to the opening is the tubal elevation (torus), caused by the shape of the underlying tubal cartilage and a collection of lymphoid tissue (the latter collection is sometimes called the tubal tonsil). Immediately posterior to the torus is a slight depression, the pharyngeal recess.

Until recently, physicians had to find the eustachian tube opening blindly. Much to the discomfort of the patient, a curved catheter was passed along the floor of the nose beyond the inferior concha and into the pharyngeal recess. It was then brought forward over the tubal elevation to enter the eustachian orifice. A nervous operator, not confident of the exact position, often used to make his own opening. Modern endoscopy has produced an instrument which visualizes the tubal opening and allows easy catheterization. The lymphoid tissue in relation to the torus may swell as a result of an upper respiratory tract infection and block the tube; this is the cause of deafness in a bad cold. Just behind the pharyngeal recess, on the roof and upper posterior wall of the nasopharynx, is a collection of lymphoid tissue, the nasopharyngeal tonsil or adenoid. Prominent in the child, this atrophies at puberty, and disappears in the aged (like most lymphoid tissue). The space between the pharynx (air) and the base of the skull (bone) can be seen on x-ray; increase of this retropharyngeal space suggests the diagnosis of an enlarged nasopharyngeal tonsil.

Oropharynx

The oropharynx, which is situated behind the mouth, lies between the soft palate and the tip of the epiglottis. Anteriorly the oropharynx is separated from the oral cavity by a pair of arches, the palatoglossal and palatopharyngeal arches. Each arch consists of a fold of mucous membrane raised by underlying muscles of the same name. The soft palate above and the dorsum of the tongue below complete the isthmus through which

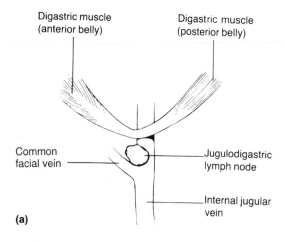

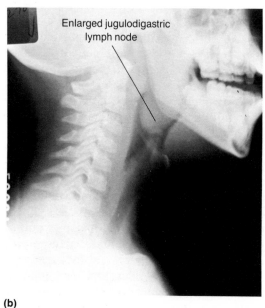

(b)

Fig. 15.5 Tonsillar lymph node: (a) diagrammatic; (b) the real thing.

food passes from the oral cavity to the oropharynx.

Palatine tonsils

On each side, between the two arches, lies a collection of lymphoid tissue called the palatine tonsil, or, in bar room parlance, *the* tonsil. A close look at the tonsil will show a number of tonsillar pits, and in its upper part even a well-marked cleft. Lying outside the

tonsil is the superior constrictor muscle, which is pierced by several arteries supplying the tonsil, the principal one being the tonsillar branch of the facial artery. The veins, likewise, pass through the pharyngeal wall to the pharyngeal venous plexus. Bleeding from these vessels is responsible for troublesome post-tonsillectomy hemorrhage. The tonsil drains to the upper deep cervical nodes, especially the tonsillar (jugulodigastric) node at the junction of the common facial vein and internal jugular vein. Such an enlarged node following a sore throat is the most common swelling in the neck. The supply to the tonsil by the fifth and ninth nerves is so diffuse that removal of tonsils under local anesthesia is performed by local infiltration rather than by blocking the main nerves (regional block). The western world is populated at the moment by people who have had their tonsils removed, but we suspect that the next generation is going to be populated by people bearing a pair of large tonsils.

We have described a lot of lymphoid tissue in this area. Waldeyer first noted this and his ring of tissue comprises the lingual tonsil, the two palatine tonsils, the adenoidal tissue (pharyngeal tonsil) and the tubal tonsil.

Laryngopharynx

Below the tip of the epiglottis lies the laryngopharynx, whose most striking feature is the posteriorly facing laryngeal inlet. The posterior bulging of the larynx leaves a recess on each side called the piriform fossa (Figs. 15.6 and 15.8). This space is worth remembering, as foreign bodies (e.g. fish bones) tend to stick there and can be difficult to find; it is also a favorite hiding place for tumors.

Besides the three main constrictor muscles which constitute the bulk of the muscular pharynx, the laryngopharynx also possesses three longitudinal muscles. All of these are inserted into the thyroid cartilage: the palatopharyngeus arising from the hard palate; the salpingopharyngeus arising from the eustachian tube and raising a fold as it runs down from the naso- to the oropharynx; and the stylopharyngeus, arising from the styloid process outside the constrictors, and passing between the upper two constrictors to join the other two muscles. These three longitudinal muscles are elevators of the laryngopharynx.

Palate

The palate consists of hard and soft parts that separate the mouth from the nose.

Hard palate (see Fig. 10.17)

This is made of the two maxillae and the palatine bones, with the former predominating. The mucous membrane, especially in the anterior part of the hard palate, is strongly adherent and cannot be easily separated from the periosteum of the underlying bone, a point of great importance in the operation of cleft palate. As one passes backward and laterally, however, mucous glands appear that separate the periosteum and the mucous membrane. Numerous small salivary glands, a well-known site for the development of salivary tumors, are also present here. The palate is supplied by the greater palatine vessels from the maxillary artery and by the greater and lesser palatine branches of the fifth nerve (V^2).

Soft palate (see Fig. 13.7)

The bulk of this structure consists of numerous mucous and serous glands. Hanging from the posterior edge of the hard palate like a curtain, the free border of the soft palate presents in the midline, the uvula or 'little tongue'. The main functional component of the soft palate is a stiff palatine aponeurosis which attaches to the posterior part of the hard palate, and which is able to change both its shape and position. The aponeurosis is the flattened out tendon of the tensor palati muscle, which arises from the upper end of the medial pterygoid plate on the lateral side of the cartilaginous part of the eustachian tube. The muscle runs inferiorly to its pulley, the hamulus, around which its tendon turns at right angles to enter the pharynx. Broadening into a triangular aponeurosis, it attaches by its anterior border to the crest of the palatine bone, where it meets the aponeurosis of the muscle of the opposite side. This aponeurosis is concave toward the mouth, but, when tensed by muscular contraction, it becomes somewhat flattened; other muscles then act on it, either elevating or depressing it. Because of the close relationship of the muscle to the cartilage of the eustachian tube, contraction opens the tube, allowing air to pass between the nose and the middle ear. The levator palati arises from the apex of the petrous temporal bone and from the medial side of the eustachian tube under the mucous membrane of the pharynx, and inserts into the upper surface of the soft palate, where it meets its fellow of the opposite side. While its main action is to elevate the soft palate, it also opens the eustachian tube.

These two muscles are responsible for abolishing the deafness which occurs on the take-off of an aircraft. When the flight attendant offers you a jelly-bean and you swallow it, the auditory tube opens, allowing air to pass from the middle ear to the nose and equalize the air pressure. The palatoglossus arises from the palatine aponeurosis and descends in the palatoglossal arch to blend with the side of the tongue. The palatopharyngeus arises from the posterior bony hard palate and the palatine aponeurosis, and descends as the palatopharyngeal arch to insert into the posterior border of the thyroid cartilage. The palatoglossus and pharyngeal muscles of each side approximate each other when they act, helping to close off the oral cavity from the oropharynx. The musculus uvulae is an apparently insignificant muscle arising from the posterior nasal spine of the palatine bone and inserting into the uvula. Though very small, paralysis of its nerve supply demonstrates an important and easily seen diagnostic sign, the displacement of the uvula toward the opposite side.

Table 15.1 Chewing and swallowing

Action	Muscles	Nerve supply
Biting off food with incisors	Masseter, temporalis, medial ptyerygoid	V³
Tearing with canines	Temporalis	
Grinding movements:		
1 Elevation	Masseter, temporalis, medial pterygoid	V³
2 Depression	Digastric, mylohyoid, geniohyoid	V³, V³ C1 (ansa cervicalis)
3 Forward	Lateral pterygoid	V³
4 Backward	Temporalis (posterior fibres)	V³
5 Molar movements	Alternate 1, 2, 3 and 4, one side to other	V³
Storing, filling and emptying vestibule	Buccinator	VII
Salivation from:		
Parotid gland		IX
Submandibular gland		VII
Sublingual gland		VII
Tip of tongue pushes bolus posteriorly	Genioglossus; superior longitudinal and transverse intrinsic muscles	XII
Elevate tongue	Styloglossus, palatoglossus	XII, pharyngeal plexus
Tense soft palate against	Tensor palati	V³
Passavant's ridge	Stylopharyngeus	IX
Hyoid moves anteriorly to open pharynx	Geniohyoid	C1 (ansa cervicalis)
Larynx raised	Thyrohyoid	C1 (ansa cervicalis)
Epiglottis tipped down	Stylopharyngeus	IX
'Stripping' wave on posterior pharyngeal wall	Palatopharyngeus	Pharyngeal plexus
Oropharynx closed off	Superior constrictor	
Larynx closed	Aryepiglottis	Recurrent laryngeal
Hyoid pulled back	Infrahyoids	Ansa cervicalis (C1, 2, 3)
Bolus pushed down by superoinferior wave	Constrictors (three) especially cricopharyngeus	Pharyngeal plexus

Nerve supply of the pharynx and palate

The nasopharynx should be thought of as fifth nerve territory, especially the second and third divisions. The nerve supply is mostly via the pharyngeal branch of the maxillary nerve, and the blood supply is a corresponding vessel from the maxillary artery. The oropharynx is mostly served by the ninth nerve, and the laryngopharynx is mostly tenth nerve territory. With regard to the motor supply, the tensor palati is innervated by a branch of the fifth nerve via the otic ganglion. All other palatine muscles are supplied by the pharyngeal plexus; in reality, these fibers come from the cranial accessory which enters the vagus and is distributed by its pharyngeal branch.

The palatine muscles are highly important in both calling out your order for a hamburger and then swallowing it; i.e. phonation and deglutition. The nasopharynx has to be sealed off in both these acts and this is accomplished by elevating the soft palate against the superior constrictor. Failure to accomplish this as a result of paralysis is evidenced by nasal speech and by the regurgitation of food through the nose. Beware laughing and drinking at the same time!

Larynx

Designed primarily as a sphincter for shutting off the air passages from the pharynx, the larynx has been modified to function as the organ of phonation.

Cartilaginous skeleton (Fig. 15.7)

Cricoid cartilage

The cricoid cartilage is the only completely circular laryngeal cartilage and it is traditionally described as being shaped like a signet ring. The posterior arch is much higher than the anterior arch and it bears a pair of arytenoid cartilages.

Arytenoids

The arytenoids are two small pyramidal cartilages, with an anterior projection for the attachment of the vocal cord (vocal process), a lateral attachment for the insertion of muscles (muscular process), and a superior process which articulates with a small corniculate cartilage. The arytenoids articulate with the sloping shoulders of the cricoid cartilage by synovial joints. Gliding down these shoulders, the space between the arytenoids is widened to a 'V'; outward rotation of the arytenoids results in a diamond-shaped opening. In man, most movement at the arytenoid–cricoid articulation is gliding rather than rotation.

Thyroid cartilage

This cartilage is large and easily palpable, especially in the male – hence Adam's apple. It consists of two laminae that meet in the midline at an angle which is more acute in the male (laryngeal prominence). The free posterior borders end in superiorly and inferiorly projecting cornua that attach to the hyoid bone and cricoid cartilage respectively.

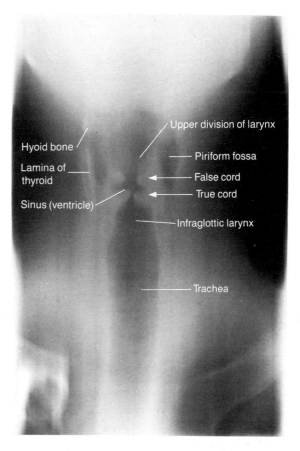

Fig. 15.6 labels: Hyoid bone / Lamina of thyroid / Sinus (ventricle) / Upper division of larynx / Piriform fossa / False cord / True cord / Infraglottic larynx / Trachea

Fig. 15.6 Tomogram of the larynx and part of the pharynx outlined by air. The air outlines the upper division of the larynx (vestibule), which reaches inferiorly to a narrowing produced by the false vocal (vestibular) fold/cord (upper arrow). The lower arrow points to the indentation produced by the true vocal cord. Between the two cords, the air passes laterally to outline the ventricle (sinus) of the larynx. Below the true cords, the air outlines the lowest division of the larynx and its continuation, the trachea. The posterior aspect of the thyroid cartilages is indistinct, the left side being more apparent. To the inner aspect of each cartilage, the air outlines the piriform fossa.

Epiglottis

The epiglottis is a pear-shaped leaf of elastic cartilage, broad above where it lies immediately behind the tongue and narrow below where it attaches to the back of the thyroid cartilage.

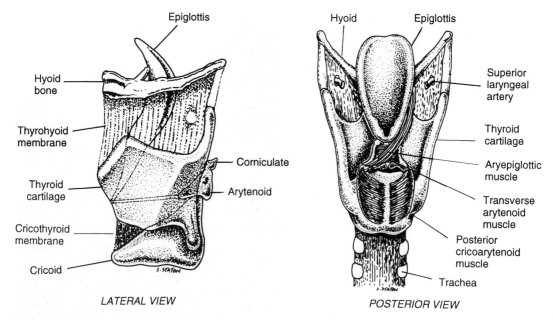

Fig. 15.7 Lateral view of the laryngeal cartilages and posterior view of the laryngeal cartilages and intrinsic muscles. For purposes of clarity, only one aryepiglottic muscle is shown.

Corniculate cartilages

This pair of small conical nodules articulates with the arytenoid cartilages and serves to prolong them.

All cartilages are of the hyaline variety, except for the epiglottis and corniculate cartilages, which are yellow elastic. *Pari passu* hyaline cartilage calcifies, but elastic cartilage never does. An x-ray of the aging neck shows this calcification. The rigidity imparted by this calcification is another reason why singers turn into frogs in their old age.

The whole laryngeal skeleton lies opposite vertebrae C3–6 in the adult; in the child it lies opposite C3–4. The higher position in the child means that you can reach the laryngeal inlet with your finger; one day you may remember this with undying gratitude as you extricate a marble blocking the laryngeal inlet and turn a blue child into a happier color.

Ligaments

These are fibrous membranes which join cartilages, but two deserve special mention.

Cricovocal (cricothyroid) ligament

This ligament stretches superiorly from the arch of the cricoid cartilage. Its free upper border attaches anteriorly to the posterior aspect of the middle of the thyroid cartilage, and posteriorly to the vocal process of the arytenoid; this free border constitutes the true vocal cord (ligament).

Quadrangular membrane

This membrane attaches inferiorly by a free border to the vocal process posteriorly and to the thyroid cartilage anteriorly, immediately above the attachment of the vocal cord. Passing upward it attaches to the corniculate cartilage posteriorly and to the epiglottis anteriorly; anteriorly it therefore is higher than posteriorly. The upper free border is the aryepiglottic fold, and the lower is the vestibular fold or false vocal cord.

Thyrohyoid membrane

The thyrohyoid membrane is a fibrous membrane joining the upper border of the thyroid cartilage to the upper border of the hyoid bone, which it reaches by passing posterior to the bone. There is a small bursa between the membrane and the hyoid bone, the subhyoid bursa, which facilitates laryngeal movement. It sometimes becomes enlarged. The membrane is pierced by the superior laryngeal artery and internal laryngeal nerve.

Muscles

The muscles which move the cartilages are intrinsic and extrinsic.

Intrinsic muscles

CRICOTHYROID

This muscle runs backward from the anterolateral outer surface of the cricoid to the inferior horn (cornu) of the thyroid lamina. Contraction tilts the thyroid cartilage forward or, if the latter is fixed, the cricoid is tilted backward. In either event the cartilages are separated and the vocal cord is lengthened.

THYROARYTENOID

This, the opponent of the above muscle, runs from the posterior aspect of the thyroid lamina to the muscular process of the arytenoid cartilage; contraction shortens the vocal cord. Some of these fibers insert into the cord itself, the vocalis muscle. Having shortened and lengthened the cords, it is time to open and close them.

POSTERIOR CRICOARYTENOID MUSCLE

This muscle runs from the posterior aspect of the cricoid cartilage on each side to the muscular process of the arytenoid. The fibers become more oblique as they descend, the upper being horizontal and the lowest vertical; the former rotate the arytenoids while the latter separates them by drawing the cartilages down the sloping shoulders of the cricoid. This movement produces a com-

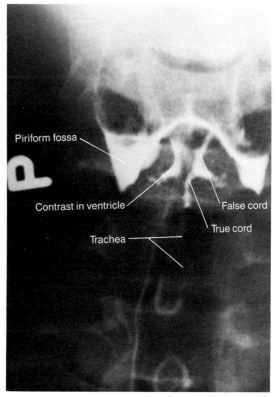

(a)

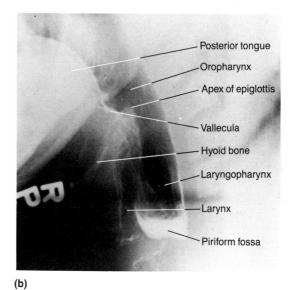

(b)

Fig. 15.8 Laryngograms. (a) anteroposterior view— contrast medium outlines the piriform fossae; (b) lateral view.

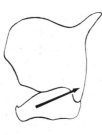
Cricothyroid muscle

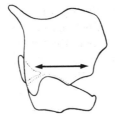
Thyroarytenoid muscle

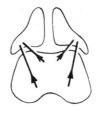
Posterior cricoarytenoid muscle

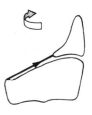
Lateral cricoarytenoid muscle

Fig. 15.9 The muscles of the larynx.

bination of 'V' and diamond-shaped glottic apertures, as described earlier. This muscle, the only dilator of the larynx, is probably the most important muscle in the body; it opens the space between the two cords, the rima glottidis. Failure of this muscle to open the space may result in an untimely demise.

LATERAL CRICOARYTENOID

The cords are closed by the lateral cricoary-tenoid muscle, which passes from the upper border of the cricoid arch backward and upward to the muscular process; it is an adductor of the cords.

The other intrinsic muscles are mainly concerned with closing the glottis and inlet of the larynx. The interarytenoid muscle is a mass of transverse fibers connecting the two arytenoid cartilages; contraction opposes them and narrows the rima glottidis. The aryepiglottic muscles arise from the muscular process on each side, cross each other behind the interarytenoid, and pass into the aryepiglottic folds to the epiglottis. Contraction of these muscles pulls the epiglottis and the arytenoid together and effectively closes the laryngeal inlet in a pursestring-like action.

Extrinsic muscles

The larynx rises during the act of swallowing, returning to its position at its completion (the reason for this will be discussed later). This movement is mediated by laryngeal elevator and depressor muscles. The elevating muscles have 'hyoid' or 'pharyngeus' as the second part of their names – mylo-, stylo- and geniohyoid. The first three attach to the hyoid bone, and the latter three to the posterior border of the thyroid cartilage – the stylo-, salpingo- and palatopharyngeus. The depressors of the larynx are the previously described strap muscles, lying anterior to the laryngeal skeleton.

Mucosal covering

The cartilages, membranes and muscles are covered by mucous membrane. The mucous membrane of the larynx is mainly columnar ciliated epithelium with scattered mucus-secreting goblet cells. However, there are two sites where there is much wear and tear and where the epithelium has to be squamous – the vocal folds and the laryngeal inlet. At the inlet the anterosuperior part of the posterior aspect of the epiglottis and the upper part of the aryepiglottic folds are covered by squamous epithelium. The vocal fold is also characterized by a virtual absence of submucosa; this implies an adherence of the mucous membrane to the underlying vocal ligaments with a minimal number of blood vessels (the submucosal layers carry the blood vessels). For this reason, the vocal cords are easily recognized as whitish structures. This anatomical arrangement, however, has even more clinical significance. In cases of laryngeal edema, the submucous tissues swell markedly; downward progression of this swelling, however, ceases at the site where epithelium is adherent to the vocal cord; consequently, the swelling above the vocal cords continues and may result in complete laryngeal obstruction with asphyxiation.

A trip to the larynx

The time has come to inspect the structures seen en route from the mouth to the larynx; this is the view you will get on examining the larynx and pharynx, and on passing a tube into the larynx. To start, place the oral cavity in direct line with the inlet of the larynx by completely extending the head at the atlanto-occipital joint. In this position, a tube passed through the mouth has a good chance of going directly into the larynx; in an emergency this may be life-saving. Remember that, in performing mouth-to-mouth resuscitation, unless the head (not the neck) is fully extended, the inflated air will pass into the esophagus and stomach; not many people can carry out respiration via their gastric mucosa. Initially identify the tongue and an area posterior to it between its base and the anterior surface of the upwardly projecting epiglottis. Note in this area median and two lateral glossoepiglottic folds of mucous membrane. The latter folds pass from the epiglottis to the side wall of the pharynx below the tonsil. These folds demarcate on each side a hollow, the vallecula; like the piriform fossa, it is an area where neoplasms may hide, and where foreign bodies may impact. On pushing the epiglottis forward (which is done with the blade of a laryngoscope) the posterior aspect of the epiglottis with its prominent tubercle is seen, with the two aryepiglottic folds running posteromedially toward the arytenoid cartilages. The cuneiform and corniculate cartilages, small cartilaginous nodules contained in these folds, are often seen as small bulges. The inlet or upper aperture of the larynx looks almost directly backward; i.e. it is set at right angles to the long axis of the tube, like the opening of the ventilating shaft on the deck of a ship. Continuing downward, the inlet of the larynx is narrowed by the projection of the vestibular folds or false vocal cords, with a space, the rima vestibuli, between them. Beyond this is the middle subdivision of the larynx, the ventricle or sinus, bounded above by the false cords and below by the true cords, the latter separated by a space, the rima glottidis. Not visible is a small diverticulum of the ventricle, the saccule, which projects from the ventricle cranially between the false cord and the lamina of the thyroid cartilage; it is very large in the gorilla, useful information for those who plan to practice on the banks of the Congo River. It is this blind-ended sac that secretes mucus and 'wets one's whistle' which has dried up by the end of a football game. The vocal folds, pale and straight edged, stand out sharply; between these folds a tube can be passed into the trachea (endotracheal tube). Below the rima glottidis is the lowest division of the larynx, which becomes the trachea at the lower border of the cricoid cartilage.

Functional movements

Sphincteric

It is very rare to inhale a piece of food. On swallowing, the larynx is lifted beneath a posteriorly bulging tongue; the epiglottis, inverted by the tongue and the bolus of food, closes like a lid over the laryngeal inlet. Additionally, the aryepiglottic and interarytenoid muscles contract and close the sphincter firmly; should further protection be needed, the rima glottidis itself can be closed. As well as during eating, the larynx also closes off during muscular exertion; air is prevented from being inspired by the closure of the vocal folds, and from being expired by closure of the false cords. This takes place whether at defecation or urination or in more enjoyable circumstances such as hitting a ball over the stands.

Phonation

Earlier it was noted that muscles can shorten, lengthen or alter the tension of the vocal cords. The expiration of air through the various positions of the cords will vary the intensity and pitch of sound. The quality of the voice depends on the resonators above the larynx, and these depend on the position of the soft palate and tongue, the final breaking up of the sound being performed by the teeth, lips and tongue (articulation).

Neurovascular supply

The vocal folds act as a dividing line between the upper and lower portions of the larynx. Above the folds, the blood supply is by the superior laryngeal branch of the superior thyroid artery, and the nerve supply by the internal laryngeal branch of the vagus, both piercing the thyrohyoid membrane (see Fig. 13.9). The lower half of the larynx is supplied by the inferior laryngeal branch of the inferior thyroid artery accompanied by the very important recurrent laryngeal nerve, which reaches it from below as it passes deep to the inferior constrictor of the pharynx. The internal laryngeal is sensory and secretomotor; except for the cricothyroid, all the muscles of the larynx are supplied by the recurrent laryngeal. The cricothyroid is supplied by the external laryngeal branch of the superior laryngeal, also a branch of the vagus. These laryngeal motor fibers actually originated with the accessory nerve which passed them on to the vagus. It is suggested by some that the external laryngeal nerve initially stimulates the cricothyroid muscles to put the cords on stretch; following this, impulses arrive by the recurrent nerve (as a result of its detour), and the stimulated muscles are now able to act on the tensed cord. In this way the cricothyroid muscle is the tuning fork of the larynx; it is tuned for sound by the time impulses reach it from the recurrent nerve. The lymphatic drainage from the larynx is important, as the nodes are often involved in laryngeal disease. The larynx drains to the deep cervical nodes as they lie along the jugular vein, the upper half into the upper nodes and the lower half into the lower. Some lymphatics, however, pass to nodes lying in front of the larynx and trachea, the prelaryngeal and pretracheal lymph nodes.

16 Lymph nodes of the head and neck

Adenoid tissue of the head and neck · Circular chain of lymph nodes · Vertical chain of lymph nodes · Clinical approach to enlarged lymph nodes

Once upon a time a student came into the anatomy lecture late; he was set to count all the lymph nodes in the body. The total turned out to be in the region of 800, of which 300 were present in the neck. It should be stated that the commonest swelling in the neck is an enlarged lymph node. As no lymph nodes are present in the scalp, a swelling of this area cannot be due to a lymph node.

Adenoid tissue of the head and neck

At the entrance of the pharynx, Waldeyer's lymphatic ring has been mentioned. This lymphoid tissue is protective and the frequent involvement in inflammatory responses confirms this.

Waldeyer's ring drains into two main areas – the main lymph node of the tonsil and the 'adenoid' nodes. The main lymph node of the palatine tonsil is situated in the angle between the internal jugular and common facial vein just below the angle of the jaw (see Fig. 15.5). Many people experience enlargement of this node following a sore throat. The 'adenoid' nodes, a few nodes lying deep to the sternomastoid muscle below the tip of the mastoid process, drain the lymph from the pharyngeal tonsils. These nodes become enlarged in diseases affecting these tonsils, a common event in young children.

Lymph nodes are roughly disposed in a circular and a vertical chain.

Circular chain of lymph nodes

Occipital nodes

These nodes, the most posterior of the circular chain, are situated between the mastoid process and the inion. If you search amongst the hair of the scalp, the usual cause for the enlargement, a septic focus, will be located.

Posterior auricular

These nodes, situated on the mastoid process behind the pinna of the ear, drain the scalp, the back of the pinna and the external auditory meatus.

Preauricular node

This node has a very precise location immediately in front of the tragus of the ear and superficial to the parotid gland. It drains the outer side of the pinna and the side of the scalp.

Parotid glands

These important nodes are situated on, in and deep to the parotid. The more superficial nodes drain the eyelids, the front of the scalp, the external auditory meatus and the tympanic cavity. The deeper glands drain the nasopharynx and the back of the nose. A swelling in the parotid gland due to an enlargement of the node draining the above areas may be mistaken, not unjustifiably, for a primary tumor of the parotid gland. Until recently it has been notoriously difficult to examine the deeper reaches of the nose and nasopharynx and to locate tumors responsible for this enlargement. Computerized tomography and endoscopy have considerably improved this situation.

Facial nodes

These nodes are situated on the buccinator, just below the orbit, and over the facial vessels as they cross the mandible. They drain the conjunctiva, eyelids, nose and cheek.

Submandibular nodes

These important nodes lie in the same fascial capsule which encloses the salivary gland. A swelling in this region (as in the case of the parotid) may be due either to an enlargement of the salivary gland or to an enlargement of the lymph nodes as a result of a lesion in their drainage area – the side of the nose, the cheek, the angle of the mouth, the whole upper lip, the outer part of the lower lip, the gums and the side of the tongue.

Submental nodes

These nodes lie in the submental triangle and drain the central part of the lower lip, the floor of the mouth, the tip of the tongue and the incisor area of the mandible. As there are very few other causes for swellings in this area, a mass in the submental triangle is nearly always an enlarged lymph node.

Vertical chain of lymph nodes

These nodes lie superficially and deeply.

Superficial nodes

These consist of the anterior cervical nodes that lie in the midline, in front of the larynx and trachea, which they drain.

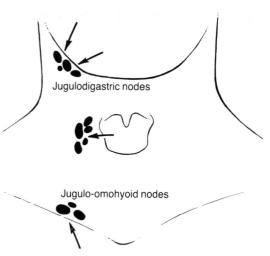

Fig. 16.1   The three areas of drainage to the deep vertical chain of lymph nodes.

Deep nodes

This chain consists of many large nodes in anterior and posterior relationship to the internal jugular vein, and reaching from the base of the skull to the root of the neck (see Fig. 15.5). Some of the nodes are situated a short distance away from the internal jugular vein and have various names applied to them; for example, those lying along the recurrent laryngeal nerve are called paratracheal nodes. Three nodes of the chain deserve mention because of their constant enlargement in various diseases: the main lymph node of the tonsil in the angle between the internal jugular and common facial vein; the main lymph node of the tongue, which lies a little lower at the bifurcation of the common carotid artery and is commonly enlarged in tongue lesions; and the supra-omohyoid node, which is situated just above the point where the anterior belly of the omohyoid crosses the carotid sheath. The supraomohyoid node is important in the lymph drainage of the tongue, as it receives some vessels from the tip that may reach it directly without traversing the other nodes; this has serious prognostic implications (see Fig. 11.19).

Final drainage of the cervical lymph nodes

The circular chain drains into the deep cervical nodes which receive lymph from the entire head and neck. From the deep cervical chain, the lymph is collected into a trunk, the jugular lymph trunk, which leaves the inferior deep cervical glands to enter the junction of the subclavian and internal jugular veins on the right side, and the thoracic duct on the left.

Clinical approach to enlarged lymph nodes

In case a deep depression ensues over what appears to be a very detailed description of the lymph nodes, reference to the clinical application and a statement of some important principles will infuse a modicum of optimism.

1   Remember that there are no lymph nodes on the cranium, which means that a cranial swelling cannot possibly be a lymph node.
2   If we divide the neck into three areas, superior, middle and inferior (Fig. 16.1), it is reasonably true to say:

a)   The nodes in the lower third of the neck (i.e. the root of the neck) usually enlarge due to diseases below the clavicle (Fig. 16.3). Most commonly the lesion is in the thorax, but pathology in the abdomen or even the testes can be responsible. A classic example is the Virchow–Troisier node from the spread of a cancer of the stomach.

b)   In the middle third of the neck, an enlarged node usually emanates from a lesion of the thyroid gland. A well-known clinical trap is a small impalpable thyroid lesion and a large lymph node; the level of the node should hint strongly as to its origin.

c)   In the upper third of the neck, the lymph nodes are nearly always en-

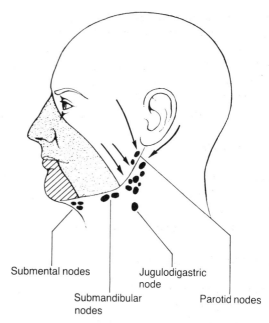

Submental nodes   Jugulodigastric node

Submandibular nodes   Parotid nodes

Fig. 16.2 The three areas of drainage from facial tissues to regional lymph nodes.

larged as a result of disease of the head and upper neck. We may subdivide this latter group further by delineating anteroposterior areas (Fig. 16.2).

i)   Anteriorly, the submental nodes drain a central wedge of tissue in the floor of the mouth by lymphatic vessels that cross the midline. This wedge includes the tip of the tongue, and the floor of the mouth, gums and lower lip opposite the four lower incisor teeth. You may puzzle about a mysterious submental swelling whose origin is revealed by an x-ray of the lower incisor teeth.

ii)   More posteriorly, the lymph nodes in the submandibular region drain the lateral part of the lower lip, all the upper lip and external nose, and the anterior two-thirds of the tongue, by lymphatics from their own side. Additionally,

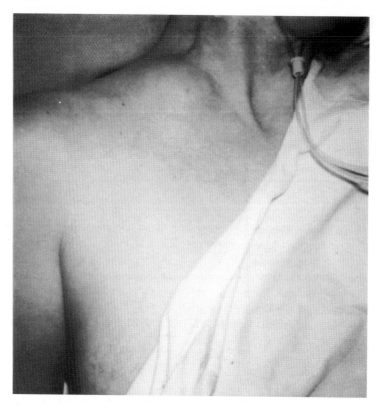

Fig. 16.3   An enlarged lymph node is seen in the root of the right side of the neck. It is a metastasis (spread) from a tumor of the lower lobe of the left lung.

these nodes receive lymph from the paranasal sinuses and anterior half of the walls of the nasal cavity. This is a bigger block of tissue than that drained by the submental nodes and is situated more posteriorly.

iii) The most posterior nodes drain all the structures not drained by the aforementioned two sets of nodes.

Of the tissues draining into the upper jugular chain, it is true to say that the more posterior the structures, the higher are the lymph nodes to which they drain; e.g. the pharyngeal tonsils, situated deep in the nasopharynx, drain to the highest jugular nodes. It is preferable to remember this rather simplified scheme than to memorize all the possible drainage areas of the head and neck.

17 Ear

External ear · Middle ear · Development of the external and middle ears ·
Internal ear

External ear

The auricle is the trumpet-shaped external appendage which gathers in sound. Its skin, firmly attached to the underlying cartilage laterally, is more mobile medially (facing cranially). There is little in the way of a subcutaneous fatty layer, which means that cellulitis (inflammation of subcutaneous tissue) is unlikely to affect the external ear. The cartilage is convoluted and continuous with the cartilage of the external meatus. It is perhaps worth mentioning one of these convolutions as it is a useful landmark: the tragus is a short triangular projection that lies closest to the face. Anterior to the tragus lies a consistently placed lymph node (preauricular), and the pulsation of the superficial temporal artery.

The muscles controlling the position of the ear are rudimentary and useful only as party tricks. The sensory nerve supply is the greater auricular to both surfaces of the lower part of the ear, and the auriculotemporal and lesser occipital to the facial and cranial surfaces, respectively, above the greater auricular territory.

The external auditory meatus, the outer one-third cartilaginous and the inner two-thirds bony, leads from the auricle to the tympanic membrane, the lateral boundary of the middle ear. The cartilage, continuous with that of the pinna, is attached to the bony meatus by fibrous tissue. The bony meatus is an incomplete ring (the tympanic plate), which is completed above by the squama of the temporal bone.

The canal is sinuous, being convex forward as well as upward. The oval shape means that an oval-shaped speculum is necessary to examine the meatus; furthermore, unless the adult ear is pulled upward and backward and the tube straightened, the instrument cannot be advanced. Note the short meatus of the child, resulting from the undeveloped small tympanic plate: failure to remember this might find you looking into the middle cranial fossa. As the otoscope is passed inward, note the thick hairs covering the cartilaginous part of the tube in contrast to the smooth skin of the bony part. Besides the hairs, the cartilaginous part has sebaceous and wax-secreting glands. This part of the meatus is a common place for inflammatory lesions, which are extremely painful; the collection of inflammatory exudate beneath the cutaneous lining is under pressure because the skin is closely adherent to the underlying cartilage or bone.

The nerve supply of the external meatus is the auricular branch of the vagus for the posterior half and the auriculotemporal nerve for the anterior half. Because of the vagal supply of the heart, lungs and many of the abdominal viscera, reflex symptoms may occur. Occasionally an elderly patient will suddenly faint as a result of cardiac syncope while you are syringing out his ear. A chronic cough in a young child may be traced to a pea in the ear. Those who have had their ears syringed will attest to the easily induced coughing or gagging. Pain from intraoral structures are often referred to the ear, the fifth nerve supplying both areas; the little boy

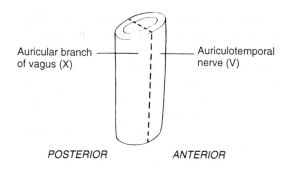

Auricular branch of vagus (X)

Auriculotemporal nerve (V)

POSTERIOR   ANTERIOR

Fig. 17.1 The nerve supply of the external auditory meatus.

with erupting teeth and patients with tongue lesions often suffer from earache. Pain often radiates up toward the temple from a lesion of the external meatus.

After straightening the meatus by pulling the auricle upward and backward, a constriction will be seen through the otoscope at the junction of the cartilaginous and bony parts. Immediately proximal to the pearly gray tympanic membrane is a depression in which foreign bodies may be stored by mischievous children.

Middle ear (see Figs. 13.2, 13.5 and 17.2)

The middle ear is a narrow slit in the petrous temporal bone, that runs more or less parallel to the tympanic membrane. Economy is the watchword as regards the size of the middle ear. It measures about 6 mm (0.25 in) wide maximally, and is only 2 mm (0.08 in) wide opposite the tympanic membrane: it is longer in the vertical than the anteroposterior direction (15 mm; 0.6 in). It is no wonder that surgery of the middle ear had to await the arrival of the operating microscope. The cavity of the middle ear is bounded as follows.

Lateral wall

The tympanic membrane or drum forms most of the lateral boundary of the ear, almost filling the notch or groove in the bony meatus; the upper part of this boundary is completed by the squamous temporal bone. Inspection of the membrane reveals a reddish streak midway between the anterior and posterior margins that runs upward and forward; this is the lateral process and handle of the malleus. Projecting from it anteroinferiorly is a reflective cone of light. Using bionic eyes and a little imagination, the long process of the incus is faintly visible behind the malleus.

The tympanic membrane is set so obliquely that its outer surface looks downward and forward as well as laterally. In the child this obliquity is such that the membrane lies almost horizontally. The membrane is also concave laterally, and the point of deepest concavity, which corresponds to the tip of the handle of the malleus, is called the umbo. The membrane consists of fibrous tissue lined by skin externally and mucous membrane internally.

The cavity of the middle ear reaches a level higher than the tympanic membrane; this area is called the epitympanic recess.

Anterior wall

The anterior wall of the middle ear leads forward to the bony part of the auditory (eustachian) tube. This tube becomes cartilaginous before it opens into the nasopharynx at the site of the tubal tonsil. The cartilaginous part of the tube is twice as long as the bony part, and contains many more mucous glands.

In the young child the tube is half the adult length, runs horizontally, is relatively wide, and opens into the pharynx by a narrow slit without a tubal elevation. This explains the great frequency with which upper respiratory tract infections pass along this short, wide horizontal tube to involve the middle ear, and of the ease with which the pharyngeal orifice can be blocked. Immediately above the auditory tube the tensor typani muscle arises from a small canal through which it runs. After turning on a pulley, it inserts into the medial side of the malleus. The muscle tenses the drum membrane and deepens loud sound.

Posterior wall

The epitympanic recess, containing the upper half of the malleus and most of the incus, leads posteriorly through an aperture (aditus) into the upper part of the important tympanic antrum. This last leads on to the mastoid air cells, contained in the mastoid process. The amount of pneumatization (air cell formation) in the mastoid process varies considerably, from one where the process is hollowed out by cells (the pneumatic mastoid), to one where the process consists mostly of compact bone and few air cells. Infections of the middle ear spread easily via the aditus into the tympanic antrum and the mastoid air cells. Mastoiditis was a scourge in all parts of the world until the introduction of antibiotics; in developing countries it still is very common. Infection in the mastoid process produces an easily visible and palpable swelling, which may become large enough to push the ear away from the side of the skull. On the posterior wall is a small pyramidal eminence, through which the tendon of the stapedius muscle emerges to insert into the stapes. The action of this muscle is to pull the stapes away from the oval window to diminish sound. The stapedius and the tensor tympani, therefore, both act as mufflers for loud sounds.

Medial wall

From above downward is the swelling of the lateral semicircular canal, the bulge produced by the facial nerve in its canal, a membrane called the fenestra vestibuli or oval window, into which fits the base of one of the auditory ossicles, the stapes, the promontory formed by the first turn of the cochlea (part of the internal ear) and covered in life by the tympanic plexus of the ninth nerve, and, lowest of all, the fenestra cochlea or round window.

Function of the middle ear

This air-containing cavity, whose pressure is held in equilibrium with the air on the other side of the drum membrane by means of the auditory tube, acts as an air-conducting mechanism for the transmission of sound. To do this, it needs three auditory ossicles, which are remarkable bones. They are the only bones in the body which are full sized at birth and grow no further. The three articulating ossicles are the malleus, which we have described before, the incus, shaped like a premolar tooth and with a long process lying behind and parallel to the handle of the malleus, and the stapes. The last articulates with the incus by a head, which attaches by two limbs to an oval base that fits snugly into the fenestra vestibuli (oval window).

When the umpire shouts into your ear that you are out, the vibrations emitted from his oral cavity enter your external meatus and impinge on your tympanic membrane. The vibrations of this membrane reach and produce intensified movements of the stapes by means of a lever-like action of the chain of auditory ossicles, acting as a mechanical transformer. The area of the tympanic membrane is eighteen or more times that of the footplate of the stapes, and, therefore, the sound is increased at least eighteenfold. The vibrations of the oval window by the stapes stimulates the internal ear; at the same time (since fluid is incompressible) compensatory outward bulging occurs in the round window (fenestra cochleae). This is not the only way in which the sound vibrations are transmitted to the internal ear – bone conduction is also important. The bony capsule of the internal ear can be stimulated by vibrations arising from immediately adjacent bone; for example, if a vibrator is placed on the side of the skull, the patient perceives sound. If a person can perceive the sound from a tuning fork placed on the mastoid process better than when placed near the external ear, he has a conductive defect. This defect could be in the ossicular chain, a not uncommon inherited defect.

Another very common problem in young children is an occlusion of the auditory tube; this lowers the pressure in the middle

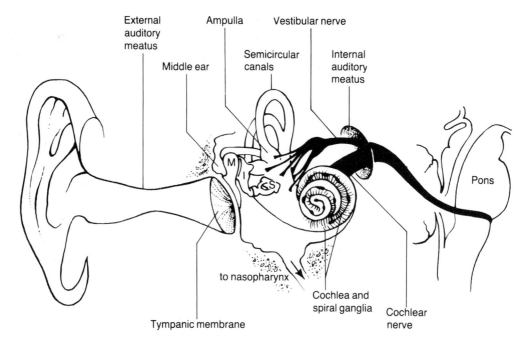

Fig. 17.2 The vestibulocochlear nerve. I, incus; M, malleus; S, stapes.

ear by causing an absorption of air. This potential vacuum is prevented by the out-pouring of sticky fluid from the mucous membrane and the formation of a 'glue ear'. This is a great culture medium for organisms and results in an impairment of the elasticity and efficiency of the hearing mechanism. A tube (grommet) placed through the tympanic drum will go a long way to equalizing the pressure and curing this condition. Destructive lesions in the middle ear, whether by infection, trauma or a surgeon, can damage the facial nerve as it passes through the middle ear.

Development of the external and middle ear

A brief note about the development of the external and middle ear is in order. The pinna is developed from a series of tubercles around the margins of the first branchial groove; failure of fusion of these tubercles may create small sinuses. The external meatus is the remains of the first external branchial groove

or cleft, which projected inward from the surface, while the middle ear is the relic of the first pharyngeal pouch, which grew out as a lateral recess of the pharynx; the auditory tube and tympanic cavity are the remains of this pouch. The tympanic membrane, therefore, is formed by the ectoderm of the external diverticulum (cleft), the endoderm of the pharyngeal diverticulum and a thin layer of mesoderm between them.

Some differences between the middle ear of the adult and that of the child have been mentioned. We should add that the tympanic antrum is a well-developed cavity at birth, but the mastoid process is not; the latter and its air cells begin to develop during the second year. Infections of the tympanic antrum in the first 2 years of life are common as a result of the wide, short eustachian tube and the frequency of upper respiratory tract infections. The absence of the mastoid process means that the facial nerve lies very superficially in the infant; an ill-placed application of forceps at delivery or a poorly placed incision behind the ear may therefore damage the nerve (see Fig. 10.23).

Internal ear

The internal ear consists of a membranous sac or labyrinth with many diverticula, lying in a very dense bony labyrinth that lies in the petrous temporal bone. The bony labyrinth consists of three portions, the cochlea lying anteriorly, the semicircular canals lying posteriorly and the connecting vestibule. The cochlea, which resembles a snail's shell, is composed of a central column, the modiolus, around which a hollow tube makes two and one-half turns. There are three semicircular canals, the superior lying in the plane at right angles to the long axis of the petrous temporal bone, the lateral canal set in a horizontal plane and responsible for the prominence on the medial wall of the middle ear, and a posterior canal set parallel to the long axis of the bone. The vestibule has two openings laterally, the fenestra vestibuli above for the base of the stapes, and the fenestra cochleae below.

The bony labyrinth contains a clear fluid, perilymph, in which the membranous labyrinth is suspended. The membranous labyrinth consists of three parts, which are filled with fluid, the endolymph. They are: the utricle and saccule (two small sacs lying in the vestibule); three semicircular ducts enclosed in the semicircular canals; and the duct of the cochlea in the bony cochlea. The auditory (eighth) nerve reaches the walls of these structures, and consists of a cochlear division to the cochlea for hearing, and a vestibular branch to the utricle, saccule and semicircular canals for equilibrium.

18 Orbit

Bony orbit · Contents of the orbit

This pyramid-shaped cavity is situated at the junction of the vault of the skull and the face. The globe of the eye occupies the anterior half of the space, while the posterior half is mostly filled by muscles, nerves and vessels, all supported by a variable amount of fat. It is interesting to note that the orbits face forward in the predator, while in the hunted they face laterally; this provides the latter with a maximal field of vision for spotting the stalker.

Bony orbit

The orbital margin consists of three bones, the frontal above and to a small extent on each side, the zygomatic laterally and the maxilla

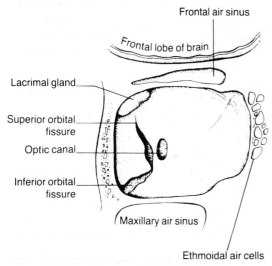

Frontal air sinus

Frontal lobe of brain

Lacrimal gland

Superior orbital fissure

Optic canal

Inferior orbital fissure

Maxillary air sinus

Ethmoidal air cells

Fig. 18.1 The bony orbit and its relations.

medially. To provide maximal peripheral vision for the embedded bulb, the lateral boundary is concave forward. To protect the globe the margins lie at least 2.5 cm (1 in) behind the root of the nose, and the supra-orbital margin projects prominently forward; these two bony parts take the bulk of the force from a blow. The bony prominences and the deeply embedded globe prevent injury to this important organ.

The cavity resembles a pear or an ice-cream cone, being widest at its margins and narrowing to an apex. This pyramidal cavity has superior, inferior, medial and lateral walls.

Superior wall

The superior wall, which doubles as the floor of the anterior cranial fossa, consists mostly of the frontal bone; fractures of the cranial fossa produce easily recognizable orbital hemorrhage.

Medial wall

Medially, the bony wall presents a small anterior fossa, bounded by anterior frontal and posterior lacrimal crests and containing the lacrimal sac. Above the fossa is a small depression, sometimes replaced by a spine, for the attachment of the fibrous pulley for the tendon of the superior oblique muscle of the eyeball. Posterior to the lacrimal bone the medial wall is formed by the orbital plate of the ethmoid bone, through which it is related to the ethmoidal sinus.

Floor

The floor shows a groove leading to a canal for the transmission of the large infraorbital terminal branch of the maxillary nerve. Medial to this is the bony attachment of the inferior oblique muscle. The floor of the orbit and the roof of the maxillary sinus are one and the same.

Lateral wall

Laterally, the zygomatic bone shows foramina for the emerging branches of the zygomatic nerve and, where it meets the roof of the orbit, a small depression for the lacrimal gland.

Clearly, the orbit is surrounded by structures which, when diseased, may involve the eye. Situated medially from above downard are the frontal, ethmoidal and maxillary air sinuses. The laterally lying zygomatic bone, despite being heavily buttressed to withstand the molar masticatory force, comes off badly in a punch to the face; the zygomatic arch and the zygomatic bone are often fractured. Depression of the orbital floor with dropping of the eyeball follows, with consequent double vision. A fractured zygomatic bone can be confidently diagnosed when diplopia (double vision) accompanies a blow to the side of the face.

The walls of the cavity converge to an apex, the site of the optic canal. This large canal transmits the large optic nerve together with its meningeal coverings and the ophthalmic artery. The extension of meningeal coverings around the nerve may seem strange until it is remembered that the eye is really an extension of the brain. This means that the subarachnoid space is prolonged along the optic nerve into the orbit; a rise of intracranial pressure is transmitted along this space and may compress the veins of the nerve, causing edema – papilledema (swelling of the optic papilla or disc), clearly visible on fundoscopy (examination of the disc by ophthalmoscope). When the eye requires removal because of infection, it is better to enucleate it from its capsule rather than sever the optic nerve; the latter would open the meninges, with a possibility of subsequent meningitis. The outer layer of the intracranial dura is prolonged through the canal and becomes the periosteum of the bony orbit.

Orbital fissures

The other openings, the superior and inferior orbital fissures, transmit important struc-

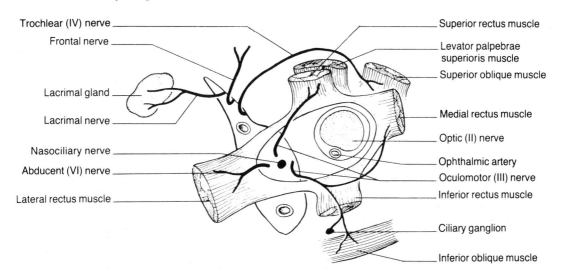

Fig. 18.2 The superior orbital fissure. The transmitted structures are seen, as well as the tendinous ring of origin for the ocular muscles.

tures to the orbit. The superior fissure, the space between the two wings of the sphenoid bone, is shaped like a retort lying on its side with slit-like lateral and bulbous medial portions. A tendinous ring for the origin of the ocular muscles surrounds the optic canal and the medial part of the fissure. The inferior fissure, situated further laterally, lies between the greater wing of the sphenoid, zygomatic and maxillary bones.

The medial walls of the orbit, separated by the nasal cavity, lie parallel to each other, while the lateral walls are so angled that, if projected backwards, they would meet at slightly more than a right angle. The long axis of the orbital cavity runs from the optic canal anterolaterally, while the visual (optic) axes are parallel to one another.

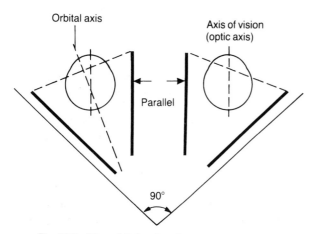

Fig. 18.3 The orbital and optic axes.

Contents of the orbit

Globe (eyeball)

This consists of three layers, an outer fibrous, a middle vascular and an inner nervous layer; the inner layer is prolonged posteriorly as the optic nerve. The posterior five-sixths of the fibrous layer is the sclera or white of the eye, while the anterior one-sixth is the transparent cornea. Interference with this transparency, as by a scar, leads to impaired vision. The cornea is almost unique in that it lacks a

blood supply, being nourished by peripheral permeation. This explains the success of corneal transplantation, where the protective elements of the blood, unlike vascular areas, are unable to reject the foreign body. The cornea is probably even more important as a refractory agent than the lens.

Optic (II) nerve

Passing posteriorly from the nervous layer, the 5 cm (2 in) long optic nerve takes a sinuous course to prevent any restraint on the movements of the globe. Its dural sheath, a continuation of the inner layer of the dura, is adherent to the structures at either end of the nerve: i.e. the optic canal and the sclera. The central artery and vein of the retina run in the nerve and are perhaps the best examples of end-vessels; occlusions produce drastic results.

Muscles of the eyeball

The muscles which move the eyeball are the four recti and the two obliques, which all insert into the sclera.

Rectus muscles

The four recti arise from the common tendinous ring and are arranged as indicated by their names: superior, medial, inferior and lateral.

The muscles, like the bony cavity, run forward as a cone that encloses the nerves and vessels for the eye, and insert into the anterior portion of the bulb of the eye, a short distance behind the corneoscleral junction. The lateral and medial recti are in the same horizontal plane, the superior and inferior in the same vertical plane.

Oblique muscles

The superior oblique muscle arises above and medial to the optic canal, passes forward along the medial wall of the orbit above the medial rectus, and forms a tendon which turns sharply backward and medially at a

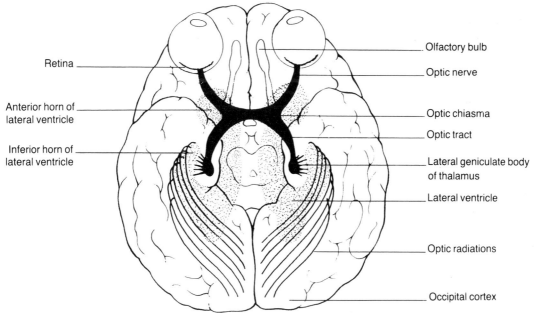

Fig. 18.4 The optic pathway.

Labels on figure:
Retina
Anterior horn of lateral ventricle
Inferior horn of lateral ventricle
Olfactory bulb
Optic nerve
Optic chiasma
Optic tract
Lateral geniculate body of thalamus
Lateral ventricle
Optic radiations
Occipital cortex

trochlea or bony spine, to which it is attached by a pulley. It inserts into the superolateral quarter of the posterior half of the globe. The inferior oblique arises from the floor of the orbit, runs laterally below the inferior rectus and inserts into the inferolateral quadrant of the posterior half of the globe. As the superior oblique acts only after it has turned at its pulley, it is clear that both obliques run in the same axes to the posterolateral quarters of the bulb.

Actions of the muscles

The globe rotates on three axes: sagittal – rotation medially or laterally; horizontal or transverse – rotation upward or downward (elevation or depression); and around a vertical axis – abduction or adduction.

Because of the pyramid shape of the orbit, the axis of the cavity runs forward and laterally. Therefore, although the recti run straight as their name indicates, they run straight but laterally. This means that the superior and inferior recti, in addition to their prime action of elevation and depression, will act as

adductors. The lateral and medial recti have only one movement, abduction and adduction respectively.

The two oblique muscles insert into the posterolateral quadrant of the bulb behind the vertical axis; therefore, in addition to their prime action of elevation or depression, they are also abductors. The action of the superior oblique muscles is conveniently remembered by naming it 'the muscle of the tramp' – down and out.

The position of the globe is kept constant by the tone of the muscles, the recti tending to retract the organ and the oblique muscles tending to protrude it. It should be appreciated that the ocular muscles, like muscles elsewhere, work together to either assist a movement or neutralize an unwanted action; for example, should you wish to look downward, the inferior rectus would be sufficient for this purpose, but, because of the axis of the orbital cavity, it will also adduct the eye. This last movement is not required or desired; the superior oblique is called into action, which assists the depression but, by its concomitant abduction action, it cancels out the unwanted adduction. Similarly,

(a)

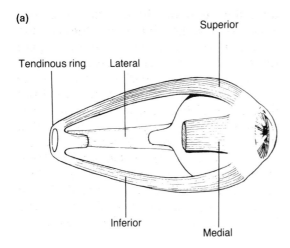

(b)

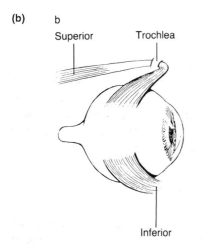

(c)

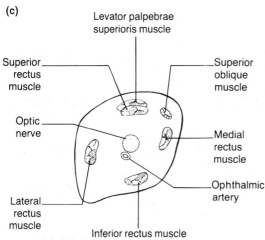

Fig. 18.5 **The muscles of the eyeball: (a) the recti, (b) the obliques, and (c) the position of the muscles in the bony orbit.**

elevation by the superior rectus produces adduction and medial rotation; the other elevator, the inferior oblique, cancels out these movements by its abductor and lateral rotator actions.

Levator palpebrae superioris

This muscle does not insert into the eyeball. Arising from the roof of the orbit anterior to the superior rectus, it passes forward on the upper surface of the latter. The thin tendon inserts into the skin of the eyelid, the superior tarsus, and the superior conjunctival fornix. The levator and the superior rectus were once probably one muscle, as they lie on top of one another and share the same nerve supply. When the superior rectus elevates the globe, the levator obligingly raises its lid and the upper part of the conjunctival sac. The portion of the levator inserted into the upper tarsus contains smooth muscle fibers (Müller's muscle) that are innervated by the sympathetic system; loss of the nerve supply to these fibres leads of drooping of the upper lid or ptosis, as seen in Horner's syndrome (paralysis of the cervical sympathetic).

Fascial sheath of the eyeball (Tenon)

This sheath forms an investment for the posterior five-sixths of the eyeball, blending with the sclera at its corneal junction anteriorly and with the sheath of the optic nerve posteriorly. Between the sclera and the sheath is a small episcleral space composed of areolar tissue and lymph spaces, which facilitates global movements. As they insert into the eyeball, the tendons of the muscle pierce the sheath, which is prolonged posteriorly onto the muscles. The sheaths of the medial and lateral recti are thick and expand laterally to attach to the orbital wall; these check ligaments prevent excessive movement. Inferiorly, thickenings of the sheaths around the inferior oblique and inferior rectus muscles fuses with the check ligaments on each side and forms a suspensory ligament that supports the eyeball. When the eye is

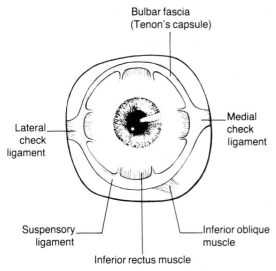

Fig. 18.6 The fascial sheath of the eyeball, thickened to form the suspensory ligament below and the check ligaments on each side.

removed, the fascia bulbi forms the socket for the prosthesis.

Nerves in the orbit

Oculomotor (III) nerve

The third cranial nerve supplies all the muscles of the eye except the superior oblique and lateral rectus. Passing forward from the midbrain through the cavernous sinus, it enters the orbit through the common tendinous ring and divides into an upper division for the superior rectus and the levator palpebrae superioris and a lower division for the medial and inferior recti and the inferior oblique. The latter branch carries an important parasympathetic contribution to the ciliary ganglion, which, after synapsing, supplies the sphincter pupillae (constrictor of the pupil) and the ciliary muscle of the eye, the muscle of accommodation or focus. These parasympathetic fibres originate from the Edinger–Westphal nucleus of the midbrain.

Trochlear (IV) nerve

The trochlear nerve is unique, as it is the only cranial nerve to arise from the dorsal aspect of the brain stem. The long slender nerve passes forward through the cavernous sinus to enter the orbit above the ring and supply the superior oblique muscle.

Abducent (VI) nerve

As its name indicates, the abducent nerve supplies the abductor, the lateral rectus. It

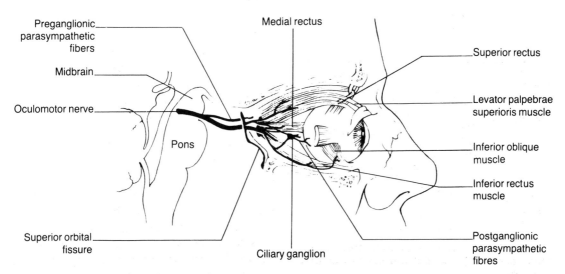

Fig. 18.7 The oculomotor nerve.

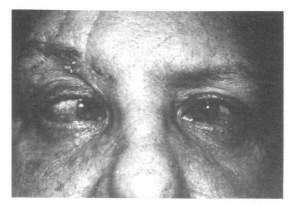

Fig. 18.8   Paralysis of the lateral rectus leads to unopposed adduction. This patient has bilateral sixth nerve lesions.

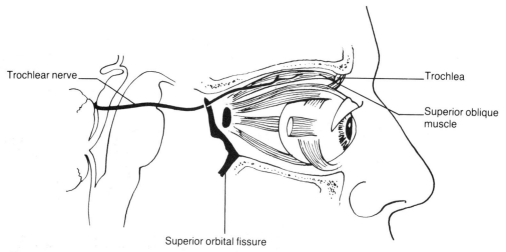

Fig. 18.9   The trochlear nerve.

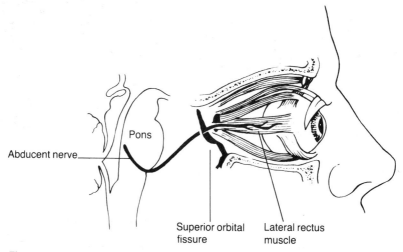

Fig. 18.10   The abducent nerve.

passes through the cavernous sinus and enters the orbit through the tendinous ring.

Ophthalmic (V¹) nerve

The most medial branch of the fifth nerve runs forward through the cavernous sinus to divide into three branches which enter the superior orbital fissure; two branches, the frontal and lacrimal, pass above the tendinous ring, while the nasociliary passes through it.

Ciliary ganglion

This pinhead-sized ganglion sits on the lateral side of the optic nerve in the posterior part of the orbit. Reaching the ganglion are:

1   Short ciliary sensory nerves travelling through the ganglion to the nasociliary nerve.

2   The parasympathetic fibers via the branch of the third nerve to the inferior oblique are the only fibers to synapse in the ganglion, the postganglionic fibers passing via the short ciliary nerves to the eye.

3   The sympathetic fibers, having synapsed in the superior cervical ganglion, reach and pass through the ganglion to the dilator pupillae muscle.

Blood vessels

Ophthalmic artery

This vessel is the major blood supply to the orbit, but it does receive some help from the infraorbital branch of the maxillary artery. Arising just before the bifurcation of the internal carotid artery in the middle cranial

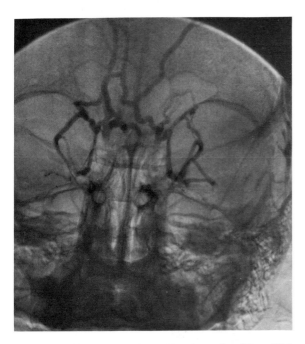
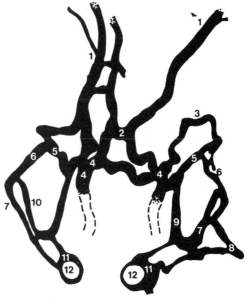

Fig. 18.11   Orbital venogram with subtraction. (From Weir and Abrahams 1978)

1 Frontal veins.
2 Superficial connecting vein.
3 Supraorbital vein.
4 Angular veins.
5 Superior ophthalmic vein (first part).
6 Superior ophthalmic vein (second part).
7 Superior ophthalmic vein (third part).
8 Inferior ophthalmic vein.
9 Anterior collateral vein.
10 Medial collateral vein.
11 Cavernous sinus.
12 Internal carotid artery.

fossa, it passes through the optic canal below the nerve. It then runs laterally either above or below the optic nerve to reach the medial side of the orbit. Its numerous branches supply the orbital structures and terminate as branches to the eyelids and scalp. The central artery of the retina pierces the optic nerve to reach the retina, where it divides into branches. These branches are unique in that they can be seen in the living eye by means of the ophthalmoscope (fundoscopy).

Ophthalmic veins

These two veins are important clinically. The superior vein accompanies the ophthalmic artery and communicates with the anterior facial vein, and the inferior vein communicates with the pterygoid venous plexus through the inferior orbital fissure. The veins drain posteriorly into the cavernous sinus; a route therefore exists for infections of the head and neck to involve the cavernous sinus, a potentially dangerous situation.

Eyelids

These are two movable fibromuscular folds that protect the globe, rest the eye from light and keep it moist. From without in, the eyelid is composed of skin, orbicularis oculi muscle, the tarsal plate with its attached levator palpebrae superioris and the conjunctiva.

The skin of the eyelid is rather loose and thin, as it does not contain any subcutaneous fat. (If it did contain fat, very obese people would have visual problems.) It easily fills up with fluid, and those at the 8:oo a.m. lecture have slight edema of the eyelids, not from being beaten up on the way to medical school but from the dependent position of the sleeping head. At the free margin of the eyelid, the skin attaches to the conjunctiva lining its inner aspect. The next layer, the orbicularis oculi, consists of two portions: the palpebral part arises mostly from the medial palpebral ligament and the lacrimal sac itself, and sweeps around to the smaller

lateral palpebral ligament; the much larger orbital portion runs over the forehead and cheek in a series of concentric loops. The palpebral part closes the lids gently and stretches the lacrimal sac, while the orbital part closes the eye tightly, and, by diminishing the volume of the conjunctival sac, is responsible for the spilling of tears.

Tarsal plates

These two dense fibrous plates in each eyelid contain some elastic fibers and are curved to conform to the shape of the eye. The plates are thickenings of the palpebral fascia, which is attached to the orbital margins peripherally. The plates are hollowed out posteriorly by sebaceous tarsal glands that are visible through the conjunctiva as yellowish streaks.

The medial palpebral ligament is a strong thickening of the palpebral fascia that joins the medial end of each tarsal plate to the bony lacrimal crest. The medial palpebral ligament, which can be seen and felt when the lids are drawn laterally, crosses anterior to the upper part of the lacrimal sac – a good guide to the latter, when drainage of the sac is required. The lateral extremities of the plates are attached to the lateral margin of the orbit by a much weaker lateral ligament.

Mucous membrane

The conjunctival sac lines the inner aspect of the eyelid (palpebral conjunctiva) and is reflected at superior and inferior fornices onto the sclera of the globe which it lines as far as the cornea, where it becomes continuous with the corneal epithelium. At the free margin of each lid it becomes continuous with the skin.

Lid margins

At each lid margin are two or three rows of large hairs (lashes); lashes of the upper eyelid are curved upward and outward, and those of the lower lid pass downward and forward.

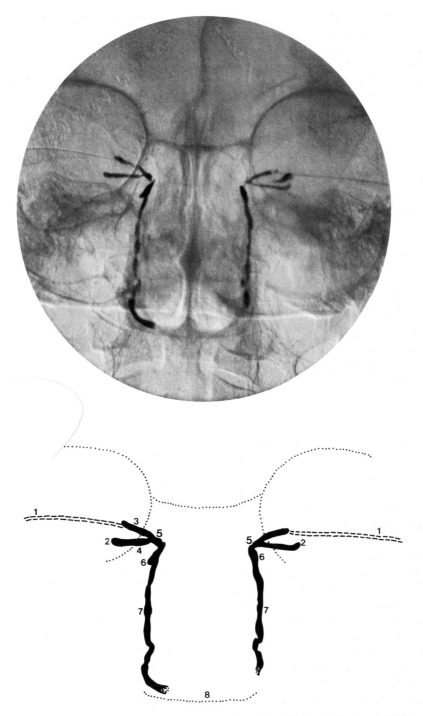

Fig. 18.12 Subtraction macrodacrocystogram. (From Weir and Abrahams 1978)

1 Lacrimal catheters.
2 Site of puncta lacrimalia.
3 Superior canaliculus.
4 Inferior canaliculus.

5 Common canaliculus.
6 Lacrimal sac.
7 Nasolacrimal duct.
8 Hard palate.

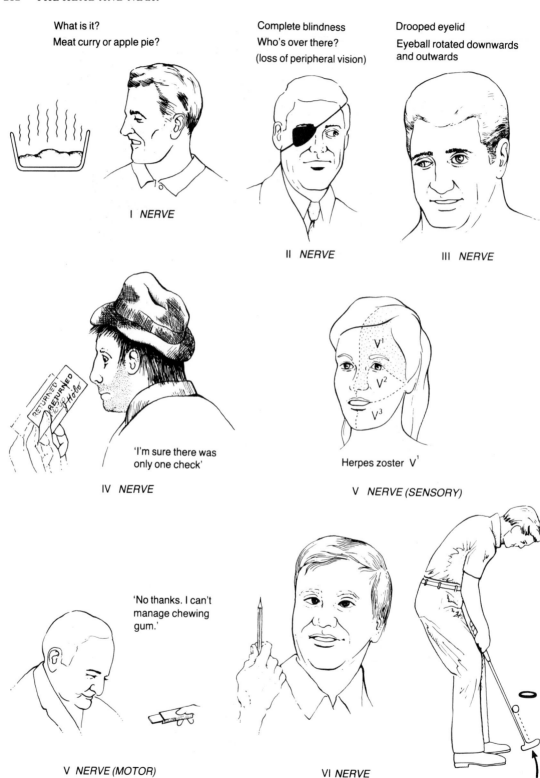

Fig. 18.13 **Summary of cranial nerves and the clinical presentation of lesions of these nerves.**

'I cannot close my eye tightly
or whistle a tune'

'What's that?

I can't hear you.'

VII *NERVE*

VIII   *NERVE (AUDITORY)*

VIII   *NERVE (VESTIBULAR
PORTION)*

'I've got palpitations.'

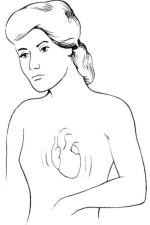

IX & X   *NERVES*

X   *NERVE*

'I've lost the shape
of my neck.'

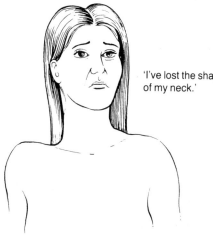

'They went that away.'
(Protruding tongue
deviates to paralysed
side)

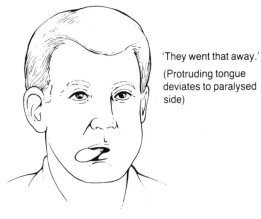

XII *NERVE*

XI *NERVE (SPINAL)*

Table 18.1   Practical cranial nerve functions – a clinical anatomist's guide

Nerve	Site of injury or disease	How to test	Abnormal signs
I. Olfactory **	Cribriform fracture	Odor to each nostril	Anosmia (no sense of smell)
II. Optic ***	1 Optic foramen fracture	Flash light in affected eye	Loss of direct + consensual pupil constriction
	2 Clot, tear of optic tract	Sudden movement from side	Blink reflex absent due to visual defect
III. Oculomotor **	Cavernous sinus fracture in this area or raised intracranial pressure	Flash light in affected eye	Direct pupil reflex absent Consensual reflex present
		Flash light in normal eye	Direct reflex present Consensual reflex absent Eye turned down and out, ptosis and dilated pupil
IV. Trochlear *	Brain stem tumor or orbital fracture	Needs specialist ocular testing	Eye fails to go down and out Patient has diplopia
V. Trigeminal **	Direct injury to maxillary sinus roof	1 Sensation to cornea 2 Sensation to cheek 3 Sensation to chin	} Numbness or paresthesia
		4 Motor – 'Clench your teeth'	Masseter and temporalis fail to contract
VI. Abducens **	Fracture through cavernous sinus, clivus or orbit Raised intracranial pressure	Follow finger from side to side	Diplopia on lateral gaze No lateral movement
VII. Facial *****	Peripherally: 1 Parotid region injury	'Smile'	1 Fascial muscle paralysis: no forehead wrinkle
	2 Fracture of temporal bone	'Whistle'	2 As 1, plus taste in anterior two-thirds of tongue and hearing abnormal
	3 Supranuclear blood clot	'Wrinkle the forehead – frown'	3 Forehead can wrinkle but otherwise facial paralysis present
VIII. Acoustic ***	Petrous temporal fracture (VII often also affected)	Clap hands near ear Tuning fork on middle of head (Weber test)	No hearing Sound travels to good ear only
IX. Glosso-pharyngeal *	Deep cut of neck Brain stem lesion	Tickle fauces with spatula	No gag reflex as sensation lost
X. Vagus **	Deep cut of neck Brain stem lesion	'Open your mouth' – look at soft palate	Uvula deviated to normal side Hoarse voice – vocal cord paralyzed
XI. Spinal accessory **	Laceration or operation in posterior triangle of neck	'Shrug your shoulders' 'Push your chin against my hand' 'Take a deep breath in'	Trapezius fails to contract } Sternomastoid muscle not seen nor felt to contract
XII. Hypoglossal *	Neck laceration under mandible involving carotid arteries	'Stick your tongue out'	Difficulty in speaking (dysarthria) Tongue protrudes to side of lesion

* = rare   **** = frequent

There are several modified sweat and sebaceous glands closely related to and opening near these hairs; when these glands become infected, the condition is called a stye (hordeolum). Situated behind the hairs are the openings of the tarsal glands which pour their oily material through small ducts; when these become blocked, a meibomian cyst forms.

Careful inspection toward the medial end of each lid margin reveals a slight elevation with an opening, the lacrimal punctum. The eyelids now become hairless and continue medially until they meet; in this angle there is a triangular space, the lacus lacrimalis, which contains a rather unattractive fold of skin, the lacrimal caruncle.

Lacrimal apparatus

The lacrimal gland, placed in a small fossa beneath the superolateral margin of the orbit, consists of an orbital part, lying on the lateral portion of the expanded tendon of levator palpebrae superioris, and a palpebral part, extending around its lateral border and visible through the superior fornix of the conjunctiva. In this respect it is similar to the submandibular gland and its relationship to the mylohyoid muscle. A dozen or more ducts lead from the palpebral part into the lateral portion of the superior conjunctival fornix.

Now let's cry. You have just been told by Dean that from now on the study of anatomy is to be voluntary. Overcome with grief, a signal is sent out from the lacrimal nucleus in the pons, via the greater superficial petrosal branch of the seventh nerve, the sphenopalatine ganglion, the maxillary nerve and its zygomatic branch, to the lacrimal nerve. The lacrimal gland is stimulated and tears are poured out of its numerous ducts. They pass into the conjunctival sac and flow medially toward the lacus lacrimalis, the eyelids acting as windshield wipers. Not wishing to make a spectacle of yourself you immediately secrete greasy material from the tarsal glands, which is expressed onto the lid margins – the

waterproofing prevents the spilling of tears. You suddenly remember that the action of the palpebral part of the orbicularis oculi, by virtue of its attachment to the lacrimal sac, is to enlarge the sac and at the same time to invert the puncta into the lacus lacrimalis. After the muscle relaxes, the lacrimal sac contracts by its own elasticity and the tears are sucked up through the two canaliculi that lead from the puncta into the sac. The lacrimal sac, lying in the lacrimal fossa on the medial side of the orbit, narrows inferiorly to become the nasolacrimal duct, which runs downward and laterally for about 2.5 cm (1 in) to open into the inferior nasal meatus about 2.5 cm (1 in) behind the nostril. The tears pour down this duct; however, you merely blow your nose at this catastrophic news and inform the Dean that you are allergic to the tobacco he is smoking. To prevent the tobacco smoke from going up your nasolacrimal duct and into your eye, the mucous membrane in the duct is thrown into several valvular folds (see Fig. 18.12).

The cornea is one of the few structures in the body that does not have a blood supply; the film of tears on the cornea is very thin, and evaporation, especially in very warm climates, would be almost instantaneous and lead to desiccation. To prevent this, a film of oil from the tarsal glands and a film of the mucus from mucous glands in the tarsal conjunctivae are secreted. Whenever the cornea becomes dry, a blinking reflex is set off which immediately replaces the film of tear, oil and mucus.

The upper lid has great mobility compared to the lower lid; the lids are closed gently by the palpebral fibers and forcibly by the orbital fibers of the orbicularis oculi. They are opened by the levator palpebrae superioris and, to a lesser degree, by the smooth muscle fibers associated with this muscle. The corneal reflex – i.e. the involuntary closing of your eye when somebody tries to poke a finger into it – requires the action of the orbicularis oculi (the seventh nerve) and corneal sensation (the fifth nerve).

Review questions on the head and neck

(answers p. 529)

86 A swelling just below the chin is most likely to be a submental lymph node. The area of drainage is from the
a) center of the upper lip
b) molar teeth
c) tip of the tongue
d) posterior one-third of the tongue
e) thyroid gland

87 The external jugular vein may be used for the passage of catheters. It can be located by a line from the angle of the mandible to the
a) suprasternal notch
b) mastoid process
c) xiphoid process
d) acromion
e) middle of clavicle

88 In a patient who has had a stroke, the food often collects between the teeth and cheeks. This is due to paralysis of the
a) masseter muscle
b) orbicularis oris muscle
c) temporalis muscle
d) buccinator muscle
e) mentalis muscle

89 A patient complains of loss of taste at the anterior two-thirds of his tongue. He may have
a) inflammation of the middle ear
b) a tumor in the external auditory meatus
c) injured the angle of his mandible
d) a tumor of his vocal cord
e) divided the hypoglossal nerve as a result of an accident

90 Which of the following does *not* lie in or pass through the jugular foramen?
a) Glossopharyngeal nerve
b) Accessory nerve
c) Superior jugular bulb
d) Superior cervical ganglion
e) Meningeal branch of vagus

91 In drinking while bending over a water fountain, nasal regurgitation of water is prevented by the contraction of several muscles under control of the fifth nerve. One of these is the
a) palatoglossus
b) palatopharyngeus
c) tensor palati (tensor veli palatini)
d) superior constrictor
e) stylopharyngeus

92 When one burns the tip of one's tongue by drinking boiling chicken soup, the pain impulse will be carried to neurons located in the
a) geniculate ganglion
b) submandibular ganglion
c) semilunar ganglion (trigeminal)
d) sphenopalatine (pterygopalatine) ganglion
e) inferior glossopharyngeal ganglion

93 The scalenus anterior is a readily found landmark for the straying surgeon. It enables him to find a nerve which runs on it. This nerve is the
a) vagus
b) ansa cervicalis
c) sympathetic trunk
d) lower trunk of brachial plexus
e) phrenic

94 It is said that the most important muscle in the body is that which opens the laryngeal glottis to allow breathing. This is the
a) cricothyroid
b) posterior cricoarytenoid
c) lateral cricoarytenoid
d) vocalis
e) thyrohyoid

95 A speck of dust is prevented from entering your eye by the blinking reflex between the
a) trigeminal and oculomotor nerves
b) C2 and facial nerves
c) facial and oculomotor nerves
d) trigeminal and facial nerves
e) lacrimal and oculomotor nerves

96 A cervical rib may cause undue tension and compression of structures passing between the anterior and middle scalene muscles of the neck. One of these is the
a) phrenic nerve
b) sympathetic trunk
c) subclavian vein
d) subclavian artery
e) thoracic duct

97 The middle meningeal artery may be involved in a head injury. It lies
a) subdurally
b) in the subarachnoid space
c) in the subpial space
d) extradurally
e) between the two clinoid processes

98 Below the floor of the pituitary fossa (sella turcica) lies the
a) cavernous sinus
b) sphenoid sinus
c) straight sinus
d) sigmoid sinus
e) internal carotid artery

99 A fracture involves the posterior cranial fossa. It may injure the
a) ninth cranial nerve entering the internal auditory meatus
b) eighth nerve entering the internal auditory meatus

c) mandibular division of the fifth nerve passing through the foramen ovale
d) twelfth nerve exiting the foramen magnum
e) middle meningeal artery entering the foramen spinosum

100 Which of the following nerves passes through the foramen magnum?
a) Hypoglossal
b) Vagus
c) Spinal accessory
d) Glossopharyngeal
e) None of these

101 The anterior facial vein communicates with and may infect the cavernous sinus via the
a) angular vein
b) ophthalmic veins
c) pterygoid plexus
d) all of the above
e) none of the above

102 In removing a tumor from the deep part of the parotid gland, the surgeon must be aware of structures lying within it. These include the
a) external carotid artery
b) external jugular vein
c) five branches of facial nerve
d) facial artery
e) stylomandibular ligament

103 Structures important in the suspension of the eyeball in the orbital cavity include the
a) orbital fascia (= Tenon's capsule)
b) superior rectus muscle
c) levator palpebrae muscle
d) suspensory ligament of the lens
e) greater tarsal plate and muscle

104 With respect to the lacrimal apparatus,
a) the lacrimal gland has a single duct, opening at the medial corner of the eye
b) the nasolacrimal duct opens into the middle nasal meatus
c) the lacrimal gland receives secretory innervation from the ciliary ganglion

d)   the lacrimal sac lies in a hollow of the maxillary and lacrimal bones
e)   the lacrimal sac is anterior to the medial palpebral ligament

105   Earache may be caused by oral lesions. Which of the following nerves would probably be responsible in this referred pain pathway?
a)   Greater auricular
b)   Auriculotemporal
c)   Auricular branch of facial
d)   Auricular branch of vagus
e)   None of the above nerves

106   Which of the following statements is true?
a)   At birth, the middle ear ossicles and mastoid air cells are fully developed
b)   Paralysis of the facial nerve may lead to hypersensitivity to sounds (hyperacusis) because the tensor tympani muscle is denervated
c)   Middle ear infection could affect the functions of both the submandibular and the otic ganglion
d)   The long process of the incus attaches to the tympanic membrane
e)   The chorda tympani lies on the stapes bone

107   A fracture through the horizontal plate of the ethmoid bone with meningeal tearing could lead to
a)   detachment of the falx cerebri
b)   loss of sense of smell (anosmia)
c)   cerebrospinal fluid leakage into the orbital cavity
d)   loss of normal nasolacrimal sac drainage
e)   none of the above

108   In the infant skull
a)   the anterior fontanelle is normally closed by 1 year of age
b)   there is no mastoid process but the motor branch of the facial nerve is superficial and protected by the styloid process
c)   there is no diploic layer
d)   no paranasal sinuses have been formed at the time of birth
e)   the tympanic bone is a complete ring at birth

109   Thrombosis of a cavernous sinus may result from spread of infection from the so-called 'danger area' of the face. This part of the face includes the
a)   occipital region
b)   chin
c)   upper lip
d)   parotid region
e)   tragus and pinna of the ear

110   In excising tumors of the parotid gland, surgeons must be aware of structures embedded within it (either superficially or deep). These include the
a)   stem of the vagus nerve
b)   internal carotid artery
c)   otic ganglion
d)   origin of the maxillary artery
e)   accessory nerve (spinal part)

111   The otorhinolaryngologist can locate which of the following between the palatoglossal and palatopharyngeal folds?
a)   Palatine tonsil
b)   Pharyngotympanic tube
c)   Valleculae
d)   Adenoids
e)   Piriform recess

112   A wrestler, with a change of heart, fondly massages the neck of his opponent; the opponent falls on to the canvas and is counted out. Surprised, he feels the victim's pulse and finds it to be very slow. He is now relieved because he remembers that
a)   the carotid sinus lies at the carotid bifurcation, situated usually at the level of the cricoid cartilage
b)   the sinus is a chemoreceptor and responds to chemical changes in the blood
c)   the sinus, when stimulated, sends afferent impulses via the ninth nerve which initiates a reflex via the tenth nerve, causing slowing of the heart
d)   the cardioinhibitory reflex is mediated by the twelfth nerve on the afferent and the ninth nerve on the efferent side
e)   he was supposed to lose and not win this fight

113 With advancing age, the internal carotid artery narrows and reduces the blood supply to the brain. This can be diagnosed by noting an increased flow through the collateral anastomosis formed by the
a) superior and inferior thyroid arteries
b) branches of the ophthalmic artery and the facial artery
c) branches of the ophthalmic artery and the superior thyroid artery
d) facial and lingual arteries
e) superficial temporal and occipital arteries

114 The branches of the external carotid artery include the
a) supraorbital
b) greater palatine
c) inferior thyroid
d) superior thyroid
e) ophthalmic

115 The carotid body
a) is located at the lower end of the common carotid artery
b) is innervated by the ansa cervicalis
c) lies at the level of the upper border of the thyroid cartilage
d) is important in the regulation of the blood pressure
e) all of the above are true

116 Enlargements of the thyroid gland commonly occur.
a) The enlargements are commonly in an upward (superior) direction
b) The enlargement is commonly in an inferior direction into the superior mediastinum, bringing it into relationship with the branches of the left brachiocephalic vein
c) The enlarged gland moves on swallowing as it is attached to the hyoid bone
d) The enlarged gland may compress the cervical sympathetic chain as it lies in the adjacent carotid sheath
e) The enlarged gland may compress the vagus nerve which lies anteriorly

117 The sternomastoid muscle
a) covers the upper end of the carotid sheath
b) receives its motor supply from the third, fourth and fifth cervical nerves

c) turns the face towards the opposite shoulder should a contracture (torticollis) develop
d) has the external jugular vein running through it
e) has three heads inferiorly

118 A tumor of the apex of the lung often invades the neck of the first rib and the structures crossing it. These include the
a) stellate ganglion
b) second thoracic nerve
c) second intercostal vein
d) vagus nerve
e) phrenic nerve

119 Branches of the vagus (tenth cranial nerve) in the neck
a) may slow the heart
b) take part in the swallowing mechanism
c) may be implicated in earache
d) may be implicated in the cough reflex
e) all are true

120 The accessory nerve – spinal portion –
a) may be seen in the anterior triangle of the neck
b) supplies strap muscles (infrahyoid muscles)
c) is important in abducting the shoulder
d) is involved in tongue sensation, posterior two-thirds
e) is involved in tongue sensation, anterior one-third

121 You may often be puncturing the subclavian vein to insert a catheter. You should find this vein
a) between the scalenus anterior and medius muscles
b) above and lateral to the subclavian artery pulsation
c) immediately above the horizontally running inferior belly of the omohyoid
d) anterior to the scalenus anterior muscle and behind the clavicle
e) deep to the lower trunk of the brachial plexus

122 The cervical sympathetic chain
a) contains fibers which cause constriction of the pupil of the eye
b) receives white rami from C5–8

c) sends gray rami to accelerate the heart
d) is located in the carotid sheath
e) joins its partner at the base of the skull to form a single ganglion

123   The hypoglossal nerve
a) lies lateral to the carotid arteries
b) crosses the loop of the lingual artery
c) is a purely motor nerve
d) hooks anteriorly around the occipital artery
e) all the above are true

124   A patient has a swelling of the parotid gland and cannot close his eye; this combination is diagnostic of cancer of the gland. He cannot close his eye because of paralysis of the
a) third nerve
b) seventh nerve
c) cervical sympathetic chain

d) ophthalmic division of the fifth nerve
e) masseter muscle

125   Persistent hiccup (spasmodic diaphragmatic contraction) may be relieved by blocking the phrenic nerve, if you know where it is. It lies
a) on the scalenus medius muscle
b) on the scalenus anterior muscle
c) on the longus colli muscle
d) superficial to the carotid sheath
e) within the carotid sheath

126   On examining the nose, the doctor finds a discharge pooling in the middle nasal meatus. It may originate from
a) the maxillary sinus
b) the ethmoidal sinus
c) the sphenoid sinus
d) a and b only
e) all of these

Part VI The Lower Limb

19　Hip bone and joint

Hip bone · Femur · Position of the lower extremity · Hip joint · Muscles acting on the hip joint

Hip bone

A study of the lower limb should commence with the hip bone; called the os coxae in Milan, the innominate bone in the public houses of London, and the good old-fashioned hip bone in the US, it forms the anterolateral portion of the bony pelvis. This last consists of the two hip bones and the posteriorly placed sacrum.

Looking as if it has been cast in one piece, it actually consists of three separate bones until the night of the sweet-sixteen party, when the three bones fuse at the cup-shaped acetabulum, from which the champagne is drunk. These three bones are the upper fan-shaped ilium, the medially placed pubis and, posteroinferiorly, the ischium.

Ilium

This consists of a lower portion, which forms part of the acetabulum, and an upper ala or wing. The wing has many important lumps and bumps to which references are frequently made. It ends above in a crest, curved convexly laterally in its anterior two-thirds and concavely in the posterior third (see Fig. 24.8). The crest ends anteriorly in an easily palpable and well-known landmark, the anterior superior iliac spine, and posteriorly in a not so easily palpable landmark, the posterior superior iliac spine. The anterior projection marks the upper attachment of the inguinal ligament. The posterior spine also is interesting. A skin dimple, better seen in women than in men (dimple of

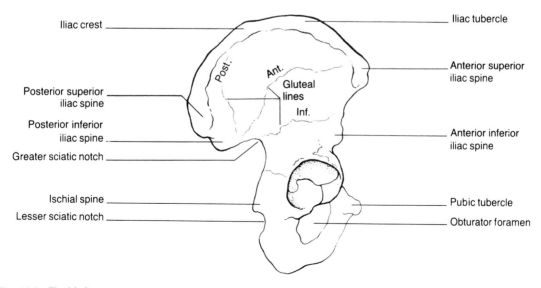

Fig. 19.1　The hip bone.

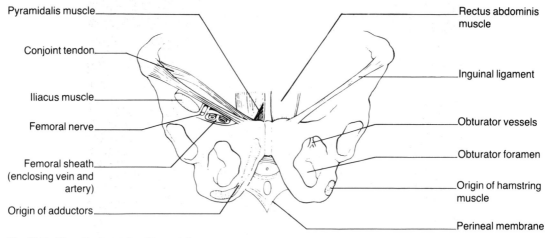

Pyramidalis muscle

Conjoint tendon

Iliacus muscle

Femoral nerve

Femoral sheath
(enclosing vein and
artery)

Origin of adductors

Rectus abdominis
muscle

Inguinal ligament

Obturator vessels

Obturator foramen

Origin of hamstring
muscle

Perineal membrane

Fig. 19.2   The attachments of the pelvis.

Venus), lies close to it, and a line joining these dimples passes through the second sacral vertebra, which indicates the middle of the sacroiliac joints. Below each superior spine lie the smaller and deeper anterior inferior and posterior inferior spines; these are difficult to palpate, and, therefore, are useless as landmarks. To the anterior inferior spine is attached one of the heads of origin of the rectus femoris muscle. About 5 cm (2 in) behind the anterior superior spine is a considerable bump, the tubercle of the iliac crest, often palpable and used as a measuring point in obstetric circles. Just behind this is the summit of the iliac crest. A line joining the two highest points of the iliac crest, the inter-cristal line, cuts across the fourth lumbar vertebra and marks the site of election for a lumbar puncture, i.e. the space between the fourth and fifth lumbar vertebrae. The inner surface of the ala, the bony component of the iliac fossa, is the site of origin of the iliacus muscle. The external surface is directed toward the buttock and is the site of origin of the three gluteal muscles. Those who wish to befriend the gluteal muscles will be interested to know the three gluteal lines. These lines run toward the greater sciatic notch: the inferior from below the anterior superior spine, the anterior from the anterior superior spine and the posterior from just in front of the posterior superior spine. The glu-

teal lines demarcate three areas for the gluteus minimus anteriorly, the medius (as the name indicates) in the middle and a portion of the maximus posteriorly. The body of the iliac bone forms the upper two-fifths of the acetabulum and meets the pubic bone at a prominence called the iliopubic eminence. The internal surface of the posterior part of the iliac bone shows a very rough area. Rather convincingly, an ear-shaped portion, the auricular surface, can be made out anteriorly, which articulates with a similar surface on the sacrum. Posterior to the auricular surface is a roughened irregular area which attaches itself to a similar area on the sacrum by a very strong interosseous sacroiliac ligament. The crest gives attachment in its anterior two-thirds to the three layers of the muscles of the anterolateral abdominal wall with their covering fascia internally and externally, while the posterior third gives attachment to muscles covering the posterior abdominal wall – i.e. quadratus lumborum internally and a portion of the latissimus dorsi and erector spinae more superficially.

Ischium

This bone forms the posteroinferior portion of the hip bone and is best described from above downwards. The upper thick portion forms two-fifths of the acetabulum and leads

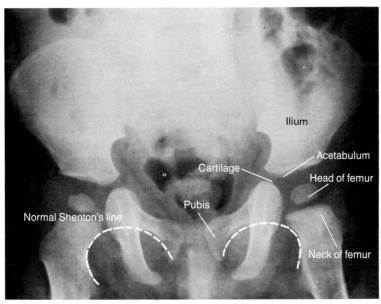

Fig. 19.3 Radiograph of pelvis of young child. The three components of the hip joint are seen (if less than 4 years old, the greater trochanter is not yet visible). The cartilage between the pubis and the acetabulum is not seen, as it is superimposed by the superior ischial ramus. Because of its obtuse angle, the neck of the femur can hardly be recognized, appearing to be a mere continuation of the shaft.

inferiorly to a rather sharp ischial spine. Attached to this is the sacrospinous ligament (see Fig. 32.3), which separates a large greater sciatic notch above from a small lesser sciatic notch below. Just below the spine is a groove for the obturator internus tendon (the tendon of one of the short muscles around the hip joint), which should remind you of the tro-chlea of the humerus; it is somewhat ridged and, in the fresh specimen, coated with fibrocartilage. Below this lies a large bump, the ischial tuberosity. This tuberosity has an area for the attachment of the hamstrings (long muscles of the posterior thigh) and, separated by a ridge, an area on which we sit. In the standing position this portion of bone is covered by the gluteus maximus, which, however, slides out of the way in the sitting position. We would never be able to sit on the bone for long were it not protected by a large thick ischial bursa and dense overlying fibrofatty tissue. Prolonged uninterrupted weight bearing may result in bursitis (weaver's bottom). Below the ischial tuberosity is the inferior ramus, which joins the same-named part of the pubis to form part of the pubic arch.

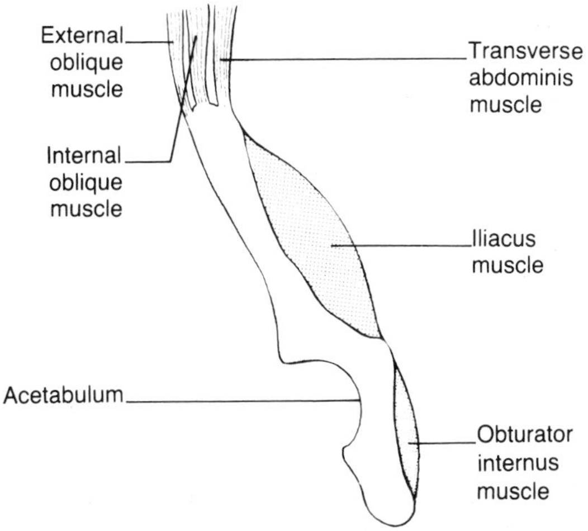

Pubis

The pubis is situated anteromedially and is composed of a flat surface, the body, to which are attached two rami, a superior and an inferior. The former meets the ilium at the

Fig. 19.4 Coronal section through the hip bone. Some of the muscular attachments to the ilium are shown.

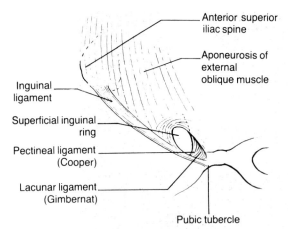

Fig. 19.5   The inguinal, lacunar and pectineal ligaments.

iliopubic eminence and the latter meets the ischial ramus to form the pubic arch. The flattened body of the pubis ends in a smooth surface medially, where it is covered by fibrocartilage and joined to the opposite pubis by ligaments and a fibrocartilaginous disc. The upper surface of the body shows an anterior crest which ends laterally in the important pubic tubercle. Apart from providing an attachment for the inguinal ligament (Poupart), it is an extremely useful landmark. From the pubic tubercle, the finger passes along a fairly smooth surface which has two borders, an anterior obturator crest, related to the obturator foramen below, and a posterior pectineal line, which runs to join the arcuate line of the ilium. Attached to the pectineal line is a prolongation of the inguinal ligament, the pectineal part of the inguinal ligament. The latter continues laterally as Cooper's ligament and merges with the periosteum. The fairly smooth area between the obturator crest and the pectineal line is for the origin of the pectineus muscle. We will later see that the large femoral vessels run over this area of the bone. The external surface of the body of the pubis gives attachment to the muscles on the medial side of the thigh, the three adductors and the gracilis. The inner aspect of the body is smooth and is separated from the urinary bladder by a pre-

vesical space (cave of Retzius), which is filled with fat. This nauseating-sounding space is important; the urologist traverses it to locate and remove the prostate by the retropubic route. The pubis commonly fractures at its two thinnest parts, the two rami, and the body of the pubis may then be displaced posteriorly and injure the bladder. It is not too early to learn that a fractured pubis may mean a ruptured bladder until disproved. Were it not for the protective fat in the prevesical space, damaged bladders would be seen even more commonly. The pubic arch, made up by the fused rami of the pubis and ischium on each side, has attached to it the two lateral margins of the double-layered triangular ligament. The arch also gives origin to many perineal structures.

Acetabulum

Fancifully likened to a vinegar cup, this socket for the head of the femur is made up by the ilium and ischium, each of which forms two-fifths, with the pubis contributing the final fifth. The socket is deep, forming a much more secure joint than its counterpart, the shoulder joint. It has a horseshoe-shaped articular area, smooth and cartilage covered, and a rough area in the floor, the acetabular fossa. The bone here is thin and, in severe accidents, the head of the femur is sometimes rammed right through into the pelvis. The fat-filled fossa is open below, where the acetabular deficiency is bridged by a ligament, beneath which pass the nerves and vessels to the hip joint.

Obturator foramen

Lying between the bodies and rami of the ischium and pubis, the foramen is closed by a

Fig. 19.6   (a) Radiograph of young adolescent male; the ischiopubic rami are fused, but the other epiphyseal lines are still visible. Radiograph of adult male (b) and female (c) pelvis; the wide subpubic angles tell you, for certain, which is the female. Shenton's line may be disrupted in hip joint pathology.

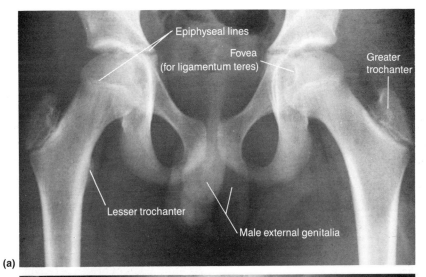

(a)

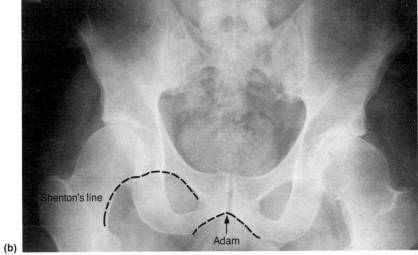

(b)

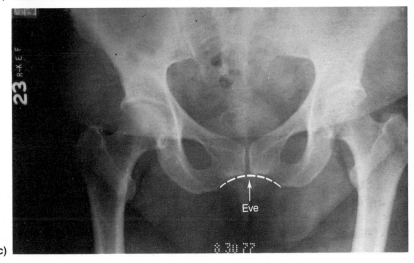

(c)

thin obturator membrane. This membrane attaches to the margins of the foramen, except above, where a small gap allows the emergence of the obturator vessels and nerve. Sometimes more than the vessels pass through; for example, a cheeky little bit of peritoneum might poke its neck through to form an obturator hernia – the alert surgeon beheads it. Surgeons are good at exploiting all possible orifices – the foramen is, at the present time, a favorite alternate route to pass an arterial graft from the abdominal vessels (e.g. aortoiliac) to the femoral vessels in the thigh. The preferred route for a graft from the abdominal to the femoral vessels is over the superior ramus of the pubis.

Femur

This large bone measures about 45 cm (18 in) long in the average person. At its upper end is the head, the ball component of the ball-and-socket hip joint. It is actually two-thirds of a ball and is beautifully smooth, except for a pitted area in its posteroinferior quarter for the attachment of the ligament of the head of the femur, the fovea. In some x-ray projections of the hip joint, this area appears as an apparent defect in the head, as if a rat has taken a nibble out of it, but it is a normal defect (Fig. 19.6a). The head gives way to a well-pitted neck, which looks as if it has had a bad case of adolescent acne – these holes are important for they carry the blood supply to the head of the femur. The neck, limited below by the intertrochanteric line anteriorly and the intertrochanteric crest posteriorly, joins the shaft in the average male at an angle of 125 degrees. In the female, with her wider pelvis and shorter femur, the angle is somewhat less, whereas in a child with its narrow pelvis, the angle is considerably more obtuse. This play on the angles might amuse the geometrically oriented student, but it is very important clinically because alterations of this angle results in diminution of hip joint movement.

Greater trochanter

This is an easily palpable, large, bony projection from the lateral surface of the upper shaft. It is separated from the skin by a bursa, the trochanteric bursa. Deeper bursae separate the trochanter from the tendons of insertions of the gluteal muscles. The summit of the trochanter gives attachment to several small deep muscles in the gluteal region, the piriformis, obturator internus and gemelli, and there is a deep pit on its medial aspect for the insertion of the obturator externus tendon.

The intertrochanteric line anteriorly gives attachment to the capsule of the hip joint and to small parts of origin of two of the vasti. The intertrochanteric crest behind has a prominence on it, the quadrate tubercle, for the insertion of another small deep gluteal muscle, the quadratus femoris.

Lesser trochanter

The lesser trochanter, into which is inserted the powerful iliopsoas muscle, lies posterior and inferior to the greater trochanter. Both trochanters have such strong muscles attached to them that they are sometimes avulsed by sudden severe muscular contractions.

Shaft

This is cylindrically shaped. Posteriorly a ridge, the linea aspera, runs down the midline; should you be curious and trace this superiorly, you will find that the lateral lip of this ridge becomes further raised as the gluteal tuberosity, while the medial lip continues as the spiral line to join the intertrochanteric line. The shaft is gently bowed anteriorly, an important point to remember when setting a fracture of the femur; failure to reproduce this bowing will result in diminished movement at the knee joint.

Position of the lower extremity

In early fetal life the upper and lower extremities are stretched out like paddles at right angles to the trunk, in a position where they could clap both hands and feet. However, as the fetus grows, the uterus cannot accommodate this cheering and clapping fetus; the limbs flex at the joints and, at the same time, rotate. The upper limb rotates so that the cephalic border of the upper limb is rotated laterally to bring the palm facing forward, whereas the cephalic border of the lower limb rotates medially, so as to bring the sole of the foot posteriorly; evidence will be offered later to show that this is not a trick.

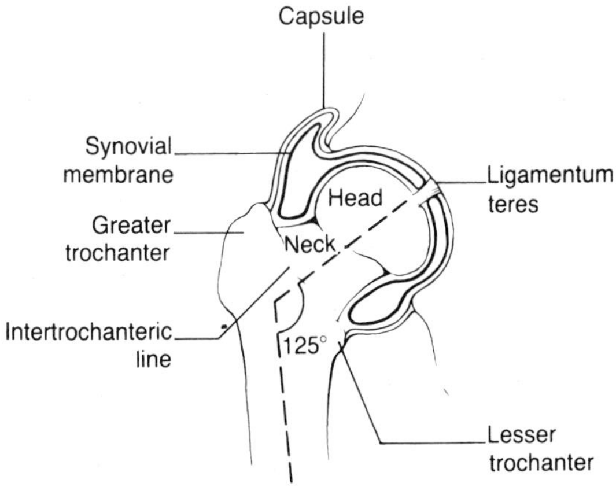

Fig. 19.7   The hip joint. The synovial membrane lines the whole nonarticular portion of the joint; here it is seen reflected onto the ligamentum teres.

Hip joint

Contrary to the shoulder joint, this joint puts stability before mobility.

Bony components

The head of the femur, comprising two-thirds of a sphere, fits into the deep cavity of the acetabulum, articulating with the smooth horseshoe-shaped part of the acetabulum. The rough nonarticular part is covered by fat, which passes inferiorly beneath the transverse acetabular ligament as it bridges the two ends of the horseshoe, to become continuous with the contiguous fat. The displacement of the fat in and out of the joint allows the head of the femur to change position without causing a vacuum; otherwise, constant clattering and clanking noises would occur, as in those who crack their knuckles. To encourage mobility, the head of the femur does not completely fit into the socket. Like the glenoid of the shoulder joint, the acetabulum is deepened by a complete ring of fibrocartilage, the labrum acetabulare. This is attached to the rim of the acetabulum, and where the latter is deficient inferiorly it continues as the transverse ligament. The labrum grasps the head somewhat beyond its equator.

Capsule and ligaments

The capsule is very strong and is attached above to the acetabular margin and transverse ligament. Below it attaches to the intertrochanteric line anteriorly, but posteriorly it does not reach as far as the crest. Here it is separated from the bone by the tendon of a muscle, the obturator externus, and by a protrusion of the articular synovial membrane which acts as a bursa for the tendon; both these structures cover the lower part of the posterior neck. The capsule is strengthened by ligaments which arise from each component bone. The fibers of these ligaments take a spiral course as a result of the rotation of the lower extremity from the fetal into the adult position (first evidence!). The strongest ligament is the iliofemoral ligament, which runs from the anterior inferior iliac spine, in a Y-shaped manner, to the upper and lower portions of the intertrochanteric line. This, one of the strongest ligaments in the body, is sometimes called the Y-shaped ligament of Bigelow. The pubo- and ischiofemoral ligaments arise from the bones just beyond the acetabulum and spiral to their insertion into the intertrochanteric line. A circular thickening of fibers, the zona orbicularis, surrounds the capsule making a sort of hourglass constriction beyond the head of the femur. The ligament of the head of the femur ('ligamentum teres' is a bad name, as it is not round) is an unimportant, sometimes rudimentary, and even absent, structure, which runs from the transverse ligament to a pit in the posteroinferior portion of the head of the femur, and carries some blood vessels.

The fibers of the capsule are reflected from their distal end onto the neck of the femur to run proximally toward the head. These retinacular fibers carry the important terminal branches of the trochanteric anastomosis to supply the head and neck of the femur. The penetration of these vessels produces the pitted neck of the femur. A fracture of the neck of the femur may disrupt these vessels, with death of the head of the femur (avascular necrosis).

Synovial membrane

This lines all the nonarticular parts of the joint; it is reflected over the internal surface of the capsule, the nonarticular fatty pad and the ligament of the head of the femur.

Movements

The muscles and ligaments are the strongest in the body. First, let's take the position of the head of the femur in relationship to gravity. As the center of gravity passes behind the center of the hip joints, it tends to rotate the trunk backward as it pivots on the femoral heads – to resist this, the anterior part of the capsule is markedly thickened.

The strength of the structures around the hip joint is such that, even after the capsule has been cut, the head of the femur is difficult to dislocate. When you observe the orthopedic surgeon dislocating the hip joint, prior to replacing it, you will notice that the surgeon, or more usually the assistant, has forearms like a weightlifter, an indication of the required effort. To permit deep embedding of the femoral head and yet provide a good range of powerful motion, the local muscles are inserted at the end of the neck lever, well away from the head; this

leaves the neck vulnerable. Fractures of the neck are therefore very common. On the other hand, because of the powerful surrounding structures and the effective ball-and-socket arrangement, dislocations and sprains are very uncommon. The neck of the femur acts as a short lever and facilitates movements of the hip joint.

The price for all this dazzling footwork is that the adoption of the erect position places the entire body weight onto the upper part of the head of the femur and, through this, to the neck. Osteoarthritis of this joint is very common – currently a fairly successful operation removes the worn out ball and socket and replaces them with a nonreactive metal or plastic material. Fractures of the neck of the femur commonly follow trivial injuries and tend to occur as one reaches a ripe old age; added difficulties occur when disruption of the retinacular vessels leads to the death of the femoral head.

The movements of the joint are: flexion, as when you sit down; extension, as when you get up; adduction crossing your legs; abduction separating your legs; medial rotation which screws the femur into its strongest

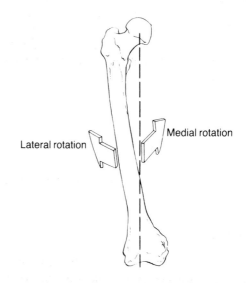

Fig. 19.8 Medial and lateral rotation of the femur. The axis of rotation is the broken line joining the head to the condyles.

position by virtue of its spiral capsular fibers; and lateral rotation which unscrews it. A combination of all these movements is circumduction. Many movements occur together; for example, extension plus medial rotation is the position of greatest stability in standing. Rotation of the thigh occurs through an axis which passes between the head of the femur and its condyles – the femoral shaft rotates backward and forward like a gate on its hinges. All muscles inserted lateral to this axis anteriorly close the gate, i.e. rotate it forward and medially; all the muscles inserted posteriorly open the gate, i.e. rotate it laterally.

Neurovascular supply

The blood supply to the hip joint comes from the surrounding vessels whose branches reach the joint via the acetabular notch. The vessels passing along the ligament of the head of the femur are relatively inconsequential (unlike those accompanying the retinacula), and by the time you are eligible for your pension, they are mostly noncontributory. The nerve supply is from the nerves supplying the three groups of muscles acting on the joint, the obturator medially, the femoral anteriorly and the sciatic posteriorly. The radiation of hip joint pain is very important clinically, as it may be referred to other joints supplied by these nerves.

Muscles acting on the hip joint

These consist of four groups, discussed below.

Flexors

While all muscles crossing the hip joint anteriorly are capable of flexing the joint, there is really only one powerful flexor – the iliopsoas muscle, a combination of the psoas and the iliacus.

Psoas major

This is a large important muscle, which arises from the bodies and intervening fibrocartilages of the lumbar vertebrae, from their transverse processes and from fibrous arches at the sides of the bodies which protect the lumbar vessels as they pass posteriorly. From this extensive origin, the muscle passes inferiorly beneath the inguinal ligament to insert into the anterior surface of the lesser trochanter. The muscle has several

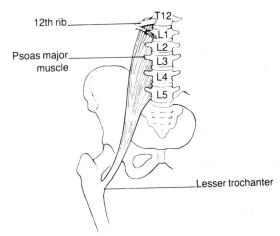

Fig. 19.9 The psoas major muscle. Note the origin reaching as high as the T12 vertebra.

important relationships. It is covered by a sheath, the psoas fascia, the upper end of which is thickened as the medial arcuate ligament and obliquely attached to the body and transverse process of the first lumbar vertebra. The sheath is open superiorly, and necrotic collections may enter it and track down to the thigh – tuberculosis of the lower thoracic spine is notorious for this, and may present as a groin swelling (psoas abscess).

The lateral border of the psoas is so well marked that it is easily seen on plain abdominal x-ray examination – the psoas line; obliteration of this line is of considerable clinical significance. Intra-abdominal disease affecting the psoas may produce spasm of the muscle and therefore flexion of

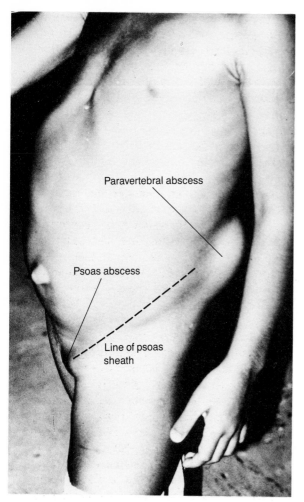

Fig. 19.10   Psoas abscess due to paravertebral tuberculosis.

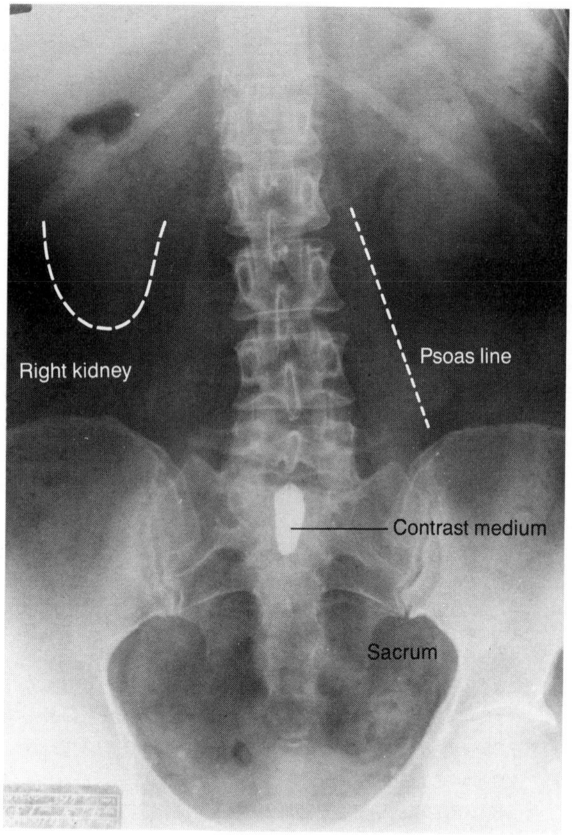

Fig. 19.11   Radiograph of abdomen, demonstrating psoas lines. Contrast medium is in the subarachnoid space and it demonstrates the lowest limit of the space (S2).

the hip joint. An acutely inflamed appendix lying on the muscle irritates it; the psoas becomes enraged, contracts and flexes the hip. Attempts to straighten the hip mean stretching an inflamed, angry psoas and cause a painful reaction – a positive psoas sign. Unaccustomed sudden strains of the psoas during sport, or following a fall, may be so severe as to avulse the lesser trochanter from the femur – easily detectable radiologically. Lesser strains are common; one test is to get the patient to flex the hip joint, a painful exercise. A better test is to have the patient already sitting up in bed (the hip is flexed at 90 degrees), and request him to lift his heel off the bed; he will either be unable to do this or grumble about your causing him extra pain. When the psoas muscle is joined by the iliacus muscle, the covering sheaths of each muscle fuse to become the fascia iliaca.

Iliacus

This muscle occupies the iliac fossa, and, from its wide origin in the fossa, it narrows to pass beneath the inguinal ligament into the thigh, where it inserts into the front of the psoas tendon and a small portion of the bone below. Beneath the converging iliopsoas muscles and separating them from the head of the femur is the psoas bursa, which may communicate with the hip joint. The iliacus muscle is not infrequently involved in intra-abdominal disease, such as cecal lesions, with resultant flexion deformity of the hip.

Adductors

This group of muscles on the medial side of the thigh consists of the three main adductors, longus, brevis and magnus, placed one on top of another, and, lying on either side, the two lesser adductors, the pectineus laterally and the gracilis medially. All of these arise from the external surface of the pubic bone, with the adductor magnus running onto the ischial bone to constitute the sciatic part of this muscle. From these origins the muscles insert into the linea aspera; the adductor longus to its middle, the deeper brevis to a deeper and higher level, and the insertion of the large magnus reaching from the highest level above to a small palpable adductor tubercle below. The adductor longus is easily identified in the living subject, as it has a visible tendinous attach-

ment in the crotch. The pectineus inserts into the upper posterior part of the femur in the same plane as the adductor longus, whilst the long slender gracilis runs down the medial side of the thigh, to insert into the antero-medial surface of the proximal tibia.

Extensors

Gluteus maximus

This muscle is the largest single muscle in the body. It arises from the sacrum and its covering ligaments and the small dorsal surface of the ilium that lies posterior to the posterior gluteal line. The large and coarse fibers (three-quarters of its bulk) run inferolaterally to insert mostly into a thickening of the deep fascia on the lateral side of the thigh, the iliotibial tract. The inferior deep quarter attaches to the gluteal tuberosity of the femur. The upper and lower borders are free; the lower border crosses, but is not responsible for, the gluteal crease, a clinically important fact. This huge muscle is a powerful extensor of the hip joint, but it is not normally called into action. If you don't believe this, carry out the 'gluteal test'. If you place your hand, for purely scientific reasons, on your companion's buttock, during gentle strolling, you will note that the gluteus maximus does not contract with each step. However, when the person is mountain climbing or, in a less adventurous frame of mind, going up some steep stairs, you will note a firmly contracting gluteus maximus. If the leg is fixed, the muscle acts at its proximal attachment to bring the lumbar spine into the line with the femur; thus, it is the muscle used to get up out of a chair. In other words, the gluteus is used mainly as an antigravity muscle during forced extension of the hip joint. For ordinary walking, the hamstring muscles (described later) suffice.

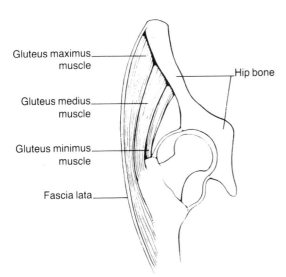

Gluteus maximus muscle

Gluteus medius muscle

Gluteus minimus muscle

Fascia lata

Hip bone

Fig. 19.12   Attachments of the gluteal muscles.

Abductors

Two other glutei, the medius and minimus, lie deep to the maximus. Both arise from the dorsal surface of the iliac bone, the medius in the middle and the minimus anteriorly – so anteriorly, in fact, that it is visible in an anterior dissection of the thigh. These muscles run directly across the lateral surface of the hip joint to insert into several ridges on the greater trochanter of the femur – this makes them pure abductors, 'the gluteal deltoid' if you wish. Because abduction of the hip joint is not a terribly useful movement (unless you are a professional waterskier) their real function occurs during walking. With one leg off the ground prior to taking a forward step, the leg on the ground holds the body weight over it, preventing its falling to the opposite side. It is the gluteus medius and minimus which contract to hold the hip bone over the femur; without them you would fall to the other side. This is seen classically in patients who have had poliomyelitis with selective paralysis of these muscles, and who develop a rolling gait (Trendelenburg). It is also seen in disorders which result in coxa vara, which sounds like the name of some mysterious female espionage agent. Coxa vara is characterized by a reduction in the angle of the neck of the femur, which results in an approximation of the greater trochanter and the ilium; the muscles, acting at a mechanical disadvantage, are rendered ineffective. Coxa vara can be caused by a congenital dislocation of the head of the femur, where the latter slips out of the acetabulum to lie on the dorsum of the ilium, by the adolescent neck of the femur slipping off the head at the epiphyseal line (slipped epiphysis) or by the neck of the femur fracturing or being bent by deformity. All result in the same waddling gait and a positive Trendelenburg sign. The latter sign is demonstrated by standing on one leg – the glutei are unable to maintain the hip bone, and the body weight above it, over the line of

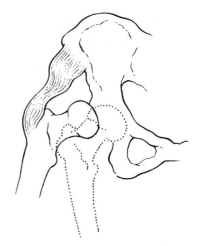

Fig. 19.13   Gluteus medius and/or minimus at a disadvantage. The approximation of the origin and insertion of the glutei (in this case due to an old dislocation of the joint) renders the muscle(s) ineffective.

the femur, and the body falls to the opposite side.

Another muscle, the tensor fasciae latae, is closely related to the glutei and has the same nerve supply, but is situated more anteriorly. Arising from the outer lip of the iliac crest, the muscle runs inferiorly to insert into the iliotibial tract. This is the second muscle inserted into this tract, the gluteus maximus being the other.

In summary, there are the adductor group on the medial side supplied by the obturator nerve; the anterolateral abductors, the gluteus medius and minimus supplied by the superior gluteal nerve; posteriorly lie the gluteus maximus, the powerful antigravity extensor, supplied by the inferior gluteal nerve and the hamstrings, the less powerful extensors used in easy walking and supplied by the sciatic nerve; and anteriorly, the iliopsoas muscle, the main flexor, supplied by the intra-abdominal lumbar plexus.

20 Thigh

Deep fascia · Superficial veins and nerves · Muscles of the anterior thigh · Femoral triangle

As in the gluteal region, the skin on the upper thigh possesses creases. These creases should be at the same level on each side; if they are at different levels in a newborn child, one should suspect congenital dislocation of the hip.

Deep fascia

The deep fascia of the thigh is very substantial, as it encloses very large muscles. Above, it attaches to the pubic tubercle, the inguinal ligament and the iliac crest. On the lateral aspect the fascia is especially thickened to form the iliotibial tract. The tensor fasciae latae and three-quarters of the gluteus maximus are inserted into the tract, which runs inferiorly to insert into the anterior aspect of the lateral tibial condyle anterior to the axis of the knee joint. This means that the gluteus maximus, a powerful extensor of the hip joint, is an equally powerful extensor of the knee joint. Because of these actions the gluteus maximus can be regarded as a major antigravity muscle. The deep fascia is a good friend to the surgeon; it is often used in strips or patches to close defects, such as herniae, elsewhere in the body.

Superficial veins and nerves

Long saphenous vein

The long saphenous vein, the longest vein in the body, starts on the dorsum of the foot and passes proximally to reach the thigh behind the medial femoral condyle. It then veers anteriorly to reach the root of the thigh, where it joins the femoral vein at a point 3–4 cm (1.5 in) below and lateral to the pubic tubercle. It is best to rely on this bony landmark, as other landmarks are often obscured by fatty collections. To reach the saphenofemoral junction the vein passes through an oval opening in the deep fascia, the saphenous opening (fossa ovalis), where it is accompanied by lymphatics connecting the superficial and deep inguinal nodes, and by superficial branches of the femoral artery. With multiple structures piercing the fascia covering the fossa ovalis, this layer is called the cribriform fascia. Before piercing the deep fascia, the vein is joined by three fairly constant tributaries – the superficial circumflex iliac from the region of the anterior spine, the superficial epigastric from the lower abdomen and the superficial external pudendal from the pubic area. Inferior to the saphenous opening, the saphenous vein is joined by two large branches from the medial and lateral aspects of the thigh. As the saphenous vein contains a far higher percentage of muscular and elastic fibers than any other vein, it is the favorite vein for use in arterial bypasses. It also contains numerous valves which assist the venous return from the lower extremity; incompetence of these valves is one of the causes of varicose veins.

Cutaneous nerves

The nerves that supply the skin of the anter-

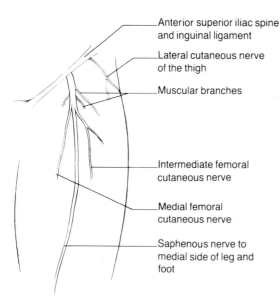

Anterior superior iliac spine
and inguinal ligament

Lateral cutaneous nerve
of the thigh

Muscular branches

Intermediate femoral
cutaneous nerve

Medial femoral
cutaneous nerve

Saphenous nerve to
medial side of leg and
foot

Fig. 20.1 Distribution of the femoral and lateral cutaneous nerves.

ior thigh are branches of the femoral, the medial and the intermediate cutaneous nerves. On the lateral aspect the lateral cutaneous nerve of the thigh, a branch of the lumbar plexus, enters the thigh medial to the anterosuperior spine and divides into branches supplying the anterolateral aspects of the thigh. Medially the obturator nerve makes its contribution.

Muscles of the anterior thigh

Sartorius

This is the longest muscle in the body. It arises from the anterosuperior iliac spine, runs downward and medially due to the rotation of the limb, and crosses the adductor muscles to insert into the anteromedial surface of the tibia. Its fanciful name is derived from its action, which brings the lower limb into the cross-legged position of the tailors of yesteryear. The femoral nerve innervates the muscle.

Quadriceps femoris

The quadriceps femoris is a group of four merging muscles, which together constitute the largest muscle in the body. Individually these muscles are the rectus femoris and the three vasti, lateralis, medialis and intermedius. The rectus femoris runs straight down from the anteroinferior iliac spine to be joined on each side by the vastus medialis and lateralis, which arise from the respective lips of the linea aspera and diverge proximally to attach as high as the intertrochanteric line. Sandwiched between the two vasti and deep to the rectus is the vastus intermedius, which arises from the anterolateral surface of the shaft of the femur. These muscles join a common tendon inferiorly, the quadriceps tendon, which inserts partly into the upper border of the patella or knee cap. Most of the fibers, however, pass over the patella as the ligamentum patellae to insert into the tuberosity of the tibia – in fact, the insertion of the quadriceps tendon can truly be said to be into the tibial tuberosity. Therefore, except for the rectus femoris, an additional flexor of the hip joint, the quadriceps are extensors of the knee joint. But what is an extensor doing on the anterior aspect of the body? All other extensors are situated on the posterior aspect. This is further proof that the leg rotated anteromedially during development.

Pectineus

This muscle cannot make up its mind whether it belongs to the adductors or to the quadriceps; in fact, it may be supplied by the nerves supplying both groups. Arising from the pubis, it inserts into the upper posterior femur and lies between the adductor longus and the termination of the iliopsoas muscle.

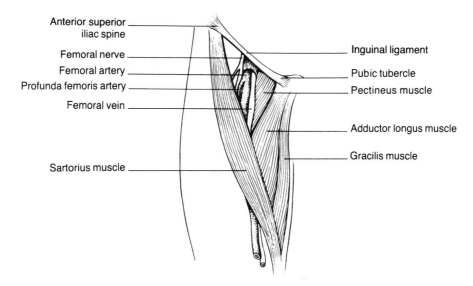

Anterior superior iliac spine

Femoral nerve

Femoral artery

Profunda femoris artery

Femoral vein

Sartorius muscle

Inguinal ligament

Pubic tubercle

Pectineus muscle

Adductor longus muscle

Gracilis muscle

Fig. 20.2   The femoral triangle.

Femoral triangle

This triangle, dearly beloved by older anatomists, is trough-like and is factually and traditionally bounded by the inguinal ligament (base), the medial border of the adductor longus and the medial margin of the sartorius. Its floor consists of the adductor longus, pectineus and iliopsoas muscles.

Contents

Below the inguinal ligament lie the femoral vein, the femoral artery and the femoral nerve mediolaterally. Proceeding distally the vein becomes posterior and the branches of the nerve anterior to the artery (the result of rotation).

Femoral vein

This large vein, the most medial structure of the three, passes behind the inguinal ligament to become the external iliac vein. Its tributaries include the saphenous vein and the profunda and its branches.

Femoral artery

This vessel (often called common femoral artery) passes beneath the inguinal ligament at the midinguinal point, and crosses anterior to the femoral vein as it passes inferiorly. It gives off three superficial branches which accompany the branches of the saphenous vein, the superficial iliac, epigastric and pudendal, as well as a deep pudendal which passes medially behind the femoral vein towards the genitalia. The very important profunda branch and its tributaries vary considerably in their origin; the profunda may come off from 1.25 to 5 cm (0.5 to 2 in) below the inguinal ligament, and its main branches, the two circumflex arteries, may arise not from the parent trunk but independently from the femoral. The profunda is really the blood supply to the muscles of the thigh, whereas the femoral artery distal to profunda's origin (often called the superficial femoral) is the blood supply to the leg. The anastomosis between the two is such that the profunda may be able to bypass a common femoral artery block by its anastomoses with branches of the iliac vessels above the block or with branches of the superficial femoral below the block. It is important, therefore, to have a patent profunda. A method of treating an obstructed common femoral artery, a common condition, is therefore directed to unblocking the origin of the profunda artery (profundaplasty), so as to take advantage of its numerous anastomoses. The femoral artery runs downward in the femoral triangle, where it lies on the adductor longus and magnus. It leaves the triangle to enter the subsartorial canal. The profunda arises on its posterolateral side and, with its accompanying vein, runs inferiorly to leave the femoral triangle at its apex where it passes behind the adductor longus. Older textbooks of anatomy often mentioned a mysterious assailant who could stab his victim through the apex of the femoral triangle, severing from before backwards the femoral artery and vein and the profunda artery and vein. The profunda artery gives off two large branches, the medial and lateral circumflex arteries, which may arise directly from the femoral artery (Fig. 35.6). The medial circumflex passes deeply on the medial side of the thigh, making for the gluteal region; on its way, it gives off very important branches of supply to the head and neck of the femur. The lateral circumflex artery runs laterally, deep to the rectus femoris, where it breaks up into ascending, transverse and descending branches. The ascending branch passes cranially toward the anterior superior spine, where it forms a very important anastomotic channel with the superior gluteal artery; this bypass can circumvent blocked common femoral or external iliac arteries. The transverse branch encircles the femur and anastomoses with the transverse branch of the medial circumflex. The descending branch runs distally to take part in the anastomosis around the knee joint; it assists the bypassing of a blocked superficial femoral artery.

Femoral nerve

The femoral nerve, lying outside the femoral sheath, passes into the thigh where it quickly breaks up into its musculocutaneous branches. The cutaneous branches are mostly expended in supplying the skin of the anteromedial thigh via the medial and intermediate cutaneous nerves of the thigh, but the saphenous branch continues distally through the femoral triangle and subsartorial canal to eventually reach the medial side of the foot. The muscular branches supply the muscles of the anterolateral thigh, the sartorius and quadriceps muscles, and possibly the dithering pectineus.

The femoral vessels are surrounded for a short distance by a femoral sheath, downward prolongations of the fascia transversalis and fascia iliaca (see Chapter 24). The sheath has a medial compartment, the femoral canal, which lies medial to the vessels and contains a pad of fat, the femoral septum, some lymph vessels and a lymph node.

Obturator nerve

If we regard the sciatic as the nerve of the posterior compartment of the thigh and the femoral as the nerve of the anterior compartment, then the obturator is the nerve of the medial compartment. It enters the thigh by passing through a small gap above the obturator membrane, the obturator foramen, in which it divides into anterior and posterior terminal divisions. These two nerves lie sandwiched between the adductors, the anterior division between the longus and brevis, and the posterior between brevis and magnus. It is a truism to say that nerves lying between muscles and near joints, supply them. The anterior division supplies the

Fig. 20.3   Cross-section of (a) midthigh (diagrammatic), (b) upper thigh (computerized tomography) and (c) lower thigh (computerized tomography). AB, adductor brevis; AL, adductor longus; AM, adductor magnus; BF, biceps femoris; F, femur; G, gracilis; GM, gluteus maximus; HS, hamstrings; LSV, long saphenous vein; RF, rectus femoris; S, sartorius; SM, semimembranosus; SV, saphenous vein; VL, vastus lateralis; VM, vastus medialis.

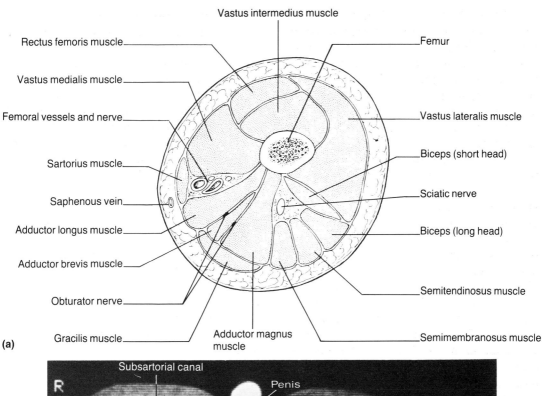

Vastus intermedius muscle

Rectus femoris muscle

Vastus medialis muscle

Femoral vessels and nerve

Sartorius muscle

Saphenous vein

Adductor longus muscle

Adductor brevis muscle

Obturator nerve

Gracilis muscle

Adductor magnus muscle

Femur

Vastus lateralis muscle

Biceps (short head)

Sciatic nerve

Biceps (long head)

Semitendinosus muscle

Semimembranosus muscle

(a)

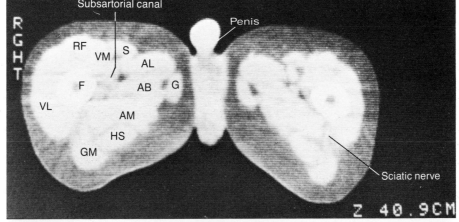

Subsartorial canal

Penis

R G H T

RF

VM   S

AL

F   AB   G

VL

AM

HS

GM

Sciatic nerve

Z 40.9CM

(b)

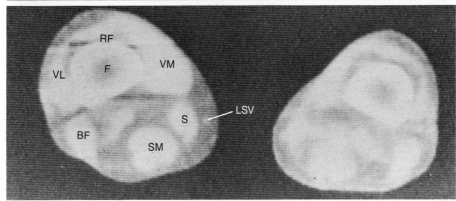

RF

VL   F   VM

S   LSV

BF

SM

(c)

adductors longus and brevis, another member of the adductor group, the gracilis, and gives a branch to the hip joint. The posterior division, having pierced and supplied the obturator externus muscle as it lies on the outer aspect of the obturator membrane, passes between the adductors brevis and magnus, supplies the latter muscle and gives off a long slender articular branch to the knee joint.

The innervation of both the hip and the knee joints by the obturator nerve is of great clinical significance. Most patients with hip joint disease complain of pain in the knee, and because of the hip joint disease they also limp. A patient limping and pointing to his knee as the site of pain often has the luxury of an x-ray examination of his knee joint; he is overjoyed when he is told it is normal. His joy turns to venom when it is later found that the trouble lies one joint higher up. The nerve supply to the gracilis is also a good example of Hilton's law (the nerve to a muscle supplies the joint on which it acts and the skin overlying the joint); the muscle, joint and the overlying skin on the medial aspect of the knee are all supplied by the obturator nerve.

Subsartorial canal (Hunter)

Leaving the relative comfort of the femoral triangle, the femoral vessels enter a narrow canal 15 cm (6 in) long, which terminates 10 cm (4 in) above the adductor tubercle. Here the vessels pass posteriorly to enter the popliteal fossa. The subsartorial canal is roofed over by the sartorius (as the name suggests) on the anteromedial aspect; the vastus medialis lies laterally, and the floor is formed by the adductors longus and magnus. Beneath the sartorius is a thick sheet of fascia which has to be incised before the vessels can be exposed. The canal contains the two femoral vessels and three nerves, the saphenous, the nerve to vastus medialis and the articular branch of the obturator nerve to the knee. The femoral artery lies on the femoral vein, but is itself crossed by the

saphenous nerve as the latter runs latero-medially to leave the canal. The nerve becomes superficial on the medial aspect of the thigh, between the sartorius and the gracilis, where it pierces the fascia in company with the normally insignificant saphenous branch of the descending genicular artery; the last may, however, become a good friend in occlusions of the femoral artery. The femoral artery passes through the tendinous fibers of the adductor magnus to reach the popliteal fossa. It is at this point that most femoral artery occlusions commence. When this was first observed, surgeons immediately busied themselves enlarging the tendinous adductor hiatus through which the vessels pass, believing that the chronic traumatic movement of the artery against the thick tendon was responsible for the commencement of the occlusive changes. However, it is inconceivable that

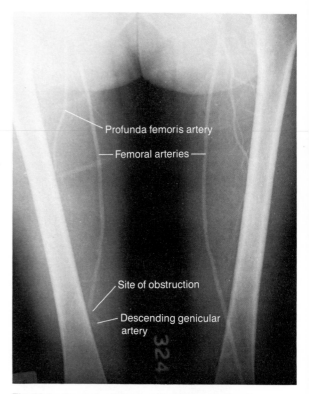

Fig. 20.4   An angiogram showing the femoral and profunda femoris arteries. The right femoral is blocked at the adductor hiatus (the usual site) and a descending genicular is seen immediately proximal to the block.

this could be a responsible factor, as everybody would then develop this block. Just above the hiatus, the femoral artery gives off a descending genicular branch to join the anastomosis around the knee joint; in the event of a block this may well become an important artery (Fig. 20.4).

Profunda femoris artery

Running behind the femoral vessels at the apex of the femoral triangle, the profunda femoris vessels pass deep to the adductor longus to lie on adductors brevis and magnus. The profunda artery gives off circumflex branches, numerous branches to the surrounding muscles and four large perforating branches. These last perforate the attachments of the muscles to the linea aspera as they pass behind the femur mediolaterally to end in the vastus lateralis.

21 Gluteal region and posterior thigh

Gluteal region · Posterior aspect of the thigh

Gluteal region

This region is heavily padded with fat, and is much more extensive than the 'artist's buttock'! Deep to the fat is the large gluteus maximus muscle, which covers several smaller muscles. Amongst these are the tensor fasciae latae anterosuperiorly and the gluteus medius and minimus a little more posteriorly. Inferior to the gluteus medius is the piriformis, a dearly beloved muscle of anatomists who feel that this is the key to the gluteal region. They justify this belief by saying that structures come out

cephalad or caudal to the muscle – true, but it does not help if you don't know the structures! Structures coming out above the muscle are the superior gluteal artery, which supplies the gluteus medius and minimus muscles, and the superior gluteal nerve that supplies the three anterosuperior muscles, the gluteus medius, gluteus minimus and tensor fasciae latae. Below the piriformis emerge numerous structures of which the largest by far is the huge sciatic nerve. Lying on the sciatic nerve is the posterior cutaneous nerve of the thigh, and lying beneath it is a nerve that supplies one of the less

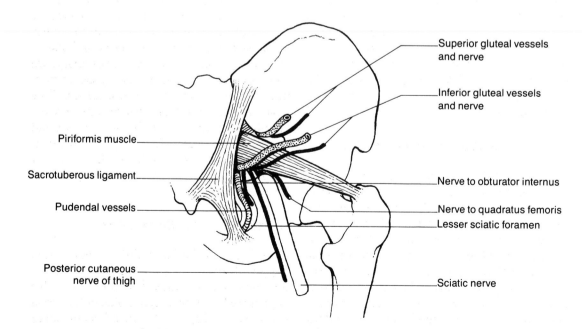

Piriformis muscle

Sacrotuberous ligament

Pudendal vessels

Posterior cutaneous nerve of thigh

Superior gluteal vessels and nerve

Inferior gluteal vessels and nerve

Nerve to obturator internus

Nerve to quadratus femoris
Lesser sciatic foramen

Sciatic nerve

Fig. 21.1 Structures passing through the sciatic foramina.

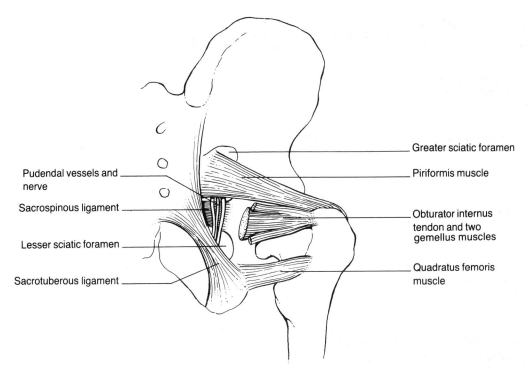

Fig. 21.2   The internal pudendal nerves and vessels, and the small muscles of the buttock. The five illustrated are all lateral rotators of the hip.

important muscles of the gluteal region, the quadratus femoris. The gluteus maximus is so large that it needs its own nerve and blood supply – the inferior gluteal artery and nerve. Most medially are grouped three structures – the nerve to obturator internus and the important pudendal vessels and nerve, the last destined for the pudendal canal en route to the perineum.

Small muscles of the gluteal region

The piriformis muscle arises inside the pelvis and passes through the greater sciatic notch to reach its insertion into the greater trochanter. Below the piriformis are several small muscles. The obturator internus also emerges from the pelvis, and inserts into the greater trochanter, but does so via the lesser sciatic notch; its tendon has attached to it above and below two small muscles fancifully called the gemelli (superior and inferior). Below this is a square-shaped muscle,

the quadratus femoris, which passes from the ischial tuberosity to the greater trochanter. The obturator externus, partially covered by the quadratus femoris, arises from the external aspect of the puboischial bone and inserts into the greater trochanter. In summary, we have a group of small muscles passing from puboischium to the greater trochanter. They cover the hip joint posteriorly and are the counterparts of the small muscles around the shoulder joint. Primarily, these small muscles help to stabilize the hip (or shoulder) joint; secondarily, they rotate the hip joint laterally.

Arterial supply of the buttock

In addition to the large gluteal vessels that supply all the muscles of the gluteal region, the medial circumflex femoral artery also appears in the gluteal region. The latter vessel supplies the hip joint and adjacent small muscles, while one of its branches runs up to the trochanteric region to anastomose with the terminal transverse branch of the lateral circumflex. This circumflex anastomosis connects superiorly with the inferior gluteal artery and inferiorly with the first perforating artery as it passes through the adductor magnus. This constitutes the cruciate anastomosis, which connects the internal iliac and femoral vessels, and which may circumvent an external iliac artery block.

Posterior aspect of the thigh

Beneath the skin and fascia in the midline lies the posterior cutaneous nerve of the thigh, with sensory branches to the skin of the posterior thigh, gluteal region and perineum. The hamstring muscles lie deep to the nerve and run from the ischial tuberosity to the upper tibia and fibula. To qualify as a hamstring, a muscle must arise from an ischial tuberosity, insert into a bone of the leg and be innervated by the tibial component of the sciatic nerve. Beneath the hamstrings lies the posterior aspect of the very large adductor magnus, which is therefore visible from the anterior and the posterior aspects of the thigh; it is duly rewarded for this effort by receiving an anterior supply from the obturator nerve and a posterior supply from the sciatic nerve. The three hamstrings are the long head of biceps laterally, and the semitendinosus lying on the semimembranosus medially. From their ischial origin the muscles diverge, the biceps passing to the fibula, and the semitendinosus and semimembranosus passing to the posteromedial tibia. The space between this diversion constitutes the upper part of the popliteal fossa. The second head of the biceps, the short head, arises from the linea aspera of the femur, and joins the long head for their common insertion into the fibula; it therefore does not qualify as a hamstring.

Sciatic nerve

The sciatic nerve, the largest nerve in the

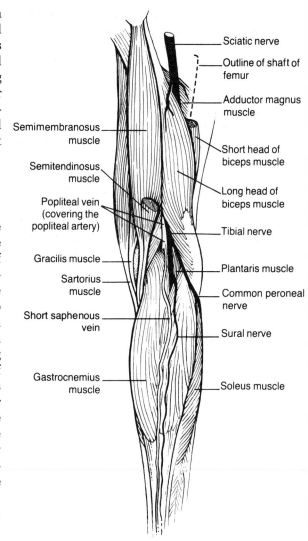

Fig. 21.3 Posterior aspect of the thigh. The femoral vessels emerge from the adductor hiatus to become the popliteal. Note the short saphenous vein emerging from behind the lateral malleolus and joining the popliteal vein. The popliteal artery is deep to the vein and the major branches of the sciatic nerve.

body, arises from the L4, 5, S1, 2, 3 nerves in the pelvis and passes through the greater sciatic notch, usually below the piriformis into the buttock. Here the nerve runs vertically downward under cover of the gluteus maximus, lying on the short muscles of the gluteal region midway between the greater trochanter and the ischial tuberosity.

Emerging from the covering gluteus maximus, it is soon crossed by the long head of biceps as this muscle passes laterally. At about the junction of the middle and lower thirds of the thigh, it divides into its two terminal branches, the tibial and common peroneal nerves. These two nerves are, in fact, really separate adherent structures; sometimes they emerge from the pelvis separately and remain so. The sciatic nerve is so large that it has its own vessel of supply from the inferior gluteal artery. It is embryologically interesting that this artery was once the major artery of the lower limb and it sometimes surprises the surgeon by retaining this embryological prominence. The nerve supplies all the hamstrings, semitendinosus, semimembranosus and long head of biceps; as the muscles arise medial to the nerve, the branches arise from the medial side of the nerve. The only branch from the lateral portion of the nerve (common peroneal) passes to the short head of the biceps — another reason for the muscle giving up its claim to being a hamstring. The point midway between the ischial tuberosity and greater trochanter indicates the site of the nerve on the body surface; above this it passes in a gentle curve medially. It is very important for injections not to be given into or near the nerve; for this reason most injections are given in the upper and outer quadrant of the anatomical buttock (gluteal region). Failure to remember this may be disastrous for patient and doctor alike!

22 Knee

Bones · Knee joint · Popliteal fossa · Bursae around the knee joint

Bones

The bones in the region of the knee are the patella, femur and tibia.

Patella

The patella (knee cap) is a triangular-shaped bone which articulates with the patellar surface of the femur. The posterior cartilage-coated articular surface has a vertical prominence, which divides it into large lateral and smaller medial surfaces to conform to the femoral surface. The patella is embedded in the quadriceps tendon, and therefore qualifies as a sesamoid bone. This tendon continues as the ligamentum patellae from the apex of the patella and inserts into the tibial tuberosity. Three of the four muscles that comprise the quadriceps are inserted mostly into the upper border and slightly into the margins of the patella; the fourth muscle, the vastus medialis, is inserted into the medial margin and only slightly into the upper margin (see reason later).

The patella starts to ossify in the third year of life. Occasionally, there are one or more small separate centers in relation to the upper and outer portion of the patella, the bipartite or tripartite patella; this may be confused with a fracture. Remember that abnormalities of ossification are nearly always bilateral, so x-ray the other side as well.

Injuries in this region may disrupt the quadriceps insertion into the patella, fracture the patella itself or rupture the tibial inser-

tion of the ligament. The last is unusual, but a partial avulsion of the tibial epiphysis is quite common in children (see below). Rupture of the extensor mechanism leads to loss of extension of the knee, and a gap can be seen or palpated above or below the patella.

Lower end of the femur

Except for its medial aspect, the lower end of the shaft has many muscle attachments. The linea aspera divides inferiorly into lateral and medial supracondylar lines that enclose the popliteal surface of the femur. The medial supracondylar line leads inferiorly to the palpable adductor tubercle, the site of attachment of the adductor magnus tendon. The end of the femur consists of two large condyles joined anteriorly by the trochlear or patellar surface, and separated posteriorly by the intercondylar notch. Each condyle has outward-projecting epicondyles for the attachments of the medial and lateral hinge ligaments of the knee joint. The two condyles are coated with articular cartilage, and a groove separates the patellar and tibial surfaces. This groove, produced by the semilunar cartilages or menisci in full extension of the knee, demarcates two articular surfaces, an anterior for the patella and a posterior for the tibial condyles. The patellar surface of the trochlea is vertically grooved so as to present a much larger lateral than medial lip. Posterior to the patellar surface, the anteroposterior axis of the lateral condyle is straight, while that of the medial condyle is curved as well as longer. You will need this knowledge to understand the movements of the knee joint.

A center of ossification appears in the lower end of the femur at or just before birth. This timing is so constant that the presence of this center is regarded as evidence of maturity; thus, in the drama of the courts of law, the cunning lawyer finally brandishes the hidden x-ray triumphantly – the presence of the epiphysis changes the crime from abortion to one of infanticide.

Upper end of the tibia

The surfaces of the two large condyles on the upper end of the tibia are almost flat and covered by articular cartilage. Between them, in the intercondylar area, are two upward projections, the anterior and posterior tibial spines. The articular surface of the femur fits into the very shallow concavity of the tibial condyles with the intra-articular discs lying peripherally. From the condyles, the tibia narrows down to a shaft. Anteriorly, this shows a large, easily palpable, tuberosity, while posteriorly its smooth surface is related to the popliteus muscle and the overlying vessels. The epiphyseal line runs horizontally just below the level of the articular surfaces posteriorly, but anteriorly it dips down to enclose the upper part of the tibial tuberosity. Into this tongue of ossifying cartilage inserts the ligamentum patellae; severe strains in young children may partially avulse the epiphysis causing a painful knee (Osgood–Schlatter disease).

Knee joint

This joint consists of the femoral condyles articulating anterosuperiorly with the patella and inferiorly with the tibial condyles, all surfaces being coated with the usual articular cartilage. The line of attachment of the capsular ligament to the femur is below the epicondyles laterally, horizontally across from the adductor tubercle posteriorly, while anteriorly it merges with the quadriceps

tendon. Inferiorly, the capsule is attached to the tibia just beyond the articular cartilage, except anteriorly where it dips down toward the tibial tuberosity to become continuous with the ligamentum patellae. Posteriorly there is a defect in the capsule through which the tendon of the popliteus passes from inside the joint to its insertion into the popliteal surface of the tibia (one of the two intra-articular tendons of the body, the other being the long head of biceps brachii). The synovial membrane of the knee joint is a rather large and complicated sac; it covers and is roughly coexistent with the internal aspect of the capsule except for certain modifications. Anteriorly there is a superior extension between the quadriceps tendon and the bone for at least 5 cm (2 in, or three fingers' breadth) above the upper border of the patella – the suprapatellar bursa. This bursa is a favorite site for joint aspiration; fullness in this region indicates fluid in the joint. From the upper part of the deep surface of the synovial membrane, a triangular fold that contains a little fat between its layers passes backward to attach by its apex to the intercondylar notch. The fold is the infrapatellar fold, and its two wings the alar folds. The tendon of the popliteus muscle, like the long head of the biceps brachii, emerges from the joint with a sheath of synovial membrane. The two cruciate ligaments that occupy the intercondylar notch appear to have been invaginated into the synovial membrane from behind and are therefore covered anteriorly and laterally. The synovial sac is a fusion of three embryologically separate joints, one patellofemoral and two femorotibial.

Ligaments

The poor fit of these bones necessitates strong ligaments. The anterior ligament is represented by the ligamentum patellae, a continuation of the quadriceps tendon. The lateral ligament is an unimpressive cord-like band that runs from the lateral epicondyle to the head of the fibula just in front of its styloid

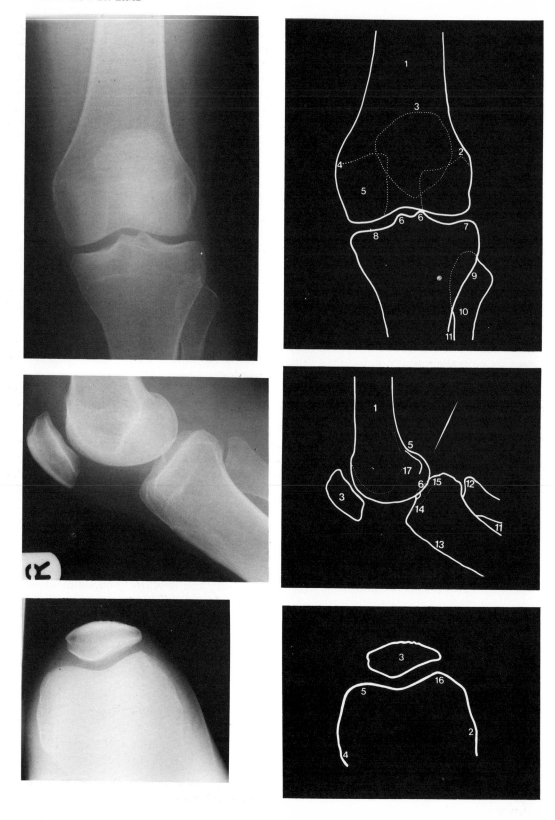

1 Femur.
2 Lateral epicondyle.
3 Patella.
4 Medial epicondyle.
5 Medial condyle.
6 Intercondylar eminences (tibial spines).
7 Lateral condyle of tibia.
8 Medial condyle of tibia.
9 Head of fibula.
10 Neck of fibula.
11 Interosseous membrane.
12 Styloid process of fibula.
13 Tibial tuberosity.
14 Anterior intercondylar area.
15 Posterior intercondylar area.
16 Lateral condyle.
17 Intercondylar notch.

process; at its insertion it splits the tendon of the biceps into two portions. The medial ligament is a strong flattened band that extends from the medial epicondyle to the medial surface of the shaft of the tibia where it attaches over a 5 cm (2 in) area. A small gap above its tibial insertion allows the genicular vessels to pass anteriorly. The medial ligament is separated from the overlying tendons, the sartorius, gracilis and semi-

tendinosus, by a bursa. The posterior ligament is an oblique ligament which is a continuation of the semimembranosus tendon insertion into the posterior aspect of the medial tibial condyle. The ligament crosses the joint and merges with the capsule. The popliteal artery lies on the ligament, which is pierced by the genicular branch of the obturator nerve and the middle genicular vessels.

Intracapsular structures

Menisci (semilunar cartilages)

These two fibrocartilaginous structures appear to have three functions:

1 They deepen the concavities of the tibial condyles to allow a better femoral fit, and in this they fail rather badly.
2 They act as buffers to diminish the pounding of the lower end of the femur on the tibia.
3 Their movements facilitate the dispersion of synovial fluid into the more remote areas of the joint.

When you read of all the troubles caused by these discs, you may wonder if these functions make them worthwhile. Your wonderment will increase when you realize that

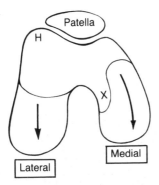

Fig. 22.2 Distal articular surfaces of the femur. Note the larger medial lip of the femoral trochlea (X). The medial condyle is longer and more curved than the lateral. Note that in this view the lateral condyle is higher (H) to prevent dislocation of the patella laterally.

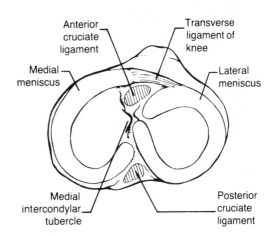

Fig. 22.3 Superior view of the proximal articular surface of the tibia.

these discs are removed with little, if any, apparent ill-effect on the patient.

The medial cartilage is 'C' shaped and the lateral one is 'O' shaped. They have thick peripheral or lateral borders which thin out medially. The peripheral borders are attached to the upper end of the tibia by small fibrous ligaments, and the extremities are anchored by anterior and posterior horns. The horns of attachment of the C-shaped medial cartilage are anterior and posterior to the horns of the smaller O-shaped lateral meniscus. The most anterior fibers of the medial cartilage pass laterally to the lateral cartilage to constitute a transverse ligament. The most striking difference in attachment between the two menisci is that the medial meniscus is firmly adherent to the medial ligament at its periphery, whereas the lateral meniscus is separated from the capsule by the intervening popliteus tendon. As the cause of most cartilage troubles is the trapping and injuring of the structure between the tibia and the femur, it will be recognized that the more mobile the meniscus, the more chance it has of avoiding the crunching femoral condyles. It is no surprise, therefore, that injuries to the medial cartilage outnumber those to the lateral cartilage by almost four to one. The disc may be injured either by being torn away at the insertion of one of its horns or, more commonly, by a piece of the cartilage being split off by the femoral condyle, with the loose flapping portion inside the joint blocking full movement between the two bones (a bucket-handle tear).

Cruciate ligaments

These ligaments are so named because they cross one another like an 'X'. The anterior ligament arises from the tibia between the attachment of the anterior horns of the medial and lateral menisci, and the posterior ligament arises posterior to both horns. The anterior ligament runs upward and laterally to the lateral condyle of the femur, and the posterior ligament runs anteriorly, crossing the anterior ligament on its medial side on its

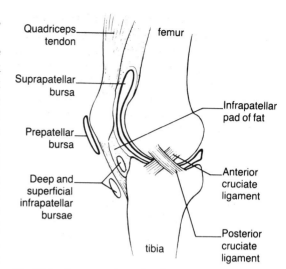

Fig. 22.4   Bursae and cruciate ligaments of the knee joint.

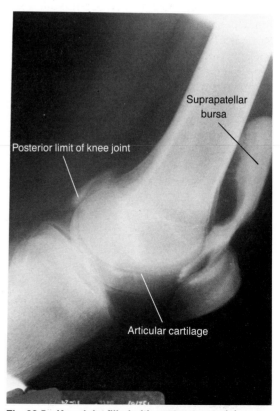

Fig. 22.5   Knee joint filled with contrast material (arthrogram). As air has also been introduced, the articular surfaces of the femoral condyles are particularly well outlined (hurrah for a leaky syringe).

way to the medial condyle of the femur. These ligaments are short and very strong, and forcibly resist any displacement of the knee joint. However, when torn – and the force required to do so is considerable – they are very difficult to replace. They resist excessive flexion and extension in a manner which will be discussed later.

Movements

The knee is a modified type of hinge joint and therefore allows extension and flexion; in the flexed position it also permits rotation.

Extension

As the joint extends, the femur rolls forward on the tibia, coming to rest when the grooves on the femoral condyles are occupied by the anterior portion of the menisci. Full extension, however, involves a terminal medial or internal rotation of the femur on the tibia. Inspection of the lower end of the femur shows that the lateral condyle is shorter and straighter than the medial condyle. When the lateral femoral and tibial condyles have come to rest in extension, there is still a portion of the medial condyle which is 'not used up' and available for further movement. At this stage, the femur pivots medially around the

anterior cruciate ligament to lock the knee joint (the screw-home movement). You can stand in this fully extended knee-locked position indefinitely without using ligaments or muscles; you can prove the lack of muscular contraction by freely waggling the patella across the femur. This terminal medial rotation is therefore an economical way of locking the joint.

Flexion

In the fully extended position, the knee must first be unlocked by lateral rotation of the femur on the tibia or, if the tibia is off the ground, medial rotation of the tibia on the femur. In less than extreme extension, flexion is a simple movement as the femur rolls back on the tibia until the posterior aspects of both its condyles are balanced on the posterior aspects of the tibial condyles.

Rotation

Rotation can be carried out only with the knee flexed; in this position, the collateral ligaments, which run from the epicondyles on the backward-projecting portions of the condyles to the tibia and fibula, are relaxed. This permits the femur and tibia to rotate internally and externally on each other through the intermediary menisci.

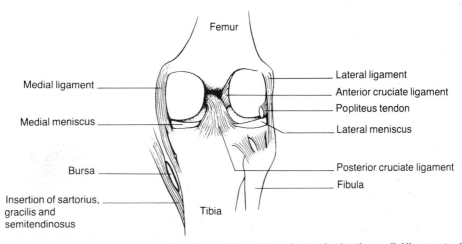

Fig. 22.6 Ligaments of the knee joint. Note that the medial meniscus is attached to the medial ligament, while the lateral meniscus is separated from the lateral ligament by the popliteus tendon.

Muscles producing movements

Extension

This is carried out by the very powerful quadriceps femoris.

Flexion

Flexion of the knee joint is carried out mainly by the hamstrings; any muscles which cross the knee joint posteriorly are additional flexors, such as gastrocnemius and plantaris muscles.

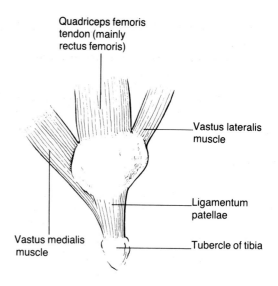

Quadriceps femoris tendon (mainly rectus femoris)

Vastus lateralis muscle

Ligamentum patellae

Tubercle of tibia

Vastus medialis muscle

Fig. 22.8 Anterior view of the extensor mechanism of the knee joint. Most of the vastus medialis is inserted into the medial margin of the patella.

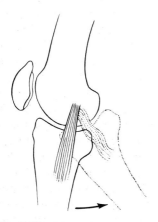

Fig. 22.7 Medial ligament relaxed in flexion.

The gastrocnemius muscle, which constitutes part of the calf of the leg, arises by two heads, a medial from the popliteal surface of the femur just above the medial condyle and a lateral which has been pushed off the popliteal surface onto the lateral aspect of the lateral condyle. The two heads are loosely joined together to insert into the tendo calcaneus. Deep to the lateral head of the gastrocnemius is the little plantaris muscle, which arises from the popliteal surface of the femur just above the lateral condyle and very soon becomes a long tendon which runs all the way down to the heel. A short muscle with a long tendon (such as palmaris longus) means two things:

1 The muscle is going out of style (atavistic), and by the time we run out of natural

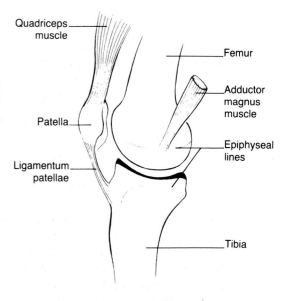

Quadriceps muscle

Femur

Adductor magnus muscle

Patella

Epiphyseal lines

Ligamentum patellae

Tibia

Fig. 22.9 Medial view of the extensor mechanism of the knee joint. Note that the epiphyseal line of the tibia runs below the tuberosity.

resources we will also have run out of plantaris.

2 The long tendon is of great use to surgeons for grafting purposes.

These three muscles form the lower boundaries of the diamond-shaped popliteal fossa.

Rotation

The rotator par excellence is the popliteus muscle. The tendon arises inside the capsule of the knee joint from a depression on the lateral aspect of the lateral femoral condyle and, carrying a synovial sheath with it, leaves the joint to be inserted into a triangular area on the popliteal surface of the tibia. In the flexed position, the tendon occupies a bony groove that is continuous with its origin. This muscle unlocks the knee joint; it rotates either the femur on the tibia or vice versa, depending on which is fixed.

Injuries to the knee joint

Menisci

During rotation, the cartilages are displaced toward the center of the joint. Should there be a sudden thrusting movement of the femoral condyles, such as occurs during a violent body check in hockey or football, the cartilage may be caught between the femur and the tibia and a portion sheared off. As one does not run with the knee completely extended or flexed, the semiflexed position which allows rotation invites this injury. The cartilages know about this and make every effort to get out of the way; the lateral is more mobile and can get out of the way fairly successfully, but the medial is more vulnerable as it is not only fixed at its two horns but also is closely adherent to the medial collateral ligament. The typical history described above, together with tenderness over the anterior portion of the cartilage and an effusion into the synovial sac, suggests the diagnosis.

Collateral ligaments

Again, the medial side of the knee is at considerable risk, as the athlete or pedestrian is usually struck on the outer side of his knee, which leads to the opening of the joint on the inside. It is rare to be struck on the inside of the knee, as the other leg protects it as it passes forward.or backward. This is one of the reasons for the larger size of medial ligament. (Another is the chronic valgus (outward) strain at the knee joint resulting from the angle between the obliquely placed femur and the straight tibia.) The upper attachment of the medial ligament to the epicondyle of the femur is much less secure than its broad attachment to the tibia. The ligament therefore tears off at its upper attachment, sometimes with a piece of epicondyle, visible radiologically, or it tears opposite the joint, where a gap can often be palpated. The tear can be proved by gently displacing the tibia laterally on the fixed femur – on the injured side the displacement is excessive. The same injury may also damage the medial cartilage as it is attached to the medial ligament. The diagnosis of ruptured ligament is important; it is better sutured than allowed to heal spontaneously by fibrous or scar tissue; such tissue has an annoying habit of stretching and producing an unstable knee.

Cruciate ligaments

The ligaments run in opposite directions, and the anterior cruciate ligament is put on tension when the femur rolls forward and separates the anterior and posterior attachment, whereas the posterior ligament tenses when the femur is flexed. The anterior cruciate is responsible for the notch at the anterior end of the intercondylar space as it tightens during full extension. It is therefore the hallmark of man in the erect position; for example, the femora of Neanderthal man, who slouched around in a crouching gait, resembling some of the more tired medical students, has no suggestion of a notch. An injury which avulses the cruciate ligaments is a violent one and is often accompanied by other injuries to the knee joint. These ligaments are difficult to repair or replace.

Patellar dislocations

The femora are separated at the hip joints by the width of the pelvis, but are in close approximation at the knee joints. As the legs are straight, it means that there is an angle between the line of each femur and tibia. The quadriceps muscle lies along the oblique axis of the femur, whereas the ligamentum patellae runs in a vertical direction. At the junction of these two different inclinations is the patella. The resultant pull of the muscle and the tendon is in a lateral direction, and this would dislocate the patella were it not for two very important facts:

1   The lateral condyle of the femur is elevated to prevent this displacement.
2   The vastus medialis, inserted well down on the medial border of the patella, pulls the patella medially. Note the horizontal direction of the lower fibers.

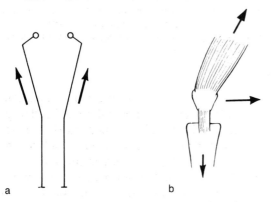

Fig. 22.10   (a) Normal valgus (outward) strain at the knee. This is exaggerated in the female because of the wider pelvis. (b) The line of pull of the quadriceps.

In some patients with either a markedly wasted vastus medialis or a congenitally underdeveloped lateral femoral condyle, the patella dislocates recurrently upon extension of the knee. Operations which cure this either build up the lateral femoral condyle by the insertion of a bone block or align the ligamentum patellae and the quadriceps by the medial transposition of the insertion of the former into the upper medial tibia.

Popliteal fossa

This fossa, a diamond shaped area posterior to the knee joint, is bounded above by the hamstrings and below by the two heads of the gastrocnemius and plantaris (see Fig. 21.3). In the superficial fascia are found the terminations of the posterior cutaneous nerve of the thigh and the short saphenous vein, the former becoming superficial and the latter passing deeply to join the popliteal vein. The fairly thick deep fascia contains fibres that run transversely to facilitate movements at the knee joint. Beneath the fascia the tibial nerve lies in the midline and the common peroneal nerve skirts the lateral boundary of the space.

Tibial nerve

This large nerve supplies both heads of gastrocnemius, popliteus and plantaris, and runs inferiorly to the lower apex of the fossa, where it disappears. At the lower border of the popliteus muscle the tibial nerve becomes the posterior tibial nerve. The cutaneous medial sural branch runs down in the midline of the calf to accompany the short saphenous vein to the lateral foot; this is a favorite nerve for replacing missing portions of other nerves. The superior, middle and inferior articular branches are sensory to the joint.

Common peroneal nerve

This nerve follows the inner border of the biceps to the neck of the fibula. Its branches join the medial sural nerve, supply the skin of the upper part of the lateral calf (lateral sural) and innervate the knee joint (superior and inferior genicular).

Popliteal vein

Immediately beneath the tibial nerve lies the large popliteal vein, which is superficial (or posterior) to the artery. The vein receives the genicular branches, as well as the termination of the short saphenous vein.

Popliteal artery

This is the deepest of the neurovascular structures (see Fig. 23.17). From above down, it is separated by a pad of fat from the popliteal surface of the femur, lies in the intercondylar notch on the oblique posterior ligament of the knee joint, and passes over the popliteus muscle, at the lower border of which it bifurcates into the anterior and posterior tibial vessels. Despite the protection offered by the popliteal pad of fat, injuries to the knee joint, such as fractures or dislocations, frequently involve the popliteal artery. This is because the latter is relatively fixed by its genicular branches as well as its close relationship to the underlying bones. The first duty of the physician in a severe injury to the knee joint is to evaluate the blood supply of the leg and foot. The artery gives off superior and inferior medial and lateral genicular branches. The drawing in your anatomy atlas may show an impressive periarticular anastomosis, but clinically it is very tenuous and blockage of the popliteal artery is notorious for its poor prognosis. A middle genicular branch pierces the oblique posterior ligament.

Bursae around the knee joint

There are many known bursae around the knee (see Fig. 22.6). The important ones are as follows:

1   The prepatellar bursa is a subcutaneous bursa lying on the lower half of the patella, which becomes enlarged from kneeling, or from other causes. 'Housemaid's knee' (almost unknown in the US) is an enlargement of this bursa.

2   The infrapatellar bursa is a bursa between the skin and the tibial tuberosity below

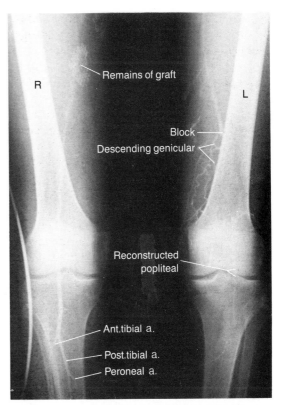

Fig. 22.11   An angiogram of the femoropopliteal artery and its branches (anterior view). Notice the block at the left adductor hiatus. To the rescue – the descending genicular artery, which refills the popliteal artery lower down.

the insertion of the ligamentum patellae. This bursa becomes enlarged in clergymen, who kneel in a more upright position than the floor scrubber; which one of these two cleanses more is debatable.

3   The suprapatellar bursa is really an extension of the knee joint. The bursa lies deep to the quadriceps and enlarges in cases of synovial effusion.

4   Posteriorly, a bursa lies between each gastrocnemius and the joint, with which each often communicates. The semimembranosus bursa lies between that muscle and the medial gastrocnemius and often communicates with the medial gastrocnemius bursa. The semimembranosus bursa (Baker's cyst) may become enlarged and usually does so at the extremes of age.

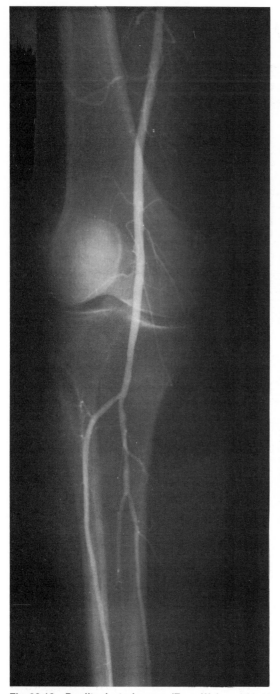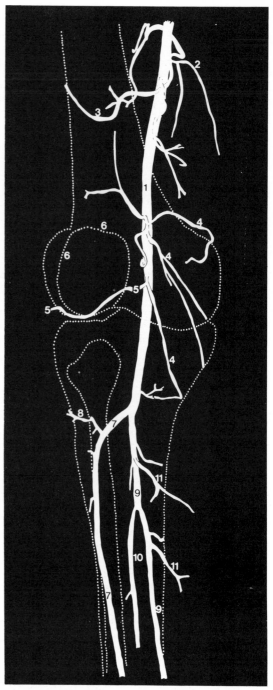

Fig. 22.12   Popliteal arteriogram. (From Weir and Abrahams 1978)

1 Popliteal artery.
2 Superior medial genicular artery.
3 Superior lateral genicular artery.
4 Inferior medial genicular arteries.
5 Inferior lateral genicular artery.
6 Patella.

7 Anterior tibial artery.
8 Recurrent tibial artery.
9 Posterior tibial artery.
10 Peroneal artery.
11 Muscular branches of posterior tibial artery.

23   Leg and foot

Bones of the leg · Ankle joint · Tarsal joints · Arches of the foot · Movements · Deep fascia · Nerves of the leg · Dermatomes and myotomes of the lower limb · Blood vessels of the leg · Dorsum of the foot · Sole of the foot · Lymph nodes of the lower limb

Bones of the leg

Tibia

The shaft of the tibia is somewhat triangular. The medial surface is large and lies in a subcutaneous position; this surface is limited anteriorly by a sharp border, often called the shin. This subcutaneous surface is exposed when the overlying skin is lost – a serious state of affairs. The lateral surface affords origin for the muscles of the leg and ends posteriorly in a sharp border, to which is attached the interosseous membrane. Posteriorly, a well-formed ridge, the soleal line, runs obliquely from the lateral condyle; from it arises part of the soleus muscle. Above the line is a triangular area for the attachment of the popliteus muscle and its covering fascia. On the undersurface of the lateral condyle is a smooth facet for the articulation with the fibula. The posterior surface below the soleal line gives origin to the deep muscles of the calf; in its upper part is the largest nutrient foramen in the body, which is sometimes mistaken for a fissured fracture on x-ray examination.

Fibula

This long, slender, nonweight-bearing bone lies posterolateral to the tibia. Its upper end is expanded into a head which bears an articular facet for the tibia; the head is surmounted by a styloid process posterolaterally. The lateral ligament and the biceps tendon insert into the lateral aspect of the head, the liga-ment splitting the tendon into two portions. Below the head, the narrow neck is closely related to the large and important common peroneal nerve. The shaft is commonly de-scribed with numerous borders. The lateral surface is related, and gives origin to, two of the peroneal muscles. On the medial surface a ridge, the interosseous border for the attached interosseous membrane, separates a narrow anterior area for the attachment of the anterolateral muscles of the leg and a poster-ior area for the origin of the tibialis posterior, one of the deep calf muscles. The posterior surface gives rise to a portion of the soleus above and a deep calf muscle, the flexor hal-lucis longus, below.

Lower ends of tibia and fibula, and skeleton of foot

The lower end of the tibia is expanded into a large weight-bearing area with a considerable medial projection, the medial malleolus. The lateral aspect of the distal extremity of the tibia has a narrow facet for the fibula, above which is a rough area for a large and strong interosseous ligament. The lower end of the fibula also has an expanded end, the lateral malleolus, which lies posterolateral to the medial malleolus. Any disturbance of the relationship of the two malleoli indicates displacement. The medial aspect of the fibula has a small articular strip for the tibia, with a rough area above for the interosseous mem-brane and a triangular facet below for the talus. When the two bones are articulated they form a mortise, into which fits the talus.

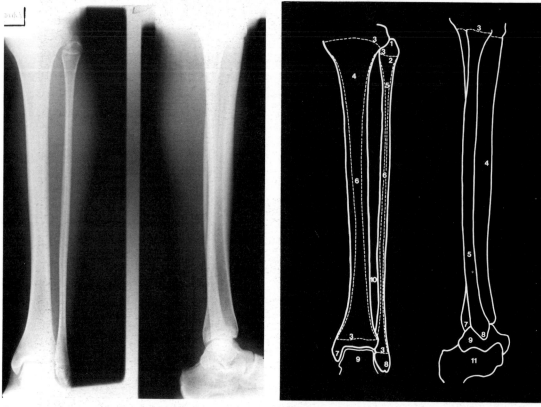

Fig. 23.1 Anteroposterior and lateral views of the tibia and fibula. (From Weir and Abrahams 1978)

1 Head of fibula.
2 Neck of fibula.
3 Epiphyseal plates.
4 Tibia.
5 Fibula.
6 Marrow cavity.

7 Medial malleolus.
8 Lateral malleolus.
9 Talus.
10 Site of interosseous membrane.
11 Calcaneus.

Talus

This bone is the most superiorly placed bone of the tarsus. It presents a body with a convex upper articular surface, continuous with large lateral and smaller medial articular surfaces on its lateral and medial aspects respectively. The convex surface articulates with the tibia, and the medial and lateral facets are firmly grasped by their respective malleoli. Note that the convex surface is broader in front than behind (see later). The body of the talus narrows anteriorly into a neck, beyond which it broadens into a head. A large facet on the head articulates with the succeeding bone of the tarsus, the navicular.

Posteriorly, there is a groove for a tendon (flexor hallucis longus) which separates posterior and medial tubercles. The under-surface of the talus articulates with the bone lying inferior to it (the calcaneus) by three facets; the posterior facet is separated from the anterior two facets by a deep groove which is occupied by a powerful interosseous ligament that unites the two bones. The talus has a facet on its anteromedial aspect, which articulates with the spring ligament (see later). This bone has no muscular attachments.

Calcaneus

This is the largest and most posterior bone of

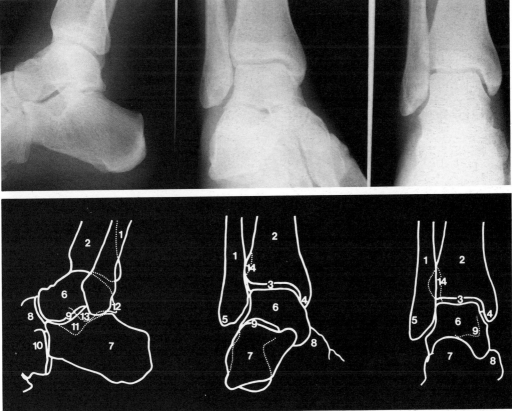

Fig. 23.2 Lateral, oblique-anteroposterior and anteroposterior views of the ankle. (From Weir and Abrahams 1978)

1 Fibula.
2 Tibia.
3 Ankle joint.
4 Medial malleolus.
5 Lateral malleolus.
6 Talus.
7 Calcaneus.

8 Navicular.
9 Talocalcaneal joint.
10 Cuboid.
11 Sustentaculum tali.
12 Posterior process of talus.
13 Lateral process of talus.
14 Inferior tibiofibular joint.

the tarsal skeleton. The posteriorly projecting part of the calcaneus, peculiar to man, is for the insertion of the tendo calcaneus; inferiorly, two large tuberosities carry the whole of the body weight. The anterior part of the upper surface articulates superiorly with the talus by three facets. Projecting medially in the region of the middle facet is a shelf of bone, the sustentaculum tali, beneath which run tendons as they pass from the leg to the foot. Anteriorly the bone ends in an articular facet for the succeeding bone of the tarsus, the cuboid. The lateral aspect has two small bosses for ligamentous attachments.

Cuboid bone

This bone articulates posteriorly with the calcaneus and anteriorly with the fourth and fifth metatarsals. Its only noteworthy feature is the large groove on its undersurface for the passage of a tendon (peroneus longus).

Navicular

The anterior aspect of the talus articulates with the boat-shaped navicular (tarsal scaphoid). The navicular bone has a large medial tuberosity that is easily palpable on the inner side of the foot.

Cuneiforms

The navicular bone articulates anteriorly with the three small wedge-shaped cuneiforms, medial, intermediate and lateral. The larger medial cuneiform is wedged upward, while the two smaller bones are wedged downward.

Metatarsus

Anterior to the cuboid and three cuneiforms are the five metatarsal bones. The first is short and stout for weight bearing, while the remaining four are slender. The heads lie distally and the bases proximally. The base of the second metatarsal is wedged between the medial and lateral cuneiforms because of the smaller size of the intermediate cuneiform bone. The fifth metatarsal is noteworthy for the easily palpable tuberosity at its base, into which inserts the peroneus brevis. An inversion injury may therefore avulse the tuberosity.

Phalanges

The four lateral toes have three phalanges, proximal, middle and distal; the big toe has two phalanges that are much stouter than those of the other toes.

Ankle joint

Bones

The bones comprising this joint are the lower ends of the tibia and fibula and the upper surface and sides of the body of the talus. The tibia and fibula both project inferiorly via their malleoli and grip the talus firmly. Note the curved articular surface of the talus, broader in front than behind.

Capsule

The capsule surrounding this joint is attach-

ed to the bones just beyond the articular surfaces.

Ligaments

Two strong ligaments protect this joint medially and laterally, the ligaments in front and behind being understandably weak.

The medial (deltoid) ligament is a stout, fan-shaped structure that is attached superiorly to the lower border and tip of the medial malleolus. Inferiorly from before backward, it attaches to the tuberosity of the navicular and sustentaculum tali and the ligament joining these two bony points, the spring ligament, and the neck and medial tubercle of the talus. The medial ligament, being thick and strong, is injured less often than the lateral ligament. The latter ligament

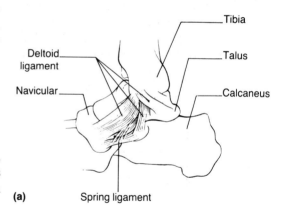

(a)

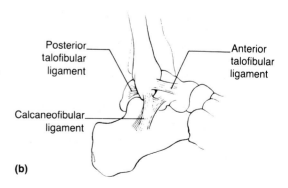

(b)

Fig. 23.3 (a) Attachments of medial (deltoid) ligament. (b) The three bands of the lateral ligament of ankle joint.

has three separate bands running in different directions from the fibular malleolus – the anterior talofibular forward to the neck of the talus, the posterior talofibular ligament from the small dimple on the medial aspect of this malleolus posteriorly to the talus and the middle calcaneofibular band inferiorly to a small protuberance on the lateral aspect of the calcaneus. Strains of this ligament are common, and, by delineating the point of tenderness, it is usually easy to diagnose which band has been stretched or ruptured.

Movements

For practical purposes, this joint is a simple hinge joint. In dorsiflexion (true extension) the toes are raised toward the leg, and in plantar flexion the toes are pointed down away from the leg. The tight grip of the two malleoli on the talus becomes somewhat looser in extreme plantar flexion, such as standing on tiptoe. In this position the narrow part of the talus engages the bones, and it is possible to rotate the talus slightly and therefore produce slight abduction or adduction, or inversion and eversion. This means that on tiptoe you are in an unstable position, whereas in dorsiflexion, with the broad anterior part of the talus engaging, you are in a much more solid position. It is therefore advisable to climb Mt Everest in the dorsiflexed position and not on tiptoe. It is also true that people sprain an ankle running down the mountain and not climbing up! When immobilizing the ankle joint, unless there is a very good reason for not doing so, always see that the foot is at right angles to the leg, which will ensure the engagement of the broader part of the talus. If the foot is placed in plantar flexion for a long period, it may not be possible to replace the broad part of the talus into the contracted mortise; this will result in a serious deformity, fixed plantar flexion of the foot.

Tarsal joints

Talocalcaneonavicular joint

This is a separate joint with its own capsule and synovial membrane. The head of the talus articulates anteriorly with the cup-shaped navicular bone, inferiorly with the two anteriorly placed calcaneal articular surfaces and medially with a ligament, the spring or plantar calcaneonavicular ligament, that joins the sustentaculum tali and the navicular tuberosity. The spring ligament, thick, powerful and partly cartilaginous, is a major support of the talus; stretching of the ligament, such as occurs in advanced flat feet, allows the head of the talus to displace inferiorly.

Talocalcaneal joint

This is a simple joint between the posterior facets of the talus and calcaneus. Between this and the previous joint is a deep groove, the sinus tarsi, narrow medially and wider laterally. Occupying the sinus is a strong ligament that unites the bones. The force required to disrupt this powerful ligament is considerable; the result is a dislocated talus.

Calcaneocuboid and cuneonavicular joints

These joints allow only a modest degree of gliding movements; the former is a somewhat saddle-shaped joint and the latter a plane joint.

Tarsometatarsal and metatarsophalangeal joints

The bases of the first three metatarsals articulate with the three cuneiforms, and, because of the different sizes of the latter bones, the joints lie at different levels, the second metatarsal being wedged between and articulating to some degree with the medial and lateral cuneiforms. This mortise locks the bone and stabilizes the anterior foot. The fourth and fifth metatarsals articulate with the cuboid.

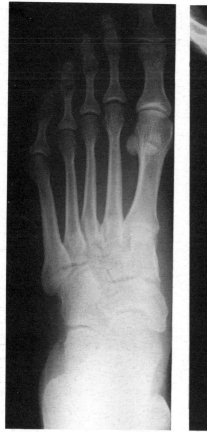
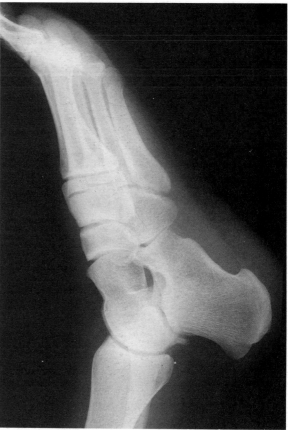

Fig. 23.4 Dorsiplantar and lateral views of the foot. (From Weir and Abrahams 1978)

The heads of the five metatarsals articulate with the proximal phalanges, and the joints possess ligaments similar to those in the hand. The first metatarsal is short and very stout, with two articular facets on the inferior surface of its head that articulate with two sesamoid bones. The small sesamoid bones are normally inconsequential, but, when fractured, the resultant sensation is as of walking on pebbles; in these cases removal may be required. In x-ray studies of the foot, don't be fooled by sesamoids which appear to be shattered into several little pieces; this may be a variation in ossification. Always compare with the other side; if different, then you are on more solid ground for diagnosing a fracture.

Interphalangeal joints

These joints are constructed on the same principles as in the fingers. Although they cannot perform the same intricate movements, they are nevertheless very important in walking.

Movements

Inversion and eversion

Inversion means turning the sole of the foot inward; eversion is the opposite. These movements cannot take place at the ankle joint which is almost a pure hinge joint; they take place at a subtalar level, the talus

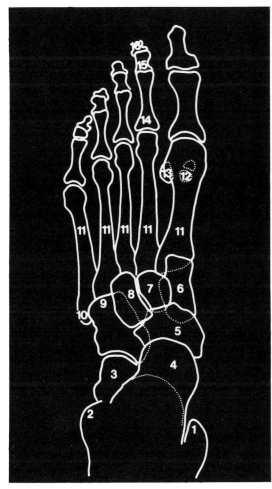

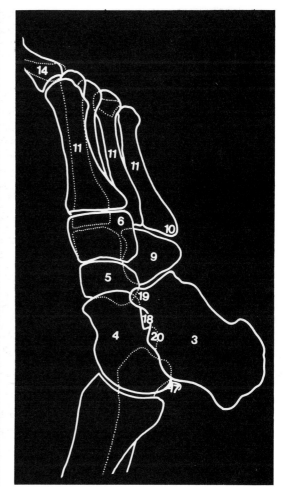

1 Medial malleolus.
2 Lateral malleolus.
3 Calcaneus.
4 Talus.
5 Navicular.
6 Medial cuneiform.
7 Intermediate cuneiform.
8 Lateral cuneiform.
9 Cuboid.
10 Styloid process of fifth metatarsal.

11 Metatarsal shafts.
12 Bipartite sesamoid bone.
13 Sesamoid bone in tendon of flexor hallucis brevis.
14 Proximal phalanx.
15 Middle phalanx.
16 Distal phalanx.
17 Os trigonum.
18 Sinus tarsi.
19 Anterior process of calcaneus.
20 Sustentaculum tali.

remaining fixed and the other bones of the foot moving in relationship to the talus. Most of the movement takes place at the talocalcaneonavicular joint, the calcaneocuboid joint moving only slightly. The cow, who lacks this movement, walks on the highest point or camber of the country road, instead of in the furrows, much to the annoyance of the motorist. The subtalar joint enables you to walk on a sloping roof or around the side of a mountain with either inverted or everted feet. (Without this movement you could not fulfil the requirements for an immigration visa to Switzerland with its dramatic mountain slopes!) The joints between the talus and navicular, medially, and the calcaneus and cuboid, laterally, lie in a horizontal line, and are sometimes

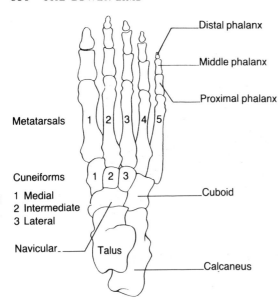

Fig. 23.5 **Superior view of the skeleton of the foot. Note that the upper surface of the body of the talus is broader anteriorly.**

Fig. 23.6 **A cow is walking along the camber of a country road; she is angry because she cannot walk in the furrow due to lack of a subtalar joint.**

collectively called the midtarsal joint. The tarsometatarsal joints lie in a somewhat irregular horizontal line because of the proximal projection of the second metatarsal; little movement takes place at these joints.

Metatarsophalangeal and interphalangeal joints

As in the fingers, the former are condyloid joints, which can flex, extend, abduct and adduct. The latter are hinge joints, which flex and extend.

Arches of the foot

Walking and running

Were the foot merely a solid platform at right angles to the leg, you would have to be satisfied with a shuffling gait. When converted into a highly sprung, mobile arch, a functional organ results.

When standing on the floor of the bathroom, note the wet imprint that indicates the weight-bearing points of the foot – the heel and the fifth and first metatarsal heads (see Fig. 23.7). This means that between these points, the foot is arched both longitudinally and horizontally. When a baby first stands, the foot appears completely flat. This doesn't mean that babies suffer from flat feet; when walking commences, the foot assumes an arched position. Ballet dancers also appear to have flat feet; their superb dancing on their toes immediately dispels this illusion. There are many factors responsible for converting the foot into a mobile, supple arch which can spring into action when the bull begins its charge.

Bones

There are two longitudinal arches, medial and lateral.

Medial longitudinal arch

The medial arch is considerably higher than the lateral. The calcaneus and body of talus form the posterior pillar, with the navicular, cuneiforms and their attached metatarsals constituting the anterior pillar. The two meet at the keystone, the head of the talus.

Lateral longitudinal arch

The calcaneus forms the posterior pillar, and the cuboid and its two metatarsals the anter-

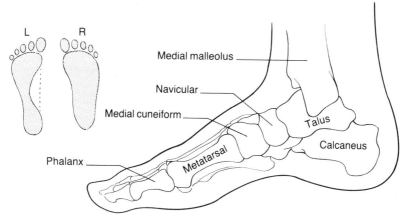

Fig. 23.7   Medial longitudinal arch. The inset shows wet imprint on bathroom floor – the right foot has lost its arch.

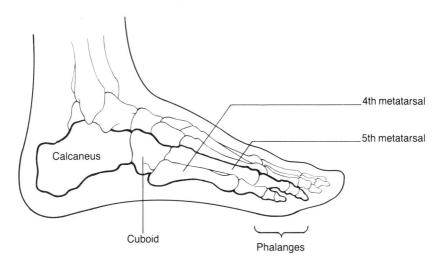

Fig. 23.8   The lateral longitudinal arch. (The bones constituting the arch are outlined in thicker lines.)

ior pillar. This arch is much lower and almost touches the ground.

Horizontal arches

The horizontal arches of each foot together constitute a complete arch; in one foot they are half arches (do not mispronounce this!). There are two arches, a tarsal arch formed by the cuneiforms and the cuboid, and a metatarsal arch consisting of the five metatarsals strung together by ligaments.

Ligaments

Ligaments play a substantial and important part in uniting the bones. When persistently stretched, however, ligaments become elongated and painful. Muscles take strain much better, and the contraction of voluntary muscle is unsurpassed in resisting distractions. The following ligaments are important.

1   The spring ligament (plantar calcaneonavicular ligament). This fibrocartilaginous band joins the navicular and

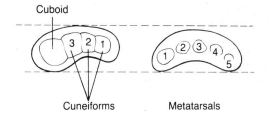

Fig. 23.9   The horizontal arches.

sustentaculum tali and supports the head of the talus; it is itself supported by the long tendons running beneath it from the leg to the foot.

2   Two plantar ligaments. The long plantar ligament is a substantial structure that runs from the plantar surface of the calcaneus across the ridge of the cuboid, which it converts into a tunnel, to the middle three metatarsals. It is an important tie beam for the longitudinal arch. The short plantar ligament is merely the calcaneocuboid ligament that lies deep to the long ligament.

3   Plantar aponeurosis. Although not a true ligament, this important subcutaneous fibrous tissue structure stretches from the calcaneus to the phalanges and acts as a strong tie.

Muscles

Muscles are probably the most important structures in the maintenance of both the longitudinal and transverse arches. They are both extrinsic muscles that arise in the calf and run to the foot, and intrinsic muscles that arise and insert in the foot.

Muscles acting at the ankle and foot joints

Anterolateral muscles of the leg

This group of muscles is enclosed by rigid structures: on each side by the tibia and fibula, deeply by the interosseous membrane and superficially by the thick deep fascia. These structures prevent substantial swelling of the muscles; swelling of the muscles from an injury or even a long unaccustomed hike may be followed by compression of the muscles by these osteofascial surroundings and lead to death (necrosis) of the muscular tissue. Physicians in charge of recruits undergoing intensive physical exercise should remember this anterior compartment syndrome (Fig. 23.13).

It is not difficult to elucidate the actions and positions of the muscles and tendons. Those tendons that act on the tarsus end in

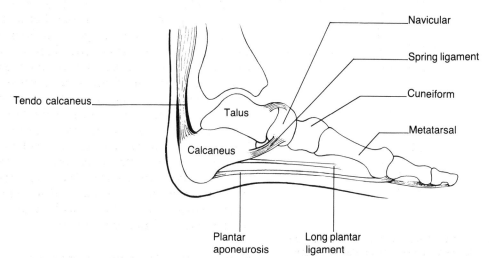

Fig. 23.10   Supports of the longitudinal arch. The long plantar ligament is shown at its posterior attachment to the calcaneus; anteriorly it inserts into the cuboid and middle three metatarsals (not shown).

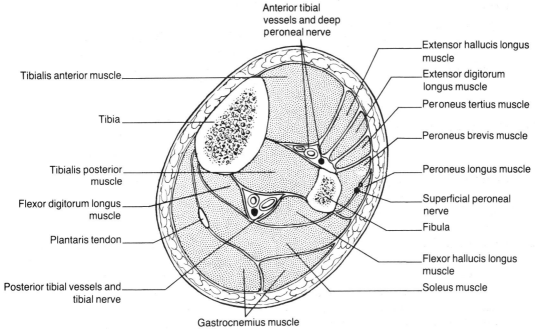

Anterior tibial
vessels and deep
peroneal nerve

Tibialis anterior muscle

Tibia

Tibialis posterior
muscle

Flexor digitorum longus
muscle

Plantaris tendon

Posterior tibial vessels and
tibial nerve

Gastrocnemius muscle

Extensor hallucis longus
muscle

Extensor digitorum
longus muscle

Peroneus tertius muscle

Peroneus brevis muscle

Peroneus longus muscle

Superficial peroneal
nerve

Fibula

Flexor hallucis longus
muscle

Soleus muscle

Fig. 23.11   Cross-section of the leg.

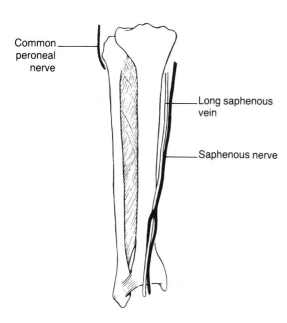

Common
peroneal
nerve

Long saphenous
vein

Saphenous nerve

**Fig. 23.12   Interosseous membrane. Note that the
membrane fibers run downward and laterally (the
opposite to that of the upper limb). The saphenous
nerve often wraps itself around the vein and is easily
damaged during surgery.**

the tarsus and lie peripherally, while those tendons that act on the digits reach them and lie centrally; i.e. evertors and invertors of the tarsus flank the digitorum muscles. The tibialis anterior muscle arises from the tibia and ends in a large tendon that runs inferiorly, and crosses and obliterates the sharp anterior border of the tibia in its lower part. It inserts into the medial side of the medial cuneiform bone and base of the first metatarsal (another muscle, the peroneus longus, inserts into these same two bones, but reaches them by another route). The peroneus tertius arises from the fibula and inserts into the fifth metatarsal bone. These two muscles are the invertors and evertors of the tarsus respectively. Between them are the muscles that proceed to the big toe, the extensor hallucis longus, and to the other four toes, the extensor digitorum longus. These last two muscles, with the peroneus tertius, arise from a narrow area on the fibula between its interosseous and anterior borders.

Peroneal muscles (longus and brevis)

These two muscles arise from the lateral

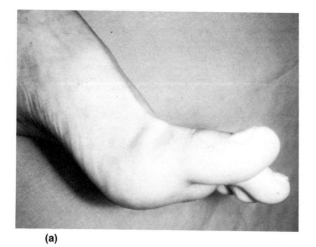

(a)

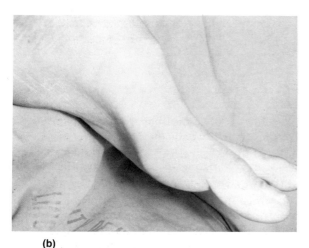

(b)

Fig. 23.13   Anterior compartment syndrome. The patient had recently returned from a long hike. He complained of pain in his anterior leg and parasthesiae (feeling of pins and needles) in the dorsum of his foot. Examination showed he could move his big toe up (dorsiflex) (a) and down (plantar flex) (b), but his other toes could not be dorsiflexed – he had a paralysis of the extensor digitorum longus. The deep fascia covering the compartment was slit – his symptoms disappeared and the toes recovered their movement.

where it develops a sesamoid bone that plays on its surface. The tendon of the longus then enters a deep ridge on the plantar surface of the cuboid and continues medially across the sole to insert into the same bones as the tibialis anterior (see Fig. 23.22). The bony ridge on the cuboid is converted into a tunnel by the long plantar ligament. These two peroneal tendons have well-formed synovial sheaths, and are held in position behind the fibular malleolus by two retinacula, one above and one below the malleolus. The condition of slipping of the peroneal tendons across the front of the fibula, due either to insufficient development of the fibular malleolus (one of its functions is to prevent this) or to rupturing or stretching of the retinacula is well known. The displacement of a superficial part of the malleolus posteriorly acts as a block, and is curative. Both muscles are powerful evertors.

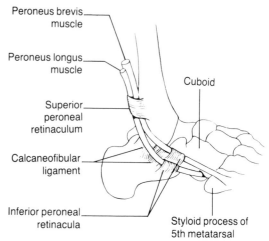

Fig. 23.14   Peroneal tendons and retinacula. The peroneus longus enters a groove in the cuboid.

Posterior muscles of the calf

These are arranged in a superficial group, the gastrocnemius, soleus and plantaris, and a deep group, the long flexors of the toes, hallucis and digitorum, and a muscle destined for the tarsus, the tibialis posterior. The two groups are separated by a thick fascial sheet.

surface of the fibula, the origin of the longus lying at first above and then posterior to that of the brevis. The two tendons pass behind the fibular malleolus, the brevis inserting into the tuberosity of the fifth metatarsal, and the longus running inferiorly to the cuboid,

Gastrocnemius

The two bellies of this muscle join at about the middle of the calf to be inserted into the tendo calcaneus. This is the largest tendon in the body and inserts into the posterior surface of the tuberosity of the calcaneus. It can, however, be ruptured by violent plantar flexion such as during a squash game.

Soleus muscle

This muscle arises from both leg bones: the oblique soleal line on the posterior surface of the upper third of the tibia and some of its medial border, and the upper third of the posterior surface of the fibula. The bellies join in a similar way to the flexor digitorum superficialis of the forearm and insert into the anterior surface of the tendo calcaneus, reaching almost to the insertion of the tendon. This muscle is responsible for more of the bulk of the calf than is the gastrocnemius.

Running lateromedially between the gastrocnemius and the soleus is the plantaris tendon. This slender tendon appears and runs on the medial side of the tendo calcaneus to insert into the calcaneus. The tendo calcaneus is surrounded by a loose sheath, often called a paratenon. In those skiing and mountaineering for the first time,

the painful creaking that emanates from the lower posterior leg at the end of the day is due to traumatic inflammation of this sheath.

The soleus and gastrocnemius muscles are often called the 'third heart' or muscular pump. These two muscles, but especially the soleus, contain large veins. Contraction (systole) of the muscles empties the blood into the deep veins, while relaxation (diastole) leaves the veins empty. A vacuum is prevented by blood passing from the superficial system via perforating veins (perforating the deep fascia) into the muscular veins. Valves in these perforating veins assure that the blood moves in the correct direction from superficial to deep veins; destruction of these valves leads to backflow from deep to superficial to produce one variety of varicose veins.

Deep layer of muscles

These muscles are the flexors hallucis longus and digitorum longus, and the tibialis posterior. In order to increase their mechanical advantage, the flexors digitorum and hallucis longus muscles arise from what appears to be the wrong bones; however, the line of pull from the fibula to the big toe is more effective than it would be from the tibia to the big toe and vice versa. Accordingly, the flexor hallucis longus arises from the fibula and the flexor digitorum longus from the tibia. The muscles give rise to tendons which cross one another in the sole as they pass their respective insertions. The tibialis posterior lies deep to these two muscles and arises from the interosseous membrane and contiguous surfaces of the tibia and fibula. Its tendon passes inferiorly and peripherally, crosses deep to the digitorum longus tendon behind the malleolus and reaches its major insertion into the tuberosity of the navicular bone. Students, from time immemorial, know that this tendon is inserted by many slips into all the bones of the tarsus except the talus. The talus is devoid of muscular attachments, but compensates for this by its numerous ligamentous attachments.

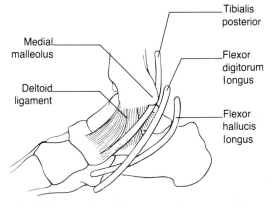

Fig. 23.15 Tendons of the deep calf muscles. Note the longus crossing the hallucis inferiorly in the sole.

Tibialis posterior

Medial malleolus

Flexor digitorum longus

Deltoid ligament

Flexor hallucis longus

Movements

Ankle joint

It is clear that all structures passing in front of the ankle joint are dorsiflexors, and all those passing behind are plantar flexors.

Dorsiflexors (true extensors)

The extensor hallucis and digitorum longus muscles are dorsiflexors, as well as extensors of the toes, whereas the tibialis anterior and peroneus tertius are dorsiflexors as well as invertors or evertors.

Plantar flexors

Of the deep muscles passing from calf to foot, the flexors hallucis longus and digitorum longus, primarily flexors of the toes, also flex the ankle joint. The tibialis posterior can also act as a flexor, although it is mostly an invertor. The most powerful plantar flexors, however, are the gastrocnemius and soleus acting through the tendo calcaneus.

Tarsal joints

Evertors

These are the three peronei; longus and brevis pass posterior to the fibula.

Invertors

The tibialis anterior and posterior are both powerful invertors.

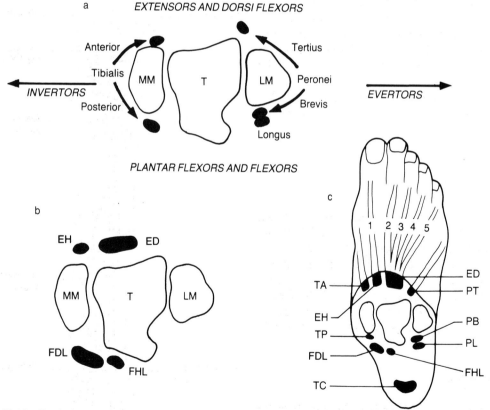

Fig. 23.16 Tendons around the ankle. The invertors and evertors are seen on the inner and outer sides of the ankle joint respectively (a). The flexors and extensor tendons to the toes lie centrally (b). (c) shows all the tendons in position. ED, extensor digitorum longus; EH, extensor hallucis longus; FDL, flexor digitorum longus; FHL, flexor hallucis longus; LM, lateral malleolus; MM, medial malleolus; PB, peroneus brevis; PL, peroneus longus; PT, peroneus tertius; T, tibia; TA, tibialis anterior; TC, tendo calcaneus; TP, tibialis posterior.

To summarize:

> *Evertors*: muscles lying on the lateral aspect of the leg, the peronei.
>
> *Invertors*: muscles lying medially, the tibialis anterior and posterior.
>
> *Dorsiflexors*: extensors of the toes.
>
> *Plantar flexors*: flexors of the toes and ankle.
>
> Additional bonuses are that all muscles with tendons passing behind the joint are additional plantar flexors, and all those with tendons passing in front of the joint are also dorsiflexors.

Many of the tendons mentioned above play an important part in maintaining the arches of the foot. The tendon of the peroneus longus runs across the inferior aspect of the tarsus from lateral to medial side; it acts as a strong pulley, elevating the lateral longitudinal arch as well as everting the foot (see Fig. 23.22). The tibialis anterior and posterior are both inserted into the middle of the tarsus and assist in its elevation. The long flexors of the calf, particularly the flexor hallucis longus, pass behind the medial malleolus and beneath the sustentaculum tali and act as major supports.

Deep fascia

The muscles of the leg are enclosed in a thick layer of deep fascia, which is so resistant that it allows little expansion of the muscles beneath it. Thickenings of the fascia in the region of the ankle joint form retinacula for stabilization of the tendons. This fascial 'stocking' is important in maintaining normal venous return through the calf muscle pump mechanism.

Retinacula

Reference has been made to the peroneal retinacula securing the peroneus longus and brevis, and to the superior and inferior extensor retinacula holding down the tendons that cross the anterior aspect of the ankle joint.

The flexor retinaculum is the thickest retinaculum that runs from the medial malleolus to the calcaneus; it restrains the tendons as they pass from the calf to the foot while also protecting the accompanying vessels.

Nerves of the leg

Tibial nerve

This large nerve passes beneath the origin of the soleus muscle, rather like the median nerve passing deep to the flexor digitorum superficialis. It lies beneath the fascial cover-

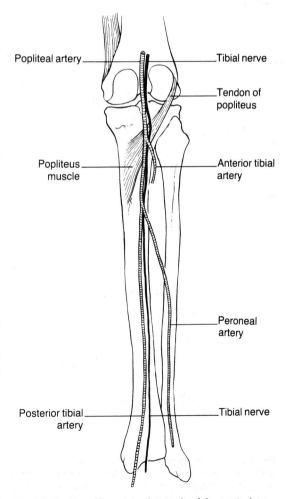

Fig. 23.17 Deep nerves and vessels of the posterior leg.

Labels: Popliteal artery; Tibial nerve; Tendon of popliteus; Popliteus muscle; Anterior tibial artery; Peroneal artery; Posterior tibial artery; Tibial nerve

ing that separates the superficial and deep calf muscles, which prevents its adherence to the overlying soleus (unlike the median nerve). The posterior tibial nerve supplies the deep muscles of the calf amongst which it lies, and becomes superficial as it approaches the flexor retinaculum posterior to the tendons of the tibialis posterior and flexor digitorum longus. As it passes deep to the retinaculum, it gives off calcaneal branches to the heel and divides into two large terminal branches, the medial and lateral plantar nerves.

Common peroneal nerve

This nerve passes along the lateral border of the popliteal fossa to reach the neck of the fibula. At this point, where it is palpable, the nerve is in danger, for it can easily be paralyzed by pressure against the bone. This may occur when the upper end of a plaster cast reaches that level (the cast must be either below or above the neck of the fibula), or when the leg rolls over the side of a metal stretcher in an unconscious patient. An old trick that was often used by pacifists in military conflicts was to hammer away with a stone against the neck of the fibula. The nerve passes over the neck of the fibula into the substance of the peroneus longus muscle, where it divides into two large terminal branches as well as giving off a small branch to the knee joint. These branches are the deep peroneal (anterior tibial is a better name) and superficial peroneal nerves (musculocutaneous is a worse name). The deep peroneal nerve enters the anterolateral compartment, and, having supplied all the muscles of that compartment, appears on the dorsum of the foot to supply the skin of the first web space; the nerve supplies mostly muscle. The superficial peroneal nerve runs inferiorly between the two peronei which it supplies, and leaves them at about the junction of the middle two-thirds and lower third of the lateral side of the leg, where it pierces the fascia. The nerve ends by supplying the skin of the dorsum of the foot, except for the first web space (deep

peroneal nerve), and the medial and lateral borders (saphenous and sural nerves respectively); the superficial peroneal supplies mostly skin. After the nerve pierces the fascia it can be palpated under the skin; anesthetizing the nerve will allow superficial surgery on the dorsum of the foot.

The saphenous nerve, having started just below the inguinal ligament, accompanies the saphenous vein and ends by supplying the skin of the medial border of the foot to the level of the ball of the big toe (see Fig. 23.12).

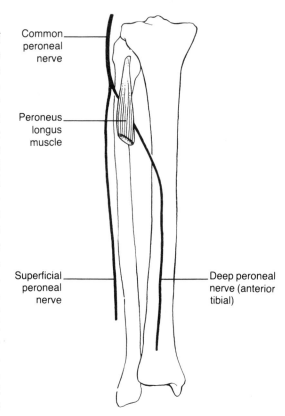

Fig. 23.18 Distribution of the common peroneal nerve.

The sural nerve accompanies the short saphenous vein and supplies the skin on the lateral border of the foot to the side of the little toe.

Dermatomes and myotomes of the lower limb

The medial and posterior rotation of the lower extremity, necessary to bring the sole into the walking position, makes a study of the dermatomes and myotomes of the lower extremity more complicated than in the upper extremity.

Dermatomes

The lower extremity develops opposite L1/2–S3 segments, the skin being stretched over the developing appendage. The anterior surface of the lower limb is supplied by segments L1–5 (see Fig. 2.14). L1 supplies the inguinal region, L3 the area of the knee, with L2 disposed between them. L5 reaches the medial part of the foot on both the dorsal and the plantar surface, and L4 is disposed between this and L3. On the posterior surface, one walks on S1 (together with the contribution from L5) and sits on S3; S2 constitutes a strip in the midline between the two, following the seam of a lady's stocking of yesteryear. The remaining S4 and 5 and the coccygeal segments supply the perianal skin in a circumferential manner.

Myotomes

The joint movements are supplied by centers which occupy four contiguous segments of the cord, two segments subserving one movement and two segments the opposite movement. The movements of the hip are served by L2 and 3 (flexion) and L4 and 5 (extension), those of the knee by L3 and 4 (extension) and L5 and S1 (flexion) and those of the ankle by L4 and 5 (dorsiflexion) and S1 and 2 (plantar flexion). In other words, the segments descend with the level of joint served. In more complex joints, movements are seldom purely unidirectional. For example, adduction at the hip joint is often accompanied by medial rotation, and abduction by lateral rotation. The same segments will be involved in these composite movements.

One can therefore work out the approximate segments of cord supply for muscles; for example, the knee is flexed by biceps femoris and the segmental supply for this muscle is therefore L5 and S1. The soleus flexes the ankle – S1 and S2 are the responsible segments. The hip is flexed, adducted and medially rotated as a composite movement; the iliacus (flexor), adductor longus (adduction) and pectineus (medial rotator) muscles are therefore supplied by the same spinal segments, L2 and L3.

Table 23.1 Segmental innervation of joint movements of the lower limb

Joint	Movements	Segments
Hip	Flexion and adduction	L2, 3, 4
	Extension and abduction	L4, 5, S1
	Lateral rotation	L4, 5, S1
	Medial rotation	L5, S1, 2
Knee	Flexion	L5, S1
	Extension	L2, 3, 4
Ankle	Dorsiflexion	L4, 5
	Plantar flexion (true flexion|)	S1, 2
Subtalar	Inversion	L4, 5
	Eversion	L5, S1

Blood vessels of the lower limb

Arterial supply (see Fig. 22.12)

The popliteal artery reaches the lower border of the popliteus, where it bifurcates into its two large terminal branches. The anterior tibial artery passes forward through the upper end of the interosseus membrane to reach the anterolateral muscular compartment, lying close to and sometimes notching the neck of the fibula on the way. The posterior tibial artery runs inferiorly and soon gives off a large peroneal artery, which may supplant or complement the continuation of the parent trunk. The posterior tibial artery runs inferiorly with its companion nerve and venae comitantes nerve deep to the fascia covering the deep muscles of the leg. The posterior tibial artery surfaces above and behind the medial malleolus, where it lies behind the tendons of the tibialis posterior and flexor digitorum longus. Passing deep to the flexor retinaculum, it divides into its terminal medial and lateral plantar arteries. This artery is very important in determining the blood supply of the leg and it can be palpated two tendons' breadth behind the medial malleolus. It is important to relax the flexor retinaculum by inverting the foot while so doing; otherwise, you may erroneously declare the artery to be absent. The peroneal artery runs inferiorly in the substance of the flexor hallucis longus and in a very close relationship to the fibula, eventually passing beyond the lower extremity of the bone to supply the heel as the lateral calcaneal artery. The peroneal artery gives off a large branch, the perforating branch, which, as the name indicates, passes through the interosseous membrane to the anterior leg. This is a fairly constant vessel and should be searched for in clinical examination. The posterior tibial and peroneal arteries supply all the muscles, bones and joints of the posterior leg.

The anterior tibial artery, having entered the anterolateral compartment of the leg, runs inferiorly, at first deeply amongst the muscles and then surfacing at the ankle joint. The anterior tibial artery then becomes the dorsalis pedis, which is crossed by the extensor hallucis brevis tendon as the latter passes to the big toe. The dorsalis pedis artery disappears at the proximal end of the first intermetatarsal space by passing deeply to join the deep plantar arch in the sole. It supplies the dorsum of the foot and the ankle region, and also anastomoses with the perforating branch of the peroneal artery. It should be stated that, in general, one artery is sufficient to keep the foot alive; therefore, it is important to be able to palpate the posterior tibial artery, two tendons' breadth behind the medial malleolus with the ankle inverted to relax the thick retinaculum, the perforating peroneal artery anterior to the lower third of the fibula, and the dorsalis pedis just proximal to the first intermetatarsal space.

Venous drainage

Superficially the anastomic network of the leg and foot connects the greater (long) and lesser (short) saphenous veins. Additionally, they communicate with an extensive system of deeper inter- and intramuscular veins and venous plexuses by valved vessels which perforate the deep fascia and are therefore called the perforating veins. The deep veins form an anastomosing network around the arteries and are called venae comitantes (comites); they are also present as intramuscular plexuses, especially in the gastrocnemius and soleus muscles (the third heart).

Short (lesser) saphenous vein

This vein commences superficially at the lateral end of the dorsal venous arch, passes behind the lateral malleolus, ascends between the two heads of the gastrocnemius muscles and pierces the popliteal fascia to end in the popliteal vein. The termination of the vein is open to considerable variation and it may, wholly or partly, end in deep veins of the thigh (see Fig. 21.3). This vein is far less liable to varicose changes than are the branches of the long saphenous.

Dorsum of the foot

Relatively thin skin covers the superficial vessels, nerves and tendons. A fairly constant dorsal venous arch lies in the distal foot, giving rise to the long and short saphenous veins at the medial and lateral ends of the arch respectively. (They are best seen on standing up after taking a hot bath.) The extensor digitorum brevis, a small and solitary muscle on the dorsum, gives rise to four tendons for the medial four toes, the tendon for the big toe crossing the dorsalis pedis. A visible, soft, compressible swelling in the proximolateral foot, it is sometimes mistaken for a tumor (usually a lipoma, a fatty tumor), especially in chubby individuals.

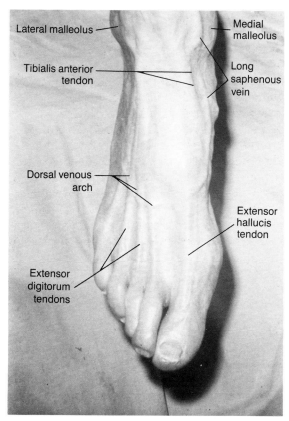

Lateral malleolus

Medial malleolus

Tibialis anterior tendon

Long saphenous vein

Dorsal venous arch

Extensor hallucis tendon

Extensor digitorum tendons

Fig. 23.19 **Veins on the dorsum of the foot. The dorsal venous arch gives rise to the long saphenous vein at its medial end (not seen is the commencement of the short saphenous vein). The long saphenous vein runs superiorly between the medial malleolus and the tibialis anterior tendon.**

Sole of the foot

Skin

Like the palm of the hand, both the epidermis and the dermis are thick, even in the newborn child. *Pari passu*, the skin of the sole becomes even thicker, and in those leaping merrily from rock to rock it becomes almost horny. The skin, as in the hand, is attached to the underlying deep fascia by fibrous bands that divide the subcutaneous fat into numerous small cushions for weight bearing, especially in the heel region. This compartmentation, while very comfortable, makes it very difficult to find a foreign body in the sole. The deep fascia of the sole is thickened, especially centrally where it is called the plantar aponeurosis. This aponeurosis extends from both calcaneal tubercles forward and splits into five bands, one for each digit (in the hand, it splits into four, as nothing is allowed to interfere with the mobility of the thumb). Each band bifurcates distally to allow for the passage of the flexor tendons, and the divisions of the band are attached to the side of the phalanges, as in the hand. The plantar aponeurosis is one of the main struts maintaining the longitudinal arch of the foot. The fascia over the medial and lateral borders of the foot is much thinner. The attachment of the skin to the underlying aponeurosis confers on it some degree of fixation.

Muscles

Traditionally, these have always been described in four layers; this has undoubtedly helped the authors of anatomical texts. It is easier and more practical to describe the muscles as medial, lateral and central. As in the forearm and hand, the muscles on the periphery reach the proximal foot (proximal phalanges) while those lying centrally reach the distal foot (middle and distal phalanges).

Muscles on the medial side

The superficial muscle is the abductor hallucis; lying anterior and deep to it is the short flexor of the big toe, the flexor hallucis brevis. This latter muscle is in two portions, its medial belly being inserted in common with the abductor into the medial side of the proximal phalanx, while its lateral head is inserted into the lateral side of the proximal phalanx in common with another muscle (adductor hallucis). The flexor hallucis brevis has a sesamoid bone in each head, which articulates with the plantar aspect of the head of the metatarsal bone. Maybe there was a time when people could abduct their big toe (e.g. hitching a ride on a dinosaur), but that time is past; the functions of these muscles have changed and they act as springs for the arches of the foot.

Muscles on the lateral side

On the lateral side are the abductor digiti minimi and, anterior and deep to it, the flexor digiti minimi brevis. These are inserted into the lateral aspect of the proximal phalanx of the fifth toe and their function is the same as the medial muscles.

Central muscles

Deep to the plantar aponeurosis is the flexor digitorum brevis destined for the middle phalanges of each toe; deep to the brevis are the flexors digitorum longus and hallucis longus tendons running to the terminal phalanges. Here there is an interesting rearrangement to allow these tendons to act both in direct lines and on more than one phalanx. For example, the flexor digitorum longus entering the foot at an angle, has its pull straightened by a muscle, the flexor accessorius (quadratus plantae), which arises from the calcaneus and inserts into the tendon. The flexor hallucis longus runs in a straight line and does not need help; more than that, as it crosses deep to the digitorum,

it gives slips to the tendons to the second and third toes, thereby extending its pull on the whole medial longitudinal arch, which is one of its main functions. Arising from the flexor digitorum longus are four lumbricals which, as in the hand, pass on the thumb or big toe side of the digitorum tendon. That being so, the first lumbrical can arise only from the side of the tendon, but the other three lumbricals arise from both the tendons between which they lie. As in the fingers, they pass dorsally to the extensor expansions on the dorsum of the toes and to the terminal two phalanges. Descending deeper, we come across the adductor hallucis, whose oblique and transverse heads insert in common with the lateral head of the flexor hallucis brevis into the proximal phalanx of the big toe. They possibly help in maintaining the transverse arch formed by the metatarsal bones. Disposing of this rather insignificant muscle, we finally reach the interossei, which, as the names indicate, lie between the bones. They have the same insertion as in the hand; they pass across the metatarsophalangeal joints to the dorsal expansion and to the two distal phalanges. They (like the lumbricals) are flexors of the metatarsophalangeal joints and extensors of the interphalangeal joints; however, they also abduct and adduct (separate and oppose). There is a difference between

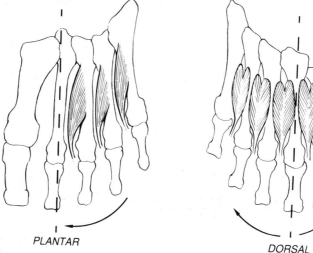

PLANTAR

DORSAL

Fig. 23.20 The interosseous muscles of the foot. The Plantar ADduct; the Dorsal ABduct.

the hand and the foot as regards the axis of movement; in the foot it passes through the second toe, while in the hand through the third finger. However, the plan of action is the same – the **d**orsally placed interossei **ab**duct (DAB) and the **p**lantar **ad**duct (PAD). Dorsal interossei arise from both bones, and the plantar interossei from a single bone. Proximal to the interossei lies the tendon of the peroneus longus.

Abduction or separation of the toes

The big and little toes have their own abductors and don't need the interossei. The abductors are therefore on either side of the second toe, abducting it away from the imaginary line passing through the second toe. The third and fourth toes have their interossei inserted into the lateral sides of the proximal phalanx, moving the third and fourth toes away from the second toe.

Adduction or opposition of the toes

The big toe has its own unimpressive adductor. The second toe cannot, of course, be adducted to itself. There remain, therefore, only three plantar interossei; these are inserted into the medial aspects of the third, fourth and fifth toes, swinging these toes toward the second toe.

Functional anatomy of the muscles of the sole

Extrinsic muscles

The long flexor muscles of the toes (hallucis and digitorum) are mainly concerned with flexing the terminal phalanges to provide a take-off in walking. The flexor hallucis longus runs below the sustentaculum tali beneath the spring ligament and gives slips of insertion to the flexor digitorum longus; it is a powerful supporter of the medial longitudinal arch. The tibialis posterior, like the tibialis anterior, is inserted into the medial side of the midtarsus; both are effective arch supports. The peroneus longus and brevis pass to the lateral side of the midtarsus and are strong evertors of the tarsus. The peroneus longus runs across the sole of the foot lateromedially and, in addition to its role as an evertor, it supports the arch very effectively. It is an important muscle for disco roller skaters!

Intrinsic muscles

Many of the intrinsic muscles run posteroanteriorly and approximate the two pillars of the arch; they are important in maintaining a mobile arch. The abductors of the big and little toes act mainly as springs, abduction

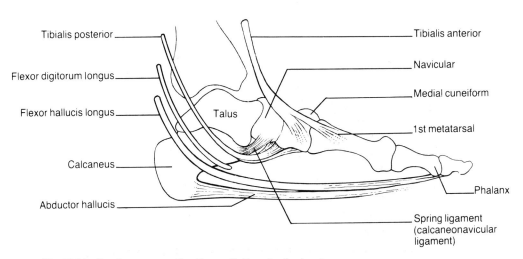

Fig. 23.21   Tendons supporting the medial longitudinal arch.

Labels: Tibialis posterior, Flexor digitorum longus, Flexor hallucis longus, Talus, Calcaneus, Abductor hallucis, Tibialis anterior, Navicular, Medial cuneiform, 1st metatarsal, Phalanx, Spring ligament (calcaneonavicular ligament)

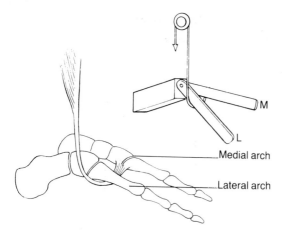

Fig. 23.22   Peroneus longus supporting the arches and acting as a pulley.

being a minor action; one day some brave anatomist will change their names. The adductor hallucis plays a part in maintaining the transverse arch of the metatarsal bones. The lumbricals and interossei, as in the hand, are flexors of the metatarsophalangeal joints and extensors of the interphalangeal joints; these movements are very important in straightening the toes. Loss of the lumbricals and interossei results in clawing of the toes; this is the condition of flexion of the inter-phalangeal joints and extension of metatarso-phalangeal joints. This results in protuberant interphalangeal joints; the overlying skin will respond by thickening and the final result will be dorsal corns and callosities. The flattened metatarsophalangeal joints will produce the sensation of walking on pebbles with bare feet. This arises because the exten-sion of the metatarsophalangeal joints leaves the head of the metatarsal bones exposed, making them bear more weight than normally. Straight toes and protected meta-tarsal heads are needed to provide a satisfac-tory weight-bearing platform. Claw feet and toes are very symptomatic; they occur mainly in diabetic patients whose neurological lesions lead to atrophy of the intrinsic muscles of the foot.

Vessels of the sole

The sole of the foot has a prolific blood supply, since it is one of the sites from which heat is lost. The posterior tibial artery gives off calcaneal branches to the heel and then divides into the medial and lateral plantar arteries deep to the flexor retinaculum. The plantar vessels pass deep to the flexor re-tinaculum and the superficial muscles of the sole, and finally run forward to supply digital arteries to the toes. Most of these arise from the medial plantar artery; the larger lateral artery mainly expends itself by forming the plantar arch. This arch lies across the bases of the metatarsals and ends medially by joining the termination of the dorsalis pedis artery after the latter has passed through the prox-imal end of the first intermetatarsal space. The deep plantar arch gives off metatarsal arteries, which anastomose with the digital arteries.

Nerves of the sole

The posterior tibial nerve gives off calcaneal branches to the heel and then divides into the medial and lateral plantar nerves deep to the flexor retinaculum. The nerves accompany the arteries; the medial plantar nerve is the larger nerve, behaving like the median nerve in the hand. It supplies the skin of three and a half toes, including the dorsal part of these toes and the muscles on the medial aspect of the foot (the 'thenar eminence of the foot'). More precisely, it supplies the abductor hal-lucis, flexor hallucis brevis, the first lum-brical and the flexor digitorum brevis. The lateral plantar nerve runs to the lateral border of the foot to supply the skin of the lateral one and a half toes, and then innervates all the remaining muscles of the sole; in this, it re-sembles the ulnar nerve in the hand.

Lymph nodes of the lower limb

Except for some nodes situated in relation to the popliteal vessels that are clinically

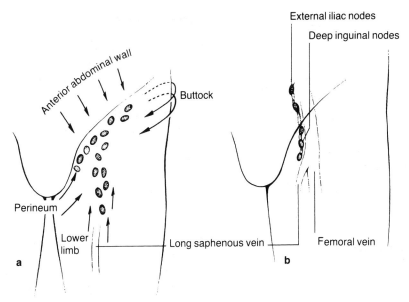

Fig. 23.23 Inguinal nodes: (a) superficial, (b) deep (see also p. 482).

impalpable and of doubtful clinical importance, the lymph nodes draining the lower extremity are concentrated at the root of the limb (as in the upper extremity).

Inguinal nodes

These are situated immediately below the inguinal ligament, and consist of a superficial and a deep group.

Superficial group

These lie in horizontal and vertical chains like a 'T'. The horizontal chain receives lymph from the superficial structures of the anterior abdominal wall below the umbilicus, from the penis, scrotum, perineum and buttock, and from the squamous mucous membrane of the terminal portions of the urethra, anal canal and vagina. The perineal structures drain to the medial members of this group of nodes, while the lymphatics from the buttock wind their way laterally to reach the lateral group of nodes. Remember that an enlarged inguinal lymph node on the medial side of the horizontal chain might come from a tumor hiding in the anal canal or vagina, or beneath the prepuce of the penis. Those unlucky enough to have had a boil (furuncle) on the butt know well the rapid appearance of a tender node in the lateral group. The vertical set of nodes surrounds the termination of the long saphenous vein and this chain receives lymphatics that drain the superficial structures of the entire lower limb, save for the area of skin drained by the short saphenous vein whose lymph goes directly to the deep nodes in the popliteal fossa.

Remember that the inguinal nodes, unlike the axillary nodes, are situated superficial to the deep fascia. It is therefore common to palpate inguinal lymph nodes in the normal individual. Many generations of medical students have been frightened out of their wits when they first palpated their own inguinal nodes; they are quickly placated when consulting the preceding generation of students, who can now reassure them.

Deep nodes

There are several nodes deep to the deep

fascia that lie on the inner aspect of the femoral vein. They receive lymph vessels running with the deep veins and all the lymph from the superficial nodes. The upper node lies in the femoral canal, the gland of Cloquet. The lymph from these nodes passes into the abdomen and to the external iliac nodes. (See also p. 482.)

Review questions on the lower limb

(answers p. 529)

127 The most muscular vein of the lower extremity, which is often used as an arterial substitute, is the
a) femoral
b) saphenous
c) popliteal
d) soleal
e) posterior tibial

128 Destruction of the superior gluteal nerve may disturb the normal gait by paralyzing the
a) piriformis
b) gluteus medius
c) gluteus maximus
d) hamstrings
e) adductor magnus

129 A child falls on a spike, injuring the upper lateral margin of the popliteal fossa; which nerve is liable to be injured?
a) Common peroneal
b) Tibial
c) Obturator
d) Sciatic
e) Femoral

130 The pull of the superficial muscles of calf may avulse a portion of bone into which they are all inserted. This bone is
a) tibia
b) fibula
c) talus
d) calcaneus
e) cuneiform

131 If the tendo calcaneus is severed in protest by an unjustly convicted prisoner he would still be able to flex his ankle by the following muscle:

a) peroneus tertius
b) tibialis anterior
c) soleus
d) plantaris
e) flexor hallucis longus

132 The proximal superficial femoral artery is often occluded by thrombus. The blood can still bypass to the lower extremity and the most important vessel that effects this bypass is the
a) internal pudendal
b) common iliac
c) superior gluteal
d) popliteal
e) descending genicular

133 Bony injuries around the knee are very common. Which structure lies closest to a fractured lower end of femur?
a) Popliteal vein
b) Tibial nerve
c) Popliteal artery
d) Common peroneal nerve
e) Saphenous nerve

134 The quadriceps femoris is composed of all the following muscles except
a) vastus medialis
b) vastus lateralis
c) rectus femoris
d) sartorius
e) vastus intermedius

135 With regard to the lower limb,
a) a femoral hernial sac would protrude just below the inguinal ligament and just lateral to the femoral artery
b) the knee jerk reflex provides a functional test for integrity of L2–4 spinal segments

c)   one stands partly on S1 and sits on L3 dermatomes

d)   the posterior cruciate ligament of the knee joint resists posterior displacement of the fibula on the tibial condyles

e)   muscles which act across two joints include the sartorius, the gastrocnemius and the soleus

136   With regard to the lower leg and foot,

a)   the ankle joint is most stable in plantar flexion of the foot

b)   the interosseous membrane transfers weight-bearing force from the tibia to the fibula

c)   the dorsalis pedis artery anastomoses with the plantar arterial arch formed by a branch of the posterior tibial artery

d)   the calcaneonavicular ('spring') ligament 'articulates' by a synovial joint with the cuneiform

e)   the lateral plantar nerve supplies cutaneous innervation to the foot just like the median nerve in the hand

137   When attempting to walk on the frozen snow, flexibility of the foot is necessary to prevent falling. In this respect, which of the following statements is true?

a)   Inversion–eversion of the foot is free at the ankle joint

b)   Inversion–eversion of the foot is free at the subtalar joint

c)   The sinus tarsi has no contents

d)   Peroneus brevis and tertius are ineffective in eversion at the transverse tarsal joint

e)   The flexor retinaculum adds substantial stability to the ankle joint

138   Muscle(s) which enable you to arise out of a chair and straighten your trunk over your lower extremities are the

a)   gluteus maximus

b)   gluteus medius

c)   psoas

d)   hamstrings

e)   none of the above

139   Enlargement of the superficial inguinal lymph nodes may be a sequel to

a)   a sore on the big toe

b)   a boil on the buttock

c)   an infected Bartholin's (greater vestibular) gland

d)   all of the above

e)   none of the above

140   A common complaint of a child with hip joint disease is pain in the knee. This is due to

a)   common innervation of both joints by the saphenous nerve

b)   common innervation of both joints by the obturator nerve

c)   the fact that the sartorius acts on the hip and the knee joints

d)   the fact that the hip joint and the knee joint are set in the same vertical plane

e)   all of the above

141   In walking along the side of a hill one's feet can compensate for the slope by inversion and eversion. These movements take place at the

a)   ankle joint

b)   talofibular joint

c)   talonavicular joint alone

d)   metatarsophalangeal joints

e)   subtalar joint

142   The longitudinal arch of the foot is supported by the

a)   plantar aponeurosis

b)   calcaneonavicular ('spring') ligament

c)   sustentaculum tali

d)   all of the above

e)   a and b only

143   A long distance runner complains of pain in his anterior leg. Being an astute physician you diagnose the anterior compartment syndrome. You do this by noticing

a)   the inverted position of the foot

b)   the complete immobility of the whole foot

c)   the plantar flexed position of the toes

d)   the separated position of the toes

e)   that the anterolateral muscles are hyperactive on electromyography

144   A patient has a cast applied to his leg for a fracture of the lower tibia. On removing the cast the patient notices his foot is turned inward and he has sensory changes. The physician informs him that the common peroneal nerve has been damaged by the ill-applied cast. The nerve

a)   crosses the head of the fibula against which it

can be rolled by the discerning examiner
b) supplies sensation to the sole
c) passes from popliteal fossa to leg superficial to the lateral head of gastrocnemius
d) is a purely motor nerve
e) passes superficial to the tendon of biceps femoris

145 **Your golfing partner has a bad swing; he strikes you on the knee with constant regularity. The accumulated blood can be removed from the synovial cavity of the knee joint by way of the following bursae which communicate with the joint:**
a) prepatellar bursa
b) deep infrapatellar bursa
c) suprapatellar bursa
d) superficial infrapatellar bursa
e) none of the above

146 **Which of the following is true?**
a) The soleus muscle is important in the pumping mechanism of the deep veins of the leg
b) The 'triceps surae' (gastrocnemius and soleus) is the only muscle group capable of plantar flexion
c) Anastomoses around the knee joint (unlike those around the elbow joint) are totally incompetent in supplying the leg and foot if the superficial femoral artery is destroyed
d) The patella is stabilized against dislocation by fibers of the vastus lateralis muscle
e) The nutrient artery of the tibia is small and situated in the distal half of the bone

147 **The superficial inguinal lymph nodes receive lymph drainage from the**
a) sole
b) buttock skin
c) labia majora
d) lower anal canal
e) all of the above

148 **For your 21st birthday you are given a set of seven tarsal bones. To assemble them, you will need to know which bones articulate with the talus. They are the**
a) navicular
b) calcaneus
c) cuboid
d) medial cuneiform
e) a and b only

149 **As the nurse is administering an intramuscular gluteal injection in the early morning, there is a power failure, and, in error, he damages the sciatic nerve. This results in**
a) loss of extension of the knee
b) total loss of sensation below the knee
c) loss of sensation of the sole of the foot
d) inability to flex the hip
e) loss of proprioception from the knee joint

150 **With regard to the greater trochanter of the femur,**
a) it receives the insertion of all of the muscles in the buttock
b) it is separated from the overlying skin by a bursa
c) it receives the insertion of the psoas tendon
d) its secondary centre of ossification appears at puberty
e) its secondary centre of ossification fuses at puberty

151 **The soleus muscle**
a) is part of what is often called the 'third heart'
b) contains few veins
c) inserts commonly with the flexor hallucis muscle
d) arises from the tibia only
e) is an example of a 'white' muscle

152 **A medical student suffers a rotation injury of the medial meniscus while dissecting on the wrong side of the body. This meniscus**
a) is smaller than the lateral
b) is more fixed than the lateral
c) is separated from the medial ligament by the popliteus tendon
d) is 'O' shaped
e) is separated from the medial ligament by a pad of fat

153 **A male patient has an aneurysm (swelling) of the aorta enlarging across the left psoas muscle involving the genitofemoral nerve. This could produce**
a) pain over the femoral triangle
b) pain in the scrotum
c) inability to elevate the testis
d) all of these
e) a and b only

Part VII   The Abdomen

24 Abdomen

Road map of the abdomen · Superficial fascia · Muscles of the abdominal wall · Fascia transversalis · Neurovascular supply of the abdominal wall · Lymphatics of the abdominal wall · Inguinal canal · Femoral canal

Road map of the abdomen

As in most US cities, the 'highways' run north–south (2×) and east–west (2×) and thus divide the abdomen into nine areas. The two vertical planes run through the mid-inguinal points. These points lie midway between the anterior superior iliac spines and the pubic symphysis. The upper horizontal line runs through the transpyloric plane; this plane lies midway between the suprasternal notch and the pubis. The lower horizontal plane, the transtubercular plane, passes through the tubercles of the iliac crests.

Superficial fascia

The superficial fascia of the abdominal wall consists of two layers: a fatty superficial layer (Camper) seen best around the pools of many hotels, and a deeper membranous layer (Scarpa). The latter layer passes inferiorly to attach to the crest of the ilium and the deep fascia of the thigh immediately below the inguinal ligament. Medial to these attachments the membranous layer passes across the symphysis pubis to fuse with the margins of the pubic arch and the base of the perineal membrane, where it forms the inferior boundary of a superficial perineal pouch (Colles' fascia). The urethra passes through the pouch. Should the urethra rupture, urine will leak into the superficial pouch and fill

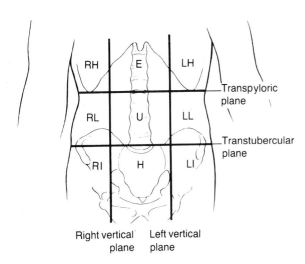

Fig. 24.1   The clinical compartments of the abdominal cavity. E, epigastrium; H, hypogastrium; LH, left hypochondrium; LI, left lower quadrant (inguinal); LL, left lumbar; RH, right hypochondrium; RI, right lower quadrant (inguinal); RL, right lumbar; U, umbilical.

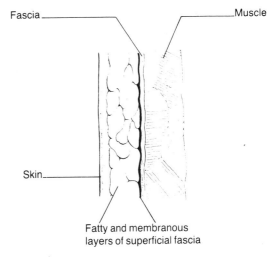

Fig. 24.2   Superficial layers of the abdominal wall.

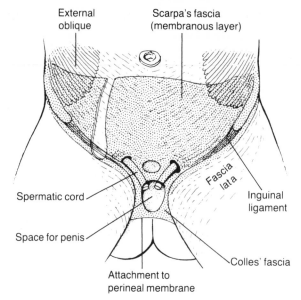

Fig. 24.3 Attachments of the membranous layer of the superficial fascia (Scarpa's fascia). This fascia is attached below to the fascia lata of the thigh and to the perineal membrane, while above it fades out in the region of the umbilicus. It is pierced by the spermatic cords and the penis with its contained urethra.

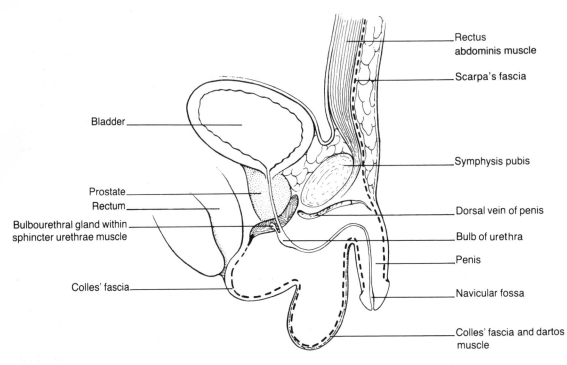

Fig. 24.4 Colles' fascia enclosing the superficial perineal pouch.

the anterior perineum, upper thighs and lower abdominal wall; the anatomical distribution of the swelling is dictated by the attachments of the fascia.

Muscles of the abdominal wall

These are disposed posteriorly, laterally and anteriorly.

Muscles of the posterior abdominal wall and lumbar fascia

The central pillar of the five lumbar vertebrae is surrounded by the sacrospinalis which lies in the groove between the vertebral spinous and transverse processes, by the quadratus lumborum which lies opposite the transverse processes and by the psoas major which covers the anterolateral parts of the bodies of the vertebrae. The sacrospinalis has origins and insertions that, fortunately, are anatomically unimportant (see p. 124). The quadratus lumborum is a flat muscle which arises from the posterior portion of the iliac crest, and runs up to the medial half of the last rib, which it steadies for the pull of the contracting diaphragm. The psoas major has been described previously (p. 301). The psoas minor, if present, is a small muscle with a long thin tendon that lies on the anterior surface of the psoas major; this thin tendon is sometimes mistaken for the genitofemoral nerve.

The lumbodorsal fascia fills the gap between the last rib and the iliac crest and is composed of three layers. The posterior diamond-shaped layer attaches to the spinous processes medially and runs laterally to meet the middle and anterior layers. The other two layers enclose the quadratus lumborum muscle. The three layers fuse laterally and form a point of origin for two of the three anterolateral muscles, the transversus abdominis and the internal oblique.

Anterolateral muscles

The fibers of the three muscles in this group, the external and internal obliques and the transversus, run in different directions – downward and forward, upward and forward, and transversely respectively. The ultimate effect of this arrangement is a strong muscular corset.

External oblique

The fibers of this muscle run downward and forward (the same direction as if the hands were in the pockets) from the lower eight ribs. The muscular fibers are inserted into the outer lip of the anterior portion of the iliac crest. Anterior to this they are replaced by tendinous fibers which join the rectus sheath; the lowest tendinous fibers, however, form the inguinal ligament (Poupart) (see Figs. 19.5, 24.9). This ligament, attached to the anterosuperior iliac spine laterally and the pubic tubercle medially, consists of the lowest part of the aponeurosis folded back on itself; by this means it forms a hammock for the spermatic cord in the male or the round ligament in the female. The insertion of the inguinal ligament into the pubic tubercle is prolonged medially along the pectineal line of the pubic bone, where it fuses with the periosteum; this prolongation is called the pectineal ligament (Cooper). Between the pubic tubercle and the pectineal line of the pubic bone, the extension of the ligamentous insertion has a free curved lateral edge, the lacunar part of the inguinal ligament (Gimbernat). The lateral curved edge is mean; it is one of the factors responsible for the strangulation of a femoral hernia. Just superior to the pubic tubercle, there is a triangular defect in the aponeurosis; this is the superficial inguinal ring. The upper and lower margins (superior and inferior crura) of the ring are thick and rigid; they are also unfriendly since they may obstruct and strangulate an inguinal hernia. The superficial ring transmits the cord/round ligament, and the ilioinguinal nerve.

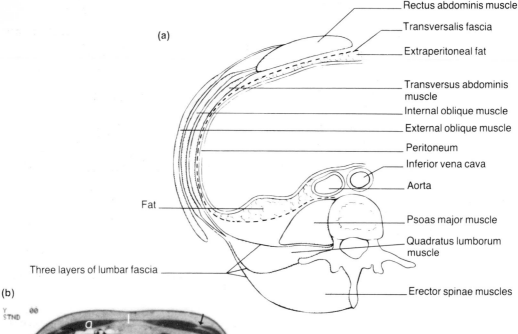

(a)

Rectus abdominis muscle
Transversalis fascia
Extraperitoneal fat
Transversus abdominis muscle
Internal oblique muscle
External oblique muscle
Peritoneum
Inferior vena cava
Aorta
Psoas major muscle
Quadratus lumborum muscle
Erector spinae muscles

Fat

Three layers of lumbar fascia

(b)

Fig. 24.5 (a) Cross-section of the abdominal cavity, illustrating the distribution of the muscles. (Note the free posterior border of the external oblique muscle.) (b) MRI scan. p, posterolateral sacrospinalis group; al, anterolateral obliques and transversus; a, anterolateral rectus abdominis; l, linea alba; →, linea semilunaris; c, crura; d, latissimus dorsi; s, superior mesenteric artery; r, left renal vein.

Internal oblique and transversus abdominis

The internal oblique and transversus abdominis muscles run between the lower six costal cartilages and the whole length of the linea alba, the anterior portions of the iliac crests, the lateral portions of the inguinal ligament and the lumbodorsal fascia.

Anterior Muscles

Rectus abdominis

This straight longitudinal muscle passes superiorly from the pubic bone and symphysis, crosses the lower costal cartilages, and inserts into the anterior surface of the fifth, sixth and seventh costal cartilages and the xiphoid process. It is wide in its upper part and, in those who are not too chubby, its lateral border is visible. The muscle is broken up into several smaller muscles by tendinous intersections; these are strong fibrous transverse bands, one at the costal margin, one at the umbilicus and one midway between, with perhaps an incomplete one lower down. These intersections increase the efficiency of the muscle and illustrate well its somatic origin, which also explains its segmental nerve supply.

Pyramidalis

This small triangular muscle is commonly absent; it arises from the pubic crest and lies in the lower rectus sheath.

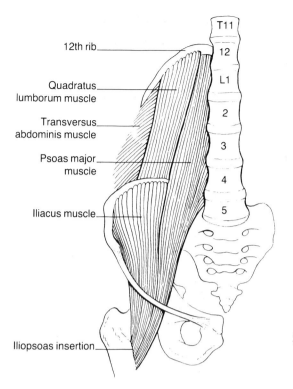

Fig. 24.6   Posterior abdominal wall

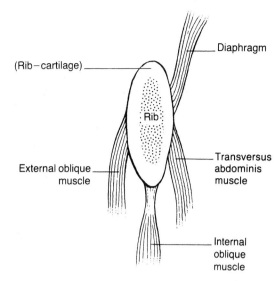

Fig. 24.7   Muscular attachments of the rib–cartilage complex.

3   Above the costal margin, the external oblique aponeurosis covers the rectus abdominis as it lies on the fifth, sixth and seventh costal cartilages.

The musculotendinous fibers making up the rectus sheath meet beyond the rectus in the midline, where they form the linea alba, a strong fibrous band, wider above than below,

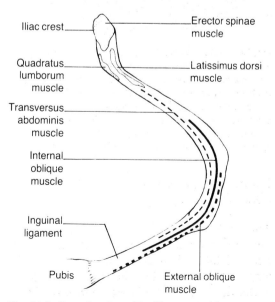

Fig. 24.8   Superior view of the iliac crest with muscular attachments.

Rectus sheath

This sheath contains the rectus abdominis and pyramidalis muscles together with the nerves and vessels supplying them. The sheath is of varying composition, as follows:

1   From the costal margin to midway between the umbilicus and pubis, it is formed by the internal oblique aponeurosis splitting into anterior and posterior lamellae to enclose the muscle. Reinforcing the lamellae are the external oblique anteriorly and the transversus posteriorly.

2   From midway between the umbilicus and pubis to the pubis, all three muscles pass anterior to the rectus muscle, leaving it with little support posteriorly. The site where the posterior sheath thins out is easily recognized and is called the semilunar fold of Douglas or arcuate line.

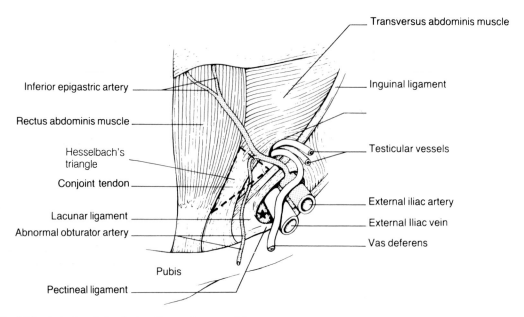

Fig. 24.9 Anterior abdominal wall seen from the INNER aspect. The ★ marks the femoral canal through which a femoral hernia passes. Note abnormal obturator artery. The dotted lines demarcate Hesselbach's triangle.

that runs from the xiphoid process to the symphysis pubis. Upper midline incisions hold sutures far better than lower midline incisions, as the sutures are placed in more substantial tissue. The skin overlying the linea alba is often the site of pigment deposition during pregnancy (linea nigra).

Actions of the abdominal muscles

These muscles provide support for the viscera, and, by their rapid contraction, protect them from injury. With closure of the glottis and contraction of the abdominal muscles, the intra-abdominal pressure can be markedly raised; this helps to empty the abdominal contents (feces, urine, babies, bad hamburgers, etc.). Paraplegics, whose abdominal muscles are paralyzed, have to use abdominal pressure with their hands to empty the bladder. The muscles flex the trunk; they initiate its forward bending, after which they relax, the strain subsequently being taken by the muscles of the back. In hyperextension of the spine, they contract to prevent a backward fall; this is one mechan-

ism by which they are injured or even ruptured. Muscles acting unilaterally flex the vertebral column to that side, or, in reciprocal contractions of each side, they rotate the spine. During inspiration, the muscular corset relaxes and affords the viscera some welcome space in the face of the descending diaphragm.

Fascia transversalis

Deep to the transversus abdominis, a layer of fascia, the fascia transversalis, encircles the abdomen. Posteriorly it covers the psoas muscle (changing its name to the psoas fascia), superiorly it runs on to the undersurface of the diaphragm and inferiorly it gains attachment to the inguinal ligament in its lateral half; deep to the medial half of the ligament, it follows the femoral vessels into the thigh to form the anterior component of the femoral sheath. Lining the iliacus muscle is the fascia iliaca; inferiorly it also inserts into the inguinal ligament in its lateral half, but deep to the medial half it passes posterior

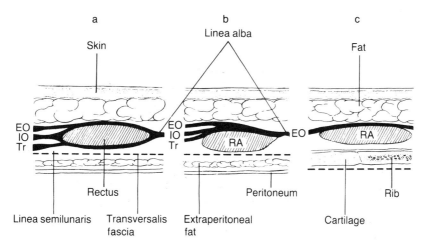

Fig. 24.10 Cross-section of the rectus sheath: (a) costal margin to arcuate line; (b) arcuate line to pubis; (c) above costal cartilage. EO, external oblique muscle; IO, internal oblique muscle; RA, rectus abdominis muscle; Tr, transversus abdominis muscle.

to the femoral vessels as the posterior component of the femoral sheath.

Deep to the fascia transversalis a layer of extra peritoneal fat covers the peritoneum, the serous membrane that lines the abdominal cavity.

Neurovascular supply of the abdominal wall

Nerves

The sacrospinalis group of muscles is supplied by posterior primary rami of the lumbar nerves, whereas the other posterior muscles, the quadratus lumborum, psoas major and minor, are supplied by the anterior primary rami of these nerves. The seventh to eleventh thoracic intercostal nerves run in the intercostal spaces as far as their anterior extremities and then pass between the costal slips of origin of the diaphragm and transverse abdominis to reach the abdominal wall. The twelfth thoracic or subcostal nerve (T12) lies below the twelfth rib and reaches the abdominal wall by piercing the transversus abdominis. The first lumbar nerve emerges from the psoas muscle and behaves

like T12. Nerves T7–12 and L1 encircle the abdominal wall in the neurovascular plane between the internal oblique and the transversus. They reach and pierce the lateral border of the rectus sheath, supply the rectus muscle and finally emerge through the anterior sheath as anterior cutaneous branches. Each nerve gives off lateral branches in the midaxillary line, which divide into anterior and posterior divisions; these branches arborize with the terminal anterior cutaneous nerves anteriorly and with the posterior primary rami posteriorly. The body wall is therefore segmentally supplied from the midline anteriorly to the midline posteriorly. The lateral division of T12 is an exception; it does not branch, but descends over the iliac crest to supply part of the skin of the buttock. Segmentally, the seventh nerve supplies the epigastric area, the tenth nerve the umbilical region, and the first lumbar nerve the area above the pubis. The segments from T7–L1 follow each other in a regular geometric pattern; this is important in spinal cord diseases, when the level of resulting anesthesia requires delineation. Intra-abdominal pain resulting from internal disease is often referred to the peripheral terminations of a somatic nerve that supplies the abdominal skin and the muscles and the peritoneum beneath them. Any injury to the skin is followed by a rapid reflex protective contraction of the muscle; this is the method of eliciting the abdominal reflexes clinically. It should be recognized that the intercostal nerves that supply the spine, diaphragm, parietal pleura and chest wall also supply the abdominal wall. If you remember this, you will not be surprised if a young child with inflammation of the pleura from an underlying pneumonia of the lung presents with abdominal pain; pneumonia in the young is a potent cause of abdominal pain and vomiting. Spinal disease may mimic abdominal disorders by the same mechanism.

Blood vessels

Deep vessels

The intercostal vessels accompany the nerves around the abdominal wall in the neurovascular plane to anastomose with the vessels lying in the rectus sheath. The latter vessels are the superior epigastric, one terminal branch of the internal thoracic which enters the sheath between the slips of the diaphragm above, and the inferior epigastric, a branch of the external iliac which enters the sheath at the semilunar fold below. These two vessels anastomose and supply the rectus muscle and overlying skin.

Superficial vessels

The superficial arteries of the abdominal wall arise mostly from branches of the femoral artery and accompany the superficial veins. The veins above the umbilicus drain superiorly to eventually reach the superior vena cava. Below the umbilicus, they pass to the femoral vein and, eventually, the inferior vena cava. A thoracoepigastric vein commonly connects these two systems and becomes important in a block of either of the two cavae. Blockage of either cava results in a visible enlargement of this vein; compression can ascertain by which route the vein is filled and the obstructed cava can be

identified (Fig. 24.11). Periumbilical veins drain to the left branch of the portal vein; obstruction of the latter, as in cirrhosis of the liver, leads to a reversed flow into the former with the appearance of distended periumbilical veins, the 'caput medusae'. Malignancies may also spread from the liver along these veins to produce an umbilical mass.

Lymphatics of the abdominal wall

Lymphatics above the umbilicus pass to the axillary nodes; those below the umbilicus drain into the inguinal nodes. A malignancy arising at the umbilicus may spread to all four

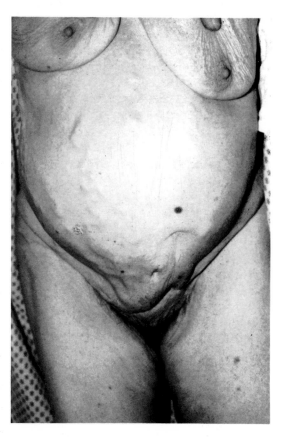

Fig. 24.11 Patient with inferior vena caval obstruction showing distended thoracoepigastric veins, which will return the blood to the heart via the superior vena cava.

sets of nodes, with a correspondingly gloomy prognosis. Remember that there are no lymph nodes in the abdominal wall (cf. scalp, Chapter 11); a swelling in the abdominal wall can never be an enlarged lymph node. There is an occasional exception in children, where one or two inguinal nodes may lie just above the inguinal ligament.

Inguinal canal

The testis is formed in the upper lumbar region of the posterior abdominal wall. It descends to the inguinal region, where it passes through the full thickness of the anterior wall to reach its permanent scrotal position; to do so it traverses the deep inguinal ring, inguinal canal and superficial ring successively. What is the reason for this journey? The testis cannot produce spermatozoa in the warm environment of the abdomen, and functions best at temperatures below this. One might conjure up the contrast between a lightly clad man standing on one of the Caribbean beaches with the cooling breezes wafting by, and the Eskimo in his frozen igloo

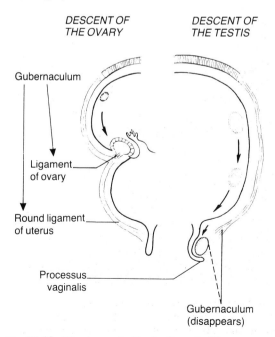

DESCENT OF THE OVARY

DESCENT OF THE TESTIS

Gubernaculum

Ligament of ovary

Round ligament of uterus

Processus vaginalis

Gubernaculum (disappears)

Fig. 24.12 The descent of the testis and of the ovary.

bundled up in heavy furs. It would appear that the former has a much higher fertility rate than the latter. However, man has adapted to his surroundings; the population of both areas shows no signs of decreasing. It is interesting to note that some animals (e.g. rodents) are rather clever; they are able to drop the testes from the abdomen into the scrotum when the breeding season arrives. Man is not so clever; he cannot easily return the testes to the abdomen, although there are times when plunges into the waters of the northerly beaches retracts them far enough to suggest a pair of swellings in the lumbar regions, the site of origin of the testes. In some animals (e.g. the elephant) the testes do not emerge from the abdomen and no scrotum has developed; it is strongly suggested that you accept this statement and not search for the elephant's scrotum. Having established the reason for the testis scurrying to the scrotum, an examination of the mechanism is instructive. The testis develops on the germinal ridge on the posterior abdominal wall and retains this retroperitoneal position during its descent. Passing inferiorly toward the inguinal region, it reaches a deep inguinal ring at about the seventh month, traverses the canal and emerges through a superficial inguinal ring at or just before birth. On its journey, it takes with it its neurovascular supply and a cushion for the ride. The latter is a process or diverticulum of the peritoneum, the processus vaginalis, which retains its original anterior relationship to the testis. Once the testis reaches the scrotum, the processus becomes pinched off at the upper pole of the testis and at the deep inguinal ring. Between these points it degenerates into a fibrous cord, the remains of the processus vaginalis. The distal part of the processus remains anterior to the testis, attaching to it by a visceral layer, and surrounding it by a parietal layer. These two continuous peritoneal surfaces are lubricated by a thin film of fluid which facilitates the up and down movement of the testis. Many mobile structures have visceral and parietal lubricated linings; for example, tendon, intestine, lung, heart and testis. The testis does not lie docilely in the scrotum; it is retracted intermittently by the cremasteric muscle. This can be seen during a cold shower in winter, or when the testis is threatened with a club; it retracts in response to cold and fright. Do not confuse this retraction with the contraction of the dartos muscle in response to cold that results in wrinkling of the scrotal skin.

The path through the anterior abdominal wall is blazed by the gubernaculum (pilot), a mass of mesenchymal tissue that precedes the descending testis. Having reached the scrotum ahead of the testis, the gubernaculum remains attached to the lower pole of the testis and the inner aspect of the scrotum for a time. After being sure that the testis is permanently in the scrotum, it degenerates and disappears by adulthood. The descent of the gubernaculum leaves a defect in the fascia transversalis, the deep (internal) inguinal ring. A defect is also present in the three muscles of the abdominal wall, but is formed so as to create an oblique passage. The ureter traversing the bladder wall, the bile duct penetrating the duodenal wall and the parotid duct piercing the buccinator muscle all do so obliquely. An oblique passage through a muscular structure implies anterior and posterior walls that are able to contract; this approximates the walls and obliterates the canal. In the case of the inguinal canal a cough apposes the boundaries of the canal and prevents the protrusion of any abdominal contents through potential areas of weakness.

The inguinal canal (see Fig. 19.5) is 3.75 cm (1.5 in) long and is situated just above and parallel to the medial half of the inguinal ligament. The contents of the canal pass from deep to superficial rings. The superficial ring is a triangular defect in the external oblique aponeurosis; its base lies on the superior border of the body of the pubis and its margins or crura are firm. The defect in the internal oblique is much larger than the defect in the external oblique. In the case of the transversus abdominis, the defect is even bigger. The fascia transversalis is pierced at the deep ring by the structures forming the spermatic cord. The internal ring is situated lateral to the inferior epigastric branch of the external iliac artery, as it runs posterior to the fascia transversalis toward the rectus sheath. The processus vaginalis passes obliquely from deep to superficial ring. Anteriorly the canal is completely covered by the external oblique, and partially, in its lateral third, by the substantial fibers of the internal oblique; the transversus abdominis does not constitute any part of the anterior wall. As the internal oblique and transversus abdominis pass medially, they join together – the conjoint tendon; this tendon passes posterior to the cord to insert into the pubic crest and pectineal line immediately behind the insertion of the inguinal ligament. The canal is formed posteriorly by the fascia transversalis completely, and the conjoint tendon in its medial third. The two rings or defects have been cleverly reinforced; the deep ring by the thick fibers of the internal oblique anteriorly, and the superficial ring by the conjoint

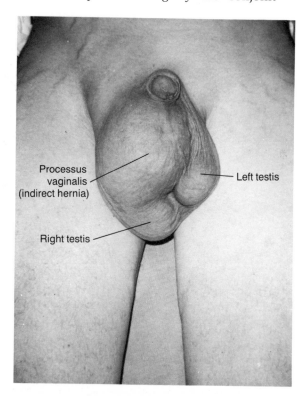

Processus vaginalis (indirect hernia)

Left testis

Right testis

Fig. 24.13 **A right indirect hernia of the complete variety. Note the right testis lying inferior to the intestine-filled processus vaginalis.**

tendon posteriorly. The arching fibers of the conjoint tendon are sometimes called the roof of the canal, and the recurved part of the inguinal ligament is designated the floor. With regard to the latter, the correct orientation of the pelvis places the pubis in a downward as well as forward inclination; in this position the inguinal ligament would form a floor or hammock for the spermatic cord/round ligament.

The ilioinguinal nerve emerges from the superficial ring to supply the skin in the area. It is not enclosed in the fascial coverings and does not form part of the spermatic cord.

Two protective mechanisms, the walls of the obliquely directed inguinal canal and the structures reinforcing the superficial and deep rings, have been described. Another particularly weak area is that which lies between the arching fibers of the internal oblique and transversus (conjoint tendon) and the inguinal ligament; it consists of fascia transversalis only. The third protec-

tive mechanism, the shutter mechanism, reinforces this area; the contraction of the arching fibers of the conjoint tendon produces flattening and therefore descent of the arch (as in the diaphragm). With this descent the conjoint tendon covers and protects the fascia transversalis – the shutter closes.

Inguinal herniae

We are now in a position to describe the inguinal herniae, which are divided into direct or indirect types. The indirect hernia results from persistence or failure of obliteration of the processus vaginalis; this leaves a diverticulum or pouch of the peritoneal cavity into which abdominal contents, usually intestine, can pass. The processus may persist in its complete fetal state and reach the scrotum or it may be obliterated somewhere along its length; a hernia therefore may or may not reach the scrotum. The management of this common hernia is clear;

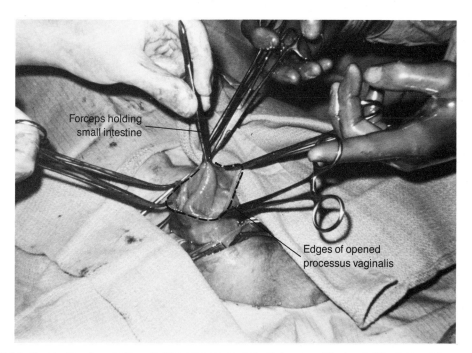

Forceps holding small intestine

Edges of opened processus vaginalis

Fig. 24.14 Surgical treatment of hernia. The processus vaginalis is open at the level of the internal ring and clamps have been placed along the edges of the opened sac. The contained small intestine is being elevated by a forceps prior to its reduction into the peritoneal cavity. Following this, the internal opening of the sac will be closed by suture.

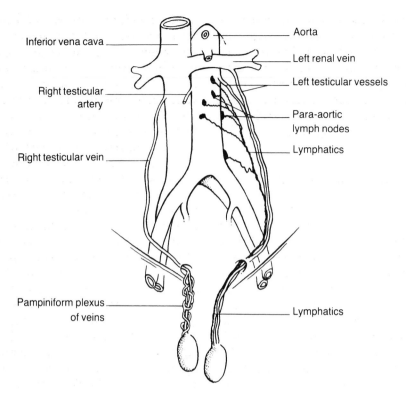

Fig. 24.15 Vascular supply of the testes.

since it is due to the presence of the persistent processus vaginalis (called the 'sac' in clinical circles), its removal will be curative. Sometimes the contents of the sac cannot return to the abdominal cavity because of constriction by either of the two rings; this is called an irreducible hernia. If the blood supply of the contained viscus is also compressed, the condition is called a strangulated hernia; this is a dangerous condition that requires an emergency operation.

The direct hernia, which usually occurs in older people, has a different genesis. *Pari passu*, all our daily difficulties unfortunately increase. All the body's muscles, and with them the expulsive acts, become weaker; coughing is less effective, bowel actions become more difficult, and the male struggles to pass urine as his prostate gland enlarges. The weak area of the inguinal canal is demarcated by a triangle, Hesselbach's triangle. This lies between the inferior epigastric

artery laterally, the rectus abdominis muscle medially and the inguinal ligament below (Fig. 24.9). It is the site where the fascia transversalis is covered only by the external oblique and through which the direct hernia protrudes as it stretches the fascia transversalis and the external oblique. The path of this hernia is directly forward through the triangle, hence its designation as a direct hernia. The shutter mechanism of the conjoint tendon is the main deterrent to the development of this hernia and it is effective until the muscles degenerate with age. As the direct hernia bulges anteriorly, it does not pass through narrow orifices; obstruction or strangulation do not therefore occur as in the indirect herniae, and surgical repair is not mandatory. Should a repair be performed, the surgeon attempts to close the shutter by suturing the conjoint tendon to the inguinal ligament; sometimes the muscles are so degenerate that these structures cannot hold sutures. In these cases, the weak area can be reinforced by fascia taken from the patient himself, or by artificial material such as polyvinyl mesh. Currently, coverage of the weak area by mesh alone is more popular than closure by the transposition of tissues. In summary, there are two types of inguinal herniae, each with a different method of management.

Contents of the inguinal canal

The contents of the canal include the structures going to and from the testis. The ductus (vas) deferens carries spermatozoa and leaves the canal at the internal ring where it passes medially into the pelvis. The veins consist of vine-like clusters (the vines of grapes – the pampiniform plexus); this arrangement delays the venous return, allows extra heat loss from the exposed scrotum and cools the testes. The rule that deep lymphatics follow arteries holds true and the vessels follow the testicular artery to its origin from the aorta at the level of L1–2. Since the lymphatics drain to the para-aortic nodes at the level of L1, an epigastric mass may be due to a tumor of the testis. The sympathetic nerves, arising originally from the tenth thoracic segment, reach the testis via the testicular artery. Acute disease of the testis may present with periumbilical pain (tenth thoracic segment); if accompanied by vomiting, an acute upper abdominal condition may be mimicked. The

cord, therefore, consists of the testicular artery, veins, lymphatics, autonomic nerves, the remains of the processus vaginalis and the vas deferens. These structures lie in the classic order – VAD, the vein anteriorly, the artery in the middle and the duct posteriorly. These structures are enclosed by a prolongation of the layers of the abdominal wall acquired during the descent of the testes; these layers are the fascia transversalis internally (internal spermatic fascia), the muscle fibers from the internal oblique muscle that are reflected onto the cord (cremasteric layer), and the fascia of the external oblique (external spermatic fascia). The cremasteric muscle consists of loops of muscle fibers reflected from the internal oblique onto the cord and which are able to elevate the cord and testis. In the infant the cremasteric muscle is so active that it can retract the testis into the groin, and this 'retractile testis' may be confused with an undescended testis. The cremaster has the same nerve supply as the skin over the femoral triangle, the genitofemoral nerve. This nerve arises from the first and second lumbar nerves and splits into two portions, the femoral part that accompanies the femoral vessels to supply the skin over the femoral triangle and the genital branch that runs with the cord to supply the cremaster muscle. The cremasteric reflex consists of retraction of the testis in response to stroking the skin over the inner portion of the upper thigh; the presence

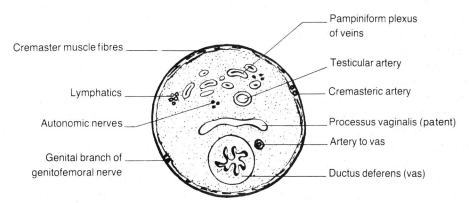

Fig. 24.16 Cross-section of spermatic cord. Normally the processus vaginalis becomes obliterated.

of the reflex proves that the cord is intact to the level of L2 spinal segment. The final coverings of the cord are the membranous and fatty layers of the continuation of the superficial abdominal wall. In the scrotum the membranous layer changes its name to Colles' fascia, and the fatty layer is replaced by smooth muscle fibers, the dartos muscle. It wouldn't do to have an insulating fatty layer lining the scrotum and preventing heat loss. The dartos muscle, which replaces the fat, is capable of wrinkling the scrotal skin. This may appear to be a fact of little interest, but it is a form of shivering. The testis prefers a cool, but not a cold, environment and shivering is a mechanism for raising the body temperature in response to cold.

Femoral canal

The external iliac vessels lie between the fascia covering the iliopsoas muscle, the fascia iliaca posteriorly and the fascia lining the transversus abdominis, fascia transversalis anteriorly. As the two fascial envelopes approach each other inferiorly, they fuse in their lateral halves and attach to the inguinal ligament. Medially both layers are carried into the thigh by the external iliac –femoral vessels as the femoral sheath. The compartments of the femoral sheath lateromedially are the artery, the vein and a space, the femoral canal (see Fig. 24.9). The canal closes shortly below the inguinal ligament by the fusion of the medial wall with the vein, and the whole sheath disappears 2.5 cm (1 in) below the inguinal ligament.

There are several reasons for the complicated arrangement. The fully flexed fetus does not have a femoral sheath; it appears only when the child begins to extend his hip, and it protects the vessels where they pass under a firm fibrous structure, the inguinal ligament. This is probably true, because surgeons avoid passing vascular grafts under the ligament, where they are very liable to be occluded. The canal is present for two reasons. Firstly, it allows the vein to distend during times of increased venous return and dilatation, such as running away from a surgeon who is trying to inject hemorrhoids below the pectinate line. Secondly, it provides a route for lymphatics to pass from the inguinal to the common iliac nodes; while most of these lymphatics pass through the canal, others run in the sheath itself.

The canal is somewhat occluded at its abdominal opening by a pad of fat called the femoral septum, and it usually contains a large lymph node (Cloquet). A portion of peritoneum protruding into the canal constitutes a femoral hernia; it enlarges by stretching the sheath ahead of it. Into this peritoneal sac may pass a small loop of intestine. Because of the unyielding structures surrounding the entrance to this femoral canal, femoral herniae are seldom large (inguinal herniae on the other hand can be extremely large). The structures bounding the abdominal opening of the canal (femoral ring) are: anteriorly the inguinal ligament; medially the lacunar projection of the inguinal ligament; posteriorly the pectineal line of the pubic bone; and laterally the femoral vein. The ligamentous components can be thought of as one continuous structure – the inguinal ligament continuing its attachment as lacunar and pectineal extensions. The content of a femoral hernia often becomes obstructed or strangulated and is difficult to free. One method of enlarging the canal is division of the lacunar part of the ligament. However, in a small percentage of patients the obturator artery is the enlarged pubic branch of the inferior epigastric artery that runs inferiorly in close relationship to the lacunar ligament and which can be severed during this procedure. Somebody must have been caught out very badly by this in the past, because this abnormal obturator artery was often called the artery of death, 'corona mortis'.

Femoral herniae are more common in the female because the brim of the pelvis is wider and the canal is therefore bigger.

25 Diaphragm

Surface relationship · Structures traversing the diaphragm · Development of the diaphragm · Nerve supply · Action of the diaphragm

The diaphragm is a double-vaulted musculotendinous sheath that forms the dome-shaped roof of the abdominal cavity and the bulging floor of the thorax. It arises by a continuous origin from the inferior aperture of the thorax; anteriorly two slips arise from the posterior aspect of the xiphoid process, laterally it shares an attachment to the lower six costal cartilages with the origin of the transversus abdominis and posteriorly two crura arise from the vertebrae. The large right crus arises from the first three lumbar vertebrae and intervening discs, the left arising from the first two lumbar vertebrae and intervening disc. The origin from the vertebrae and the ribs is completed posteriorly by fibers that arise from the thickened fascia covering the psoas and quadratus lumborum muscles – the medial and lateral arcuate ligaments.

From these attachments the fibers arch upward and medially to insert into a trefoil central tendon, one leaf lying anteriorly and two posteriorly.

Surface relationship

The upper limit of the diaphragm is much higher than is generally thought; it reaches the right fourth rib and the left fourth space in the expired position (both living and cadaveric), the liver being responsible for the higher position on the right side. A useful working rule is that it reaches just above the

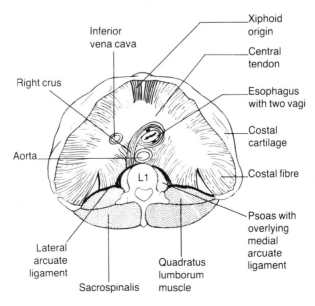

Inferior vena cava
Xiphoid origin
Central tendon
Right crus
Esophagus with two vagi
Costal cartilage
Aorta
Costal fibre
L1
Psoas with overlying medial arcuate ligament
Lateral arcuate ligament
Quadratus lumborum muscle
Sacrospinalis

Fig. 25.1 The diaphragm – an inferior view.

nipple line; any wound below the nipple line should be considered to have penetrated the abdominal cavity until this fact is disproved.

Structures traversing the diaphragm

A number of important structures pass through openings in the diaphragm on their way to or from the thorax.

1   The aorta passes anterior to the T12 vertebra beneath a thickening of fibrous tissue that unites the two crura, the median arcuate ligament. The commencements of the thoracic duct, the cisterna chyli and the azygos vein lie to the right of the aorta.

2   The esophageal opening lies more anteriorly at a higher level (T10), and slightly to the left of the aortic opening. Fibers from the right crus surround the esophagus, one of the factors that prevents gastroesophageal reflux. On the esophageal wall is the portosystemic venous communication; in the event of portal vein obstruction this anastomosis becomes varicose. The two vagi collect themselves from the interlacing esophageal plexus and enter the abdomen as well-behaved anterior (left) and posterior (right) gastric nerves, in front of and behind the esophagus. It is at this point that surgeons like to sever the nerve and reduce the gastric acid in cases of peptic ulceration, instead of chasing the numerous branches of the plexus. The esophageal hiatus quite frequently allows abdominal contents to pass illegally into the thorax, the condition of hiatus hernia.

3   The caval opening situated in the central tendon at the level of T8 transmits the inferior vena cava as well as part of the right phrenic nerve.

There are numerous other openings of the diaphragm through which structures pass from chest to abdomen and vice versa. Amongst these structures are the superior

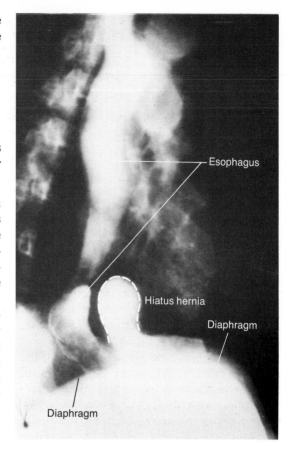

Fig. 25.2   The fundus of the stomach is seen parked illegally in the esophageal hiatus – a hiatus hernia.

epigastric and musculophrenic branches of the internal thoracic artery, the lower five intercostal nerves and vessels as they pass from the thoracic to the abdominal wall, and the sympathetic trunk and its branches, the greater, lesser and least splanchnic nerves.

It is interesting to study the function of the large openings during the contraction of inspiration. The caval opening is pulled open by the tendon, thereby enlarging the diameter of the cava and increasing the venous return. On the other hand, the right crus of the diaphragm closes the esophageal opening to prevent gastric acid from burning a hole in your lower esophagus, while the aortic opening remains passive and does not permit changes of caliber.

Table 25.1 Openings in the diaphragm – a summary

Vertebral level	Site	Major structure	Accompanying structures	Effect of inspiration
T8	Right trefoil of central tendon	Inferior vena cava	Right phrenic nerve (part) Lymphatics from liver	Opens (to increase venous return)
T10	Left of midline Sling of right crus	Esophagus	Left gastric vessels Anterior and posterior gastric nerves Lymphatics	Closes (to prevent acid regurgitation)
T12	Posterior to median arcuate ligament	Aorta	Thoracic duct Azygos vein	*No effect*

Development of the diaphragm

This is interesting (even for the student). The septum transversum is a mesodermal mass originally situated in the cervical region before being pushed down by the developing heart and lungs to its adult position. The history of this descent is confirmed by the retention of its nerve supply from the third, fourth and fifth nerves, the phrenic nerve. The upper part of the septum forms part of the fibrous pericardium, the middle part of the diaphragm, and its lower part gives rise to the connective tissue stroma of the liver. The anteromedial part of the diaphragm, including the central tendon, is formed from the septum transversum. The rest of the diaphragm is completed posteriorly by the mesoderm that surrounds the alimentary tube, and laterally by the inward projections of the pleuroperitoneal folds. Between the latter folds and the periesophageal mesoderm lie the pericardioperitoneal canals through which the pleural and peritoneal cavities communicate; the fusion of the folds and the mesoderm closes the canals. At times this fusion does not occur and abdominal contents are able to pass into the chest; this constitutes the commonest form of congenital diaphragmatic hernia (Bochdalek). This hernia is clinically manifest on the left side, as a right-sided hiatus would be blocked by the liver. The presence of abdominal contents in the chest is one cause of respiratory embarrassment at birth; by feeding the baby and distending the abdominal contents, these symptoms are exacerbated.

As the central tendon of the diaphragm and the pericardium arise from the septum transversum and are inseparable, a needle can be passed through both structures and blood aspirated in cases of acute cardiac tamponade (a collection of blood in the pericardial cavity). The needle is passed directly cranially between the xiphisternum and the seventh costal cartilage. At the present time, this is the most popular method of tapping the pericardial space; the alternative method of passing the needle through the fourth and fifth spaces is more likely to strike vessels on the anterior surface of the heart.

Nerve supply

The story of the development of the diaphragm is reflected in its nerve supply. The septum transversum is supplied by the phrenic nerve. The lateral parts of the diaphragm that arise from the costal cartilages receive their nerve supply from the intercostal nerves which course through this part of the body wall (T7–12). The posterior part of the diaphragm, the crura, is formed from the mesoderm lying behind the esophagus and is supplied by the neighboring intercostal nerves as well as the phrenic.

The fibers of the phrenic nerve are both afferent and efferent. The afferent fibers theoretically register sensation from the dia-

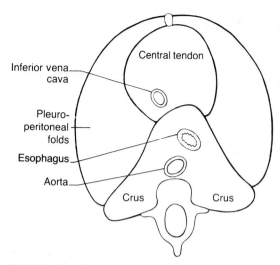

Inferior vena cava

Pleuro-peritoneal folds

Esophagus

Aorta

Central tendon

Crus   Crus

Fig. 25.3   Development of the diaphragm.

phragm in the sensory cortex. The cortex, however, is more accustomed to register sensation from skin, a far more highly developed sensory organ. Sensations from the diaphragm are therefore interpreted as coming from the shoulder region, which is clinically important, as conditions which irritate the diaphragm often produce shoulder pain (dermatomes C3, 4, and 5), see p. 215. For example, in suspected cases of tubal pregnancies, where tubal rupture liberates a large quantity of blood into the peritoneal cavity, placing the patient in a steep Trendelenburg position (head down, feet up) may elicit shoulder pain as the blood gravitates toward, and stimulates, the diaphragm. The nerve is also afferent to all the structures beginning with 'p', the **p**eritoneum over the abdominal and the **p**leura over the thoracic surfaces of the diaphragm, the **p**arietal and fibrous **p**ericardium, and the mediastinal **p**leura. Sensation from the peripheral diaphragm is mediated by the intercostal nerves, which also supply the chest and abdominal walls; conditions involving the peripheral diaphragm commonly present as abdominal pain. For example, a lower lobe pneumonia which inflames the overlying pleura of the peripheral diaphragm may present with abdominal pain (T7–12) and a temperature; in a young child who is unable to give a good history, the thoracic condition may be confused with an intra-abdominal condition.

Action of the diaphragm

On contraction the diaphragm straightens its arch and descends, pushing the abdominal viscera ahead of it. By increasing the intra-abdominal pressure, it contributes importantly to such functions as defecation, micturition and parturition. The descent acts like the withdrawn barrel of a syringe; it increases the vertical diameter of the thorax, and, to prevent a vacuum, air is inhaled.

26 Peritoneal cavity

**Extent and contents · Vertical disposition of peritoneal reflections ·
Transverse disposition of peritoneal reflections**

Extent and contents

The peritoneum is a smooth shiny membrane which lines the abdominopelvic cavity as a parietal layer and is reflected onto the viscera as a visceral layer. Small amounts of exudate allow easy movement of the viscera, the same mechanism as in the pleura, pericardium, testes and tendon sheath. The peritoneal cavity extends from the diaphragm (fourth/fifth space) to the pelvis (pubis anteriorly and coccyx posteriorly). A wound below the nipple line may enter and damage the contents of the abdominal cavity. By the same token, a shot in the buttock while you are leaving the field of battle to have a cup of tea may also enter the cavity; wounds seemingly confined to the buttock should be viewed with suspicion. Fortunately, most of the viscera are protected by the bony pelvis and the ribs with their costal cartilages; in

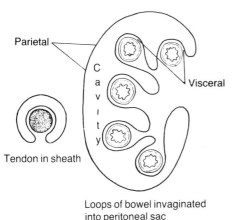

Fig. 26.1 **Visceral and parietal peritoneum (an analogy to a tendon sheath is drawn).**

Parietal

Visceral

C
a
v
i
t
y

Tendon in sheath

Loops of bowel invaginated into peritoneal sac

between these areas, the erect person is dangerously exposed.

Before describing the disposition of the peritoneum, a superficial visit to the intra-abdominal structures is worth while. Unpaired viscera lie in the upper abdomen; the liver occupies most of the right side, the spleen a smaller part of the left side, and below and between them lies the pancreas. The paired organs are the two kidneys and, above them, the suprarenals. The gastrointestinal tract is a conduit which starts with the entry of the esophagus into the abdomen and ends with the exiting of the anal canal.

The esophagus, after only a short course in the upper abdomen, enters a large fusiform bag, the stomach, which lies horizontally across the upper abdomen. The stomach narrows to give way to the succeeding C-shaped duodenum that is situated mostly behind the peritoneum. The duodenum is followed by the jejunum, which gives way imperceptibly to the ileum. The jejunum and ileum constitute the small intestine and have a mesentery. The ileum then enters a sacculated larger tube, the large intestine, the first part of which is the cecum with its attached appendix. The large intestine then runs as three sides of a square, the ascending colon cranially, the transverse colon transversely and the descending colon and sigmoid colon caudally; the last finally becomes the rectum and anal canal, which opens at the external anal orifice.

At this point, we should review the embryological events in the gastrointestinal tract. This was once a simple tube suspended

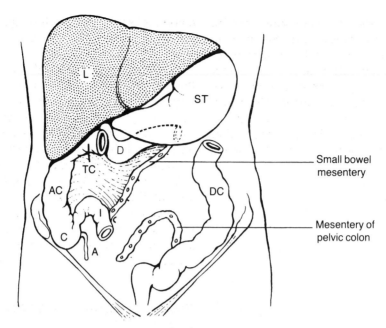

Fig. 26.2 Simple layout of alimentary tract. The mesenteries of small intestine and pelvic colon are seen in section. A, appendix; AC, ascending colon; C, cecum; D, duodenum; DC, descending colon; I, ileum; L, liver; ST, stomach; TC, transverse colon.

posteriorly by a mesentery from the posterior abdominal wall through which the blood vessels reached the tube. The same layers were reflected anteriorly around the upper part of the tube to the anterior abdominal wall. Certain events changed this simple arrangement. Briefly, the rapid enlargement of the stomach and its displacement by the growing liver pushes the organ to the left and rotates it onto its right surface so that it runs from left to right. As this happens, the duodenum also rotates on its right side and its mesentery merges with that of the posterior peritoneum – the duodenum becomes a retroperitoneal structure. As the duodenum ends, the mesentery is reconstituted as the small bowel mesentery. Another major event is the rotation of the intestine. The intestine grows so fast during the fourth to ninth weeks of fetal life that it cannot be contained in the abdominal cavity; the midgut emerges through the umbilical orifice and grows outside the abdomen. When the abdominal cavity enlarges sufficiently to hold the proliferating intestine, the bowel returns in an

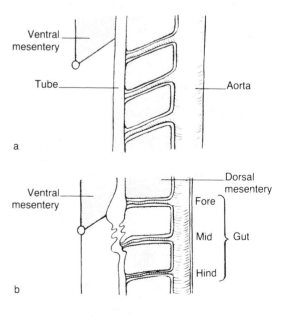

Fig. 26.3 (a) Primitive alimentary tube with dorsal mesentery carrying arterial supply. (b) A later stage, when the fore-, mid- and hindgut have developed.

orderly manner, the details of which should be sought in embryology texts. Suffice it to say that the small bowel retains its mesentery and the ascending colon loses it; the latter becomes covered by peritoneum on three sides, but lies bare posteriorly. The transverse colon retains its mesentery, but the descending colon loses it in a manner similar to that of the ascending colon. The pelvic

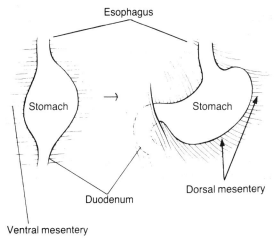

Fig. 26.4 Rotation of the stomach and duodenum onto its right side with resultant loss of the duodenal mesentery.

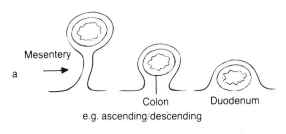

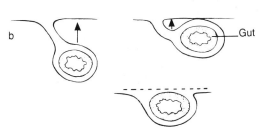

Fig. 26.5 Two ways of becoming retroperitoneal. Note that in (b) zygosis takes place and the fused peritoneum eventually disappears posteriorly.

colon picks up the mesentery again (pelvic or sigmoid mesocolon), only to lose it when the colon becomes the rectum.

The liver bud arises as a diverticulum from the foregut and grows into the ventral mesentery, which it separates into a sickle-shaped (falciform) ligament anterior to the liver, right and left triangular ligaments above and behind the liver, and a lesser omentum between the liver and the stomach. The spleen develops in the dorsal mesentery behind the stomach and divides this ligament into two portions, a gastrosplenic ligament between the stomach and the spleen, and the lienorenal between the spleen and the left kidney.

Vertical disposition of peritoneal reflections

A hand passed headward along the inner aspect of the anterior abdominal wall will be stopped by the reflection to the right of the anterior layer of the coronary ligament (see Fig. 28.3). At the right border the peritoneal reflection returns to the left as the posterior layer of the coronary ligament; the right triangular ligament is the point of fusion of these two layers of the coronary ligament. If the hand is passed to the left of the falciform ligament, it will meet the left triangular ligament; the reflection runs to the left as the anterior layer and returns to the right as the posterior layer. The posterior layers of the triangular ligaments approach each other and then attach to a fissure in the liver, the fissure for the ligamentum venosum. From the fissure the two peritoneal folds pass inferiorly to the stomach as the lesser omentum. After enclosing the stomach, the two layers run caudally for a variable distance as the anterior two layers of the greater omentum; they then turn back on themselves as the posterior two layers of the greater omentum and reach the transverse colon. These four layers of the greater omentum are difficult to separate in the adult because they tend to fuse and become fat laden; with age this omentum

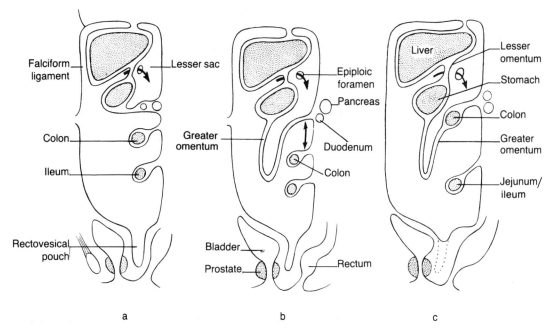

Fig. 26.6   The development of the mesenteries and peritoneal sacs as seen in longitudinal section. (a) Greater omentum not yet developed. (b) Developing greater omentum; colon not incorporated in omentum as yet. (c) Fusion of two layers (straight arrow) in (b) shows returning layers of greater omentum fusing with mesocolon to form the adult arrangement. The lower end of the rectovesical pouch has been obliterated.

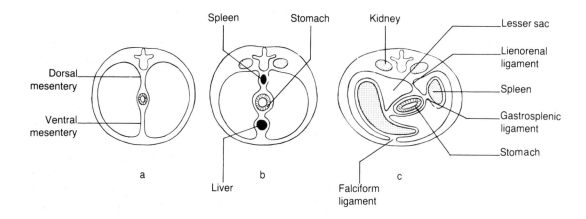

Fig. 26.7   The development of the mesenteries and peritoneal sacs as seen in transverse section. (a) Primitive gut with dorsal and ventral mesentery. (b) The liver has developed in the ventral mesentery and the spleen in the dorsal mesentery. (c) The disproportionate enlargement of the liver has rotated the stomach onto its right side. The spleen is displaced well to the left. Note that the lesser sac lies posterior to the stomach.

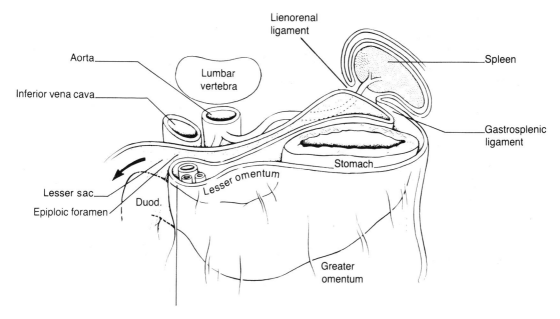

Fig. 26.8 Lesser and greater sacs. The lesser sac (omental bursa) lies posterior to the stomach and leads (via arrow) into the greater sac.

becomes markedly fatty and this is one of the causes of a pot belly. When the two posterior layers of the greater omentum reach the transverse colon, they enclose it and pass posteriorly as the two layers of the transverse mesocolon, the mesentery of the transverse colon (this is the only part of the colon which does not lose its mesentery). On reaching the posterior abdominal wall, the superior layer of the transverse mesocolon passes superiorly on the posterior abdominal wall to the diaphragm, where it becomes continuous with the diaphragmatic peritoneum. This peritoneum is then reflected onto the liver as the posterior layers of the coronary and triangular ligaments at which point this journey commenced. The inferior layer of the transverse mesocolon passes inferiorly on the posterior abdominal wall, from which it is quickly reflected as the two layers of the mesentery of the small intestine. Returning from the mesentery to the posterior abdominal wall, the peritoneum runs inferiorly until its reflection as the two layers of the pelvic mesocolon that surrounds the pelvic colon.

When the pelvic colon loses its mesentery, it becomes the rectum; the peritoneum then passes anteriorly onto the pelvic organs.

Transverse disposition of peritoneal reflections

These reflections are best appreciated in transverse sections of the abdomen. As the two peritoneal layers enclosing the stomach are reflected to the left, they reach and enclose the spleen, after which they are reflected onto the left kidney. Tracing the peritoneum enclosing the stomach to the right, it will be seen that the anterior layer passes onto the posterior abdominal wall after it has covered the duodenum as this flopped onto its side and became retroperitoneal. The posterior layer turns leftward at its right extremity to become continuous with the upper layer of the transverse mesocolon. A cavity is formed behind the two layers of lesser omentum and stomach; this is the lesser sac or omental bursa. If we stood in this lesser sac, which lies between the lesser

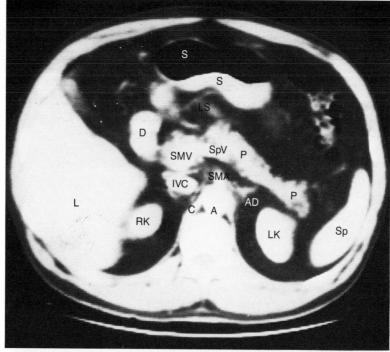

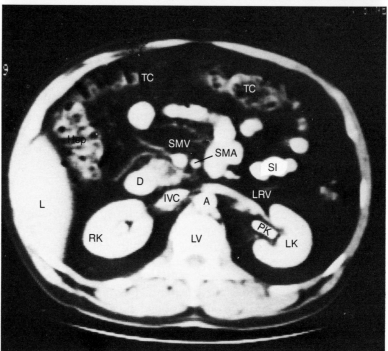

Fig. 26.9 The anatomist's dream – a living transverse section. The gastro-intestinal tract has been outlined by radio-opaque barium. A, aorta; AD, left adrenal gland; C, crus diaphragm; D, duodenum; Hep, hepatic flexure, colon; IVC, inferior vena cava; L, liver; LK, left kidney; LRV, left renal vein; LS, lesser sac (omental bursa); LV, lumbar vertebra; P, pancreas; PK, pelvis of kidney; RK, right kidney; S, stomach; SI, small intestine; SMA, SMV, superior mesenteric artery and vein; Sp, spleen; SpV, splenic vein; TC, transverse colon.

omentum and stomach anteriorly and the peritoneum of the posterior abdominal wall posteriorly, and we walked to the right, we would emerge into the greater sac through a foramen, the epiploic foramen (Winslow).

There was a time when the above was of interest only to surgeons and anatomists. Anatomists would decorate their practical examinations with numerous cross-sections, including those of the abdomen. As a clinician would be most unlikely to see such a view of a patient, the exercise was considered merely a means of humoring the anatomy faculty and a necessary evil in passing the examination. Lo and behold! Computer assisted tomography arrived and we are now able to look at transverse sections of the abdomen.

27 Gastrointestinal tract

Esophagus · Stomach · Duodenum · Small intestine · Large intestine

Esophagus

The abdominal esophagus is a continuation of the thoracic esophagus. This tube passes through the esophageal hiatus of the diaphragm, where it is surrounded by fibers of the right crus, one of the possible mechanisms that prevents esophageal reflux. After a short abdominal course of 1.25 cm (0.5 in), where it lies behind the liver, the esophagus enters the stomach. It is here that the right and left vagi have now become the posterior and anterior gastric nerves respectively. This lower esophagus is also an important site of portosystemic venous anastomosis (see p. 430).

Stomach

The stomach, the largest part of the gastrointestinal tract, was originally a simple tube bulging anteriorly and posteriorly (see Fig. 26.4). Rotation onto its right side has resulted in a viscus that runs from left to right posterior to the left lobe of the liver and the anterior abdominal wall successively (see Fig. 26.8). The stomach ends by passing to the right of the midline to become the duodenum. The part of the stomach behind the anterior abdominal wall can be percussed, and a well-dilated organ may be visible. The patient with a large tympanitic (resonance on percussion) epigastric swelling is the proud bearer of a dilated stomach, and may soon prove it by vomiting.

The stomach is divided into several parts.

A line projected to the left from the esophageal entry marks the lower limit of the fundus of the stomach. The fundus lies against the diaphragm and through this it is related to the heart, left lung and pleura – a stab through the left chest below the level of the nipple has every chance of passing into the fundus of the stomach. The cardiac relationship is very important; many patients with angina (i.e. cardiac pain on exertion) are worse after a heavy meal (the dilated stomach impairs the cardiac function). Below the fundus lies the body of the stomach, which passes to the right to become the pyloric antrum. The division between the body and the pyloric antrum is a niche and a notch. The niche, the incisura angularis, is an indentation on the lesser curvature or upper border of the stomach. The notch is a less distinct indentation on the greater curvature or lower border of the stomach. To the left of these indentations is the body; to the right is the pyloric antrum. The latter narrows into a pyloric canal and ends in the pyloric sphincter. An ill-defined area at the entrance of the esophagus is called the cardia. The shape of the stomach (like people's noses) varies; a common configuration is a J-shaped stomach (Fig. 27.1). The stomach can be so mobile that the greater curvature may lie in the pelvis; an inguinal hernia may even contain a portion of stomach. While the stomach can distend markedly with contents (as it should do at a banquet), its two ends, cardiac and pyloric, are relatively fixed. The cardiac end is situated beneath the seventh costal cartilage about 2.5 cm (1 in) to the left of the midline,

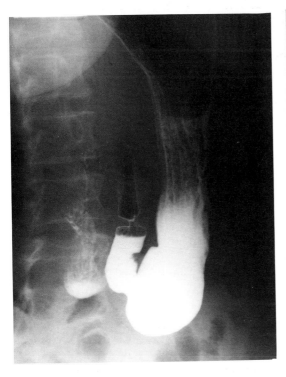
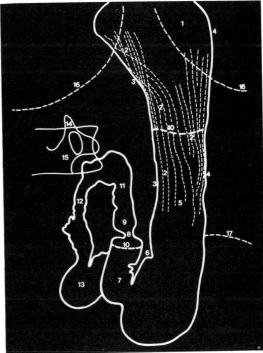

Fig. 27.1 Right anterior oblique barium study of stomach and duodenum. (From Weir and Abrahams 1978)

1 Fundal gas bubble.
2 Longitudinal ridges of mucous membrane (rugae).
3 Lesser curvature.
4 Greater curvature.
5 Body of stomach.
6 Incisura angularis.
7 Antrum.
8 Pyloric canal.
9 Duodenal cap (bulb).
10 Contrast medium fluid level.
11 First part of duodenum.

12 Second part of duodenum (descending).
13 Third part of duodenum (horizontal).
14 Superior articular process of L3.
15 Pars interarticularis of L3.
16 Breast shadow.
17 Iliac crest.

and the pyloric end is about 2.5 cm (1 in) to the right of the midline opposite the first lumbar vertebra. Posteriorly, the stomach lies comfortably on what is called the stomach bed. As tumors or ulcers of the stomach often penetrate posteriorly to involve these structures, it is important to know the contents of the bed. These include the pancreas, the upper half of the left kidney and left suprarenal gland, left crus of the diaphragm, left celiac ganglion and, through the greater sac, the gastric surface of the spleen. By the same token, diseases of any of these structures may involve the stomach.

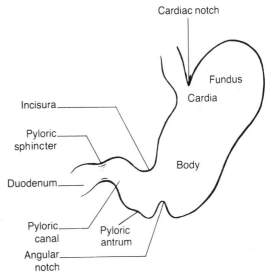

Fig. 27.2 Divisions of the stomach.

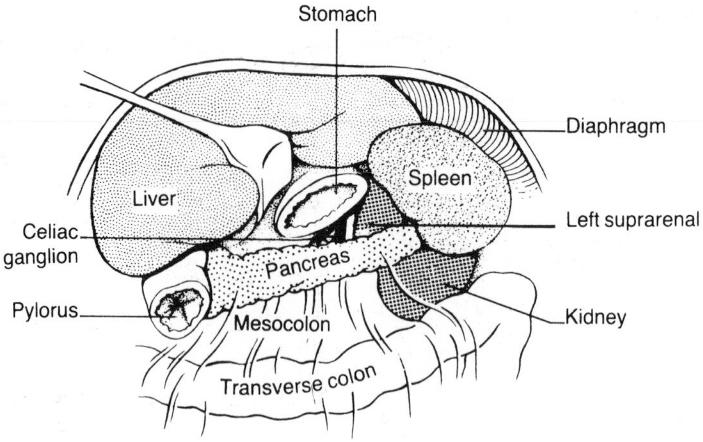

Fig. 27.3 The stomach bed.

Structure

The whole gastrointestinal tract is composed of an outer longitudinal layer and an inner circular layer. The stomach, however, has a third oblique layer, which is mostly confined to the fundus. The longitudinal layer is continuous over the duodenum, but the circular layer becomes markedly thickened at the pyloroduodenal junction as the pyloric sphincter. The sphincter is easily palpable at surgery, and in certain conditions (e.g. congenital pyloric stenosis) it may so enlarge as to be palpable through the abdominal wall; the swelling will lie 2.5 cm (1 in) right of the first lumbar vertebra. The mucous membrane of the stomach is thrown into numerous folds (rugae). Along the lesser curve these folds form furrows, the magenstrasse (road of the stomach), along which most of the food passes; interestingly, most gastric ulcers are situated along this roadway.

Duodenum (Figs. 27.4, 28.13 and 28.14)

The succeeding portion of the gastrointestinal tract is called the duodenum. It is shaped like a 'C' loop that encloses within it the head of the pancreas; diseases of either the duodenum or the pancreas, therefore, may involve one another. The duodenum is divided into four parts that bristle with clinical implications; these parts are 5 cm. 7.5 cm, 10 cm and 2.5 cm (2, 3, 4 and 1 in) long respectively.

The first part runs upward to the right and almost directly backward to touch (as does the pylorus) the undersurface of the right lobe of liver – duodenal or pyloric ulcers may therefore penetrate the liver. The first 1.25 cm (0.5 in) has the same peritoneal relationships as the stomach; it is lined by the anterior and posterior layers of the lesser omentum, which continue inferiorly as the greater omentum. To the right these two layers become continuous with each other as the free edge or border of the lesser omentum, the anterior boundary of the epiploic foramen. Running in this free edge are three very important structures (see later). The anterior layer that covers the anterior wall of the first 2.5 cm (1 in) of the duodenum passes on to completely cover the rest of the duodenum, and to make it an entirely retroperitoneal structure for the rest of its course. Running posterior to the first 2.5 cm (1 in) of the duodenum is the common bile duct on its way to the second part of the duodenum, a large branch of the hepatic artery and the portal vein.

The duodenum continues with a fairly sharp turn caudally; in so doing, its second part is crossed by the transverse colon as this passes to the left. The latter divides the second part

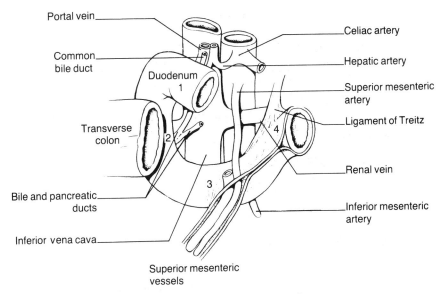

Fig. 27.4   The duodenum. The pancreas has been removed.

of the duodenum into upper (supracolic) and lower (infracolic) parts. About halfway down the second part, on the posteromedial wall, is the common entrance of the common bile and pancreatic ducts. Obstruction of the duodenum below this site will force bile and pancreatic juice back into the stomach, shortly to appear in an emesis basin! Duodenal blocks above the opening of the ducts will not produce such dramatic effects; the presence of bile in the vomit is therefore of considerable clinical significance. This part of the duode-

num also lies on the renal pelvis, enlargements of which may displace the duodenum.

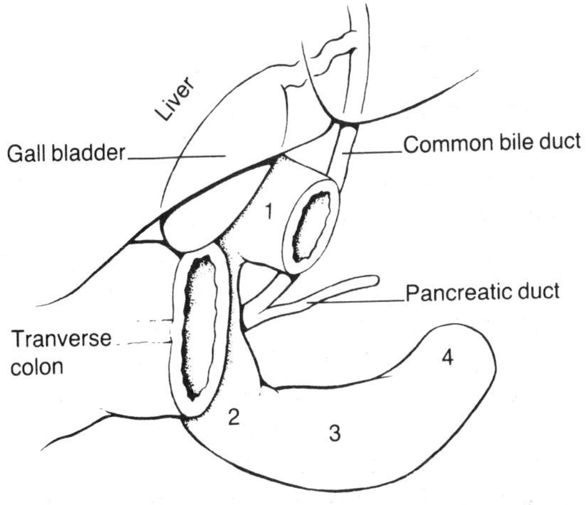

The third part of the duodenum runs transversely across the abdomen sandwiched between large vascular structures. Posteriorly lie the vena cava and aorta; the relationship to the aorta is particularly important, since aneurysms (pathologic swellings) of the aorta come into very close relationship to the duodenum and sometimes burst into it with dramatic results. Anteriorly, the third part of the duodenum is crossed by the superior mesenteric vessels as they enter the small bowel mesentery. Sometimes, when the small intestine hangs from its mesentery in an exaggerated manner, the vessels may actually compress the third part of the duodenum – the superior mesenteric artery syndrome. This syndrome occurs mostly in slender young girls, who have usually lost a lot of weight with a resulting visceroptosis (drooping of the viscera from loss of the supporting fat – the kidneys and spleen may also be affected). Usually, the cure is to send them to a camp specializing in French pastry; weight gain builds up enough fat to support the intestines and prevent their drooping. Of course, surgeons have an operation to cure this condition!

Fig. 27.5   The duodenum.

At the end of its transverse run, the fourth part of the duodenum turns upward to the left of the vertebral column and finally bends forward to end at the duodenojejunal junction. This junction is supported by an important ligament, the suspensory ligament of Treitz, a fibromuscular structure that runs between the duodenojejunal junction and the crus of the diaphragm. Its purpose is debatable; it is possible that it kinks the duodenojejunal junction and holds up the contents of the duodenum so as to allow maximal digestion and absorption.

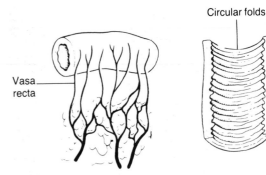

Fig. 27.6   The jejunum.

Structure

The duodenum is the widest part of the small bowel (see Fig. 27.1). The proximal first part is smooth because it is devoid of folds; this part is easily seen on x-rays where it appears as a triangular duodenal cap. Submucous glands (Brunner) are present, which are quite large and sometimes visible on barium studies. Toward the end of the first part large folds appear, which are characteristic of most of the small intestine. These are the circular folds or valvulae conniventes, which run circularly around the small intestine; they increase the surface area of the intestine without increasing its length and they also allow distension of the bowel. The entrance of the pancreatic and bile ducts is marked by a projection on the posteromedial surface of the second part, the ampulla of Vater. This ampulla has on its tip a little papilla. There is often a little fold running longitudinally toward the ampulla and identifying the ampulla. This fold is said to be the surgeon's friend in the sometimes difficult task of finding the ampulla; most surgeons can find the ampulla, but not the fold! It is also a site much sought after by the endoscopist.

Small intestine

This part of the intestine is some 6 m (22 ft) long and the proximal portion is named the jejunum, the distal portion the ileum. This division is arbitrary since one imperceptibly becomes the other. It is, however, important clinically, radiologically and surgically to recognize the differences in the intestine as one proceeds distally. The circular folds are most numerous in the upper small bowel, diminishing distally to become infrequent in the ileum. Minute finger-like projections, villi, are important in the absorptive process, and they are also more numerous in the proximal small bowel. The lymphoid tissue accumulates progressively distally, where it collects itself into small aggregations; some of these aggregations become confluent to form large patches, Peyer's patches. These are most numerous in the terminal ileum, where they lie on the antimesenteric border, the

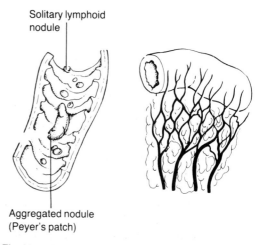

Fig. 27.7   The ileum.

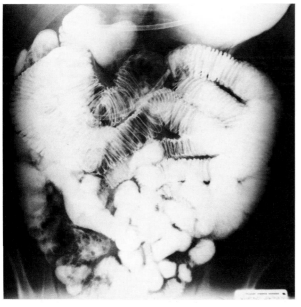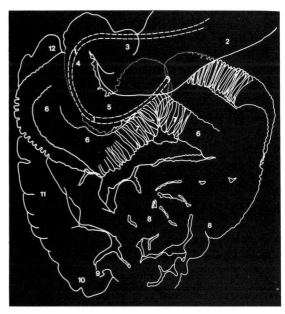

Fig. 27.8 Small bowel enema – double air contrast. (From Weir and Abrahams 1978)

1 Catheter in C-shaped duodenum.
2 Contrast medium in stomach antrum.
3 First part of duodenum.
4 Second part of duodenum.
5 Third part of duodenum.
6 Jejunal coils.

7 Valvulae conniventes (plicae circulares).
8 Ileum.
9 Ileocecal valve.
10 Cecum.
11 Ascending colon.
12 Right colic (hepatic) flexure.

side of the gut opposite the mesenteric attachment. These patches become involved in various diseases, such as typhoid and lymphoma, and they are crucial in the antigenic response of the gastrointestinal tract; sometimes they are visible radiologically. The vessels in the upper mesentery form a small number of arcades which end in long straight vessels, the vasa recta, which supply the intestine. Toward the ileum, the number of arcades increases and the terminal vessels are much smaller; mostly, they cannot be seen because of the accumulation of mesenteric fat, which becomes more marked as one proceeds distally. The jejunum is much thicker than the ileum, and some liken this to handling a shirt sleeve inside a coat sleeve; this sensation is due to the marked accumulation of circular folds. Finally, the caliber decreases as one descends and the terminal ileum is the narrowest part of the small in-testine, the site of obstruction for foreign bodies travelling illegally down the gastro-intestinal tract. It is important to remember that some 130 cm (5 ft) of small bowel lies suspended from its mesentery in the pelvis; pelvic inflammation may therefore cause small bowel symptoms or obstruction.

Meckel's diverticulum

The vitellointestinal duct with its companion artery forms the apex of the protruded midgut loop during early embryonic life. Normally disappearing, the duct or its artery may persist in whole or in part and create a lot of mischief. Persistence of the intestinal end of the duct results in a diverticulum. The characteristics of this diverticulum are best summed up as the story of the twos – it is about 2 in (5 cm) long, situated 2 ft (60 cm) from the ileocecal valve,

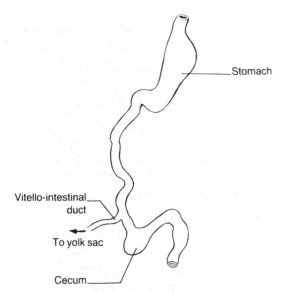

Fig. 27.9   The primitive gut. The proximal end of the vitellointestinal duct leading to the yolk sac is seen. Mostly, it disappears; if it doesn't – see Figs. 27.10 and 27.11.

times the diverticulum is connected to the umbilicus by a fibrous strand, or a strand only may exist; the strand, the remains of the vitelline artery, may cause intestinal obstruction by ensnaring the bowel or by causing it to twist (volvulus).

Mesentery of the small intestine

Although the small bowel is 6 m (22 ft) long, it is suspended from a mesentery whose base measures only about 15 cm (6 in) (see Fig. 26.2). The mesenteric attachment runs from the left side of the second lumbar vertebrae to the right quadrant, crossing the vertebral column and the vessels anterior to it. The mesentery contains the superior mesenteric vessels and its branches together with numerous lymphatics, lymph nodes and autonomic nerves.

present in 2 percent of people, contains two types of mucosa (intestinal and gastric) and these facts are true two-thirds of the time. The gastric mucosa may lead to ulceration in the diverticulum, the commonest cause of lower intestinal bleeding in the youngster. Some-

Large intestine

This part of the bowel is characterized by a relatively wide-bored lumen and the presence of teniae coli, sacculations and appendices epiploicae. The cecum is the

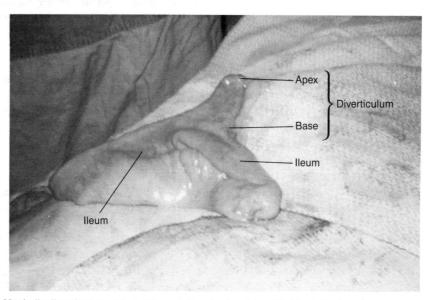

Fig. 27.10   Meckel's diverticulum. A broad-based diverticulum found incidentally at surgery. Removal of this anomaly is mandatory.

widest part of the tube and the distal sigmoid the narrowest; with the contents of the former fluid and the latter solid, obstructive symptoms are commoner in lesions of the sigmoid. The teniae coli are three longitudinally running bands that consist of the congregated outer longitudinal layer of the colon. Over each end of the large bowel, the appendix and the rectum, the teniae are lost and the longitudinal muscle becomes more uniformly spread over the circumference. As the teniae are shorter than the length of the bowel, the colon has a sacculated appearance. The appendices epiploicae are tabs of fat that lie under the serosal covering, sometimes in relation to the teniae. Most numerous in the sigmoid colon, they sometimes become very fat and twist on themselves to produce acute abdominal symptoms (torsion of the appendices epiploicae). The large intestine lacks circular folds and villi and is almost reduced to the status of a mere conduit. It is liberally provided with goblet cells which secrete mucus to lubricate the stool during its onward passage. The cecum, the first part of the large intestine, is a blind diverticulum situated below the entry of the ileum into the large intestine.

Ileocecal valve

The ileocecal entrance is marked by a valve that consists of two lips, upper and lower, which are prolonged into two folds, the frenula. When cecal distension distracts the frenula, the valve narrows and the ingress of ileal contents is discouraged; it is analogous to pulling on the angles of the mouth and narrowing the orifice. It is important to appreciate that in some people the valve is deficient and cecal contents may regurgitate into the ileum. In cases of distal large bowel obstruction (a common condition) a competent valve will prevent the backed-up large bowel contents from passing retrograde into the small bowel; this creates a dangerous situation where the increasing large bowel distension may eventually lead to rupture of the cecum. In the case of an incompetent valve, the large

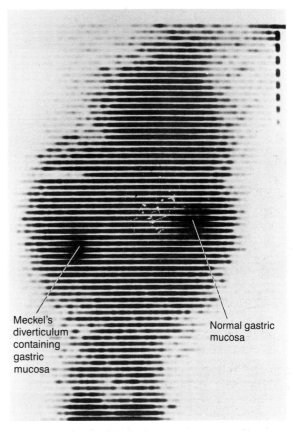

Meckel's diverticulum containing gastric mucosa

Normal gastric mucosa

Fig. 27.11 A little girl (aged 9) had severe bleeding per rectum. This technetium-99m scan (which shows selective pickup by gastric mucosa) shows normal gastric mucosa on the left, and similar mucosa in a diverticulum on the right.

bowel contents pass into the small bowel and the large bowel is decompressed; this buys time and prevents cecal rupture. It is strange for the competent to be punished!

Appendix

Since acute appendicitis is the commonest acute abdominal condition in males and, in many parts of the world, in females, the appendix is an important structure. While in the child the appendix comes off the apex of the cecum, the disproportionate growth of the right wall of the cecum displaces the appendix and its orifice to the posterolateral left side. The appendix, which varies from

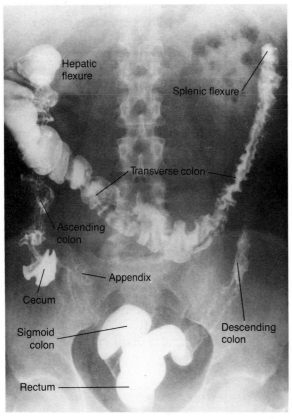

Fig. 27.12 The colon. The barium enema outlines the colon. Note the appendix pointing toward the pelvis and the transverse colon drooping toward the false pelvis.

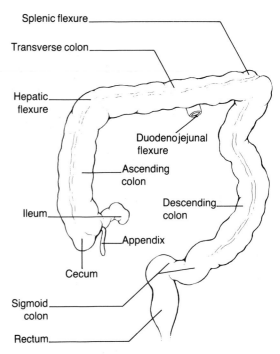

Fig. 27.13 The colon.

2.5 to 22.5 cm (1 to 9 in) in length, may point in any direction: upward and to the left in relationship to the terminal ileum; inferiorly over the brim of the pelvis in close contact to the tube, ovary or bladder; along the right side of the cecum; or behind the cecum in the retrocecal fossa. Of these positions the retrocecal is the commonest, followed by the pelvic. It is easy to see how an inflamed pelvic appendix, lying next to the tube, ovary or bladder, can, and often does, mimic pathologic conditions in these organs. The appendix is supported by a mesentery, which is a prolongation of the posterior leaf of the small bowel mesentery of the terminal ileum. In the mesentery of the appendix lies the appendicular artery, which is very nearly an end-artery. Thrombosis of the artery in acute

appendicitis often leads to necrosis of the appendix, and perforation of an acutely inflamed appendix is a common end-result. The nearer to the appendix that the artery lies, the more likely the thrombosis, and the quicker the perforation in cases of appendicitis. The appendix may come into relationship with neighboring muscles and, when inflamed, may irritate these muscles; in the retrocecal position, it may lie on the iliopsoas and in the pelvic position on the obturator internus. Pain is produced on extending the hip or rotating it externally (psoas and obturator signs respectively). Inflammation of the appendix may be felt as referred pain to the periumbilical region (T10).

Structure

The appendix contains a large amount of lymphoid tissue and is sometimes even called the abdominal tonsil (it is lucky that ear, nose and throat surgeons don't know

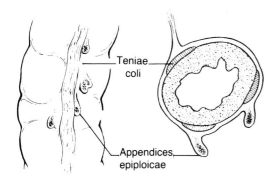

Fig. 27.14 The colon showing three teniae coli and appendices epiploicae.

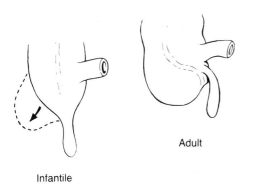

Fig. 27.15 The appendix. Growth of the right wall of the cecum pushes the appendix to the left.

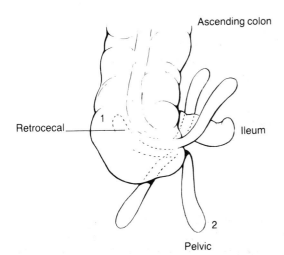

Fig. 27.16 The various positions of the appendix.

about this). This lymphoid tissue, like all lymphoid tissue, swells up markedly when infected and may obstruct the narrow blind appendix. There are often deficiencies in the muscular wall of the appendix, which may allow the protrusion of small portions of the mucosa (diverticuli).

Cecum

The cecum, the thinnest-walled part of the large intestine, has a blind lower end that is fully covered with peritoneum. It lies on the iliacus and is anteriorly related to the anterior abdominal wall. It can sometimes be playfully gurgled by the examining hand, a point of clinical importance; when distended, it may be visible. The cecum is as short as it is broad, and its posterior peritoneum is reflected posteriorly onto the iliac fossa to create a small recess behind the organ, the retrocecal fossa (see Fig. 27.17).

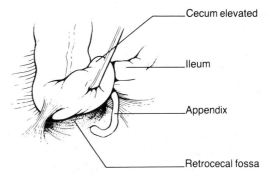

Fig. 27.17 The retrocecal fossa. The commonest position of the appendix is in this fossa.

The appendix often hides in this fossa, and for this reason one is sometimes fooled by the relative lack of tenderness of the inflamed appendix, protected as it is by the overlying cecum.

Colon

Above the ileocecal valve, the cecum loses its complete peritoneal investment and becomes the ascending colon. Because it is covered with continuous peritoneum in front and on the sides it is fixed in position. The ascending colon is 10–12.5 cm (4–5 in) long, and when it reaches the liver it bends forward and to the left as the hepatic flexure. Posteriorly the ascending colon lies on the muscles of the posterior abdominal wall, the iliacus, the quadratus lumborum and the origin of the transversus abdominis, and, just before reaching the liver, the lower pole of the right kidney. Some people believe that this last relationship accounts for kidney infections originating in the colon, but to believe in this is to believe in rocking horse manure.

From the right colic flexure, which indents the liver, the gut then makes a wide U-shaped curve of about 45 cm (18 in) before reaching the splenic flexure, where it indents the spleen. The hepatic and splenic flexures are fixed points, and the transverse colon swings downward from them to various levels – in some people it may even reach the pelvic brim. It is separated from the anterior abdominal wall by the layers of the greater omentum, and slung from the posterior abdominal wall by the transverse mesocolon. This last carries the neurovascular supply to the transverse colon and separates the abdomen into an upper supracolic compartment containing the stomach, spleen, liver and pancreas, and a lower infracolic compartment containing the small intestine. The transverse mesocolon is attached to the posterior abdominal wall while crossing the second part of the duodenum and to the inferior border of the pancreas.

The splenic flexure lies at a much higher level than the hepatic flexure; the latter has been pushed down by the liver. The splenic flexure is fixed by a fold of peritoneum that joins the colon to the diaphragm, the phrenicocolic ligament, on which the spleen sits. The intestine then turns downward and backward to become the descending colon, which has lost its mesentery in the same way as the ascending colon and is similarly fixed. As it runs downward it lies on the same structures as the ascending colon. On reaching the brim of the pelvis, it turns medially across the psoas muscle to pick up a mesentery and changes its name to the pelvic or sigmoid colon.

Both the ascending and the descending colon are separated from the muscles of the posterior wall by nerves which traverse this space before they enter the anterolateral abdominal wall, the subcostal and the two branches of the first lumbar nerve with their accompanying vessels.

The sigmoid colon hangs from a generous mesentery (mesocolon) whose root is shaped like an inverted 'V'. The mesocolon first runs proximally along the medial border of the psoas and then distally presacrally as far as its third or middle piece. In this area the sigmoid loses its mesentery and changes its name to rectum. The mesentery may be so long as to allow the two limbs of the bowel to twist as a volvulus and become obstructed. It is also common for the part of the loop of the convoluted bowel to lie in the right lower quadrant (iliac fossa), where inflammations of the intestine may mimic appendicitis. The appendices epiploicae are at their longest in this area and may also twist on themselves – torsion of the appendices epiploicae, a surgical emergency.

28 Liver, spleen, pancreas and biliary system

Liver · Spleen · Pancreas · Biliary apparatus and gallbladder

Liver

The liver is the largest gland in the body and is proportionately bigger in the child than in the adult. It is the major cause for the intestines having to seek refuge outside the abdominal cavity in early fetal life and is also partly responsible for the pot belly of young children. Situated in the right upper quadrant and extending across the midline to the left upper quadrant, it is largely protected by the ribs and cartilages; however, liver injuries are becoming increasingly common as the highways become more crowded. The liver is surrounded by a capsule (Glisson's capsule), which gives it a fairly definite shape in the cadaver. However, in life, it is a soft and fairly mushy organ, which is easily ruptured and less easily sutured. It is wedge shaped, with the base of the wedge to the right and the apex to the left.

The superior surface is moulded to the diaphragm; it therefore reaches to the fifth rib on the right and the fifth space on the left. Through the diaphragm, it is related to the lung and pleura on each side and the pericardium in between. Liver abscesses may burst into the chest through the diaphragm, and intrathoracic diseases may involve the liver.

The right lateral surface is often needled for liver biopsies. Remember that, in the midaxillary line, the lung reaches the eighth and the pleura the tenth ribs. A needle passed above the tenth rib will traverse the lung and/or pleura, depending on the level of insertion, before passing through the diaphragm into the right lobe of the liver. If infection is suspected it is best to stay below the 10th rib; if not, the 8th intercostal or lower will suffice (see p. 521).

The inferior margin extends from right to left where its apex reaches approximately the level of the apex beat of the heart. The inferior edge lies just below the right costal margin, but a substantial portion of the liver appears in the epigastrium before disappearing under the left costal margin. Students often feel this normal epigastric mass in thin people and mistake it for some mysterious tumor. In the child, and in patients where the diaphragm is pushed down by an enlarged lung, such as

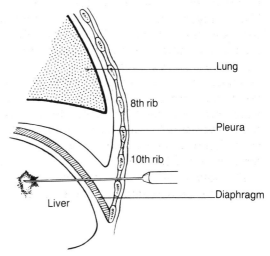

Fig. 28.1 Coronal section through the right lung, pleura, diaphragm and right lobe of the liver (posterior view). The superior and right lateral surfaces are related, through the diaphragm, to the pleura and lung. Note the safe route (below pleura and lung) to aspirate a liver lesion or obtain a biopsy. Don't go deeper than 8.8 cm (3.5 in) or you may obtain a voluminous sample of venous blood from the inferior vena cava!

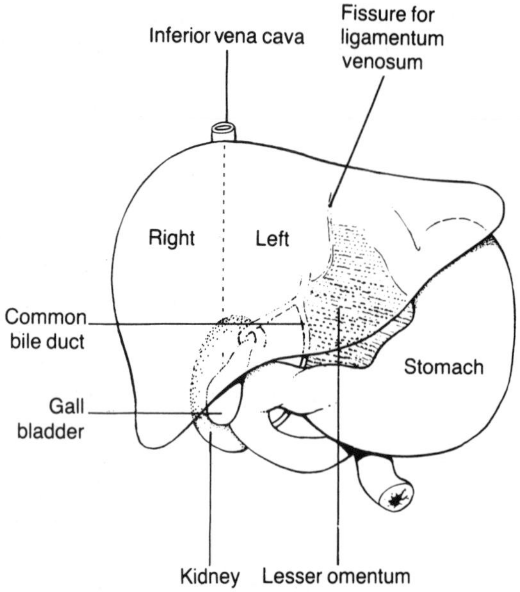

Inferior vena cava

Fissure for
ligamentum
venosum

Right   Left

Common
bile duct

Stomach

Gall
bladder

Kidney   Lesser omentum

Fig. 28.2   Structures lying beneath the posteroinferior surface of the liver. Note the lesser omentum (shaded) attaching between the fissure for the ligamentum venosum above, and the lesser curvature of the stomach and upper border of the first 2.5 cm (1 in) of the duodenum below. The inferior margin of the liver has to be elevated to see the proximal stomach and esophagus, superior pole of the right kidney and proximal portions of the biliary system. Note the line of division between the 'real' right and left lobes.

occurs in emphysema, the inferior border of the liver is palpable. Another reason for palpating a normal right lobe of the liver is the presence of a Riedel's lobe, a normal inconsequential mass of developmental origin (Figs. 28.4 and 28.5).

The anterior surface lies between the superior and inferior margins. It is related on the right and left sides to the ribs and cartilages (the diaphragm, pleurae and lungs intervening), and between them to the abdominal wall.

The posterior and inferior surfaces merge into one another. A prominent feature in this area is a transverse slit, the hilum in the liver, which is traversed by the vessels, nerves and lymphatics to and from the interior of the organ. In front of the hilum is situated, on the right, the fossa for the gallbladder, and, on the left, the attachment of the falciform ligament with its contained ligamentum teres. These two structures demarcate the quadrate lobe between them. Behind the hilum lie the inferior vena cava on the right and the fissure for the ligamentum venosum on the left; they demarcate between them the caudate lobe, which extends to the right as the caudate pro-

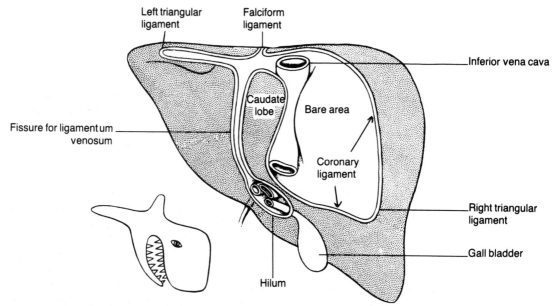

Left triangular
ligament

Falciform
ligament

Inferior vena cava

Caudate
lobe

Bare area

Fissure for ligamentum
venosum

Coronary
ligament

Right triangular
ligament

Gall bladder

Hilum

Fig. 28.3   Posteroinferior surface of liver. The reflections of the ligaments are seen. (For those who find it exasperating and may have to reproduce these reflections on demand, draw a shark with tail and dorsal fin (inset); extract his teeth and eye, and you will impress any examiner.)

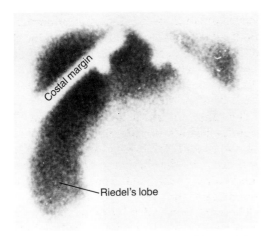

Fig. 28.4   Liver scan demonstrating Riedel's lobe. Clinically, it may simulate an ominous tumour.

cess. This is the commonly described 'H', the cross-bar being the hilum, the two anterior limbs being the gallbladder and the ligamentum teres, and the two posterior limbs the vena cava and ligamentum venosum. This surface covers and hides many important structures. Starting from the right, it covers the upper part of the right kidney and the suprarenal gland. Portions of these two organs lie against the bare area of the liver uncovered by peritoneum. In front of them lie the hepatic flexure of the colon and the junction of the first and second parts of the duodenum. Passing to the left, the liver covers the inferior vena cava as it emerges from behind the peritoneum to lie against the bare area, and the esophagus and proximal stomach.

Ligaments

The falciform ligament, the remnant of the ventral mesentery, carries the obliterated left umbilical vein (ligamentum teres) in the lower border of its sickle-shaped fold. In its heyday (in utero) it joined the left branch of the portal vein, and then bypassed the liver, a relatively inactive organ in the fetus where the placenta does all the work, via the large ductus venosus to the left hepatic vein. After fetal life, the liver becomes a dynamo, and the left umbilical vein and the ductus venosus

regress; the vein becomes the ligamentum teres, and the ductus the ligamentum venosum.

Now follow the ligamentum teres as it leaves the umbilicus to pass between the two layers of the falciform ligament, which passes upward and to the right. When the falciform ligament reaches the postero-superior surface of the liver, the two peritoneal layers diverge. Following the left layer, it passes to the left and returns as a left triangular ligament. The right layer passes to the right and returns to the left again as the right triangular ligament. The two layers forming the right ligament are so widely separated that a considerable area of the liver is devoid of peritoneum – the bare area. The ligament is often called the coronary ligament, the triangular ligament being the portion where the two layers actually meet. The posterior layers of each triangular ligament then happily reunite and pass inferiorly in the fissure for the ligamentum venosum, finally curving to the right to reach the hilum of the liver. The attachment to the fissure and the hilum of the liver constitutes the lesser omentum, which ends inferiorly as the layers gain attachment to the lesser curvature of the stomach.

Lobes

Before the advent of modern liver sugery, the right lobe was demarcated from the left by the falciform ligament. When surgeons attempted to remove one or other lobe, they were discouraged by the excessive bleeding accompanying the procedure. The anatomists came to the rescue. By injection methods they showed that there was a line across which the right and left hepatic arteries and ducts, as well as the tributaries of the right and left branches of the portal vein did not communicate. This line, running from the gallbladder fossa to the inferior vena cava, demarcates the right and left lobes. By separating the two lobes along this line, hepatic lobectomy has now become a much

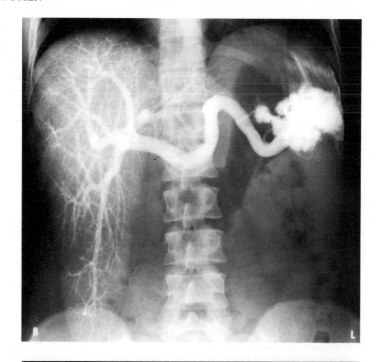

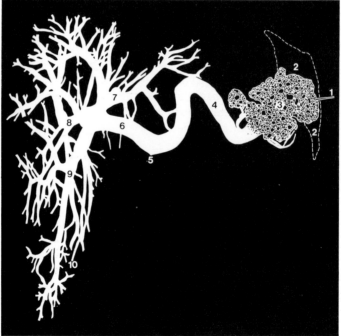

Fig. 28.5   Splenoportogram. (From Weir and Abrahams 1978)

1 Needle in splenic pulp.
2 Intraperitoneal contrast medium (leakage).
3 Contrast medium within splenic pulp.
4 Splenic vein.
5 Point of entrance of superior mesenteric vein.

6 Portal vein.
7 Left portal vein.
8 Right portal vein.
9 Abnormally large segmental vein to Riedel's lobe.
10 Riedel's lobe.

more acceptable procedure, the surgeon having only to contend with hepatic veins which cross this line of demarcation.

Spleen

If you remember the odd numbers 1, 3, 5, 7, 9 and 11, you will know quite a lot about the spleen. It is 1 × 3 × 5 in (2.5 × 7.5 × 12.5 cm) in size, weighs 7 oz (500 mg) and lies against ribs 9–11. If you have to choose between remembering these not too difficult numbers, the last two are the most important. The spleen is placed along the long axes of ribs 9–11, and most spleens worthy of the name show indentations from these ribs in the formalin-hardened cadaveric organ. The large diaphragmatic surface is related to these ribs, but separated from them by the diaphragm, pleura and lung. Traumatic fracture of one of these ribs should lead to the diagnosis of ruptured spleen until proved otherwise. The leaking of blood from this vascular organ irritates the diaphragm on which it

lies, and shoulder pain following an injury to the lower chest/upper abdomen is very suggestive of splenic injury. It is a truism to say that the spleen has to enlarge to approximately twice its size before it can be felt beyond the costal margin; it lies much deeper than generally thought, the medial edge being only 3.75 cm (1.5 in) from the vertebral column.

Visceral surface

Anteriorly the posterior surface of the stomach rests on the gastric surface of the spleen, posteriorly the spleen lies on the kidney, and inferiorly the splenic flexure of the colon indents the lower pole. On the gas-

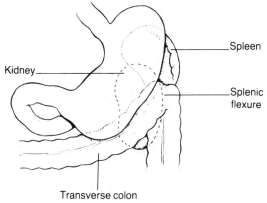

Fig. 28.7   The viscera abutting on the spleen. The stomach rests anteriorly, the kidney posteriorly and the colon inferiorly. The tail of the pancreas has been omitted.

tric surface lies a fairly large hilum, through which the vessels and nerves enter the spleen and against which the tail of the pancreas abuts. The surgeon is very wary when removing the spleen to avoid taking a portion of either the stomach wall or the pancreas with the specimen – these additional structures may puzzle the pathologist examining the removed spleen. The spleen is covered by a thin layer of peritoneum and is easily torn, with severe resultant hemorrhage. The anterior border is notched, and people with long sensitive fingers (such as internists and housemen) imagine that they can feel these

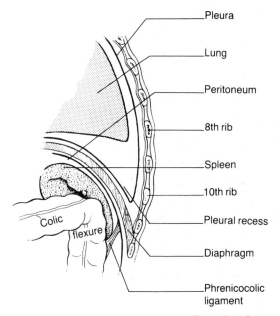

Fig. 28.6   The spleen and its relations. The spleen is seen related to ribs 9, 10 and 11, the diaphragm and the pleural space intervening. The visceral surface is indented by the colic flexure, and the lower pole rests comfortably on the phrenicocolic ligament.

notches in an enlarged spleen. Normally, however, the spleen resembles the size and shape of a closed fist, the knuckles representing the splenic notches.

Ligaments (see Figs. 26.7 and 26.8)

The two layers of the lesser omentum have been noted passing inferiorly and splitting to enclose the stomach, after which they continue as the anterior two layers of the greater omentum. On the left side of the stomach, the two layers come off as the two layers of the gastrosplenic ligament. These layers run posteriorly to the hilum of the spleen; the posterior layer then continues backward to the kidney as one layer of the lienorenal ligament, while the anterior layer encloses the spleen and continues posteriorly as the left layer of the lienorenal ligament. When both layers of the lienorenal ligament reach the kidney, they separate and become continuous with the parietal peritoneum, the left becoming continuous with the peritoneum of the greater sac and the right passing to the right as the posterior lining of the lesser sac. Note that the spleen is actually in the greater sac. The layers forming the ligaments serve as routes for the neurovascular and lymphatic supply to and from the spleen.

Sometimes there are accessory spleens; they lie in close relationship to the hilum of the spleen, in the gastrosplenic ligament or in the greater omentum. When removing the spleen for diseases that are destroying the patient's blood, the surgeon is careful to remove these accessory spleens, in case they get delusions of grandeur and enlarge to replace the excised organ.

Pancreas

(see Figs. 26.9, 27.3, 28.8 and 28.9)

This mixed endocrine and exocrine organ lies against the posterior abdominal wall and is likened to a little animal; it has a head, neck, body and tail. (Those with a sweet tooth have described it as like a bar of Swiss Toblerone chocolate, which is of similar

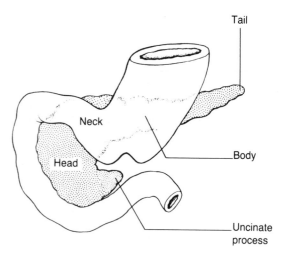

Fig. 28.8   Component divisions of the pancreas.

length and triangular in shape.) The head lies in the epigastrium, with the rest of the organ extending into the left hypochondrium, where it ends by touching the spleen. It is very deeply placed and usually not accessible to clinical examination; radiologic techniques now allow direct visualization of the pancreas. Diseases of the pancreas are common and difficult to treat.

Head

The head lies in the C-shaped duodenal loop with a flange projecting in front and behind; diseases of the head therefore commonly produce duodenal obstruction. The lowest part of the head extends to the left as an uncinate process. The head lies on the vena cava and the two large renal veins. Most importantly, the posterior aspect of the head is grooved by the bile duct; sometimes the duct actually embeds in the substance of the pancreas. Most elderly patients presenting with painless jaundice have a tumor of the head of the pancreas blocking the bile duct as it runs to its duodenal exit.

Neck

The neck is a narrow part of the gland, behind

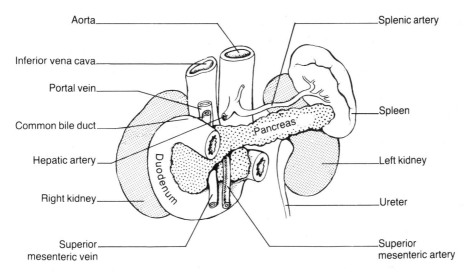

Fig. 28.9 Relations of the pancreas and duodenum. The superior mesenteric vessels cross the uncinate process of the pancreas and the third part of the duodenum.

which the portal vein is formed; one of the symptoms of tumors of the neck of the pancreas is obstruction of the portal vein.

Body and tail

The body of the pancreas runs across the aorta, vena cava and left renal vein as they lie on the vertebral column and reaches the upper part of the left kidney and suprarenal gland. Finally, the tail turns forward into the lienorenal ligament to touch the apex of the visceral surface of the spleen. It is common for patients with lesions of the body of the pancreas to present with backache. Backache and diabetes of recent origin are characteristic symptoms of a pancreatic lesion which involves the paravertebral structures and destroys the endocrine secretions. Anteriorly, the transverse mesocolon is attached to the neck and inferior border of the body of the pancreas. With the body of the pancreas lying in the supracolic compartment, it forms part of the stomach bed (see Fig. 27.3). It is common for peptic ulcers to penetrate the pancreas from the stomach or first part of the duodenum; the patient usually notices this by telling you that the original epigastric ulcer pain now passes through to the back. It is also often difficult to tell whether epigastric tumors start in the stomach and invade the pancreas, or vice versa.

Pancreatic duct

Originally there were two ducts serving the dorsal and ventral embryological pancreatic buds, which developed from the dorsal aspect of the duodenum and from the primitive bile duct respectively. The adult ductal system is made up by a fusion of the ducts, but with drainage areas that differ from the original. The main adult duct (Wirsung) starts near the tail of the pancreas, and, enlarging on the way, runs to the right in the posterior part of the gland. When it reaches the duodenal end of the pancreas, it joins the common duct to open into the duodenum. An accessory duct (Santorini), which drains the lower part of the head and uncinate process, runs superiorly across the main duct to open into the duodenum proximal to the main duct.

Biliary apparatus and gallbladder

Bile produced in the liver cells is discharged into small canaliculi lying between these

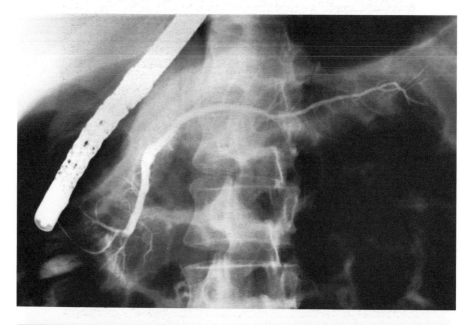

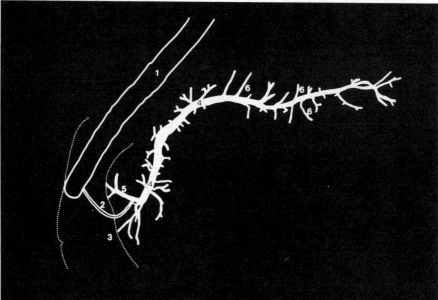

Fig. 28.10 Endoscopic retrograde cholangiopancreatogram. (From Weir and Abrahams 1978)

1 Fiberoptic endoscope in second part of duodenum.
2 Cannula inserted into main pancreatic duct (Wirsung) via ampulla.
3 Gas in second part of duodenum.

4 Main pancreatic duct.
5 Accessory duct (Santorini).
6 Intralobular ducts in 'herring-bone' pattern.

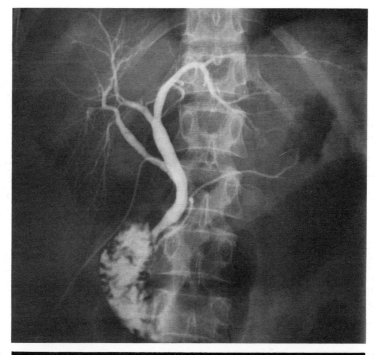

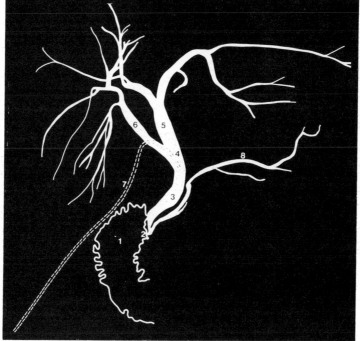

Fig. 28.11   Peroperative cholangiogram. (From Weir and Abrahams 1978)

1 Contrast medium in descending part of duodenum.
2 Hepatopancreatic ampulla (Vater).
3 Common bile duct.
4 Common hepatic duct.

5 Left hepatic duct.
6 Right hepatic duct.
7 Catheter (T-tube).
8 Contrast reflux in pancreatic duct.

cells. These canals lead to small ductules and eventually into the right and left hepatic ducts. The ducts join just outside the hilum of the liver to form the common hepatic duct. After a short course, the common hepatic duct is joined by the cystic duct from the gallbladder, and continues as the common bile duct. The common bile duct, 7.5 cm (3 in) long, runs in the free edge of the mesentery of the lesser omentum, behind the first part of the duodenum, enters the groove or actually embeds in the pancreas, and reaches the posteromedial surface of the second part of the duodenum. Here it is joined by the pancreatic duct to open into a common ampulla, which terminates in a small papilla. The common duct therefore has a part above, a part behind and a part below the duodenum. It can be obstructed in any of these positions, but most commonly the lowest part is blocked by a growth in the

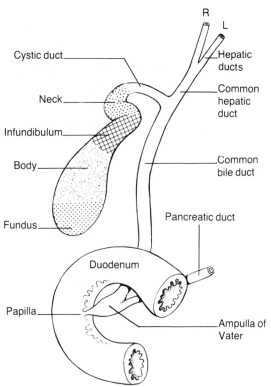

Fig. 28.13   The biliary apparatus.

head of the pancreas or a gallstone. There are variations in the manner by which the pancreatic and common bile ducts terminate, but normally there is a separate sphincter mechanism for the common bile duct (Boyden) proximal to an effective joint sphincter with the pancreatic duct (Oddi).

The gallbladder lies in a fossa on the under-surface of the right lobe of the liver. The fundus just peeps out beyond the liver edge, and can be located at the junction of the ninth costal cartilage and outer border of the rectus sheath (linea semilunaris); this is the site of maximum tenderness in acute inflammation of the gallbladder. Any pear-shaped swelling in this area which moves on respiration (being thrust down by the diaphragm push-ing the liver) can be confidently called a gallbladder; if this isn't so you may obtain a refund from the authors. The fundus leads to the body, which lies against the liver surface, with which it shares a common peritoneum

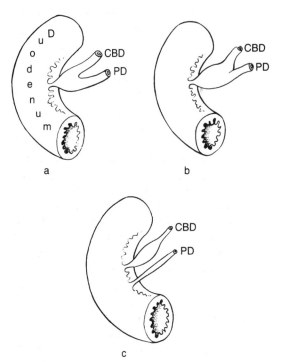

Fig. 28.12   The terminations of the common bile duct (CBD) and the pancreatic duct (PD). (a) and (b) are common, and (c) is less common. It is easy to see how a gallstone can block both ducts in (a) and (b), and produce serious sequelae.

(sometimes this peritoneum droops as a mesentery, making the surgeon's work easy). Beyond the body the viscus narrows into a neck, the most dependent part of which lies in close relationship to the duodenum and is often called Hartmann's pouch. The rela-

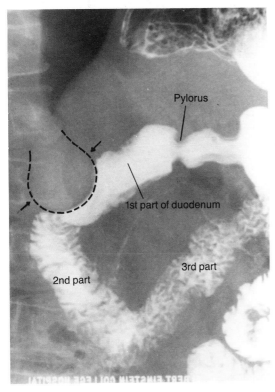

Fig. 28.14   The duodenum and the gallbladder. The indentation of the first part of the duodenum by an enlarged gallbladder (arrows) is well demonstrated in this barium study.

tionship of the gallbladder to the duodenum is so close that, in the cadaver, the first part of the duodenum is nearly always stained with bile. It is often difficult to determine whether the patient with pain in the right upper quadrant is suffering from a duodenal ulcer or inflammation of the gallbladder. Hartmann's pouch is a common site for the impaction of large gallstones – sometimes a false passage is formed between the pouch and the duodenum and the stone escapes into the gastrointestinal tract to cause bother further on. The neck of the gallbladder narrows into a

2.5-cm (1 in) long cystic duct, which runs inferiorly to join the common hepatic duct; below this junction the duct continues as the common bile duct. The cystic duct is remarkable for its numerous spiral valves (the valves of Heister); the function of these valves is difficult to determine, but some feel it keeps the narrow duct patent (obviously this is a myth; narrower ducts than this don't have these folds). Nevertheless, they fortunately tend to prevent gallstones from passing down the cystic duct into the common duct to produce serious problems; they also irritate the surgeon who sometimes has trouble passing a catheter down the duct.

Structure

The mucous membrane of the gallbladder has a honeycomb appearance, which, like the small intestine, enlarges the absorptive surface. The function of the gallbladder, apart from acting as a reservoir, is to absorb various substances and so concentrate the bile. The bile is ejected after the food reaches the duodenum, especially if it is a fatty meal. In fact, the test for determining the functional ability of the gallbladder is initially a dose of iodinated contrast medium to outline the gallbladder (cholecystogram), followed, as a reward, by bacon and eggs to stimulate contraction. For this contraction, the gallbladder is provided with a well-defined single-layer muscle coat; the hepatic and bile ducts, on the other hand, are mostly composed of fibrous and elastic tissue, with very little in the way of muscle fibers. Since all colicky pains result from excessive muscle spasm, the syndrome of biliary colic due to a stone in the ducts does not find an easy explanation. The only substantial muscle fibers found in these ducts are situated at the lower end of the common bile duct and at the common entrance of the pancreatic and bile ducts into the duodenum, the sphincter of Oddi. This sphincter holds up the column of bile until the fatty meal reaches the duodenum; it relaxes to allow the passage of the now-concentrated bile.

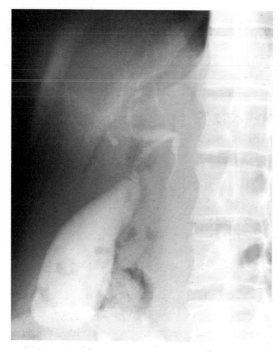 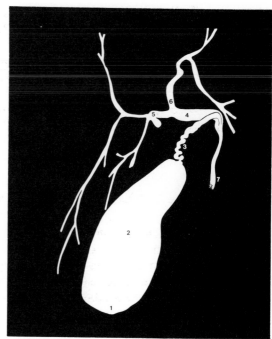

Fig. 28.15   Oral cholecystogram. (From Weir and Abrahams 1978)

1 Fundus of gallbladder.
2 Body of gallbladder.
3 Cystic duct.
4 Common hepatic duct.

5 Right hepatic duct.
6 Left hepatic duct.
7 Common bile duct.

29 Neurovascular supply of abdominal viscera

Arterial supply · Portal circulation · Blood supply of viscera · Lymph drainage of viscera · Nerve supply

Arterial supply

The abdominal aorta gives rise to three large unpaired vessels, the celiac artery arising opposite T12, the superior mesenteric opposite L1 and the inferior mesenteric opposite L3. The celiac artery supplies the foregut (esophagus to ampulla of Vater), the superior mesenteric the midgut (ampulla to the junction of the middle and left thirds of transverse colon), and the inferior mesenteric the hindgut (from the left third of the transverse colon to halfway down the anal canal).

Celiac trunk

This short wide trunk emerges from the abdominal aorta above the pancreas where it is surrounded by the celiac plexus, which is partially distributed along the branches of the artery. The trunk gives rise to three large arteries, hepatic, splenic and left gastric. For reasons that are important only to examiners, the splenic is the largest branch. Sometimes the celiac artery is absent, in which case the three branches arise directly from the aorta.

Hepatic artery

This large artery runs downward and to the right behind the peritoneum of the lesser sac to reach the first part of the duodenum. Here it turns forward and then upward to enter the free edge of the lesser omentum, in which it runs to enter the hilum of the liver as its terminal branches, the right and left hepatic arteries. In the free edge of the lesser omentum, it lies to the left of the common bile duct. Students are often given mnemonics to remember what lies on which side in the free edge of the lesser omentum, but it should be obvious; the artery comes from the celiac artery on the left and the duct is destined for the duodenum on the right. Near the duodenum the hepatic artery gives off the relatively small right gastric, which anastomoses with the left gastric along the lesser curvature of the stomach. As the hepatic turns up into the free edge of the lesser omentum, a large gastroduodenal branch arises, which passes behind the first part of the duodenum to reach the head of the pancreas. Behind the first part of the duodenum, the gastroduodenal is commonly involved in duodenal ulcers which penetrate posteriorly and erode this vessel. The destruction of this artery is heralded by the passage of a large quantity of bright red blood upward and downward in the gastrointestinal tract. The gastroduodenal terminates by dividing into two branches. The first, the superior pancreaticoduodenal runs between the pancreas and the duodenum to supply both organs and anastomoses with the inferior pancreaticoduodenal from the superior mesenteric below – by this means an anastomosis between the vascular supply of the fore and midgut is established (Fig. 29.4). The second branch, the right gastroepiploic runs between the two layers of the greater omentum to meet the left gastroepiploic and form the gastroepiploic arch along the greater curvature of the stomach.

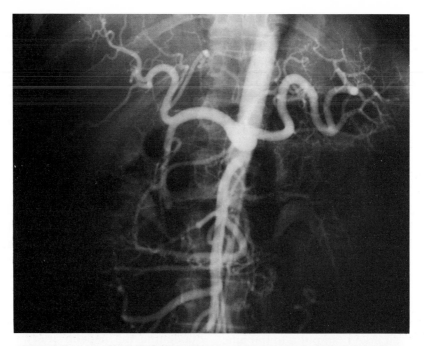

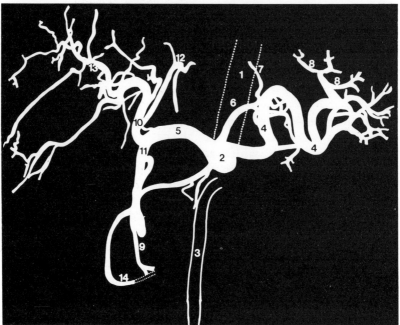

Fig. 29.1   Celiac arteriogram. (From Weir and Abrahams 1978)

1 Abdominal aorta.
2 Celiac trunk.
3 Superior mesenteric artery.
4 Splenic artery.
5 Common hepatic artery.
6 Left gastric artery.
7 Esophageal branches of left gastric artery.

8 Splenic artery branches.
9 Superior pancreaticoduodenal artery.
10 Hepatic artery proper.
11 Gastroduodenal artery.
12 Left hepatic artery.
13 Right hepatic artery.
14 Right gastroepiploic artery.

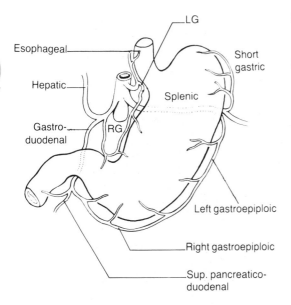

Fig. 29.2   The arterial supply of the stomach. LG and RG, left and right gastric arteries.

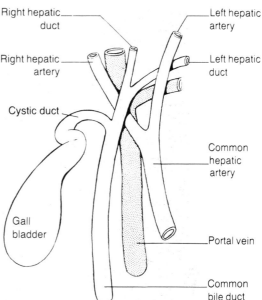

Fig. 29.3   The arrangements of the structures in the free edge of the lesser omentum on their way to and from the hilum of the liver.

Splenic artery

This vessel runs to the left in a wavy manner, the crest of the waves lying above and the troughs behind the upper border of the pancreas. After this dillydallying it reaches the lienorenal ligament, in which it passes forward to the hilum of the spleen where it divides into approximately five large branches. Before so doing, it gives off the short gastric (vasa brevia) and the left gastroepiploic branches which continue in the gastrosplenic ligament; the former branches supply the fundus of the stomach, and the latter enters the two layers of the greater omentum to complete the gastroepiploic arch for the supply of the greater omentum and both surfaces of the stomach.

Left gastric artery

This artery, the smallest of the three branches of the celiac, runs upward to insinuate itself between the two layers of the lesser omentum. Running inferiorly, the left gastric supplies the cardiac area of the stomach and anastomoses with the right gastric along the lesser curvature. Just before turning downward, an important esophageal branch is given off; this supplies the intra-abdominal esophagus, and anastomoses with the esophageal branches of the thoracic aorta.

Superior mesenteric artery

Arising a little below the celiac artery and behind the neck of the pancreas, the superior mesenteric artery runs down and seemingly emerges from a split in the pancreas. This illusion arises because the vessel is sandwiched between the uncinate process posteriorly and the neck of the pancreas anteriorly. The artery reaches the third part of the duodenum, where it enters the mesentery of the small intestine; both the mesentery and the artery cross the gut as they run inferiorly (the superior mesenteric artery syndrome has been referred to on p. 385).

Branches

Numerous jejunal and ileal branches arise from the left side of the artery. These branches form arcades until their terminal

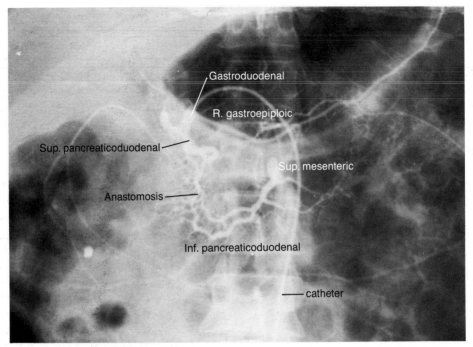

Fig. 29.4   The anastomosis between the pancreaticoduodenal arteries. This links the fore- and mid-gut supply and can be a life saver.

branches, which tend to be straight (vasa recta), especially in the jejunum.

From the right side of the vessel the following branches are given off:

1   *The inferior pancreaticoduodenal* arises as the superior mesenteric artery crosses the duodenum. It then runs to the right to anastomose with the superior pancreaticoduodenal branch of the gastroduodenal. In cases where the celiac artery origin becomes blocked, this is the route by which blood will flow retrograde to fill the celiac trunk.

2   *Middle colic.* This large branch is given off as the artery passes behind the transverse mesocolon. Entering the transverse mesocolon, the vessel divides into right and left branches, which run to each colic flexure, where they each anastomose.

3   *Right colic.* Running to the right, this artery lies anterior to all the structures in the retroperitoneum. It supplies the ascending colon.

4   *Ileocolic.* This branch runs to the right and downward, and divides into: an ascending branch which anastomoses with the right colic; cecal branches to the anterior and posterior surface of the cecum, the latter giving off the appendicular artery; and, finally, a descending branch which enters the root of the mesentery to anastomose with the terminal branch of the parent trunk, both of which supply the terminal ileum.

The superior mesenteric artery arises from the aorta at a fairly acute angle which puts it in much the same mainstream as the aorta itself; it is therefore easy for clots (emboli) from higher up in the arterial tree to enter this artery. Surgeons may have to approach the origin of the vessel to remove the offending clot. They find themselves in an unenviable position, exposing an artery behind the pancreas which is hidden between the splenic vein in front and the left renal vein behind.

A few operations like this will tame most surgeons.

Inferior mesenteric artery

Arising where the duodenum crosses the aorta, the vessel runs downward to the left across the left common iliac vessels, where it changes its name to superior rectal (hemorrhoidal). Its first branch is the upper left colic, which runs upward to the left to divide into branches which join the middle colic above and the lower left colics below. The succeeding branches are the lower left colics (sigmoidal), a series of three or four vessels that supply the descending and pelvic colon. These arteries anastomose with each other, with the upper left colic above and with the superior rectal below; this last anastomosis is a miserable one.

Portal circulation

The veins carrying the portal blood differ from most other veins by being devoid of valves and by beginning and ending in capillaries. They carry all the products of digestion from the gastrointestinal tract, as well as blood from the spleen, pancreas and gallbladder.

The inferior mesenteric vein drains the upper anal canal, rectum and hindgut, at first running with the artery of the same name. Leaving the artery as the latter curves toward the aorta, the vein runs vertically upward just behind the duodenojejunal flexure and joins the splenic vein deep to the body of the pancreas.

The large splenic vein, closely adherent to the body of the pancreas and well below the splenic artery, runs to the right as far as the neck of the pancreas, where it is joined by the superior mesenteric vein to form the portal vein (Figs. 28.5 and 29.8). Because of its close relationship to the pancreas, diseases of this organ may thrombose the splenic vein with consequent enlargement of the spleen. The superior mesenteric vein lies to the right of its artery in the root of the mesentery, and crosses the duodenum to join the splenic vein behind the neck of the pancreas. The two mesenteric veins therefore outflank their accompanying arteries. There is, however, much variation in the formation of the portal vein, and it is worth remembering that, in general, veins are variable.

The portal vein, commencing behind the neck of the pancreas, runs upward to enter the free edge of the lesser omentum, where it forms the third member of the portal triad. Running superiorly with the hepatic artery anterior and to the left and the bile duct anterior and to the right, the portal vein divides into right and left branches which enter the hilum of the liver. The right branch is joined by the cystic vein and the left branch by the left umbilical vein of old, the ligamentum teres. Pancreaticoduodenal, a small right gastric and large left gastric veins join the lower end of the portal vein. The left gastric vein reaches it by first accompanying the left gastric artery and then continuing to the right with the hepatic artery; it is this vein which bears the brunt of portal vein occlusion, as a result of which its tributaries become grossly enlarged. The bulk of the blood supplying the liver comes from the portal vein. Despite the fact that it is deoxygenated blood, relative to arterial blood, the total oxygen content reaching the liver by the portal vein is much greater than that via the hepatic artery. The latter vessel supplies the supporting structures and capsule of the liver. Some consider that it is possible to ligate the hepatic artery and rely on the portal vein to supply the needs of the liver. Blockage of the portal vein by a diseased liver is sometimes treated by diverting the blood to the systemic system (portosystemic shunt) – this usually has marked side effects.

Portosystemic communications

The commencing tributaries of the portal vein anastomose with tributaries of the

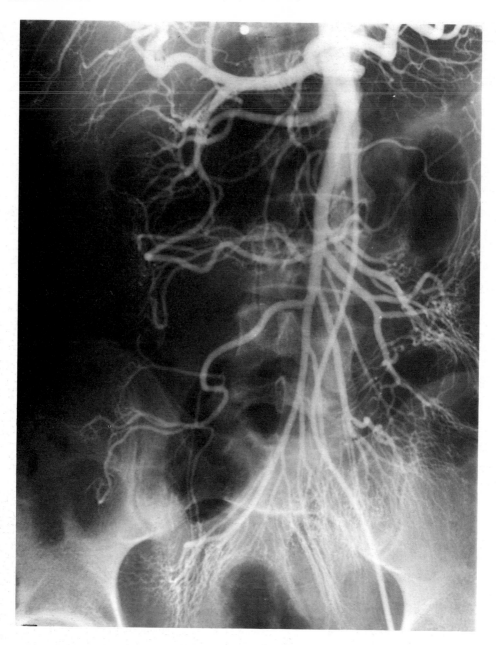

Fig. 29.5   Superior mesenteric arteriogram. (From Weir and Abrahams 1978)

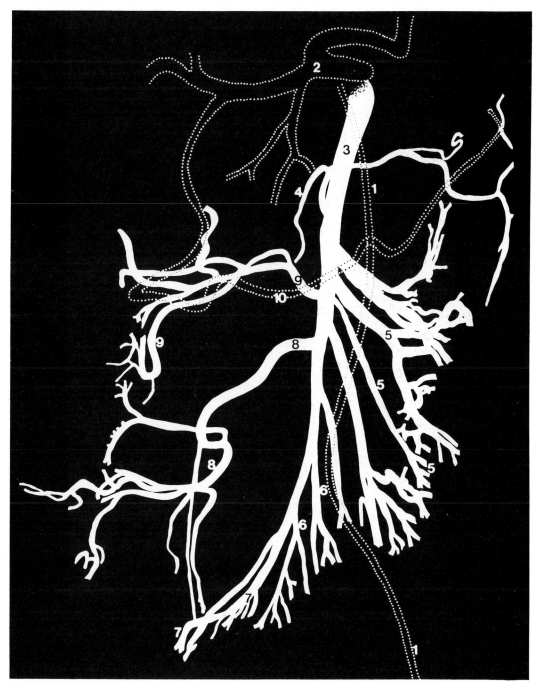

1 Catheter via femoral route.
2 Celiac trunk and branches (dotted).
3 Superior mesenteric artery.
4 Inferior pancreaticoduodenal artery.
5 Jejunal branches.

6 Ileal branches.
7 Ileocolic artery.
8 Right colic artery.
9 Middle colic artery.
10 Left gastroepiploic artery (branch of celiac trunk).

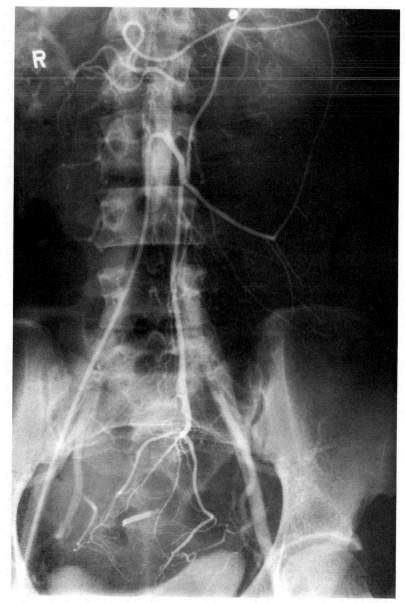

Fig. 29.6   Inferior mesenteric arteriogram. (From Weir and Abrahams 1978)

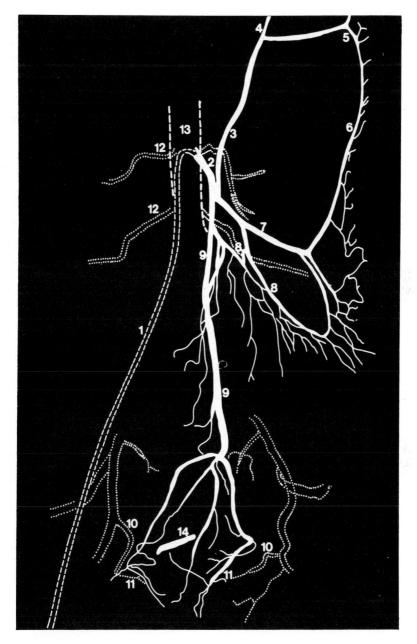

1 Percutaneous catheter into inferior mesenteric artery.
2 Inferior mesenteric artery.
3 Left colic artery.
4 Ascending branch of left colic artery.
5 Descending branch of left colic artery.
6 Marginal artery (Drummond).
7 Sigmoid artery (inferior left colic).
8 Anastomosis between sigmoid and superior rectal
   arteries.

9 Superior rectal artery.
10 Middle rectal artery – branch of internal iliac artery.
11 Anastomosis of superior and middle rectal arteries.
12 Reflux in lumbar arteries.
13 Reflux in aorta.
14 Intrauterine contraceptive device (IUCD).

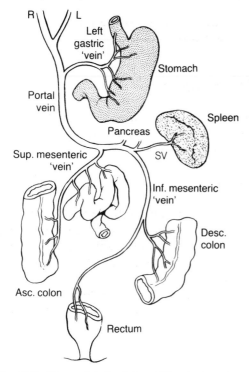

Fig. 29.7 The portal circulation. SV, splenic vein.

systemic circulation. Portal vein occlusion, such as occurs in cases of cirrhosis of the liver, results in the blood at increased pressure backing into the systemic capillaries (portal hypertension). These capillaries, unaccustomed to this high pressure of blood, become dilated and often burst.

1 The most important communication involves the tributaries of the left gastric and azygos veins on the abdominal esophagus. Rupture of these veins (esophageal varices) is the first or second commonest cause of violent upper gastrointestinal bleeding, depending on the country (Fig. 29.9).

2 The terminal branches of the inferior mesenteric vein (the superior rectal veins) join the inferior rectal veins, which are branches of the iliac veins – dilata-

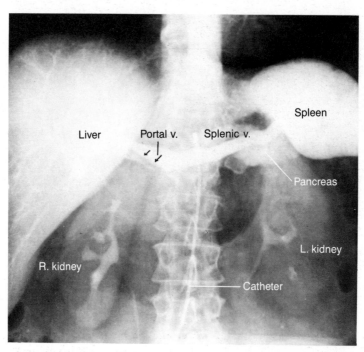

Fig. 29.8 Radiograph of a study of the splenic and portal veins. Contrast medium injected via the aorta has reached the venous phase of the study; the spleen and splenic vein, the body and tail of the pancreas, the portal vein (two arrows) and the liver are outlined. Note the contrast medium being excreted via the kidneys (pyelogram).

tions of this anastomosis may result in hemorrhoids, which are varicosities of the superior rectal veins. However, there are very few people aged over 40 who do not have hemorrhoids of some degree, and it is difficult in cases of portal hypertension to determine whether the condition is in fact due to the liver disease.

3 Small veins run along the ligamentum teres and falciform ligament to join the superior and inferior epigastric veins; enlargement of the former veins produces a characteristic picture of dilated periumbilical veins, the caput medusae.

4 Lastly, small branches of the colic and splenic veins anastomose in the retroperitoneal region with branches of the renal vein and of the body wall. This cannot be appreciated clinically, but it is noticeable surgically as it leads to hemorrhagic dissections.

Blood supply

We are now in a position to study the supply of some individual organs and relate them to clinical situations.

Liver

The blood supply to the liver via the portal vein and hepatic artery has been described above. The blood is now collected via systemic veins which commence in the central veins of the lobules of the liver. The confluence of the central veins forms hepatic veins, a right and left for the respective lobes and a

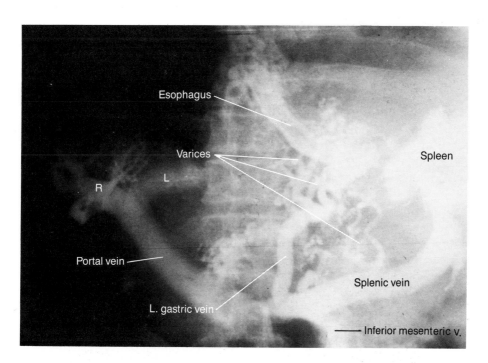

Fig. 29.9 In a patient with portal hypertension, contrast medium has been injected into the spleen, outlining part of the spleen, as well as the splenic, inferior mesenteric and portal veins, the last dividing into right and left branches. The superior mesenteric vein is not outlined in this study. The left gastric vein is hugely dilated, and the emptying of this vein into the azygos veins on the lower end of the esophagus has led to esophageal varices (dilation and tortuosity of the anastomotic veins). Some of these have ruptured into the esophagus to produce a violent hematemesis (vomiting of blood); the esophagus is thus outlined.

shared middle vein. The hepatic veins have hardly any extrahepatic course; they immediately join the inferior vena cava as it embeds in the liver substance on its upward journey. These hepatic veins are important for several reasons.

1 Blockage will produce one form of portal hypertension (Budd–Chiari syndrome), quite a common condition in the Caribbean.

2 These veins act as supports for the liver. It is sometimes claimed that the liver is supported by ligaments and intra-abdominal pressure, but when you open the abdomen at surgery, the liver doesn't fall into your lap; cutting the ligaments does not accomplish this either. The main supports of the liver are the hepatic veins.

3 Liver trauma is very common with the rising auto accident rate. A terrible injury is the avulsion of the hepatic veins from the vena cava by an acceleration or deceleration injury. The bleeding occurs in an almost inaccessible area behind the liver and comes from the right atrium of the heart, since there are no valves to prevent the backflow of blood. It is not surprising therefore that this injury has a high mortality rate. If one can get the patient to surgery in time, a catheter can be pushed up the vena cava to just below the right atrium and inflated to control the backflow. This allows time to sever the right triangular ligament and rotate the liver to locate the bleeding site.

Gallbladder

Most textbooks mention the cystic artery as the only blood supply of the gallbladder, but when the gallbladder is being removed from its liver bed you will notice the surgeon controlling considerable bleeding; the gallbladder is supplied from both the liver bed and the cystic artery, This explains the rarity of gallbladder necrosis during inflammation; the inflamed appendix, on the other hand, because of its dependence on virtually a single artery, almost always undergoes necro-

sis and perforation. This dictates the different approaches in the treatment of these two conditions.

Stomach

The stomach can be divided into three regions of arterial supply (see Fig. 29.2): the left gastric for the cardiac region; the splenic for the fundus; and the hepatic for the body and pylorus. In such an active organ, this profuse supply is not surprising. Once upon a time, it was thought that bleeding from peptic ulcers could be controlled by the ligation of the branches going to the stomach; no matter how many branches were ligated, the stomach merrily went on bleeding.

Duodenum

The duodenum is supplied by the two pancreaticoduodenal vessels, the upper arising from the hepatic and the lower from the superior mesenteric (see Fig. 29.5). This anastomotic route is taken when the origin of either the celiac or the superior mesenteric artery stenoses, a common concomitant of advancing age. Since the arch between these two arteries runs in the groove between the pancreas and duodenum, it means that two organs share the same blood supply and one cannot be removed without the other.

Small intestine

The jejunal and ileal branches of the superior mesenteric artery supply this very active portion of intestine. The vasa recta of the jejunum, longer than those of the ileum, have minimal anastomoses once they leave the arcades, and disruption of one of these vasa recta may lead to necrosis of a portion of the gut. This is the danger of traumatic lacerations of the small bowel mesentery.

Appendix

The importance of the appendicular artery being virtually an end-artery has already been stressed.

Colon

The colon is supplied in its so-called right half by the superior mesenteric, and its left half by the inferior mesenteric. The branches have been detailed, but it is important to note the marginal artery (of Drummond). The mar-

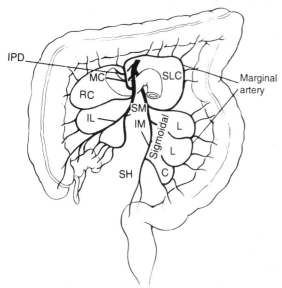

Fig. 29.10 Arterial supply of colon. Note all the branches of both mesenteric arteries are linked by the marginal artery. IL, ileocolic; IM, inferior mesenteric; IPD, inferior pancreaticoduodenal; LLC, lower left colics (sigmoidal); RC, right colic; SH, superior hemorrhoidal; SLC, superior left colic; SM, superior mesenteric.

ginal artery runs in close proximity to the large bowel along its whole length from ascending to pelvic colon. It is formed by the various branches of supply to the gut being joined near their terminations by a longitudinal vessel, from which branches are given off. The marginal artery is weakest and sometimes deficient where the superior and inferior mesenteric distributions meet just proximal to the splenic flexure; diminution of the blood supply in this region may lead to death of the colon in that area, a condition known as ischemic colitis. There is a second dangerous area where ischemic colitis may occur: at the rectosigmoid, the lower left colic branches of the inferior mesenteric join the

superior hemorrhoidal continuation of the inferior mesenteric in an unimpressive anastomosis. In the elderly, the origins of the superior and inferior mesenteric arteries may both become blocked, and the marginal artery may fill from below via the superior hemorrhoidal by way of the hemorrhoidal branches of the internal iliac and keep the colon alive. Surgeons can transpose large segments of colon as far as the neck if it is necessary to bypass the esophagus – the bowel will depend on the marginal artery for its blood supply. The right colon – a term including the cecum, the ascending colon and the right half of the transverse colon – has a much better blood supply than the left colon; the former is an actively absorbing area (mainly water), while the latter acts mainly as a passive conduit. It follows that tumors of the colon, the second commonest tumor in males, spread much more easily on the right than the left side, with a correspondingly worse prognosis.

Lymph drainage

Lymph capillaries are, in most areas, a network, and blockage of one lymph channel leads to the detouring of the lymph to an adjacent channel; for this reason, cancer cells have very little trouble travelling widely via the lymphatics.

Stomach

As it is an aphorism that superficial lymphatics follow veins and deep lymphatics follow arteries, a knowledge of the arterial supply of the stomach is tantamount to a knowledge of the lymphatic drainage. It is popular to draw various lines and divide the stomach into three sections, but there is so much overlap that it is doubtful whether this is really practical. Certainly it can be said that the area of the stomach supplied by the splenic artery drains via the lymphatics accompanying that artery to the lymph nodes in the hilum of the spleen, then to nodes

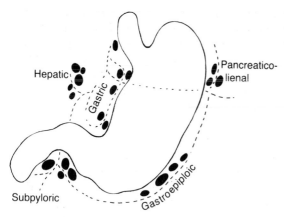

Hepatic

Gastric

Pancreatico-
lienal

Subpyloric

Gastroepiploic

Fig. 29.11 Lymph drainage of the stomach.

situated along the upper border of the pancreas, and eventually to the celiac nodes around the celiac artery. The cardiac area, which is left gastric artery territory, drains along the left gastric artery to reach the celiac nodes. The rest of the stomach drains via branches of the hepatic artery, and there are large nodes situated along these arteries. We should mention those situated along the hepatic artery as it runs up to the liver; enlargement of these nodes may cause blockage of the adjacent bile duct to produce jaundice. There is also a well-known node situated at the bifurcation of the gastroduodenal artery below the first part of the duodenum whose enlargement may produce duodenal obstruction. It is clinically interesting to note that there is seldom lymphatic spread from a tumor of the stomach across the pylorus into the duodenum; this is probably due to an interruption of the lymphatics by the thickened circular muscle that forms the pyloric sphincter and to an infolding of the longitudinal muscle at this site. At the opposite end of the stomach the spread of tumors onto the lower esophagus is only too common and carries a poor prognosis.

Duodenum and pancreas

Nodes that lie along the pancreaticoduodenal vessels drain the duodenum and the head of the pancreas. The body of the pancreas drains to nodes situated along the splenic artery, while the tail drains to nodes in the splenic hilum.

Spleen

The spleen drains to nodes that are situated on the splenic hilum and which drain to the celiac nodes.

Liver

The liver drains mostly by vessels that accompany the hepatic artery; a large cystic node lies in relationship to the gallbladder, and other nodes lie in relationship to the hepatic artery in the free edge of the lesser omentum. Some vessels pass via the falciform ligament to retrosternal nodes and through this to diaphragmatic nodes, while others reach the latter nodes via the inferior vena cava. In summary, the lymph from the dome of the liver drains upward into the chest, while the rest of the organ drains eventually to the celiac nodes.

Small and large intestine

The villi of the small intestine are noteworthy for their large central lymphatics which are white (lacteals) because of the contained absorbed emulsified fat. Lymphatic channels pass to lymph nodes situated near the intestine, along the larger branches of the superior mesenteric artery, and at the origin of the main artery, sometimes called proximal, intermediate and main nodes respectively. The same system of drainage (apart from the villi) occurs in the large intestine. The lymphatics draining the intestine eventually collect into an intestinal trunk, which joins a trunk from the stomach to form a gastrointestinal trunk; this empties into the anterior aspect of the cisterna chyli.

Cisterna chyli

This lymph sac lies between the aorta and the

right crus of the diaphragm, and is separated from the latter by the vena azygos (when present). Apart from the gastrointestinal trunk, the cisterna receives the right and left lumbar lymph trunks, tributaries from the parietes. The sac narrows superiorly to become the thoracic duct.

Nerve supply

The nerve supply is mediated by the sympathetic and parasympathetic systems. On the efferent side the parasympathetic system is responsible for inpouring secretions, for stimulating peristalsis and for opening sphincters and passing out contents (babies, urine and feces, amongst others). The sympathetic system, on the other hand, causes diminished activity of the viscus, closes sphincters, and is a powerful vasoconstrictor. The parasympathetic is therefore the active and opening agent, the sympathetic delaying and closing. On the afferent side, much of the sensation from the viscera travels along the sympathetic system. It is true to say that the abdominal viscera appreciate only changes in tension, namely distension or contraction. The excessively contracting intestine causes colic, and the distending liver stretches its capsule to produce a constant ache. Other sensations, such as touch, temperature and pinprick, that are superbly appreciated in other areas, are not appreciated at all by the viscera. On observing a patient with a colostomy (an opening of the colon onto the abdominal wall), note that the bowel can be cut or pinched severely, as is your fancy, and the patient will carry on reading the newspaper. This is unfortunate, as many curable diseases remain painless until they are far advanced; on the other hand, who would want to be reminded of the constant activities going on in the abdomen? The visceral peritoneum surrounding the organs localizes sensations poorly (autonomic supply), whereas the parietal peritoneum that lines the anterior abdominal wall (somatic nerve supply) appreciates a larger variety of sensations and can localize extremely well. Although afferent sensations from viscera are relatively poorly localized, they do cover fairly characteristic areas. The foregut afferent fibers end in the seventh and ninth segments of the cord and this sensation is appreciated by the sensory area of the brain as arising from the seventh to ninth somatic segments, the area from xiphisternum to above the umbilicus. The midgut relates to the tenth segment, which is the periumbilical area, whereas the large gut relates to the eleventh and twelfth segments, which is the hypogastrium, the area between the umbilicus and the pubis. Furthermore, most viscera tend to localize toward the midline, the site of their original development.

Let us exemplify the above discussion by presenting the symptoms and signs of acute appendicitis, the commonest cause of an acute abdomen. When the inflamed appendix protests by means of contractions, the pain is appreciated as periumbilical abdominal colic although the inflamed appendix lies in the right lower quadrant; the appendix localizes as midgut to the tenth spinal segment through its autonomic nerve supply. As the inflammation of the appendix proceeds, it involves the parietal peritoneum against which it is lying in the right lower quadrant. This is a much cleverer peritoneum, and it localizes well (it is blessed by a somatic nerve supply). The pain therefore shifts from the umbilicus to the right lower quadrant; further pressure on the inflamed parietal peritoneum, as on an inflamed finger, is resented and is appreciated as tenderness.

Intra-abdominal sympathetic system

These nerves emanate from the lateral horns of the thoracolumbar sympathetic outflow and pass into the sympathetic chain and its ganglia.

Celiac ganglia

These ganglia are two nodular masses that lie on each crus of the diaphragm with the celiac artery medially and the adrenal glands laterally. The ganglia are joined by and emit numerous branches that form the celiac or solar plexus, so called because of the branches which radiate from the ganglia like the rays of the sun. This is a popular place for a boxer to strike his opponent and it is said by the ringside commentators to lead to all sorts of autonomic nerve disturbances; however, a

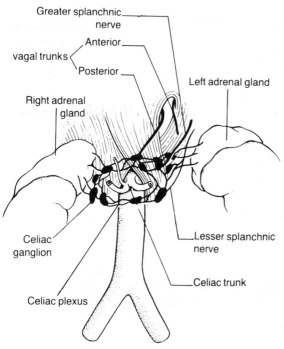

Fig. 29.12 The celiac ganglia and suprarenal glands.

hard whack into the upper abdomen, replete as it is with many organs, is in itself sufficient to produce the long count. The celiac plexus is prolonged downward in front of the aorta as the preaortic plexus, which becomes the intermesenteric plexus and eventually ends below the bifurcation of aorta on the fifth lumbar vertebra as the hypogastric plexus. The latter plexus was called the presacral nerve until someone realized that the sacrum started below the fifth lumbar vertebra. Offshoots of the plexus are common, and a particularly large one follows the renal artery as the renal plexus.

The nerves forming the celiac ganglia and its extensions are mostly of sympathetic origin, the parasympathetic being a lesser contributor. The thoracic trunk gives off the greater splanchnic (fifth to ninth ganglia) and the lesser splanchnic (tenth and eleventh ganglia) branches, both of which pierce the diaphragmatic crus to join the celiac ganglion on each side; the least splanchnic branch from the twelfth ganglion on each side also pierces the crus, but joins the renal plexus.

Sympathetic chain

The intra-abdominal or lumbar sympathetic chain contains four or five ganglia, usually four because of the frequent fusion of two ganglia. The thoracic chain passes under the medial arcuate ligament as the lumbar chain to reach and lie on the anterior border of the psoas muscle. The lumbar chain leaves the abdomen for the pelvis by passing behind the common iliac artery to join the hypogastric plexus. On the anterior border of the psoas the right chain is overlapped by the vena cava and the left chain by the aorta. The trunks are attacked by surgeons who wish to sever the vasoconstrictor fibers to the vessels of the lower extremity.

Branches

The chain receives two white rami from the upper two lumbar nerves, but sends off gray rami to all five lumbar nerves. These rami can be seen passing with the lumbar vessels beneath the fibrous arches from which the psoas muscles arise. The chain also gives off branches to reinforce the intermesenteric and hypogastric plexuses.

Intra-abdominal parasympathetic system

Most of the parasympathetic supply to the abdomen emanates from the two vagi. They were left in a state of apparent disarray as an esophageal plexus on the lower end of the

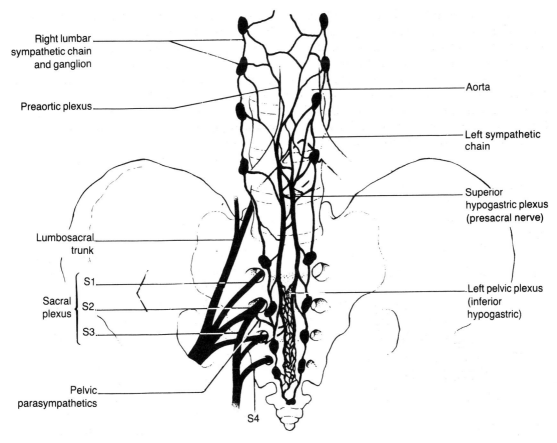

Fig. 29.13 Sympathetic chains and plexuses and the lumbosacral plexus. Each chain contributes to the preaortic plexus. The superior hypogastric plexus is dividing into right and left pelvic plexuses. Note the parasympathetic pelvic branches from S2 and 3 joining the pelvic (inferior hypogastric) plexus.

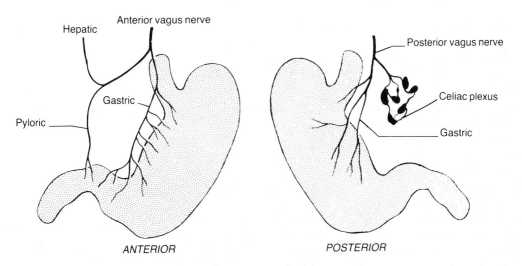

Fig. 29.14 The nerve supply of the stomach.

esophagus. As the esophagus passes through its hiatus into the abdomen, the fibers regain their control and rejoin to form left and right nerves. The rotation of the stomach onto its right side results in the left vagus becoming anterior and the right becoming posterior. Occasionally, the grouping is not into two distinct nerves and there are three or more branches. The anterior vagus runs caudally on the esophagus and is mostly expended on the anterior surface and lesser curve of the stomach. Before it reaches the stomach, a hepatic branch passes in the lesser omentum to the liver and gallbladder and a pyloric branch runs down to partly control the pyloric sphincter. The posterior branch, mostly destined for and distributed with the celiac plexus, sends a gastric branch to the posterior surface of the stomach. The vagal supply to the stomach is secretomotor – stimulation produces acid secretion and the emptying of the stomach by churning the food and passing it through the relaxed sphincter. The medical treatment of a peptic ulcer is to paralyze the vagi to produce less acid. When the medicine man fails, the surgeon severs the vagi (truncal vagotomy) and reduces the amount of acid, but he unfortunately also paralyzes the stomach and prevents its emptying; he therefore has to add an operative procedure to assist the emptying of this somewhat atonic bag. More recently, the secretory fibers passing to the stomach are severed, leaving the pyloric branches intact and thus retaining the pyloric pump (selective vagotomy).

From the celiac plexus, the vagal (and sympathetic) fibers are distributed via the blood vessels along the gut, approximately as far as the splenic flexure. The bowel below this is supplied by pelvic splanchnic nerves (S2, 3, 4 outflow) via the inferior mesenteric artery.

30 Kidney and suprarenal gland

Kidney·Suprarenal gland

Kidney

In the days when one studied botany, the bean was called kidney shaped; now, in studying anatomy, the kidney is called bean shaped. Both are true, and these paired organs, each about 10 cm (4 in) long and 5 cm (2 in) wide, lie opposite the twelfth thoracic to the third lumbar vertebrae. Each kidney is tilted in two directions: the upper pole is nearer the midline than the lower, and it lies obliquely on the posterior abdominal wall with its medial border facing somewhat anteriorly. Because of the latter it would appear that the urine emanating from the kidney runs slightly uphill. The right kidney lies at a slightly lower level than the left, its ascent having been prevented by the liver. A palpable right kidney may be normal, but a left kidney should not be palpable abdominally – if it is, it usually means enlargement of that kidney.

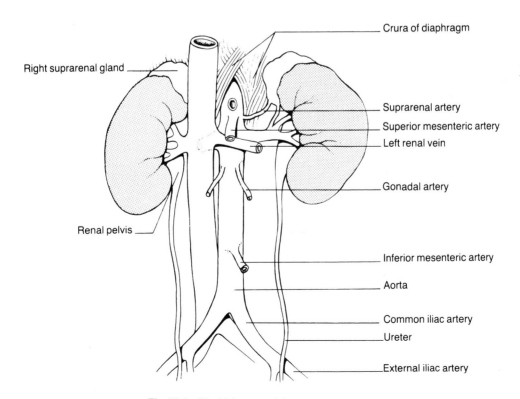

Fig. 30.1 The kidneys and the suprarenal glands.

Anterior relations

The kidneys are covered by different organs on each side. The right kidney is capped by the right suprarenal gland. The second part of the duodenum lies along its medial border with the hepatic flexure of the colon crossing both kidney and duodenum. The left kidney is capped less successfully by the left suprarenal gland, which has slipped down its medial border. The upper part is covered by the spleen and the stomach, where it forms part of the stomach bed. The pancreas runs across the middle of the kidney to reach the spleen; below this it is related to the colic flexure as the latter turns down to become the descending colon. Small portions of the lower part of each kidney look into the greater sac below the attachment of the transverse mesocolon and are related to loops of small intestine. With all these organs in front of the kidney, it is no surprise that the urologist approaches the kidney from behind.

Posterior relations

Superiorly the kidney lies on the diaphragm and its crus. Lower down it lies on the three muscles that form the posterior abdominal wall, mediolaterally the psoas, quadratus lumborum and the origin of transversus abdominis. In the dissecting room three dents may be seen on the posterior surface of the kidney; these are caused by the transverse processes of the first three lumbar vertebrae. The pleura is separated from the kidney by the diaphragm. Sometimes the origin of the diaphragm posteriorly is somewhat deficient and the kidney may actually be in direct contact with the pleura. Inflammation surrounding the kidney, such as a perinephric abscess, may irritate the nearby pleura and produce a so-called sympathetic pleural effusion. The subcostal nerve, the two branches of the first lumbar nerve and their accompanying vessels run posterior to the kidney as they encircle the abdominal wall.

Hilum

The hilum of the kidney lies on its medial border, and transmits the renal artery and vein, the ureter and the autonomic nerves. The large structures are arranged in the old VAD order with the ureter lying posteriorly.

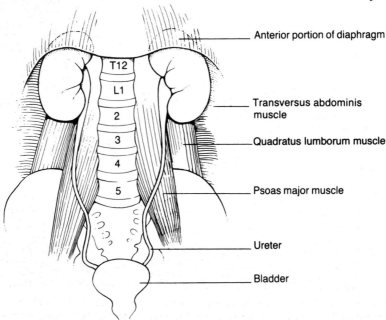

Anterior portion of diaphragm

Transversus abdominis muscle

Quadratus lumborum muscle

Psoas major muscle

Ureter

Bladder

Fig. 30.2   The kidneys and ureters lying on the posterior abdominal wall.

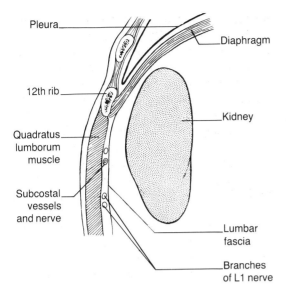

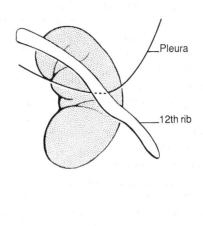

Fig. 30.3 The posterior relations of the kidney. The anterior layer of the lumbar fascia separates the kidney from the nerves and the vessels lying on the quadratus lumborum; this prevents damage to the nerves and vessels during removal of the kidney (surgeons call it the friendly fascia).

Fig. 30.4 The pleural relationship of the kidney. This close posterior relationship explains why pleural effusions may complicate kidney infections. Surgeons have to remove the twelfth rib to get at the kidney and they remember this relationship well (surgeons call it the perilous pleura).

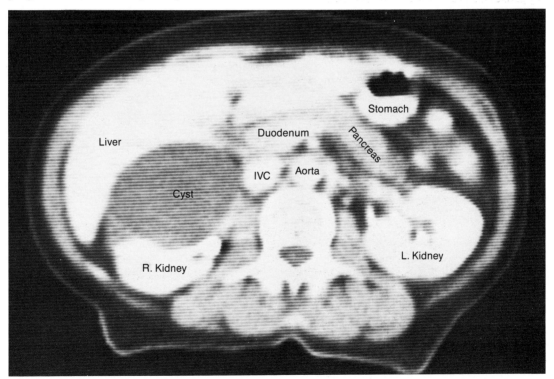

Fig. 30.5 Computerized axial tomogram of abdomen. Note the pancreas crossing the left kidney. The stomach has contrast medium in it. The aorta and vena cava are well seen, with the superior mesenteric artery(s) coming off the former behind the pancreas. Note the large cyst of the right kidney.

The ureter is a tube carrying the urine from the kidney, and it expands at its upper end into a pelvis (basin), which lies partly within and partly without the kidney substance. The urine is collected by tubules that empty into renal papillae. These papillae project into minor calyces, three or four of which join to form a major calyx. The two major calyces are so constituted as to empty into the pelvis.

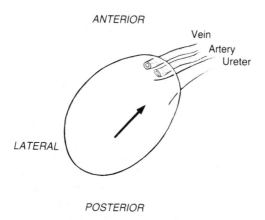

Fig. 30.6   The hilum of the kidney. The structures entering and leaving the hilum are arranged in the VAD fashion. Note the oblique tilt of the kidney.

Capsules of the kidney

The kidney is surrounded by a fibrous capsule which, in the healthy kidney, can be easily stripped. Outside this capsule lies a considerable layer of perinephric fat that is partly responsible for holding the kidney in position. The fat is condensed peripherally to form the perinephric fascia (Gerota). Loss of the fat is one of the reasons for a floating kidney, a condition where the organ is excessively mobile. This used to be a prolific source of surgery in the bad old days when these floating kidneys were replaced in their proper position (nephropexy) by the more sporting type of surgeon.

Ureter

Each expanded pelvis narrows into the ureter, a 25 cm (10 in) long fibromuscular tube that runs downward on the psoas muscle opposite the tips of the lumbar transverse processes to reach the brim of the pelvis. Here the ureter lies anterior to the bifurcation of the common iliac artery or the sacroiliac joint, and then passes into the true pelvis on its way to the bladder. The ureter has three physiologic narrowed areas where stones tend to impact: at the pelviureteric junction, where it crosses the bifurcation of the common iliac artery and at its entry into the bladder. The relationship of the ureter to the tips of the transverse processes is important; a radiologic opacity in this position commonly represents a calculus (stone). Like all well-behaved ducts it is the most posterior structure on the posterior abdominal wall (VAD), lying behind all the visceral vessels, i.e. mesenteric, colic and gonadal. The surgeon recognizes the ureter by its position and by its spontaneous peristalsis.

Neurovascular supply of the kidney and ureter

The renal arteries arise from the aorta and are virtually end-arteries (see Fig. 30.1). Any interference with these arteries is serious as the kidney has little collateral supply.

The veins lie in front of the arteries, and for embryological reasons the left is much longer than the right. Neophyte urologists removing their first kidney (nephrectomy) are only too happy if it is the left kidney, as the pedicle (handle) is much longer and there is little risk of pulling the renal vein off the vena cava. The left vein, a cross-branch between the two subcardinal veins, is embryologically more than a renal vein, and it takes upon itself the drainage of both the adrenal and the testicular or ovarian veins. On the right side, where the right posterior cardinal vein becomes the vena cava, the latter receives the veins of the gonad and adrenal, as well as the renal vein. Tumors of the kidney may grow along the renal veins and, in doing so, may block its tributaries; the only one that you would notice would be the left testicular vein. Varicosity (dilatation) of the testicular vein (scrotal varicocele) appearing initially in an older person is diagnostic of this situation.

The kidney and ureter receive their afferent supply by sympathetic nerves that arise from segments T10 to about L2, the upper segments supplying the kidney, the lower the ureter. These nerves relay the pain occurring in renal colic; for example, the patient's pain starts in the lumbar region where the kidney is situated, and spreads forward toward the umbilicus (T10). When the pain passes down toward the groin (L1) and even to the thigh (L2) it means that the stone, as commonly happens, has left the renal pelvis and passed down the ureter. Its passage through the narrow opening into the bladder can be diagnosed by a further change in the site of the pain (see p. 474).

Besides afferent fibers, the renal plexus contains sympathetic fibers which have a considerable effect on the vessels of the kidney and, therefore, the flow of urine. In the past, periarterial renal sympathectomy (removal of the sympathetic fibers from around the renal artery) was quite a common operation for the treatment of kidney diseases, but it now belongs in the annals of 'operations I have done and regretted'.

Congenital anomalies of the kidney and ureter

These anomalies are quite common and clinically important. The kidney develops in two areas. Inferiorly the ureter develops as an outgrowth (bud) of the lower end of the mesonephric or wolffian duct, grows cranially, dilates to form the pelvis, divides to form the calyces, and terminates in the collecting tubules. These collecting tubules join the secreting tubules that are derived from the metanephros, the secreting part of the kidney that develops from the mesonephric ridge. After fusion of these two parts, the kidney

'ascends' by differential body wall growth until it comes into contact with the suprarenal gland. As it is passing cranially, it waves cheerily to the downward travelling testis/ovary which has developed immediately medial to it from the germinal ridge and is descending to its rightful position in the scrotum/pelvis, carrying its vessels with it; the vessels therefore cross the ureter. During ascent, the kidney is supplied by vessels that arise successively from the iliac arteries and the aorta, each succeeding vessel being followed by the disappearance of the vessel below it until it reaches the adult renal artery. Some of these vessels, especially thos eto the upper or lower poles, may not disappear and are then often called accessory arteries. Some authorities, however, consider these accessory vessels to be vessels of supply rather than additional vessels. Urologists are wary of this, and they temporarily clamp the 'accessory' vessel to watch the color of the kidney before they consider dividing it. In the past, such 'aberrant' vessels crossing the ureter were thought to be responsible for initiating a ureteric obstruction. Sometimes this does happen, but the concept is not as easily accepted as it was in the past, when surgeons were happily operating on all patients whose upper ureters were blocked, and severing the vessels as they came across them.

It should be easy to elucidate the anomalies.

1　The collecting system may not unite with the secreting system, and cysts form where this failure occurs – congenital polycystic kidneys. These are large kidneys, not popular with insurance companies.

2　A kidney may not ascend to its adult position, and the unwary surgeon may explore a patient who has an undiagnosed mass in the pelvis or the lower abdomen, remove it, only to find that it is a functioning kidney. This is not ideal, but if the patient has another kidney, it is not disastrous; should she belong to group 3 below, it is

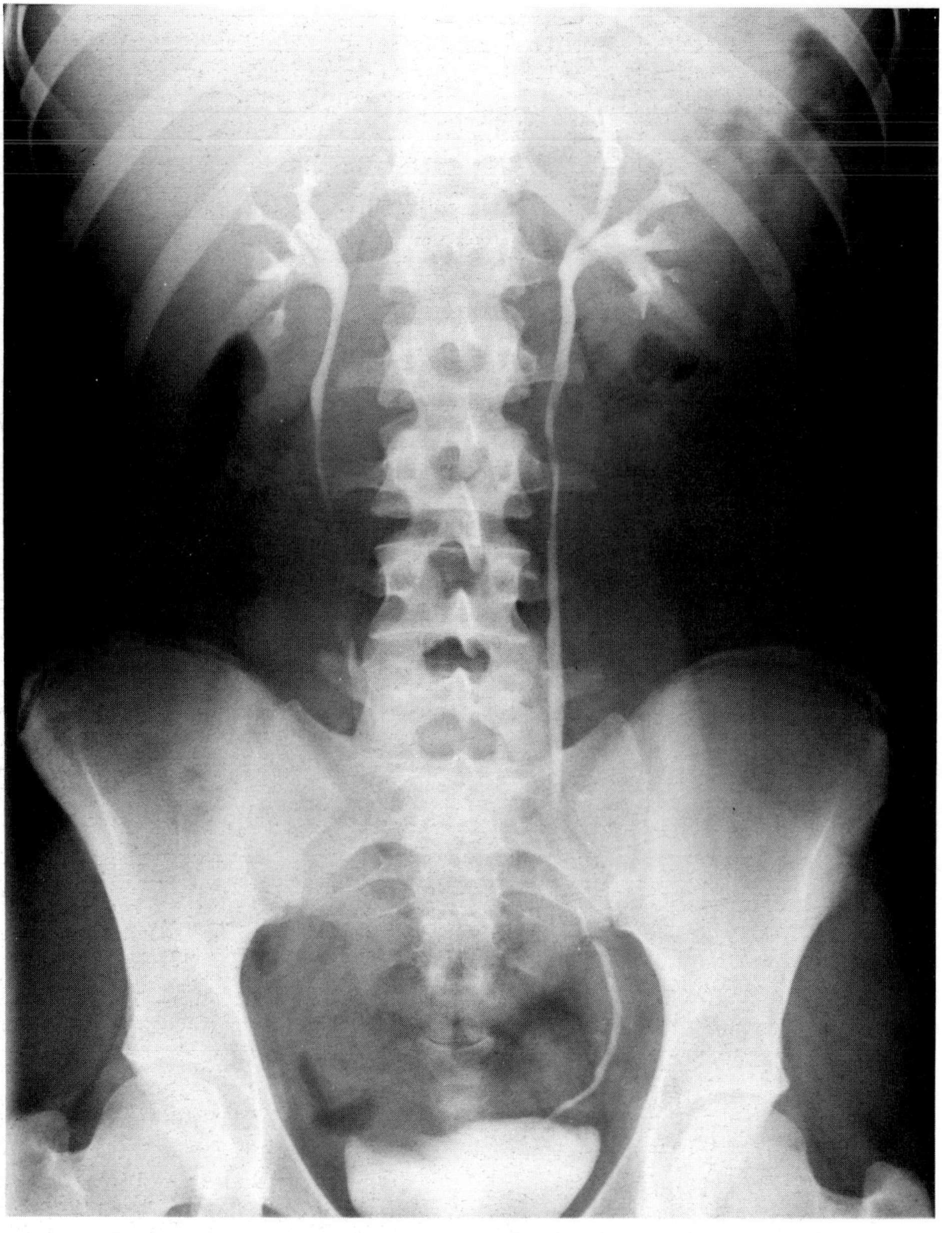

Fig. 30.7   An intravenous pyelogram outlining the calyceal system of both kidneys, most of the left and right ureters and the bladder.

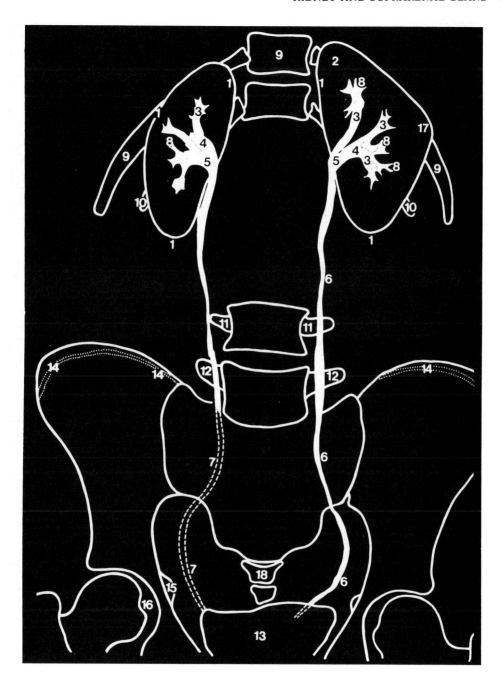

1 Renal outline.
2 Superior lobe, renal cortex.
3 Minor calyx.
4 Major calyx.
5 Pelvis of kidney.
6 Left ureter.
7 Position of right ureter.
8 Position of renal papilla.
9 Eleventh thoracic vertebral body and rib.

10 Twelfth rib.
11 Transverse process of L4.
12 Transverse process of L5.
13 Bladder.
14 Epiphyseal line.
15 Ischial spine.
16 Fovea.
17 Splenic hump.
18 Coccyx.

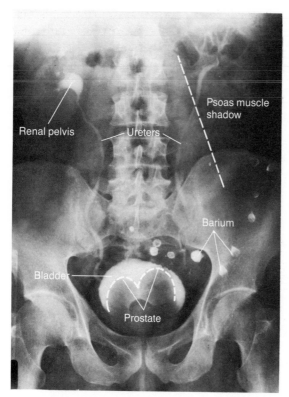

Fig. 30.8 Radiograph of an intravenous urogram. The injected contrast medium is excreted via the kidneys. The right ureter is outlined throughout its length, the left through most of its length – the lower 2.5 cm (1 in) is not seen. The circular shadows are inspissated blobs of barium retained in colonic diverticula from a previous study. A grossly enlarged prostate fills most of the bladder.

bad news for all. Therefore, always investigate the kidneys in any unexplained abdominopelvic mass.

3   One ureter and/or kidney may fail to develop and the patient may have only one kidney. No one would dream of removing the kidney of a patient unless he is sure that there is another kidney.

4   Uncommonly, the lower poles of the two kidneys become fused across the great vessels, the so-called horseshoe kidney. Its ascent is prevented by the root of the inferior mesenteric artery and it is thus confined to the lower lumbar region.

5   Abnormalities involving the ureter usually concern the ureteric bud, and it

may split into two portions as it ascends to produce a bifid ureter or pelvis. The split may be complete and the two ureters may enter the bladder separately, but more commonly the second completely separate ureter opens lower down into the urethra or vagina, one of the uncommon causes of urinary incontinence.

6   One of the rarest abnormalities occurs when the ureter becomes heavily involved with the complicated development of the inferior vena cava and passes behind it, the retrocaval ureter. It usually becomes obstructed by the vein.

Suprarenal (adrenal) gland

These two endocrine glands are situated in close relationship to the kidney, the right one being perched on the upper pole of the kidney, whereas the left one has fallen slightly off its perch and has slid down the medial border to lie between the renal vessels and the upper pole. Each gland lies on a crus of the diaphragm with the celiac ganglia and their plexuses between them. The left gland lies behind the lesser sac, forming part of the stomach bed superiorly, and being crossed by the pancreas in its lower part. The right gland is covered by the liver and the vena cava as the latter lies embedded in the liver.

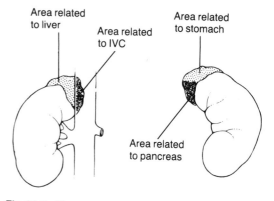

Fig. 30.9 The suprarenal glands.

Development

These glands have an interesting development: the outer golden cortex is derived from mesoderm, and the inner vascular medulla originates from the ectoderm of the neural crest. Both parts have entirely different functions, the medulla producing norepinephrine (noradrenaline) and the cortex producing steroid hormones. Their relationship to the kidneys is purely fortuitous, as they are enclosed in a completely separate capsule from the kidney. This is fortunate – otherwise a kidney removal would mean an adrenal removal, with serious consequences.

Neurovascular supply

The textbooks are not too generous with their arterial supply and mention three arteries giving off branches of supply – the phrenic above, the aortic in the middle and the renal vessel below. In fact, at surgery, the gland has an incredibly vascular supply by numerous small, mostly unnamed, vessels.

On the other hand, the veins are single, the right draining into the vena cava by an extremely short trunk, and the left usually draining into the left renal vein. These veins are important clinically. They are wide and it is possible to pass a catheter via the inferior vena cava into these veins to collect secretions. This may determine whether the adrenal has pathology, and on which side. Recently, the catheter has been left in the vein to cause venous thrombosis and destroy the gland, thereby obviating the need for surgery.

The suprarenal has a prolific nerve supply from the celiac ganglia, and it is thought that many of these fibers are preganglionic and that the postganglionic neurons are replaced by the cells of the medulla. In the past, hypertension was treated by severing these branches and denervating the gland, but better methods now exist.

31 Retroperitoneal vessels and nerves

Abdominal aorta · Inferior vena cava · Lumbar plexus · Summary of posterior abdominal wall structures

Abdominal aorta

This large trunk, the continuation of the thoracic aorta, enters the abdomen posterior to a fibrous band, the median arcuate ligament, which joins the two crura of the diaphragm. From the level of the disc between T12 and L1, the aorta runs downward to end in front of the fourth lumbar vertebra, about 1.25 cm (0.5 in) below and to the left of the umbilicus, where it bifurcates into its two terminal common iliac branches. Although lying deeply, it may surprise you to note visible aortic pulsations, even in fairly obese individuals. The reason for this is the forward displacement of the vessel by the normal anterior curvature of the lumbar spine (lumbar lordosis); as this lordosis is often exaggerated in women, the pulsation is even more noticeable in the gentler sex.

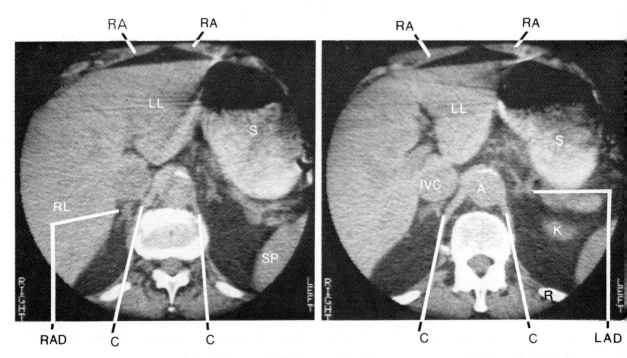

Fig. 31.1   Adrenal views. A, Aorta; C, Crura of diaphragm; IVC, Inferior vena cava; K, Upper pole left kidney; LAD, Left adrenal gland; LL, Left lobe of liver; R, Rib; RA, Rectus abdominis m.; RAD, Right adrenal gland; RL, Right lobe liver; S, Stomach; SP, Spleen. (Reproduced from Weir and Abrahams: An Atlas of Radiological Anatomy, 2nd Edition, Churchill Livingstone, 1986).

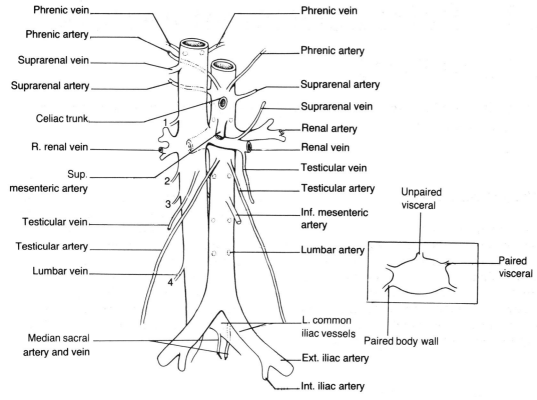

Fig. 31.2 The aorta, inferior vena cava and their branches. Numerals 1–4 depict the four right lumbar veins.

Anteriorly the aorta is placed behind all the abdominal viscera. Above, it lies behind the peritoneum of the lesser sac, flanked on each side by the crura of the diaphragm which support the celiac ganglia and suprarenals. At first covered by the liver, it passes inferiorly posterior to the body of the pancreas. Sometimes it is difficult to tell whether an epigastric mass is due to an enlargement of the pancreas or of the aorta, since the pulsations of the aorta are transmitted through the pancreas. Below the pancreas the third part of the duodenum crosses the aorta transversely. One of the more dramatic ways of leaving this planet occurs when an aneurysm (bulging of a weakened portion of the wall) of the aorta, or the graft replacing it, ruptures into this part of the alimentary tract. Below the dueodenum, the aorta is covered by the peritoneum and numerous loops of small intestine

Posteriorly the aorta lies on the vertebral column and its ligaments, with the left lumbar veins passing behind it on their way to the inferior vena cava; backache is one of the early symptoms of pathologic enlargement of the aorta, and is often mistaken for renal pain.

To the right of the aorta lies the vena cava, into which an aneurysm may rupture and to its left lies the psoas muscle with the sympathetic trunk running on it and partially hidden by the aorta.

Branches

These are distributed ventrally, laterally and dorsally. The three ventral unpaired branches, the celiac and the superior and inferior mesenteric arteries, have been discussed in Chapter 29.

The lateral branches are distributed to the three paired glands. The suprarenal and renal arteries have been described (Chapter 30),

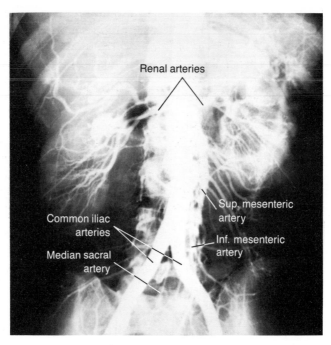

Fig. 31.3 Aortogram showing the course of the aorta and some of its branches.

and the third lateral branch is the gonadal. The testicular/ovarian arteries have similar courses, running inferiorly on the posterior abdominal wall anterior to the ureter, which they supply, but behind the colic vessels. The testicular vessels reach the internal ring, at which point they join the other constituents of the spermatic cord to pass down the inguinal canal and supply the testes. The ovarian vessels do not reach the internal ring, but cross the external iliac vessels at the pelvic brim to enter the lateral end of the broad ligament, via the infundibulopelvic or suspensory ligament of the ovary, to supply the ovary and tube.

Posterior

The posterior branches, the phrenic and lumbar, supply the roof and posterior walls of the abdomen respectively. The phrenic vessels arise at the same level as the celiac artery and pass upward and laterally to supply the diaphragm, and, to a lesser degree,

the esophagus, adrenal gland and liver. The latter blood supply is held by some to be an important collateral to the liver when the hepatic artery is occluded. The lumbar arteries are four pairs only, as the aorta fails to reach the fifth lumbar vertebra; the fifth pair of vessels arise from the internal iliac arteries. The lumbar arteries pass under the protection of the fibrous arches of origin of the psoas muscle, after which they run laterally to supply the posterior abdominal wall on their way to the anterolateral abdominal muscles. Posterior branches are given off to the spinal muscles, which supply the spinal cord en route. One of the complications of clamping the aorta during surgery is the interruption of the blood flow to the spinal cord, with disastrous results; it happens uncommonly and, at the present time, unpredictably.

The median sacral artery, the continuation of the aorta, runs inferiorly in front of the fifth lumbar vertebra and sacrum; while a considerable vessel in a crocodile, in man it is of little consequence.

Inferior vena cava

This vessel is a much longer and wider structure than the aorta, running from the fifth lumbar vertebra to the eighth thoracic vertebra, where it enters the right atrium of the heart. It is formed by the junction of the two common iliac veins in front of the fifth lumbar vertebra, where it is crossed by the right common iliac artery. Running up on the vertebral column to the right of the aorta, it lies in much the same plane as the aorta until the upper lumbar vertebrae when it passes forward to reach a deep groove in the posterior aspect of the liver. After this it penetrates the central tendon of the diaphragm to reach the right atrium. The complicated development of the vena cava may lead to occasional deviations from normal, with a vena cava being left sided instead of right sided, or being double. The forward inclination of the cava allows branches of the aorta such as the right renal, middle suprarenal and phrenic to pass to the right behind it. The cava partially covers the right suprarenal gland, accounting for the short course of the latter's vein.

Like the aorta, the cava lies deeply and is deep to, from above downward, the posterior aspect of the liver, the lesser sac at its epiploic foramen, the first part of the duodenum and portal vein, the head of the pancreas containing the bile duct, the third part of the duodenum, the right gonadal artery, the posterior peritoneum of the infracolic compartment with the root of the mesentery and its contents, and, finally, the right common iliac artery. The aorta lies on the left side of the cava, and the sympathetic trunk lies behind it. The proximity of the vena cava to the portal vein at the epiploic foramen is useful to the surgeon – who sometimes anastomoses the two vessels to bypass a cirrhotic liver that is obstructing the portal vein (portacaval shunt).

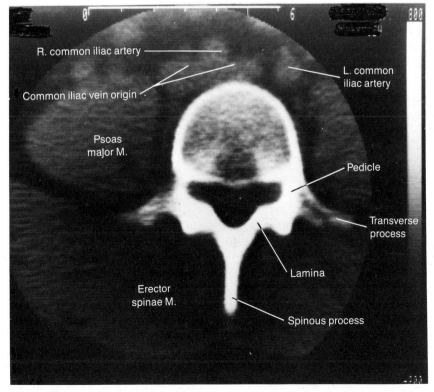

Fig. 31.4   Computerized tomogram of the posterior abdominal wall, showing the common iliac vessels.

Tributaries

The lumbar veins open into the posterior aspect of the cava, the veins of the left side having to cross behind the aorta. The first and second veins may not open directly into the cava, but may join the ascending lumbar vein or either azygos venous system. The right gonadal vein opens just below the entrance of the renal vein; the left gonadal vein, it must be remembered, enters the left renal vein, as the left vena cava left it in the lurch. The renal veins join the cava, each vein lying anterior to its own artery; the longer left vein receives the suprarenal and gonadal vein. Reference has been made to the right suprarenal vein (Chapter 30) and the hepatic veins (Chapter 29). It should be noted, however, that the complex developmental patterns of these veins lead to anomalies in some 10–20 percent of normal individuals. The right gonadal vein, for example, may open directly into the right renal vein.

Lumbar plexus

This plexus consists of the ventral rami of L1, 2 and 3 lumbar nerves with contributions from T12 and L4. The plexus assembles itself in the posterior fibers of the psoas muscle and its branches appear in the following manner:

1. One branch, the obturator nerve, appears on the medial border of the psoas muscle and runs along the side wall of the pelvis.
2. One branch, the genitofemoral nerve, pierces the psoas to appear on its anterior surface.
3. All the other branches appear at the lateral border of the psoas.

The subcostal nerve, the twelfth thoracic nerve, lies below the twelfth rib on the posterior abdominal wall. It passes laterally to enter the neurovascular plane of the abdominal wall and surfaces in the hypogastrium above the pubis.

The first lumbar nerve divides into iliohypogastric and ilioinguinal branches,

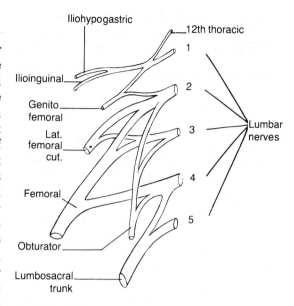

Fig. 31.5   The lumbar plexus.

which run, as does the subcostal nerve, across the quadratus lumborum to enter the anterolateral abdominal muscles. The superior nerve, the iliohypogastric, ends (as the name indicates) in the hypogastrium, supplying the area in the region of the pubis. The ilioinguinal nerve enters the inguinal canal, runs through it outside the cord and emerges through the superficial ring to supply the skin on the medial side of the scrotum/labium and adjoining thigh. Patients often complain of pain in this area when a hernia develops and presses on the nerve. It is also the site of pain from a calculus passing down the ureter (L1). Don't be caught out by the patient complaining of groin pain, where you cannot demonstrate a hernia; she may have a stone in the ureter.

The lateral cutaneous nerve arises from L2 and 3, runs across the iliacus muscle toward the anterior superior iliac spine, where it penetrates the inguinal ligament near its lateral bony attachment to emerge onto and supply the upper lateral thigh. There is an interesting entrapment neuropathy, meralgia paraesthetica, where the nerve is compressed between the two portions of the inguinal ligament – it produces a burning pain on the outer aspect of the thigh. Blocking the nerve by a local anesthetic agent close to the anterior spine produces a large area of anesthetic skin on the lateral thigh, a favorite site for taking skin grafts.

The large femoral nerve arises from the second, third and fourth anterior rami, and appears at the lateral border of the psoas. Lying between the psoas and the iliacus, which it supplies, it stays in that groove to enter the thigh. Since the nerve lies behind the iliac fascia, it enters the thigh outside the femoral sheath.

The genitofemoral nerve arises from L1 and 2, pierces the psoas muscle and its fascia, and runs down anterior to the muscle with the external iliac vessels. The femoral branch, lying within the femoral sheath, pierces this sheath in the thigh on the lateral side of the artery and supplies the skin over the femoral triangle. The genital branch accompanies the structures passing to the testis via the inguinal canal and supplies the cremaster muscle and other scrotal structures. This explains the cremasteric reflex; stroking the skin of the thigh is accompanied by retraction of the testis by the contracting cremasteric fibers.

Summary of posterior abdominal wall structures

At dissection, surgery or even end-of-course examinations it is often very difficult to relate all the posterior abdominal structures correctly. A useful guide is to think of these structures as belonging to four separate layers that are stacked one upon another in the following order, from deep to superficial:

1   Vertebrae and posterior wall muscles (e.g. psoas, quadratus and transversus abdominis).
2   The lumbar plexus.
3   The great vessels (aorta, vena cava) and urinary tract.
4   The genital tract (e.g. gonadal vessels).

Life then becomes much simpler if you need to distinguish the psoas, genitofemoral nerve, ureter and ovarian artery.

32 Pelvis

Bony skeleton · Orientation · Composition

Bony skeleton

The bony skeleton has been partially reviewed in a study of the lower limb (Chapter 19). To the two os coxae must be added the bones completing the pelvis – the sacrum and the coccyx.

Sacrum

This bone is triangular shaped with its base above and its apex below. It represents five fused sacral vertebrae, including their representative spinous, costal and transverse processes. The bodies of the five vertebrae can still be made out anteriorly since they are separated by grooves (lineae transversae). Posteriorly the spinous processes, articular tubercles and transverse processes are visible mediolaterally. The four intervertebral foramina diminish in size from above downward. The base or upper surface of the first sacral vertebra articulates with the fifth lumbar vertebra at the lumbosacral angle. On each side of the lumbosacral articulation lies the upper surface of the fused transverse and costal processes, the alae, which represent the upper surface of the lateral masses, those parts of the bone that lie lateral to the foramina. The anterosuperior margin of the body of the first sacral vertebra projects forward as a promontory, a palpable landmark on vaginal examinations.

At the lower end of the sacrum, the laminae and the spinous process of the fifth sacral vertebra are missing; the gap is called the sacral hiatus. The articular processes of the fourth sacral vertebra project downward as sacral cornua which articulate with the cornua of the succeeding bone, the coccyx, at

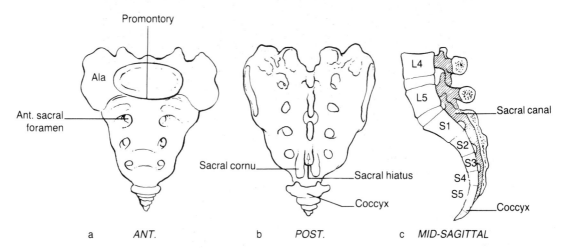

Fig. 32.1   The sacrum and coccyx: (a) anterior view; (b) posterior view; (c) midsagittal view.

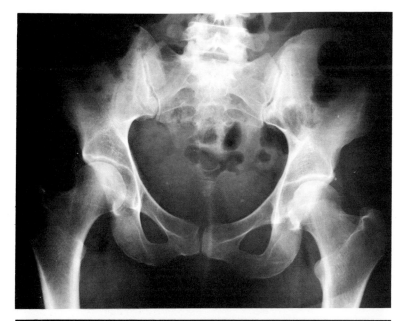

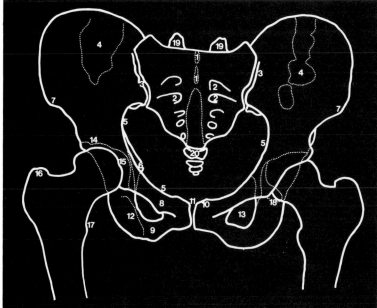

Fig. 32.2   Anteroposterior view of the pelvis. (From Weir and Abrahams 1978)

 1 Sacral spinous crest.
 2 Anterior sacral foramina.
 3 Sacroiliac joint.
 4 Gas in bowel overlying iliac fossae.
 5 Pelvic brim.
 6 Ischial spine.
 7 Anterior superior iliac spine.
 8 Superior ramus of pubis.
 9 Inferior ramus of pubis.
10 Pubic tubercle.

11 Pubic symphysis.
12 Ischial tuberosity.
13 Obturator foramen.
14 Acetabular rim.
15 Fovea of femoral head.
16 Greater trochanter.
17 Lesser trochanter.
18 Acetabular fossa (teardrop).
19 Articular facets of S1.
20 Four coccygeal segments.

a fibrous joint. The sacral hiatus is a very important site since it allows the passage of a needle into the sacral canal extradurally to anesthetize the contained nerves (caudal anesthesia – used in childbirth).

The sacral canal is the downward continuation of the lumbar vertebral canal and contains the five anterior and posterior sacral roots and coccygeal nerves as they run inferiorly in a test-tube-like extension formed by the dura and arachnoid membranes, which end at the level of the second sacral vertebra. The pia mater is prolonged downward to perforate the two outer layers of spinal membranes and reaches the coccyx as the filum terminale. The sacral canal also contains vessels, lymphatics and fat.

The anterior surface of the middle three sacral segments gives attachment to the piriformis muscle, while the posterior surface affords origin to the large muscles of the back of the trunk, the sacrospinalis, multifidus and gluteus maximus.

Coccyx

This bone consists of from three to five (usually four) rudimentary vertebrae. The first coccygeal vertebra articulates with the fifth sacral vertebra at a joint which commonly fuses. The transverse processes of the first coccygeal vertebra sometimes fuse with the fifth sacral vertebra and so add further sacral foramina. Two cornua project upward to meet the sacral cornua. The joints between the coccygeal vertebrae fuse to a varying degree. During pregnancy, however, the joints become mobile enough to increase the pelvic outlet, while with old age the degree of fusion increases. Falls on the coccyx are often extremely painful, and the radiologic studies are difficult to interpret.

Orientation

When holding the articulated pelvis, it is

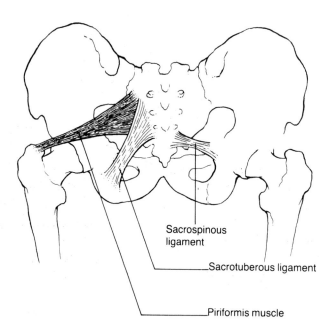

Fig. 32.3   The sacrotuberous and sacrospinous ligaments – posterior view.

Sacrospinous ligament

Sacrotuberous ligament

Piriformis muscle

difficult to appreciate that there is a distinct downward and backward tilt of the pelvis. Push the articulated pelvis against the wall so that the anterior superior iliac spine and the upper margin of the symphysis are in the same vertical plane – the anterior aspect of the pelvis now looks downward and backward. If you pass a sword above the symphysis, it will pass through the same plane as the ischial spines and the tip of the coccyx. Unless this is appreciated, it will be difficult to understand the mechanisms of obstetrics.

Composition

The pelvis has an anterior wall composed of the pubic bone and its two rami, with a urogenital membrane (triangular ligament) filling the space between the diverging inferior rami. Laterally, it is formed by the ischium and its two rami and by the fusion of all the bones in the acetabular region, the obturator membrane completing the side wall. The posterior wall is formed by the sacrum and coccyx and by two ligaments, the sacrotuberous and sacrospinous.

Sacrotuberous ligament

This large ligament lies on the posterior aspect of the pelvis and runs from the medial margin of the ischial tuberosity to the posterior iliac spines and the sides of the sacrum and coccyx below these spines.

Sacrospinous ligaments

This triangular ligament is attached by its apex to the ischial spine and by its base to the last segment of the sacrum and first segment of the coccyx. It is often regarded as a fibrous degeneration of the posterior part of the coccygeus muscle.

These two ligaments convert the two sciatic notches into foramina through which intrapelvic structures pass to the buttock or perineum. During pregnancy, most ligaments soften and allow some extra bony movement,

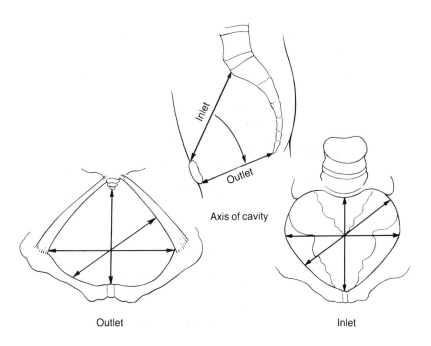

Fig. 32.4 The diameters of the inlet, outlet and axis of the pelvic cavity.

which accounts for the waddling gait of some pregnant women.

Inlet of the pelvis

The inlet is bounded by the promontory and anterior margin of the sacral alae posteriorly, on each side by the arcuate and pectineal lines, tubercle and crest of the pubic bones, and anteriorly by the symphysis pubis.

Cavity of the pelvis

The anterior wall measures 5 cm (2 in) and the posterior 15 cm (6 in).

Outlet of the pelvis

The outlet is diamond shaped and bounded in front by the symphysis pubis, on each side by the conjoined ischiopubic rami forming the pubic arch, and posterolaterally by the sacrotuberous ligaments with the coccyx projecting into the posterior angle.

Diameters of the pelvis

The measurements of the pelvis are very important in determining whether there is disproportion between the fetal head and the pelvis. The diameters measure as listed in Table 32.1. From this it can be deduced that the fetal head engages the inlet in the transverse diameter and emerges through the outlet in the anteroposterior diameter; this implies that it descends in a spiral manner. Clinically, however, the actual measurements are not as vital as the relationship between the pelvis and the fetal head. It is these proportions that matter in each delivery. Genetic and racial variations account for a large range in the normal pelvic measurements. The differences in the male and the female pelvis are usually obvious, the pelvis of the female being 1.25 cm (0.5 in) larger in all diameters. The pelvis of the female is wider and shallower with the ischia everted, the male tending to be narrower and deeper and funnel shaped. The inlet to the female pelvis is kidney shaped, whereas in the male it is heart shaped. The sacrum is wider, shorter and less curved in the female, to allow more room. The bones of the male pelvis are thicker and rougher. The differences are better summed up in a simple line diagram (Fig. 32.5).

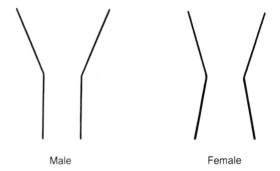

Male   Female

Fig. 32.5   The shape of the pelvis in each sex.

These bones must now be padded before they are in a fit condition to contain the pelvic viscera.

Obturator internus

The obturator internus muscle arises from the inner aspect of the bones forming the margin of the obturator foramen as well as the fascia closing this foramen. At the upper end of its fascial origin it leaves a little gap for the ob-

Table 32.1   Diameters of the bony pelvis

| | Transverse | | Oblique | | Anteroposterior | |
	cm	(in)	cm	(in)	cm	(in)
Inlet (brim)	12.5	(5)	11.25	(4.5)	10	(4)
Cavity	12.5	(5)	12.5	(5)	12.5	(5)
Outlet	10	(4)	11.25	(4.5)	12.5	(5)

turator vessels and nerve to pass from pelvis to thigh. The muscle narrows into a tendon which exits from the lesser sciatic foramen into the buttock to be inserted into the medial side of the greater trochanter of the femur. As the tendon turns at a sharp angle from pelvis to thigh, it leaves ridges on the bone, which is lubricated by an intervening bursa. (See also Figs. 21.1 and 21.2.)

Piriformis (see Fig. 35.1)

The piriformis muscle arises from the anterior surface of the lateral mass of the sacrum opposite its middle three segments, and leaves the pelvis through the greater sciatic foramen to be inserted into the greater trochanter close to the obturator internus.

Below the piriformis the small coccygeus muscle runs from the fifth sacral vertebra and adjoining coccyx to the ischial spine (see Fig. 32.9 and 35.1).

Having padded the bones with muscle, the latter will be covered by a thick sheet of pelvic fascia.

Parietal fascia

This sheet of fascia lines the bony pelvis and its muscles. Nearly all the structures that leave the pelvis lie to its outer side, which

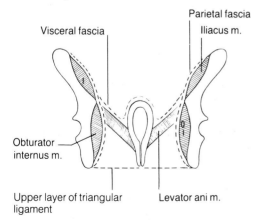

Fig. 32.6   The visceral and parietal fasciae. The viscus depicted could be bladder, uterus or rectum, as all are supported by the levator ani. The fascia overlying the viscus is thin and easily stretched, as all these organs are very distensible.

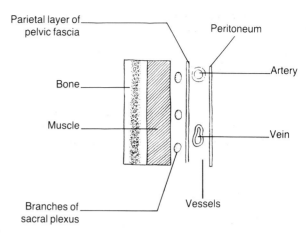

Fig. 32.7   The linings of the pelvic wall.

spares them piercing and weakening the fascia, while the structures destined for the pelvis lie on its inner side. The superior border of the fascia is attached along the pelvic brim; anteriorly it is attached to and fuses with the periosteum of the posterior aspect of the pubis, while posteriorly it crosses the anterior aspect of the piriformis and sacral plexus of nerves to line the hollow of the sacrum behind the rectum. This posterior sheet of fascia is often called Waldeyer's fascia. Laterally the fascia covers the obturator internus. Inferiorly it reaches and attaches to the ischiopubic ramus, where it is reflected from one side to the other as the upper layer of the two-layered triangular ligament (urogenital diaphragm), the inferior layer of which is called the perineal membrane.

Visceral fascia

The visceral fascia arises from the parietal fascia laterally, and is attached anteriorly and posteriorly to the back of the pubis and the ischial spine respectively. The visceral fascia passes medially and downward onto the upper surface of a muscle which forms the pelvic hammock or floor of the pelvis, the levator ani, from which it is reflected onto the pelvic viscera. Thickenings of the fascia run from the side wall of the pelvis to the viscera, and constitute ligaments of support.

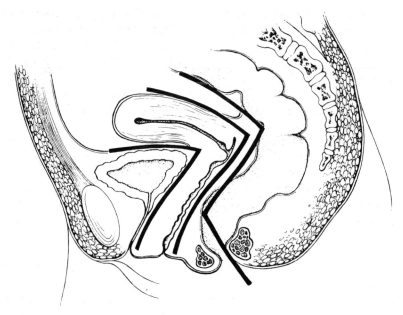

Fig. 32.8   The antiprolapse positions of the three pelvic viscera. They form a series of four sevens, the anorectum being a double seven.

From without in, therefore, the arrangement of the structures in the pelvis is bone (the pelvic wall), parietal muscles (obturator internus, piriformis and coccygeus), the sacral plexus of nerves, the parietal pelvic fascia, the blood vessels, the visceral pelvic fascia and, finally, the peritoneum. Note that as the muscles and the majority of sacral nerves are going to leave the pelvis, they lie outside the parietal pelvic fascia. The only nerve that pierces the fascia to enter the pelvis is the obturator nerve; it does not have to pierce it twice, however, as it exits above the upper margin of the obturator fascia, as the latter obligingly loops downward at the obturator foramen so as to leave a small gap

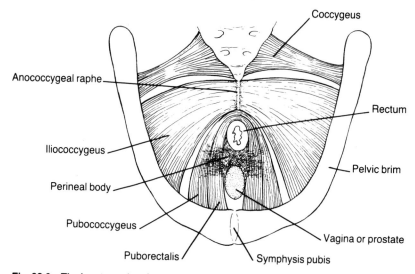

Fig. 32.9   The levator ani and coccygeus muscles.

for the nerve. Most of the blood vessels supply the pelvic viscera, but those that do not will have to pierce the fascia.

Levator ani muscle

This muscle arises from the pubis and ischial spines and a thickening of the parietal layer of the pelvic fascia between them. The two muscles pass downward and medially from each side to form a pelvic diaphragm, the support of the pelvic viscera. The muscle is divided into several portions, each of which lies in a different plane.

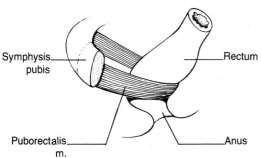

Symphysis pubis

Rectum

Puborectalis m.

Anus

Fig. 32.10 The puborectalis muscle sling. The change of angles of the rectum at the anorectal junction is partly maintained by the puborectalis. This angle is most important for fecal continence.

1 The *iliococcygeus* arises from the posterior part of the thickened fascia and ischial spine, and is inserted into the coccyx and a fibrous thickening between it and the anus, the anococcygeal raphe or ligament.

2 The *pubococcygeus* arises from the back of the pubis and anterior part of the thickened fascia, and passes horizontally backward beside the rectum and above the previous muscle to insert into the coccyx and anococcygeal raphe.

3 The *puborectalis* arises from the back of the pubis, and passes backward and downward to the side of the rectum, where it meets the muscle of the other side to form a loop or sling behind the rectum. This helps to constitute the anorectal angle where the rectum changes direction and name, and passes directly backward as the anal canal.

This powerful muscle can be easily felt as a large ridge by a finger in the rectum; its presence is a major factor in the maintenance of anal continence. Stretching or destruction of the levator ani is one of the factors responsible for prolapse of the pelvic viscera. The levator ani is supplied by the second, third and fourth sacral nerves.

33 Pelvic organs

Bladder · Seminal vesicles · Uterus · Rectum · Anal canal · Prostate gland · Pelvic peritoneum

In embryonic life one tube (the cloaca) enters the pelvis and emerges at the cloacal membrane. It later splits into an anterior portion, the lower end of the urinary system (urogenital sinus), and a posterior portion, the lower end of the alimentary tract. The reproductive organs insinuate themselves between the two systems.

Bladder

The bladder is a distensible hollow organ that acts as a reservoir for urine. It has a forward-pointing apex, a posteriorly directed base, two inferolateral surfaces and a superior surface, and has been likened to a small rowboat. When empty, the bladder is confined to the pelvis, but as it fills with urine it rises above the brim of the pelvis, dissects the peritoneum off the anterior abdominal wall and presents as a suprapubic swelling. A distended bladder is the second commonest cause of a suprapubic swelling, and before any exotic diagnoses are made the bladder should be emptied. Remember also that in a child the pelvis is a teeny weeny affair, and there is little room for the bladder; for this reason, up to about the age of 8 years, the bladder is an abdominal organ. It is normal to palpate a distended bladder above the pubis in a child. Advantage is taken of this in the child where samples of urine can be obtained by suprapubic needle aspiration. Normally holding about 300 ml (0.6 pint), the adult bladder can (with a stiff upper lip) accommodate almost 1 liter (1.76 pint). Distended bladders are very liable to traumatic rupture, and this is a danger when the alcoholic celebrant drives home via a lamp post.

The apex of the bladder points forward and ends in a median umbilical ligament, a relic of the urachus which ran to the umbilicus. It is quite common for the lower part of this ligament to contain a urachal lumen continuous with the cavity of the bladder. Rarely, the whole urachus remains open as a form of congenital umbilical urinary fistula.

The superior aspect or fundus of the bladder is covered by peritoneum that is reflected onto it from the anterior abdominal wall. The rest of the bladder lies below the level of the peritoneum, except for the very uppermost part of the posterior wall. Lying

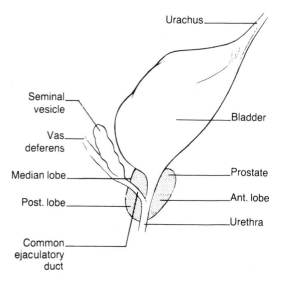

Fig. 33.1 The bladder, prostate gland and seminal vesicles – lateral view.

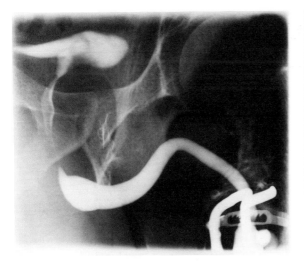
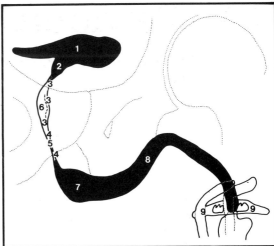

Fig. 33.2   Male urethrogram. (From Weir and Abrahams 1978)

1 Contrast medium in urinary bladder.
2 Bladder neck.
3 Prostatic urethra.
4 Membranous urethra.
5 Site of sphincter urethrae (external sphincter).

6 Site of colliculus seminalis (verumontanum).
7 Bulbous urethra.
8 Spongy urethra (penile).
9 Penile clamp (Knudson's).

on the superior surface are coils of intestine, and in the female the uterus leans forward to partially cover the bladder. A full bladder will displace the uterus backward and it is customary, for this reason, to examine patients vaginally with the bladder empty; otherwise, an erroneous diagnosis of a displaced uterus will be made.

The inferolateral surfaces are supported on each side by the sloping levator ani muscles, and are separated from the symphysis pubis by a fat-filled retropubic space (the cave of Retzius or prevesical space). One of the surgical approaches for removing the prostate is via this retropubic route. The fat pad does constitute a protective buffer against injuries to the bladder; however, fractures of the pubis often lacerate the bladder, and a patient with a fractured pelvis must be investigated to exclude this injury.

With regard to the bladder base, the seminal vesicle is somewhat adherent and separates the bladder from the rectum; in the female, the base is related to the anterior aspect of the cervix and anterior vaginal wall.

The bladder is loosely surrounded by fascia which allows marked distension. The median umbilical ligament has been mentioned above. There are two medial umbilical ligaments which are the remnants of the umbilical branches of the internal iliac arteries. After reaching the bladder in postfetal life as the blood-containing superior vesical arteries, the umbilical arteries become fibrous strands that pass to the umbilicus as the medial umbilical ligaments. These three umbilical ligaments can be seen as vertical folds that ridge the peritoneum of the anterior abdominal wall. Also visible, lateral to the medial ligaments, are the peritoneal folds raised by the inferior epigastric vessels, the lateral umbilical folds or ligaments.

Structure

The wall of the bladder is composed of interlacing muscle fibers that run in different directions. The superior surface is the only part of the organ that has a serous covering of peritoneum. The mucous membrane of the bladder is transitional epithelium that is

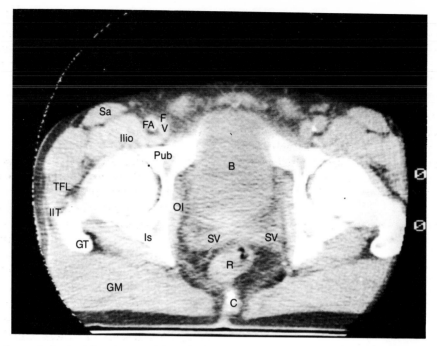

Fig. 33.3 Computerized axial tomogram of the pelvis. B, bladder; C, coccyx; FA, FV, femoral artery and vein; GM, gluteus maximus; GT, greater trochanter of femur; Ilio, iliopsoas; IIT, iliotibial tract; Is, ischium; OI, obturator internus; Pub, pubis; R, rectum; Sa, sartorius; SV, seminal vesicle; TFL, tensor fascia lata.

thrown into numerous folds to allow maximal distensibility while conserving space (as in the intestine); it is also urine proof in that it does not permit absorption of urine (urine can be absorbed by other mucous membranes). The mucous membrane of the posteroinferior aspect of the bladder is smooth, firmly adherent to the underlying muscle, and pale. This area is called the trigone and it is easily recognizable at cystoscopy (inspection of the interior of the bladder by a telescope); it is bounded on each side by the two slit-like orifices of the ureter and inferiorly by the internal urethral orifice. The ureters can be seen discharging urine through their orifices; a lifetime of carrying out cystoscopies will result in the growth of transitional epithelium over the cornea (examine your neighborhood urologist's eyes). The slit-like orifices bear testimony to the oblique passage of the ureters through the bladder wall. This constitutes the valve-like mechanism that closes off the intramural part of the ureter and prevents the reflux of urine up the ureters during contraction of the bladder. A small proportion of people suffer from failure of this mechanism; the result is vesicoureteric reflux. There are no glands in the bladder; mucus in a urine specimen comes from the urethra or from a dirty urinal. A bulge, caused by the underlying prostate, is visible at the lower portion of the trigone, the uvula vesicae. Small in the adolescent, the prostate enlarges as time passes and eventually projects into and may obstruct the bladder (Fig. 30.8).

Seminal vesicles

The seminal vesicles (see Figs. 33.1 and 33.4) are a pair of blind-ended tubes that are coiled on themselves to produce a structure about 5 cm (2 in) long; when teased out, each tube is about 12.5 cm (5 in) long. The function of the vesicle is to produce energized, fructose-rich fluid in which the spermatozoa may frolic. They do not form a reservoir for the sperm.

The seminal vesicles lie between the rectum and the posterior surface of the bladder; the peritoneum covers their upper surfaces, dipping down slightly onto their posterior surfaces before being reflected onto the rectum. It is theoretically possible, but in practice difficult, to palpate normal seminal vesicles via the rectum; however, enlarged diseased vesicles are easily palpable. The vesicles end below in narrow ducts which join the terminations of the vasa deferentia to form the common ejaculatory ducts; these latter ducts run through the substance of the prostate to empty into the urethra (see Fig. 33.20).

Ductus (vas) deferens

This structure, last seen leaving the inguinal canal by hooking around the inferior epigastric artery at the internal ring, runs extraperitoneally on the lateral wall of the pelvis. Passing inferiorly the ductus crosses the ureter to reach the base of the bladder, behind which it runs on the medial side of the seminal vesicle. At its lower end the vas dilates markedly as an ampulla, after which it narrows again to join the seminal vesicle. The ampulla lies between the rectum and the bladder medial to the seminal vesicle. The ductus deferens, with its thick wall and narrow lumen, feels exactly like a whipcord, or, in modern usage, a fine insulated electric wire. (See also Figs. 24.9, 24.16 and 34.3).

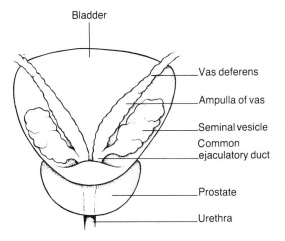

Fig. 33.4 **Base of bladder, vasa deferentia, seminal vesicles and prostate gland – posterior view.**

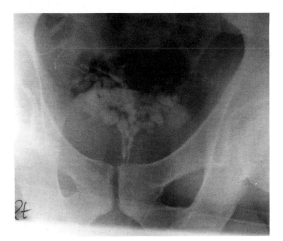

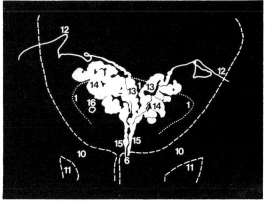

Fig. 33.5 **Seminal vesioculogram (vasogram). (From Weir and Abrahams 1978)**

1 Contrast medium in urinary bladder.
6 Site of colliculus seminalis (verumontanum).
10 Superior ramus of pubis.
11 Obturator foramen.
12 Ductus deferens (vas deferens).

13 Ampulla of ductus deferens.
14 Seminal vesicle.
15 Ejaculatory duct.
16 Phlebolith.

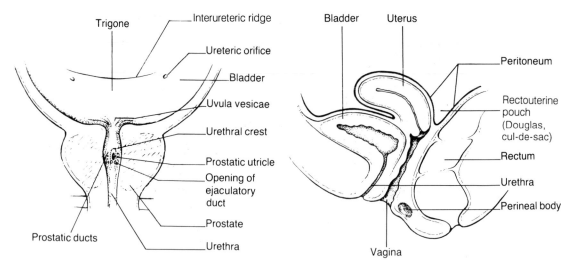

Fig. 33.6   The interior of the bladder base and the prostatic urethra.

Fig. 33.7   The female pelvis – sagittal section.

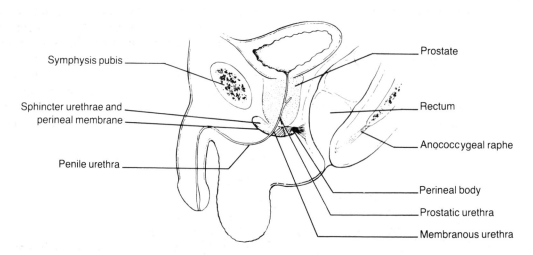

Fig. 33.8   The male pelvis – sagittal section.

Uterus

This muscular organ is shaped like a pear with its broad end, the body, superiorly, and its narrow end, the cervix, inferiorly. When unripe, the pear measures 7.5 cm (3 in) long, 5 cm (2 in) broad, and 2.5 cm (1 in) thick, but when ripe it enlarges sufficiently so as to reach the epigastrium – a remarkable achievement. Two tubes entering the upper part of the body demarcate its uppermost portion as the fundus. The cervix projects so far into a hollow muscular tube, the vagina, so as to leave a space between it and the vaginal wall, the fornix. The fornix is artificially divided into anterior, lateral and posterior fornices, of which the last is by far the deepest. The cavity of the body, slit-like in

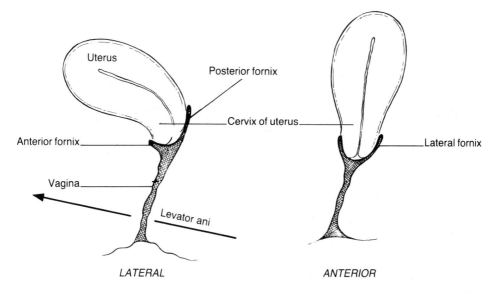

Fig. 33.9 **The uterus and vagina (lateral and anterior views).**

the virgin, leads to the canal of the cervix, which has an internal os above and an external os below.

Position of the uterus

Normally the uterus lies in an anteverted and somewhat anteflexed position. Anteversion means that the uterus and cervix are set almost at right angles to the axis of the succeeding canal, the vagina. Anteflexion implies that the uterus is also bent on itself. It is common for the uterus to be retroverted and sometimes even retroflexed. It is easy to determine the position by vaginal examination; a full bladder will push the uterus backward and mimic retroversion (empty the bladder before vaginal examination). The

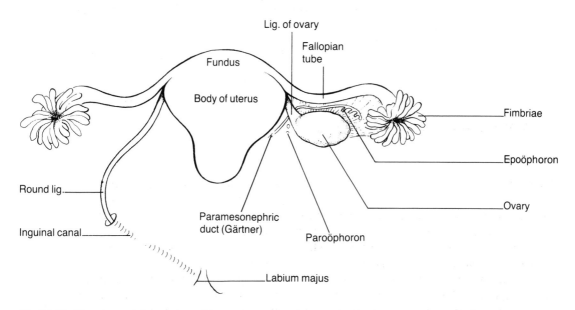

Fig. 33.10 **The uterus, tubes and ovaries.**

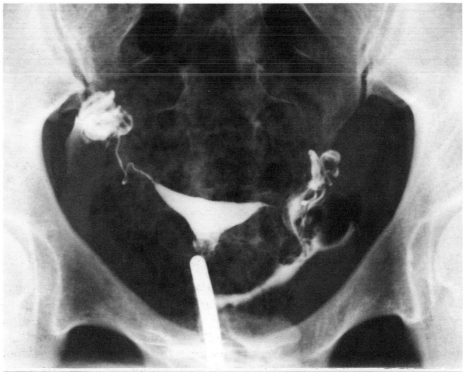

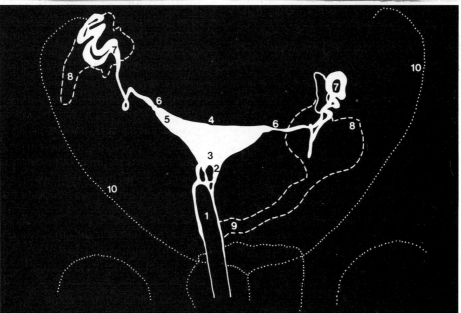

Fig. 33.11   Hysterosalpingogram. (From Weir and Abrahams 1978)

1 Vaginal cannula.
2 Cervix uteri.
3 Body of uterus filled with contrast medium.
4 Fundus of uterus.
5 Cornu.

6 Isthmus of uterine tube (fallopian tube).
7 Ampulla of uterine tube.
8 'Overspill' into peritoneal cavity.
9 'Overspill' in rectouterine pouch (Douglas).
10 Pelvic brim.

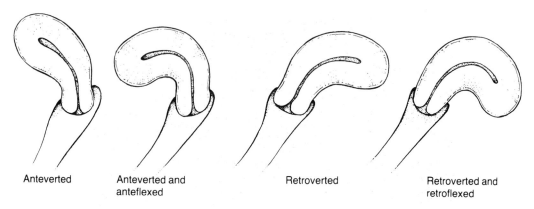

| Anteverted | Anteverted and anteflexed | Retroverted | Retroverted and retroflexed |

Fig. 33.12   Different positions of the uterus.

anteverted position is a happy position. If the uterus lies in a neutral position, in the same axis as the vagina, the intra-abdominal pressure will tend to push it down the vagina – in cases of severe prolapse it may exit from the vagina. The retroverted uterus may carry other problems, amongst them backache. However, 20 percent of normal happy women have a retroverted uterus.

Peritoneal attachments of uterus

From the superior surface of the bladder the peritoneum passes onto the anterior surface of the uterus, over the fundus, down the posterior wall and over the cervix and upper part of the posterior vaginal wall, before passing onto the rectum. Laterally, the anterior and posterior layers pass toward the walls of the pelvis.

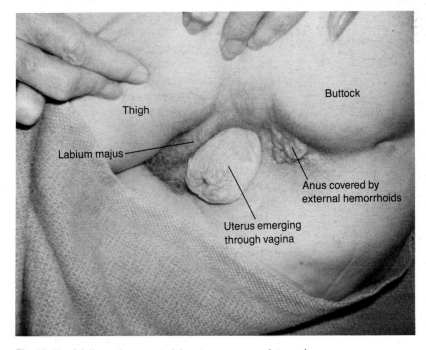

Fig. 33.13   A failure of support of the uterus – a complete prolapse.

Anteriorly, the uterus lies on the bladder, to which the cervix has a particularly close relationship; a disaster occurs when a tear of the cervix extends into the bladder during delivery, a vesicovaginal fistula. Posteriorly, the uterus is separated from the rectum by a large pouch, the rectouterine pouch of Douglas or cul-de-sac, which usually contains small and large intestine. The cervix can be palpated through the wall of the rectum, and during labor the degree of its dilatation can be ascertained via this route.

Supports of uterus

The position of the uterus in relationship to the vagina has been mentioned above.

The levator ani muscle or pelvic diaphragm is a most important support, and where deficient anteriorly for the passage of the vagina, it is strengthened by the fascia of the urogenital diaphragm (see Chapter 34). A major cause of prolapse in older women is the divarication of the levatores ani by multiparity.

The ligaments of the uterus are:

1 The round ligament of the uterus is a fibromuscular structure that extends subperitoneally from each cornu of the uterus to the deep inguinal ring. After traversing the inguinal canal it attaches to the fibrofatty tissue of the labium majus. It is derived from the gubernaculum ovarii, as is the ligament of the ovary.

2 The uterosacral ligaments are fibromuscular structures that pass posteriorly on either side of the rectum from the cervix to the middle of the sacrum. Note that the uterosacral ligaments pull the lower part of the cervix backward while the round ligaments pull the uterus forward. It is along these uterosacral ligaments that lymphatics from the cervix reach the sacral nodes.

3 The transverse ligaments (cardinal, Mackenrodt) are really thickenings of the visceral layer of pelvic fascia that run laterally from the cervix to the side wall of the pelvis, forming a hammock.

4 The broad ligament is not really a ligament, but two folds of the peritoneum that line the uterus, front and back, and which pass laterally to the side wall of the pelvis. It can be likened to a sheet that has covered a ski-jumper in flight. At the inferior extremity of the ligament, opposite the cervix, the layers pass backward and forward to line the pelvic floor. The ligament does not actually face forward and backward; because the uterus is anteverted, the ligament faces anteroinferiorly and posterosuperiorly.

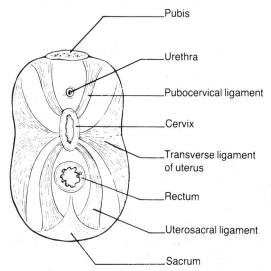

Fig. 33.14 The ligamentous supports of the uterus.

Pubis

Urethra

Pubocervical ligament

Cervix

Transverse ligament of uterus

Rectum

Uterosacral ligament

Sacrum

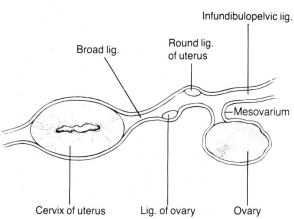

Fig. 33.15 The broad ligament and its relations.

Infundibulopelvic lig.

Broad lig.

Round lig. of uterus

Mesovarium

Cervix of uterus

Lig. of ovary

Ovary

In the past, when retroversion was thought to be the cause of many ills, a common operation was to pull the uterus forward by shortening the round ligaments (ventrosuspension); many gynecologists of yesteryear had their large cabin cruisers named Ventrosuspension. This was largely unsuccessful as the round ligament is generally too slack.

Uterine tube (salpinx, fallopian tube)

This tube runs laterally from the cornu of the uterus for about three-quarters of the length of the broad ligament (see Fig. 33.10). The tube has a funnel-shaped end, the infundibulum, that is surrounded by fimbriae (fingers); one of the fimbriae is particularly tenacious and attaches itself to the ovary, the ovarian fimbria. This encourages the discharging ovum to pass into the grasping tentacles of infundibulum. The ovum should find its initial passage easy, because the distal end of the tube is dilated into an ampulla; more medially the tube narrows to form the isthmus before it enters the muscular wall of the uterus as the intramural part. To encourage the ovum on its voyage, the surface epithelium of the tube is of a ciliated columnar type.

Ovary

Each ovary lies against the side wall of the pelvis attached to the posterior leaf of the broad ligament by a double fold of the peritoneum called the mesovarium (see Fig. 33.15). The peritoneum does not cover the surface of the gland but fuses with the covering germinal epithelium. The ovary is supported by a fibromuscular ligament, the ligament of the ovary, through which it attaches to the uterus just behind the attachment of the tube and round ligament. The ovarian and round ligaments were originally one continuous structure passing through the inguinal canal, the female gubernaculum. This gubernaculum encountered the cornu of the uterus, halting the progress of the ovary and confining it to a pelvic position.

As it lies on the side wall of the pelvis, the ovary is related through the peritoneum to the obturator internus and its overlying obturator nerve. This explains why ovarian pathology may cause pain radiating down to the knee. The ovary is somewhat surrounded by the uterine tube, as the latter curls downward and medially to come into relationship with the ovary at its fimbriated end. At this point, the peritoneum of the posterior leaf of the broad ligament has fused with the fimbriae except for a minute opening that permits entry of the ovum. It should be realized, therefore, that the tube actually opens into the peritoneal cavity and a communication exists between the external atmosphere and the peritoneal cavity via the vagina, uterus and tubes. This can be translated into practice. Pelvic peritonitis may occur under very poor hygienic conditions in younger girls (pneumococcal peritonitis following vulvovaginitis in children). It also allows a test to ascertain the patency of the tube; air or gas insufflated through the cervix can be heard bubbling into the peritoneal cavity. If you are more inquisitive, you can inject dye through the cervix and, by means of a laparoscope (an endoscope to view the peritoneal cavity), watch it emerge through the tube. This normal intraperitoneal passage can, very rarely, result in conception within the peritoneal cavity (abdominal pregnancy).

Contents of broad ligament

The ligament of the ovary passing backward to the ovary, the round ligament passing forward to the groin, and the uterine tube lie between the layers of the broad ligament. There are some important vestigial remnants of the mesonephric tubules and ducts (Gartner) which are present, some situated above the ovary, the epoophoron, and some lying medial to the ovary, the paroophoron. While sounding like inconsequential structures, they are not uncommonly the sites of cystic enlargements and are then called paraovarian cysts or cysts of the broad ligament. The other contents of the ligament, to be de-

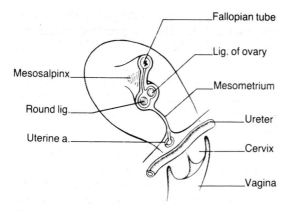

Fallopian tube

Lig. of ovary

Mesosalpinx

Mesometrium

Round lig.

Ureter

Uterine a.

Cervix

Vagina

Fig. 33.16   Sagittal section of the broad ligament.

scribed later, are the uterine and ovarian vessels, nerves and lymphatics (p. 470).

Rectum

After the pelvic colon loses its mesentery, it becomes the rectum, which is approximately 12.5 cm (5 in) long and ends at the tip of the coccyx where it becomes the anal canal (see Figs. 32.8 and 32.10). Rectus means straight, and the tube is certainly straight in the Tibetan flying fox; in man it is anything but straight, and is curved in several directions. It is curved in an anteroposterior direction

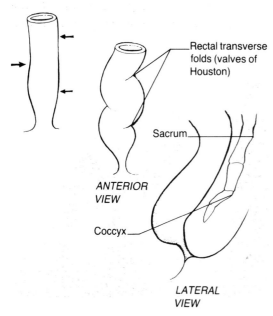

Rectal transverse folds (valves of Houston)

Sacrum

ANTERIOR VIEW

Coccyx

LATERAL VIEW

Fig. 33.17   The rectum – it is anything but straight!

to fit into the hollow of the sacrum, a most important point in preventing its descent or prolapse (see Fig. 32.8). The upper and lower thirds are straight, but the middle third curves to the left as the ampulla, which is the usual dilatation of any duct before it reaches its external opening. This left-sided curvature produces indentations into the rectum, of which two lie on the left with one, at a level between them, on the right. These projections contain not only mucous membrane (cf. the circular folds of the upper bowel, Chapter 27) but also the circular muscle of the gut. The projections are the rectal valves of Houston (New Yorkers think that these valves are aptly named after the Texan city). The exact function of these valves is uncertain, but it is thought that they can help distinguish between flatus and feces. The structure of these projecting valves means that a biopsy will include muscle; a specimen that requires this thickness taken in another part of the rectum may risk a perforation. Some small cancers can be missed behind these valves. At the tip of the coccyx, the rectum passes posteriorly as the anal canal; the angle at this anorectal junction is maintained by the sling of the puborectalis muscle (see Fig. 32.10). The three teniae coli, the hallmark of the large intestine, become a single layer as they spread fairly equally around the wall of the rectum (as in the appendix). The peritoneum of the rectum gradually passes onto the bladder anteriorly, so that the upper third of the rectum is covered by peritoneum in front and on either side, the second third of the rectum in front only, while the lower third lies below the peritoneal reflection.

Rectal examination

As a rectal examination is performed on nearly every patient, it should be recognized what the finger in the rectum feels in front, on the sides and posteriorly. As the finger is introduced, the firmly contracted sphincters are noteworthy; these are followed by a capacious cavity, the ampulla, around which is the thick shelf of the puborectalis.

Posterior structures

As the rectum lies in the hollow of the sacrum, this bone and the coccyx are easily palpable. Lesions of the bone itself or of the presacral space are easily appreciated. In a patient with coccydynia (pain in the region of the coccyx) movement of the coccyx is easily performed per rectum and may reproduce the patient's symptoms.

Anterior relations

MALE

The uppermost part of the finger can appreciate a diseased seminal vesical only, a normal vesical being difficult to recognize. An empty bladder is difficult to palpate, but one can easily feel a full bladder, one of the poorest ways of diagnosing retention of urine. The prostate is easily palpable, and the parts amenable to examination will be discussed later; prostatic pathology is readily diagnosed by rectal examination. The rectum slides easily over the posterior prostatic surface due to the intervening rectovesical fascia (Denonvilliers).

FEMALE

The cervix can be palpated, and this is one of the routes for monitoring the degree of cervical dilatation during childbirth; above this is the posterior fornix of the vagina. A retroverted uterus is easily palpable through the rectal wall.

In both sexes it is vital to recognize that the palpating finger can reach the cul-de-sac or pouch of Douglas, the lowermost part of the peritoneal cavity. Swellings can be appreciated, as can peritoneal tenderness. Striking examples of severe peritoneal tenderness are where the acid secretion from a perforated duodenal ulcer flows down into the pelvis to irritate the pelvic peritoneum, or a ruptured,

leaking ectopic pregnancy pours blood into the pouch. Laterally it is possible to feel the ovary in the female, especially if it is abnormal.

Just as the uterus has mechanisms which prevent it from being thrust into the open world, so has the rectum. This is held in place by:

1. The attachment of the puborectalis part of the levator ani to the wall of the rectum.
2. The visceral layer of pelvic fascia reflected from the upper surface of the levator ani (see Fig. 32.6).
3. The fatty tissue of the pelvis, and inferiorly the dense lobulated fat of the ischiorectal fossae.
4. The neurovascular supply of the rectum in the pelvis (referred to later).
5. Small slips of muscle (rectourethralis) that pass from the anterior rectal wall to the triangular ligament and perineal body and which require severance before the rectum can be removed.
6. The curve of the sacrum. Young children may develop rectal prolapse; as you wave your magic proctoscope over the protruding rectum over the next few years, the prolapse disappears. What really happens is that the straight sacrum of the youngster produces a straight tube which

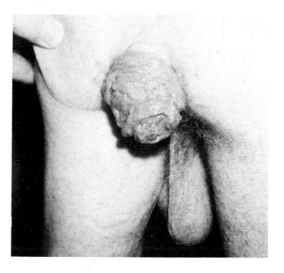

Fig. 33.18 A patient with a rectal prolapse.

is pushed out of the pelvis by the intra-abdominal pressure. As the sacral curve develops, the rectum occupies this curve and avoids the direct intra-abdominal pressure; in fact, the surgical treatment of prolapsed rectum is to fix the viscus into the sacral hollow.

Anal canal

The rectum becomes the anal canal below the tip of the coccyx, at which point it bends sharply backward. The canal, 3.75 cm (1.5 in) long, exits at the external anal orifice. The upper two-thirds of the canal is the lower end of the endodermal tube or cloaca, while the lower third is formed by the infolding of the skin at the proctodeum. These two meet in the vast majority of the population, but occasionally they do not and the result is an imperforate anus.

Structure

The double development can often be appreciated by noting the color change on procto-scopy. The bright intestinal red gives way to relative cutaneous pallor; Hilton called this junction the 'white line'. There are many differences in the anal canal above and below the white line, for it acts as an anatomic watershed.

1   Arterial supply. The blood supply above the line comes from the superior hemorrhoidal (inferior mesenteric), below the line from the inferior hemorrhoidal branch of the internal iliac. This anastomosis can circumvent a blocked inferior mesenteric and keep a pelvic colon alive.
2   Venous drainage. The white line is one of the sites of portosystemic anastomosis. Portal vein obstruction can theoretically lead to hemorrhoids; i.e. dilatation of the systemic veins. Theoretic, because hemorrhoids are so common that it is difficult to incriminate cirrhosis as the etiologic agent in cases of portal hypertension.
3   Nerve supply. Above the white line the canal is supplied by the autonomic system, below by the much smarter

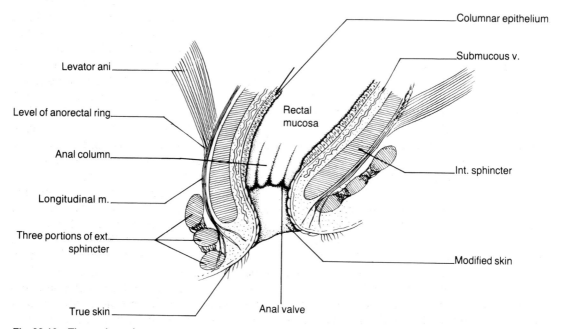

Fig. 33.19   The anal canal.

somatic inferior hemorrhoidal nerve. This means that injections above the line for hemorrhoids are painless, whereas an attempted injection below the line will require four strong men to control the patient. Lesions above the line, such as cancers, are unfortunately painless and lead to delay in diagnosis; below the line they are symptomatic.

4 Lymph drainage. The lymphatics following the arterial supply above the line pass upward to the inferior mesenteric group, whereas those below drain to the superficial inguinal nodes. Beware of the inguinal swelling whose source originates in the anal canal.

5 Tumors above the line arise from columnar epithelium (gut) and behave differently from squamous tumors below the line (skin).

Proctoscopy

When an instrument is passed into the anal canal, it must be passed anteriorly in the direction of the umbilicus, because the anal canal passes upward and forward before it inclines backward as the rectum. Pushing an instrument upward and backward may prevent this entry.

Mucous membrane

The anal canal is lined in its upper part by the pinkish columnar epithelium. Where the columnar epithelium meets the squamous epithelium, there is an area of transitional epithelium (not as in the bladder), where the columnar epithelium is heaped into several layers. The squamous epithelium is modified skin because it does not have skin appendages (e.g. hair, sweat glands, sebaceous glands). The hairs at the entrance of the anal canal, where the modified skin becomes true skin, prevents creepy crawly things from passing into the anal canal. An irregular ring, the pectinate or dentate ring, is visible about 1.8 cm (0.75 in) from the anal verge. This was once thought to represent the site of the tat-

tered remnants of the membrane that separated the two original portions of the anorectum; however, it is formed differently. Vertical columns, the anal columns of Morgagni, run vertically down to the dentate line. These columns contain submucous veins that are the terminal tributaries of the superior hemorrhoidal veins; in hemorrhoids these veins become enlarged. The columns themselves are joined together at their lower ends by folds of mucous membrane, the anal valves (Ball); the ring of valves is the dentate line. The valves joining the columns form small crypts, into which open the ducts of some eight to ten small mucous anal glands. Inflammation of these glands produces cryptitis, and sometimes the pus can be seen in the crypts.

Anal sphincters

Anal continence is a very precious thing; the external anal orifice is always closed and in a state of tonic contraction. There are two sphincters, one internal and one external. The internal sphincter, the thickened circular muscle of the lower bowel wall, is under control of the autonomic system, whereas the external sphincter is striated muscle with a somatic supply. The latter has been divided rather artificially into three portions, one lying on top of another, and perhaps one should call them superficial, intermediate and deep (profundus) to avoid confusion as they have been given bad names. Nevertheless, the external sphincter consists of a ring of muscle under voluntary control which surrounds the bowel wall; the deepest part of the muscle (the profundus) merges with the puborectalis. The internal sphincter occupies the upper two-thirds of the anal canal and the external sphincter the lower two-thirds; the middle third of the anal tube is therefore surrounded by both muscles. The puborectalis sling, the profundus part of the external sphincter and the circular muscle of the internal sphincter constitute the anorectal ring, a thick ridge of muscle that is easily palpable on rectal examination, and part of (the somatic portion) which the patient can contract on command. Complete continence depends on the integrity of this ring; minor degrees of incontinence may occur when small portions of the internal or external sphincter are damaged. From the patient's point of view, an intact puborectalis probably is, socially, the most important muscle in the body, as it maintains continence day and night by continuous contraction.

Prostate gland

This gland belongs in the domain of the male, and has no exact female homologue (see Figs. 33.1, 33.4, 33.6 and 33.8). It is always said to be shaped like a chestnut, sometimes a rather old chestnut. Like the cecum, it is a little broader than it is long, measuring about 3.75 cm (1.5 in) broad and 3.15 cm (1.25 in) long. The base is attached to the neck of the bladder, the muscle wall of the bladder being continuous with that of the prostate. The organs are separated merely by a small vein-containing groove. The apex of the gland rests on the upper surface of the urogenital diaphragm (triangular ligament). The posterior wall is palpable per rectum and the two inferolateral walls are supported by the levator ani, here often called the levator prostatae. The anterior border lies behind the symphysis pubis and is separated from it by a fat-filled space (cave of Retzius), but attached to it by puboprostatic ligaments. A sheath of fascia lying on the levator ani is reflected onto the prostate as a false capsule.

Lobes of prostate

While many of the lobes merge imperceptibly, a differentiation is important clinically. The lobes are demarcated by the structures passing through the gland. The urethra, the continuation of the bladder, runs through the prostate as the prostatic urethra. The ejaculatory ducts, formed by the union of the ducts of the seminal vesicles and the vasa,

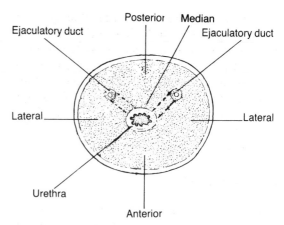

Fig. 33.20   The lobes of the prostate gland as seen in cross-section.

enter the posterior surface of the prostate to run through its substance and enter the prostatic urethra. This demarcates the following lobes: the anterior lobe lies in front of the urethra, and, as it contains very little glandular tissue, it is seldom the site of pathology. The lateral lobes lie on either side of the urethra and may become enlarged to compress and narrow the urethra or bulge into the bladder (see Fig. 30.8). The most important lobes, however, are the median and posterior lobes. The median lobe lies between the prostatic urethra anteriorly and the two ejaculatory ducts as they pass on each side toward the urethra. This triangular-shaped area projects into the urethra and bladder to raise bulges, the urethral crest and uvula vesicae respectively. The posterior lobe lies behind the ejaculatory ducts of the urethra and contains a large amount of glandular tissue.

It is important to realize that the posterior and lateral lobes are, and the median lobe is not, palpable per rectum; the median lobe lies anterior to the ejaculatory ducts and the posterior lobe. Therefore in patients with urinary symptoms the palpation of a normal-sized prostate does not exclude the prostate as the cause; it may only be possible to diagnose pathology of the median lobe by seeing the lobe bulging into the urethra or bladder by perurethral endoscopy or a radiologic study.

The intrusion into the bladder disrupts the internal sphincter and allows urine to trickle into the urethra to produce the earliest symptoms of prostatism – a call to action in the middle of the night, nocturia. The median lobe is the site par excellence of benign prostatic hypertrophy, and the posterior lobe that of carcinoma of the prostate, both conditions reserved for men of advancing years.

Prostatic urethra

This tube leads from the bladder to the tip of the penis where it exits as the external urethral orifice. The urethra is divided into several parts (Fig. 33.2). The first is the 3.75 cm (1.5 in) long prostatic urethra, which is the widest part of the tube. As it tunnels through the prostate it presents a midline posterior ridge, the urethral crest or verumontanum, which is produced by the bulging of the prostate. In the center of the crest is a small depression, the utricle, where the two müllerian (paramesonephric) ducts have been dashed in their attempts to form a vagina. This very small unimportant depression is the male vagina, and, despite its size, it has a long name – the uterus masculinis. On each side of the utricle, each common ejaculatory duct opens into the urethral crest. On each side of the urethral crest is a sulcus into which the dozen or so prostatic ducts open. The prostate is a fibromuscular gland, which produces secretions that drain into the prostatic urethra. One of the treatments of prostatic infective disease is to massage the prostate through the rectum to empty its contents into the urethra. This is only partly successful because the massage tends to merely move the prostate around and make it a bit dizzy.

Pelvic peritoneum

After discussing the pelvic organs it is time to trace the pelvic peritoneum (see Fig. 26.6). In the male, after lining the lower inner surface of the anterior abdominal wall, it is reflected at the level of the pubis onto and along the upper surface of the bladder. It reaches the seminal vesicles, the upper parts of which it covers, and dips down to form a small pouch after which it passes on to the rectum. It partially lines the upper two-thirds of rectum before completely surrounding the pelvic colon as a mesentery. The peritoneum extends laterally to cover the pelvic contents and becomes continuous with the abdominal peritoneum at the pelvic brim.

In the female, having passed across the dome of the bladder, it is reflected onto the anterior and posterior surfaces of the uterus, over the posterior fornix, and onto the rectum where it continues as in the male. Laterally the folds of peritoneum hang down over the uterine tubes as the layers of the broad ligament.

34 Perineum

Penis · Male urethra · Female urethra · Vagina · Perineal muscles · Female
perineum · Scrotum · Testis and epididymis · Perianal and ischiorectal spaces

The perineum is that part of the pelvic outlet which lies inferior to the pelvic diaphragm, the levator ani and coccygeus. It is artificially divided by a line between the anterior parts of the ischial tuberosities that demarcates anterior urogenital and posterior anal triangles, as seen in the lithotomy position (*lithos* = stone, *otomy* = cut; i.e. the position once used to remove bladder stones).

Anococcygeal body

The fibers of the iliococcygeus, part of the levator ani, interdigitate in front of the coccyx to form a raphe which extends from the tip of the coccyx to the anorectal junction (see Fig. 32.9). In this raphe is a fibromuscular mass of tissue, the anococcygeal body, into which several muscles insert, including a portion of the external sphincter.

Perineal body

The perineal body is a similar but larger body, of special significance in women, lying between the posterior end of the perineal membrane and the anal canal. Into this fibromuscular mass pass, amongst other perineal muscles, fibers of the external anal sphincter.

Urogenital diaphragm (perineal membrane, triangular ligament

The outlet to the pelvis is partially closed by a ligament that is composed of superior and inferior layers, through which the vagina and urethra pass from pelvis to perineum.

Attached on each side to the conjoint ischiopubic rami, the membrane has free anterior and posterior borders where the layers fuse; the latter border is much longer due to the divergence of the rami. The inferior fascial layer is commonly known as the perineal membrane, and is the strong platform that supports the external genitalia.

Perineal pouches

The superficial perineal pouch
(see Fig. 24.4)

The membranous layer of the subcutaneous tissue of the abdominal wall (Scarpa's fascia) passes inferiorly across the pubis and inguinal ligaments to be attached to the fascia lata of the upper thighs laterally, while in the midline it is reflected over the emerging penis and spermatic cords. The fascia then lines the scrotum, and ends posteriorly by attaching to the posterior border of the urogenital diaphragm.

Deep perineal pouch

This space lies between the two layers of the urogenital diaphragm.

Penis

The penis consists of three parts, a bulb and two crura. The bulb of the penis has an expanded posterior portion that is attached to the undersurface of the perineal membrane in the region of the perineal body. Passing forward, the bulb narrows, to become the corpus spongiosum of the penis; this expands at its distal extremity to form the conical glans. Each crus is attached at the angle between the insertion of the perineal membrane and the ischiopubic ramus. Each crus passes forward as a cylindrical corpus cavernosum to meet and adhere to its fellow and to the corpus spongiosum. These three structures then pass forward beyond the subpubic angle, where they form the penis. All three corpora are composed of erectile tissue, cavernous spaces into which arterioles open directly, and each is surrounded by an inelastic fibrous membrane, the tunica albuginea. This tunica is loosely surrounded by the continuation of the Colles' fascia, here called the fascia of the penis. The skin of the penis is hairless, and on the glans it is prolonged forward as a fold, the prepuce, which invests the glans and which is the structure removed in the operation of circumcision. For obvious reasons, the subcutaneous tissue of the penis is devoid of fat.

Male urethra (see Figs. 33.2, 33.8)

The prostatic urethra, 3.75 cm (1.5 in) long, continues through the perineal membrane as the membranous urethra. This is the most rigid part of the urethra, and certainly the most difficult through which to pass a catheter; apart from the narrowness, the surrounding sphincter urethrae muscle acts as an impediment. Having escaped from the perineal membrane, the urethra now enters the bulb of the penis and passes through the corpus spongiosum as the penile urethra, which averages 15 cm (6 in) in length. The urethra therefore consists of 15 cm (6 in) penile, 1.25 cm (0.5 in) membranous, and 3.75 cm (1.5 in) prostatic portions. At the beginning of the penile part of the urethra (bulb of penis), there is a dependent dilatation called the intrabulbar fossa; as urethral secretions and purulent collections tend to pool there, it is a common site for stricture formation. A further dilatation occurs in the glandular portion, the navicular fossa or fossa terminalis, after which the urethra passes

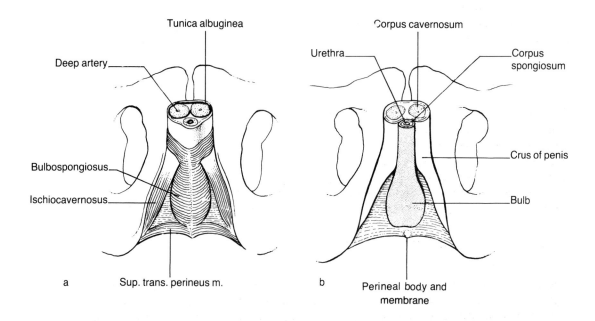

Fig. 34.1 (a) The muscles of the male superficial perineal pouch. (b) The constituent parts of the penis.

through the glans to open at the external urethral orifice. The external opening of the urethra (like most ducts) is its narrowest part; if a catheter can be passed through the external orifice, it can be passed into the bladder. The navicular fossa lies anteriorly, so it is usual to pass the catheter in a posterior direction and avoid getting lost in this pit. The urethra is lined by transitional epithelium throughout except for the terminal portion where it is lined by hard-wearing squamous epithelium. The mucous membrane contains numerous glands, persistent infection of which is difficult to cure.

The stream of urine is a fascinating and useless study. The shapes of the urethra differ. The prostatic urethra is horseshoe shaped because of the posterior bulging of the prostate; the membranous urethra is stellate shaped; the penile urethra is horizontal in section but the external meatus is a vertical slit. It is no wonder that the male passes his urine as a spiral stream. Apparently this delays the separation of the stream into droplets, otherwise many more men would miss their train while waiting for the last drop to fall (Newton's law).

Female urethra

The female urethra is only 3.75 cm (1.5 in) long, which makes catheterization easy but bladder infections much more common. From the bladder the urethra passes downward and forward to empty through a small orifice. Because of the close adherence of the urethra to the anterior vaginal wall it is possible to palpate stones or other foreign objects in the urethra via the vagina. More important is the proclivity for vaginal injuries to involve the urethra with a resultant vesicovaginal fistula.

Vagina

The vagina is an empty canal whose walls run downward and forward at right angles to the anteverted uterus to open in the perineum

(see Figs. 33.7, 33.9 and 33.12). The anterior wall is 7.5 cm (3 in) long, the posterior 10 cm (4 in). The upper end of the vagina is attached to the circumference of the cervix so that the latter projects into it to form a moat between the vagina and the cervix, the fornix. The large size of the posterior fornix and the reflection of pelvic peritoneum from the cervix onto the posterior fornix before it passes on to the rectum deserve re-emphasis; this means that an examining finger is separated from the pelvic peritoneum by the posterior vaginal fornix only. As this pouch is the lowest part of the peritoneal cavity, intraperitoneal collections tend to gravitate into it. A needle can also be passed through the posterior fornix to sample the peritoneal contents; a modern-day advance is the ability of the gynecologist to look into this cul-de-sac by passing an instrument (culdoscope) through the posterior fornix. In the bad old abortion days, the sleezy attendant, while attempting to pass a knitting needle through the cervix to stimulate the inside of the uterus, passed it through the posterior fornix into the peritoneal cavity and punctured the intestine, resulting in peritonitis.

Vaginal examination

The examining fingers will palpate the projection of the cervix 7.5–10 cm (3–4 in) from the vaginal introitus. If the cervix points backward, it means that the uterus is anteverted; if forward, the uterus is retroverted. Anteriorly the urethra cannot normally be palpated but an impacted foreign body may be appreciated. Superiorly a distended bladder is easily recognized. Posteriorly the perineal body may be palpable inferiorly and at a higher level large compressible fecal masses may be present in the rectum. Finally, at the uppermost extremity, the posterior fornix, covered by the parietal peritoneum of the cul-de-sac, can be reached; this last may contain abscesses, collections of blood (hematomata) or neoplastic masses. Laterally a normal ovary may be palpable through each lateral fornix, especially if it prolapses close to the fornix, but this is not always easy. More easily appreciated is a cyst or tumor of the ovary, or other pelvic masses. It will be pointed out later that the ureter comes within 6 mm (0.25 in) of the lateral fornix and those with very sensitive fingers may be able to feel a thickened ureter or a stone in the ureter. Finally, with the fingers in the vagina, and the other hand on the anterior abdominal wall (bimanual examination), the body of the uterus can be palpated between the two hands and its size estimated.

Perineal muscles

Muscles of the superficial perineal pouch

The superficial transverse perineal muscle has the same origin and insertion as the deep, and, like it, is of minimal significance.

The ischiocavernosus muscle arises from the ischial ramus and passes forward over the crus to be inserted into the commencement of the corpus cavernosum. It aids in the support of the erect organ.

In the male the bulbospongiosus muscle arises from the perineal body and the median raphe anterior to it. The posterior fibers pass forward and laterally over the bulb to be inserted into the perineal membrane. The anterior fibers insert into the dorsal surface of the penis. The old name was the compressor urethra or accelerator urinae; it empties the urethra.

Muscles of the deep perineal pouch

Lying between the two layers of the perineal membrane and surrounding the membranous urethra is the sphincter urethrae (external sphincter), Fig. 33.8. It is the muscle which enables you to stop what you are doing and run for the train. Most of its fibers attach to the pubic ramus, encircle the urethra and mingle with fibers of the other muscles at the perineal body. The small deep transverse perineal muscle arises from the ischiopubic ramus and runs medially to the perineal body.

All perineal muscles are supplied by branches of the pudendal nerve (S2, 3 and 4).

Female perineum

The passage of the vagina and urethra through the perineal membrane makes it a less substantial structure in the female. The two crura of the clitoris are attached at the angle between the insertion of the membrane and the ischiopubic ramus. The crura become the corpora cavernosa and form the body of the clitoris. The bulb of the vestibule, a mass of the erectile tissue, is split by, and lies on each side of, the vagina. The two masses join together anteriorly at the commissure of the bulbs, which is prolonged forward as the clitoris with its terminal expansion, the glans. The muscles are somewhat modified. The ischiocavernosus covers each crus as in the male, but the bulbospongiosus is split into two portions as it runs forward from the perineal body. Each half covers the bulb of the vestibule, and inserts near the clitoris. The two parts of this muscle act as a perineal vaginal sphincter; the levator ani is the sphincter at a pelvic level.

Bulbourethral glands

In the female, each gland (also called Bartholin's gland) is situated at the posterior aspect of the bulb of the vestibule under cover of the bulbospongiosus muscle; its long duct opens into the vagina at the junction of the anterior two-thirds and posterior one-third of the lateral wall. The gland commonly becomes infected when its duct blocks; a swelling situated at the junction of the anterior two-thirds and posterior one-third of the vaginal outlet is characteristic of Bartholin's gland enlargement. In the male, these glands, sometimes called Cowper's glands, are situated between the layers of the urogenital diaphragm on either side of the urethra,

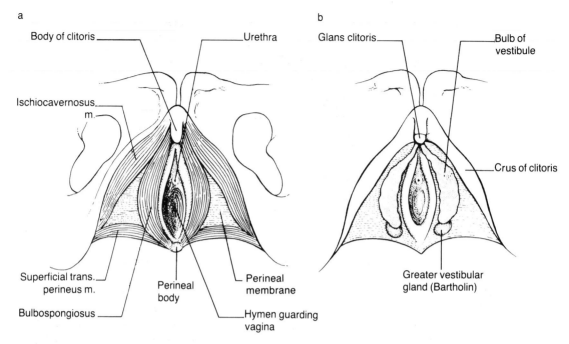

a

Body of clitoris

Urethra

Ischiocavernosus m.

Superficial trans. perineus m.

Perineal body

Perineal membrane

Bulbospongiosus

Hymen guarding vagina

b

Glans clitoris

Bulb of vestibule

Crus of clitoris

Greater vestibular gland (Bartholin)

Fig. 34.2   (a) The muscles of the female superficial perineal pouch. (b) The constituent parts of the clitoris.

embedded in the sphincter urethrae muscle. Their ducts pierce the perineal membrane to enter the bulbous urethra. Their functions are uncertain but they assist lubrication during intercourse. They become easily involved in the urethritis accompanying gonorrhea.

Scrotum

This pouch of skin between the upper thighs is divided by a septum into two compartments that house each testis. The subcutaneous layer is devoid of fat at the behest of the testes, as they dislike insulation and warmth. Instead, the fat is replaced by the dartos muscle that is supplied by sympathetic fibers. The dartos can wrinkle the skin; in low temperatures it can cause the pouch to shiver, which is one of the mechanisms for raising the temperature. It is probable that the spartan swimmers of the northern climes have wrinkled scrotums to rival that of Methuselah. Deep to the dartos are a succession of layers that have been carried down as the

testes passed through the abdominal wall on their way to their scrotal sac. From superficial to deep, these are the membranous layer of the superficial fascia (Colles' fascia), the fascia covering the external oblique (external spermatic fascia), that covering the internal oblique (cremasteric fascia containing a considerable number of cremasteric muscle fibers) and the prolongation of the transversalis fascia (the internal spermatic fascia). Inside these layers are the testes, each surrounded by its two layers of tunica vaginalis.

Testis and epididymis

Each testis is surrounded by the remains of the lower part of the two layers of the processus vaginalis, now called the tunica vaginalis. As in all mobile organs of the body, the visceral layer is firmly applied to the body of the testis, while the parietal layer lines the innermost layer of the scrotum. As with all two-layered serous membranes (cf. pleura, peritoneum, pericardium), disease processes may lead to an increased amount of the

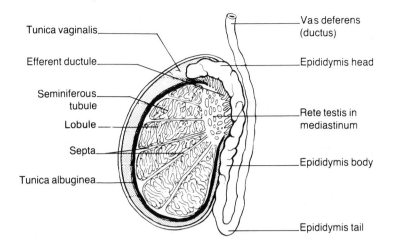

Tunica vaginalis

Efferent ductule

Seminiferous
tubule

Lobule

Septa

Tunica albuginea

Vas deferens
(ductus)

Epididymis head

Rete testis in
mediastinum

Epididymis body

Epididymis tail

Fig. 34.3   Testis and epididymis – lateral view. The testis is seen in section.

normal lubricating fluid. Such an accumulation of this fluid in the tunica vaginalis is called a hydrocele. The two layers become continuous toward the posterior margin of the gland, where a small gap allows the neurovascular and ductal structures to pass to and from the testis. The tunica vaginalis should serve as a reminder that the testis is not such a docile organ as it appears, and it is capable of moving from the base of the scrotum to the inguinal canal, especially in children. Inside the tunica, the testis is surrounded by a thick capsule, the tunica albuginea. At the posterior border of the gland the tunica thickens into the mediastinum testis from which fibrous septa pass forward to divide the testis into over 200 compartments. In each compartment lie two or more convoluted seminiferous tubules, each of which, when teased out, is 60 cm (2 ft) long. As they pass toward the mediastinum, they take a straighter course and are called the straight tubules. On reaching the mediastinum, they anastomose with one another in a network, the rete testis. From this network, a dozen or so efferent ductules emerge from the upper pole of the testis to enter the epididymis.

The epididymis is applied to the posterolateral surface of the testis, which identifies to which side the testis belongs. Each epididymis consists of a head, body and tail. The head, the enlarged upper portion, consists of a collection of lobules, each lobule being the coils of an efferent ductule. The body and tail consist of a duct, into which the lobules empty, which is so remarkably coiled on itself that, when uncoiled, it equals the length of the small intestine (6 m, 20 ft). At the lower pole of the testis, the duct in the tail becomes the ductus deferens. The relatively thin-walled ductus passes up on the medial side of the epididymis and, at the level of the upper pole of the testis, assumes its characteristic thick muscular wall and small lumen.

The testis and epididymis arise from embryologically differing structures, and are slightly separated by a small space or sinus, into which dips the tunica vaginalis. In an imperfectly developed testis, the separation

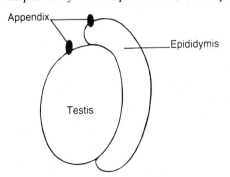

Appendix

Epididymis

Testis

Fig. 34.4   The testis and epididymis with their appendages.

of testis and epididymis is often excessive. Certain of the tubules of the testis and epididymis persist as vestigial structures. One of these structures is attached to the upper pole of the testis, the appendix of the testis; a smaller one attaches to the epididymis, the appendix of the epididymis. Although looking rather innocuous, the appendix of the testis is well known as a bothersome structure in young boys at and around puberty, when the appendix sometimes undergoes a twist (torsion of the appendix testis). This presents as acute testicular pain, which is often referred to the abdomen (the original site of the testis) and vomiting. In the younger child suffering from abdominal pain and vomiting, torsion of the testis itself may be mistaken for an acute abdominal condition, the more urgent abdominal symptoms overshadowing those from the testis. A torsion of the appendix epididymis is a rare event.

The structures running to and from the mediastinum of the testis also bear witness to the abdominal origin of the organ, and these structures, which comprise the spermatic cord, start and end in the abdomen. They have been mentioned in the discussion of the inguinal canal, but are worth a gentle reminder. Do not be caught out by the lymph vessels of the testis, which drain to the para-aortic glands at the origin of the testicular artery in the epigastrium. An epigastric mass in a young man may be the first evidence of a tumor of the testis. By the same token, testicular pain is often felt in the abdomen as the sympathetic supply emanates from the ninth and tenth thoracic segments. The testis develops in a similar way to the adjacent kidney. Its productive part (sperm-producing tubules), like the secreting part of the kidney (urine-producing glomeruli), is developed retroperitoneally and separately from the ductal structures, which later join to provide the excretory apparatus for transport of the spermatozoa or urine. Remnants of these ducts can be found in either system – the term 'urogenital apparatus' is therefore an apt description of the situation.

Perianal and ischiorectal spaces (ischioanal fossae)

These spaces lie between the anal canal and the side wall of the pelvis, the latter consisting of the ischial tuberosity and the obturator internus covering the bone above the tuberosity. As the longitudinal muscular layer of the large intestine descends, it becomes fibrous and gives off a lateral extension which divides this space into two areas, in which abscesses commonly occur.

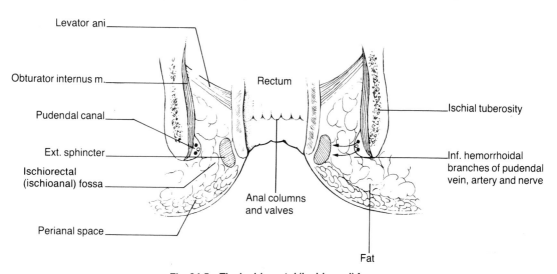

Fig. 34.5   The ischiorectal (ischioanal) fossae.

1 The lower or perianal area, situated beneath the skin, is composed of coarse fat broken up into small loculi and admirably suited to act as a cushion.

2 The upper part of the space is filled with delicate soft fat which is not for sitting. This fat fills a space and also allows distension of the gas- or stool-filled rectal ampulla. The space is inappropriately called the ischiorectal fossa, bounded by the fascia overlying the obturator internus laterally, the anal canal medially, the sacrotuberous ligament posteriorly and the urogenital diaphragm anteriorly; its more modern and sensible name is the ischioanal fossae.

Above the attachment of the sharp edge of the sacrotuberous ligament to the ischial tuberosity, lies the pudendal canal (Alcock). This canal lies in the extreme lateral portion of the ischioanal fossae and contains the neurovascular structures passing from the lesser sciatic foramen forward to the perineum.

35 Vessels, nerves and ureter in the pelvis

Sacral plexus · Coccygeal plexus · Pelvic autonomic system · Common iliac artery · Veins of the pelvis · Vertebral system of veins · Structures entering the pelvis from above · Lymphatic drainage of the viscera

Sacral plexus

This large plexus lies in front of the piriformis muscle and deep to the pelvic fascia. The plexus consists of S1, 2 and 3 anterior rami, with considerable assistance from L4 and 5 via the lumbosacral trunk, which runs across the ala of the sacrum, and a smaller assist from S4, which dillydallies between the sacral and coccygeal plexuses. Most of the distribution is to the lower limb and perineum, but there is some intrapelvic distribution. (See also Figs. 21.1 to 21.3.)

The intrapelvic distribution consists of branches of supply to the piriformis, levator ani and coccygeus muscles, and the sacral outflow of the autonomic parasympathetic system from S2, 3, (4).

The majority of nerves to the lower limb pass below the piriformis. Only one passes above the muscle, the superior gluteal nerve, which supplies the gluteus medius and minimus and the tensor fasciae latae. The sciatic nerve, the largest nerve of the body, derives its fibers from all the roots of the sacral plexus, namely L4, 5, S1, 2 and 3. It normally passes below the piriformis and crosses the ischium to lie on the small lateral rotator muscles of the thigh midway between the ischial tuberosity and the greater trochanter. The nerve derives its name from its relationship to the ischium, having been abbreviated from ischiatic to sciatic. It is really composed of two loosely bound portions, the tibial and common peroneal nerves. Occasionally the common peroneal nerve takes a violent dislike to its partner and

separates from it, piercing or passing above the piriformis. Running with and lying anterior or deep to the sciatic nerve is the nerve to quadratus femoris; lying on or posterior to the sciatic is the posterior cutaneous nerve of the thigh. The inferior gluteal nerve comes out close to the sciatic nerve and supplies the gluteus maximus, a muscle large enough to demand its own nerve supply. Two small branches passing out of the pelvis are the nerve to the external sphincter from S4, and a perforating cutaneous which pierces the sacrotuberous ligament to supply the skin of the lower buttock.

Two branches that lie well to the medial side of the sciatic nerve make a brief foray into the buttock, the pudendal nerve and the nerve for the obturator internus. These nerves emerge through the greater sciatic foramen and cross the ischial spine before leaving the buttock by passing through the lesser sciatic foramen. The pudendal nerve is the second largest branch of the sacral plexus (Fig. 35.1 to 35.3). After exiting the greater sciatic foramen and crossing the ischial spine, it passes into the perineum within the pudendal canal (Alcock), where it lies close to the ischial tuberosity and ramus in the lateral part of the ischioanal fossa (see Fig. 34.5). Soon after entering the canal it gives off an inferior rectal (hemorrhoidal) branch, which runs across the ischioanal fossa to supply the external sphincter, its overlying skin and the anal canal as far up as the pectinate line. It then divides into its two terminal branches, the dorsal nerve of penis/clitoris and the perineal nerve. The close relationship of the

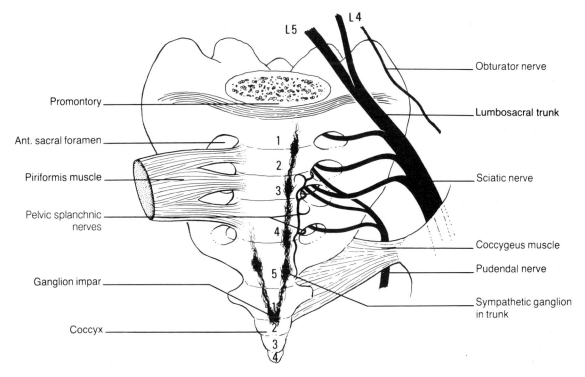

Fig. 35.1 The sacral plexus. The two major branches are shown.

pudendal nerve to the ischial spine is important. As this bony point is palpable on vaginal examination, the nerve can be blocked and the perineum anesthetized, which is often performed to reduce the discomfort of delivery (Fig. 35.4).

The perineal nerve supplies the skin and muscles of the superficial and deep perineal pouch, the bulbospongiosus, ischiocavernosus and superficial and deep transverse perineal muscles, the sphincter urethrae, the bulb of the penis/clitoris and the overlying skin.

The dorsal branch of penis/clitoris pierces the posterior margin of the urogenital diaphragm to lie between its two layers. The nerve then pierces the inferior layer of the diaphragm and, after supplying the corpus cavernosum, continues forward via the suspensory ligament of the penis/clitoris onto the dorsum of the organ to innervate the skin, prepuce and glans. Note that the dorsum of the penis refers to that organ in the erect position!

Coccygeal plexus

The coccygeal plexus is constituted by the lower part of S4 joining with S5 and C1 nerves. The plexus is distributed to the sacrococcygeal joint, coccyx and overlying skin.

Pelvic autonomic system

Sympathetic trunks

These trunks were left stranded behind the common iliac vessels. Picking them up again, they run down the anterior surface of the sacrum, medial to the foramina, until the first piece of the coccyx where they join to form the ganglion impar. There are three or four ganglia along the trunk. The fibers running in the sacral sympathetic trunk are mostly preganglionic; they join the sacral nerves after synapsing in the ganglia by postganglionic fibers, which are distributed to the lower

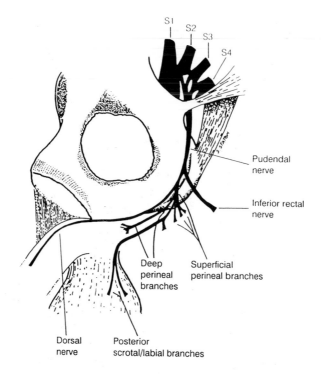

Fig. 35.2   The origin, course and branches of the pudendal nerve.

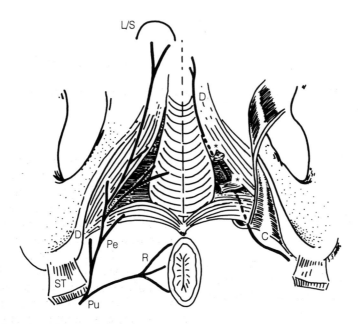

Fig. 35.3   The branches of the pudendal nerve in the anal and urogenital triangles. The dorsal nerve (D), discontinued on the reader's left, is shown on the right piercing the fascia covering the superficial perineal pouch (c), continuous with the membranous layer of superficial abdominal fascia, often called Colles' fascia in the perineum, then entering the deep perineal pouch by piercing the superficial layer of the deep perineal pouch, which has been partly removed(*) to demonstrate this. The nerve exits the deep pouch to reach the penis/clitoris. On the left the distribution of the perineal division (Pe) is shown. ST, sacrotuberous ligament; R, rectal branches; L/S, labial scrotal branches; Pu, pudendal.

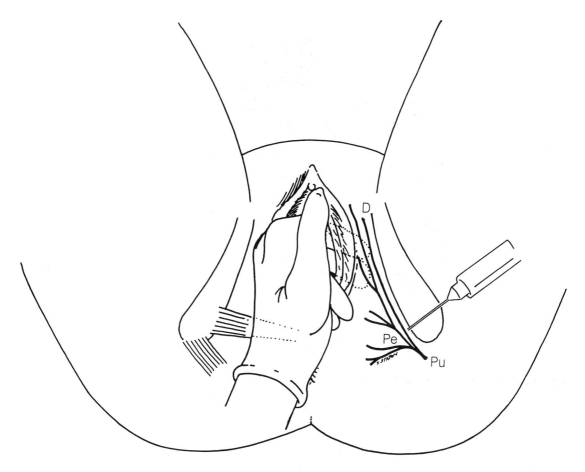

Fig. 35.4 Pudendal nerve block. The needle is directed toward the fingers as they palpate the ischial spine. Pu, pudendal nerve; Pe, perineal branch; D, dorsal branch.

limb and perineum. (See also Figs. 2.18–2.20 and 29.13).

Sacral parasympathetic nerves (pelvic splanchnics)

Sacral parasympathetic nerves arise from the S2, 3 (4) nerves – they are the parasympathetic contribution to the sacral autonomic plexuses.

Autonomic plexuses (see Fig. 29.13)

The aortic plexus continues inferiorly to the front of the fifth lumbar vertebra, where it becomes the superior hypogastric plexus or presacral nerve, the latter obviously incorrectly named. This plexus consists of fibers from the preaortic plexus with a contribution from the lower lumbar sympathetic chain. It divides into two narrow networks that pass down on each side of the rectum, the right and left hypogastric nerves, which are joined in front of the lower part of the sacrum by the pelvic parasympathetic nerves and branches from the sacral sympathetic ganglia to form the right and left inferior hypogastric or pelvic plexuses. These pelvic plexuses, which contain scattered small ganglia, are distributed by the visceral branches of the internal iliac arteries to the visceral organs. The autonomic plexuses are distributed as follows.

1 Prostatic plexus. A considerable part of the autonomic plexus is reserved for this musculoglandular organ. Offshoots of the plexus pass as cavernous nerves to the erectile tissue of the penis/clitoris; they are responsible for producing vasodilatation of the vessels of the corpora and thereby erection. This can occur reflexly, stimulation of dorsal skin of the penis/clitoris resulting in reflex vasodilatation. The reflex involves S2, 3, 4 spinal segments; the skin is supplied by the pudendal nerve (S2, 3, 4), and the vasodilatation is produced by the parasympathetic S2, 3, (4); hence the former name of nervi erigentes for these fibers.

2 Bladder. The parasympathetic supply is responsible for contraction of the muscular wall, inhibition of the internal sphincter and evacuation of urine. For this reason bladder retention is often treated by drugs which stimulate the parasympathetic supply, which may turn on the tap. The bladder and perineal skin are supplied by the same nerve roots, S2, 3 and 4, the former being autonomic and the latter somatic. Painful stimuli from the bladder base are often referred to the tip of the penis; the passage of a stone from ureter to bladder is often heralded by the onset of such pain. Operations on the perineal skin, such as hemorrhoidectomy, are often accompanied by reflex retention of urine. Furthermore, the rectum and bladder (the original cloaca) share the same nerve supply; many an elderly man can tell you that he cannot empty his bladder until his rectum has been disimpacted, and fecal impaction is a well known, but often forgotten, cause of retention of urine. There is a close and sophisticated relationship between the passage of wind and water; during defecation the urinary stream shuts off.

3 Ducti deferentia and seminal vesicles. The provision of the fluid for the sperm is a function of the sympathetic system (seminal vesicles, semen, sympathetic and sex (all the S's)). Most of the fluid in which the sperm frolic is provided by the prostate and seminal vesicles. The sympathetic outflow which activates these organs originates in the region of L1 spinal segment. Surgeons are very careful not to remove the L1 ganglion during excision of the sympathetic chain (sympathectomy) in young males; the prostate and seminal vesicles would be unable to contract and empty, and sterility would result. Note that it would not produce impotence; **potency** is a function of the sacral **para**sympathetic (pelvic splanchnics).

4 Left colon. The aortic plexuses send their autonomic supply to the colon via the vessels, except for the left side of the colon. The latter receives its supply via an offshoot of the pelvic plexuses, which ascends anterior to the aorta and joins the distribution of the inferior mesenteric artery to the descending and sigmoid colon.

Afferent sensation

The course of the nerves carrying afferent sensations, particularly pain, is not well known. At one time it was thought that the superior hypogastric sympathetic plexus carried the bulk of afferent sensation from the female genital organs; unlucky ladies who had their presacral plexuses severed in intractable cases of dysmenorrhea (painful periods) can testify to this misinformation. It appears, however, that most of the sensation from the bladder does pass through the sacral autonomic nerves.

Common iliac artery

The abdominal aorta bifurcates opposite the L4 vertebra into two common iliac arteries (see also Fig. 31.2). Each passes down across the L5 vertebra to reach the front of the sacroiliac joint, where it divides into its two

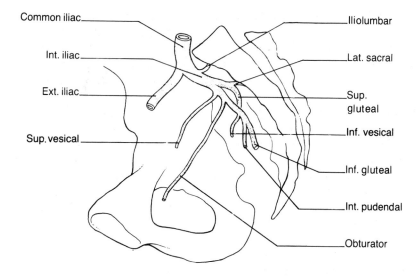

Fig. 35.5 The internal iliac artery and its branches.

terminal branches, the internal iliac for the intrapelvic structures, perineum and gluteal region, and the external iliac for the lower limb. The ureter crosses the bifurcation of the common iliac artery, a fact gratefully remembered by the surgeon in locating this sometimes elusive structure. The sympathetic trunk passes from abdomen to pelvis behind the artery. The inferior mesenteric vessels lying in the sigmoid mesentery cross the left common iliac vessels to become the superior rectal. The right artery crosses the left common iliac vein and sometimes causes partial obstruction. In a patient with varicose veins of the left leg only, which is a most unusual occurrence (most varicose veins are bilateral), suspect this obstruction and exclude it by a venogram; sometimes a deep groove across the left common iliac vein can be seen in the cadaver.

External iliac artery

This artery runs inferiorly on the psoas muscle to the midinguinal point where it becomes the femoral artery. On the right, the ileocecal part of the intestine may lie in front of the artery, and on the left the sigmoid

colon. The vas deferens/round ligament hooks around its inferior epigastric branch and crosses the lower end of the external iliac artery, as it makes its way toward the urethra/uterus. The inferior epigastric artery is a large branch that arises at the medial margin of the internal ring behind the fascia transversalis. It runs superiorly as the lateral border of Hesselbach's triangle to pierce the fascia at the arcuate line. Here it enters and runs in the rectus sheath where it anastomoses with the superior epigastric artery. Besides being a landmark around which the vas deferens/round ligament hooks as it leaves the company of the other cord structures to pass into the pelvis, it forms an important anastomotic channel between the external iliac and the subclavian artery – via the internal thoracic, a friendly connection in times of trouble. The deep circumflex iliac artery runs laterally to the anterior superior iliac spine, where it forms an important anastomosis with numerous other vessels. On its way, it gives off an ascending branch which runs in the neurovascular plane of the abdominal wall toward the costal margin and which is often severed during the operation of appendectomy.

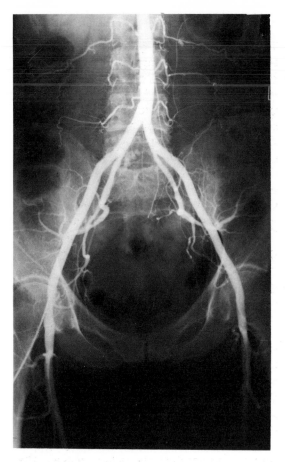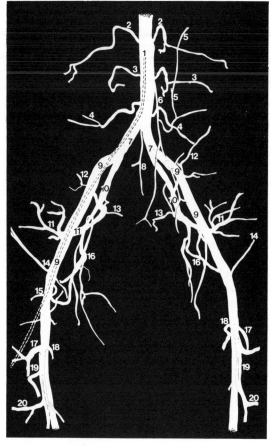

Fig. 35.6 Aortic and iliac arteriogram. (From Weir and Abrahams 1978)

1 Catheter tip in aorta via retrograde femoral route.
2 Third lumbar artery.
3 Fourth lumbar artery.
4 Fifth lumbar artery.
5 Left colic artery.
6 Inferior mesenteric artery.
7 Common iliac artery.
8 Median sacral artery.
9 External iliac artery.
10 Internal iliac artery.
11 Superior gluteal artery.
12 Iliolumbar artery.
13 Lateral sacral artery.
14 Deep circumflex iliac artery.
15 Superficial circumflex iliac artery.
16 Inferior gluteal artery.
17 Lateral circumflex femoral artery.
18 Medial circumflex femoral artery.
19 Profunda femoris artery.
20 Perforating artery.

Internal iliac artery

This large, but short, trunk arises opposite the sacroiliac joint, and passes backward in front of its vein to divide into branches for the lower limb, buttock and perineum, as well as segmental branches. Those branches that leave the pelvis pierce the fascia, except for the obturator vessel which passes through the small gap above the obturator membrane.

Branches for the lower limb

The two gluteal branches pass amongst the branches of the sacral plexus to reach and supply the gluteal region, the superior exiting above, and the inferior below, the piriformis. The obturator runs along the lateral wall of the pelvis in the reversed VAN position; the nerve lies at the highest level. Having passed through the small gap above the obturator foramen, it ends by supplying the adductor or medial aspect of the thigh. Near the pubis it gives off a small pubic branch, which anastomoses with a small descending branch of the inferior epigastric. At times the obturator artery arises from the inferior epigastric instead of the internal iliac; to reach the inferiorly placed obturator foramen it passes in close relationship to the femoral canal, where it is endangered in an operation for femoral hernia (Fig. 24.9).

Branches for the perineum

The internal pudendal artery, with the pudendal nerve, passes through the greater sciatic foramen into the buttock, which it quickly leaves through the lesser sciatic foramen. It supplies the musculocutaneous structures of the perineum, including the erectile tissues. Its terminal branch is the dorsal artery of the penis; on the dorsal surface of this organ the artery runs with, and lies medial to, the nerve on each side, the deep dorsal vein of the penis lying in the midline. The internal pudendal artery provides the increased flow necessary for erection, and patients with common/internal iliac artery occlusion may complain of impotence; this can sometimes be relieved by dis-obliterating the common and internal iliac arteries. Attempts to circumvent this problem have also been made by anastomosing one of the branches of the femoral artery to the internal pudendal. The results have been somewhat unpredictable, varying from success to either a complete flop or a persistent priapism (erection).

Visceral branches

The two vesical arteries run forward. The superior vesical supplies the dome and upper portion of the bladder, after which it becomes the medial umbilical ligament. The inferior vesical supplies the lower portions of the bladder and the structures lying in relationship to it, namely the prostate, ducti, seminal vesicles and lower ureters; the main artery to the ductus arises from this vessel. The middle rectal is often absent, especially in the female; if present, it supplies only the muscular layers of the rectum, most of its supply being reserved for the prostate. In the female these prostatic branches are replaced by the uterine and vaginal arteries.

The uterine artery is a large vessel that crosses the floor of the pelvis and passes beneath the base of the broad ligament, where it lies above the ureter in close relationship to the cervix. At this point it turns up between the two layers of the broad ligament to run in a corkscrew manner toward the fundus where it anastomoses with the tubal branch of the ovarian artery. The corkscrew allows unwinding of the artery during pregnancy. The relationship of the ureter to the artery (water under the bridge) is what makes the gynecologists sweat during hysterectomy. The relationship of both structures to the cervix is so close that the operator must be very careful not to include the ureter while ligating the uterine artery. In fact, in the bad old days, the surgeon avoided the risk by carrying out a supracervical hysterectomy, avoiding the cervix and its dangerous relations. The vaginal artery often comes from the uterine as it turns upward, but may be a separate branch of the internal iliac.

Segmental branches

The somatic segmental branches follow the line of the intercostal and lumbar arteries. With the aorta terminating at the fourth lumbar vertebra, the fifth lumbar artery comes off the internal iliac in the form of the iliolumbar. The lumbar branch passes poster-

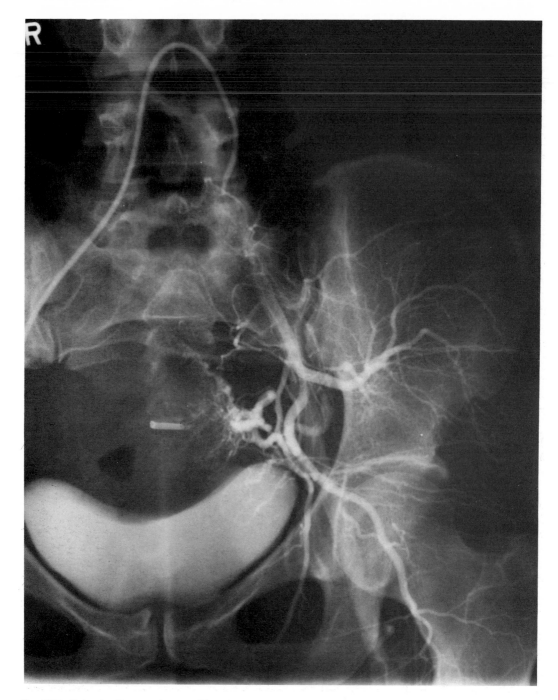

Fig. 35.7 Selective iliac arteriogram. (From Weir and Abrahams 1978)

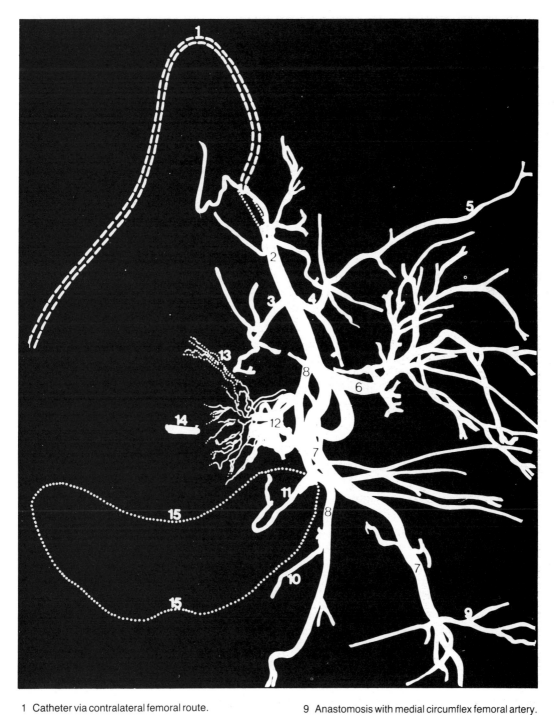

1 Catheter via contralateral femoral route.
2 Internal iliac artery.
3 Lateral sacral artery.
4 Iliolumbar artery.
5 Iliac branch of iliolumbar artery.
6 Superior gluteal artery.
7 Inferior gluteal artery.
8 Obturator artery.

9 Anastomosis with medial circumflex femoral artery.
10 Pubic branch of obturator artery.
11 Superior vesical arteries.
12 Uterine artery.
13 Uterine venous plexus.
14 Intrauterine contraceptive device (IUCD).
15 Bladder outlined by contrast medium.

iorly and supplies the lower reaches of the cauda equina on its way to supply lumbar muscles, while the iliac branch passes beneath the psoas to reach and supply the iliacus muscle, ending in the anastomosis around the anterior superior iliac spine. The lateral sacral artery runs inferiorly on the lateral side of the anterior sacral foramina. Branches enter the foramina to supply the sacral canal and emerge through the posterior foramina to supply the structures overlying the sacrum and coccyx.

Anastomosis around the anterior superior iliac spine

Many vessels meet here, including the deep circumflex iliac (external iliac), the superficial circumflex (femoral), the ascending branch of the lateral femoral circumflex (profunda), the superior gluteal and the iliac branch of the iliolumbar (internal iliac). By this means, blocks of common and external iliacs and common femoral arteries may be circumvented to supply the pelvis and lower extremity.

Veins of the pelvis

The pelvic viscera drain into venous plexuses which surround the walls of each organ. Prostatic, uterine and rectal plexuses are particularly noteworthy and drain into the internal iliac veins. The ovarian veins drain (as do the testicular veins) to the vena cava or right renal vein on the right and to the renal vein on the left. This last drainage embarrasses surgeons, who occlude the vena cava below the renal veins to prevent the dangerous condition of pelvic clots dislodging and migrating to the lungs (pulmonary emboli). Caval occlusion may control the right, but not the left ovarian vein; clots can still pass via the left renal vein to the vena cava above the iatrogenic occlusion. A higher occlusion of the vena cava above the entry of renals cannot be performed, as the kidneys would die. The left ovarian vein can be

separately ligated but it means a different and dangerous operative procedure. A fairly common condition is a varicocele, seen more often on the left, and due to varicosities of the pampiniform or cremasteric plexus. A tumor of the left kidney that grows along the left renal vein from the left kidney may block the spermatic vein and cause a varicocele. The latter, appearing for the first time in an older man, is very suggestive of a kidney neoplasm.

Vertebral system of veins (Batson's plexus) (Fig. 35.8)

The vertebral system of veins consists of an internal plexus in the vertebral canal (including the sacral canal), which drains through the intervertebral foramina to an external system, lying anterior and posterior to the vertebral column from sacrum to atlas. The external plexus drains into, from above down, the vertebral veins of the neck, the intercostal veins of the thorax, the lumbar veins of the abdomen, and the iliolumbar and lateral sacral veins of the pelvis. The visceral plexuses in the pelvis communicate with the unvalved vertebral plexus, providing a route for the spread of lesions from the pelvis through the vertebral column to the skull; prostatic and breast tumors commonly involve pelvis, spine and skull through these channels.

Structures entering the pelvis from above

Ureter

This structure was last seen crossing the bifurcation of the common iliac vessels; it then passes slightly laterally toward the ischial spine and then curves forward to enter the the upper outer angle of the trigone of the bladder. An opacity seen near the ischial spine can be considered to be a calculus in

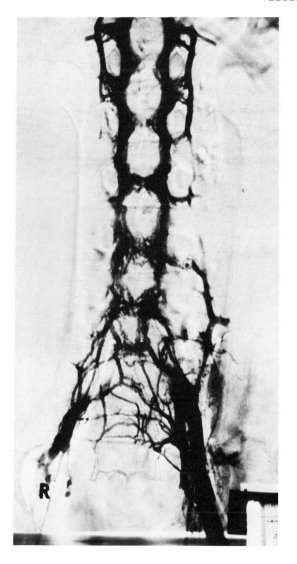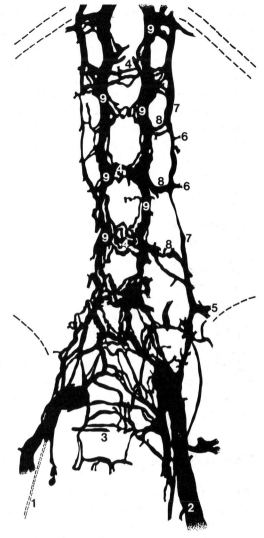

Fig. 35.8 Lumbar vertebral venogram. (From Weir and Abrahams 1978)

1 Catheter in common iliac vein.
2 External iliac vein.
3 Sacral venous plexus.

4 Basivertebral veins.
5 Iliolumbar vein.
6 Segmental lumbar veins.
7 Ascending lumbar vein.
8 Intervertebral veins.
9 Longitudinal vertebral venous plexi.

the ureter until proved otherwise. A very important structure crosses the ureter in each sex, the ductus deferens in the male and the uterine artery in the female. In the pelvis the ureter receives blood from the inferior vesical artery (see Fig. 35.5); abdominally it is supplied by the renal artery superiorly and by the gonadal vessel, which crosses it in the lumbar region, inferiorly.

Gonadal vessels

The ovarian artery is a branch of the aorta that passes inferiorly, crosses the ureter but lies behind the colic vessels, and enters the pelvis by running over the proximal part of the external iliac vessels (see Fig. 31.2). Passing down on the side wall of the pelvis, it enters the lateral extremity of the broad ligament, the infundibulopelvic ligament. In the broad ligament it anastomoses with the uterine artery, and supplies the ovary and tube.

The testicular artery runs the same course as the ovarian artery until it reaches the pelvic brim, which it skirts. It then passes superior to the external iliac artery and, as a component of the spermatic cord, enters the deep inguinal ring to reach the testis via the inguinal canal.

Lymphatic drainage of the viscera

The pelvic nodes are grouped near the pelvic brim in relationship to the external and common iliac arteries, and in the pelvis in relation to the internal iliac and its branches (see Figs. 2.22 and 2.23). The lymphatic drainage, important clinically, is detailed below.

1 Lymphatics from the bladder and male internal genital organs all pass through the internal iliac nodes.
2 Those from the ovary, like the testis, pass to the para-aortic nodes.
3 The lymphatics of the uterine tubes and fundus of the uterus drain with the ovarian lymphatics, but the body of the uterus drains to the internal and external nodes, and, by a single vessel along the round ligament, to the superficial inguinal nodes. The cervix drains mainly with the uterine arteries to the internal iliac nodes, but a few vessels pass to the external iliac and sacral nodes. As the uterine lymphatics travel so widely, this organ is sometimes called the 'lymphatic gypsy' of the pelvis.
4 The lower ends of the squamous-lined structures in the perineum, the vagina and rectum, drain to the superficial inguinal nodes (see also p. 349).

Review questions on the abdomen

(answers p. 529)

154 A bull gores the anterior abdominal wall of a matador with relish and revenge. A loop of intestine appears as the matador lifts his cape. The bull recognizes it as transverse colon because

a) it has appendices epiploicae attached to it
b) it has no circular folds
c) it has omentum attached to it
d) it has longitudinal bands running along it
e) all of the above are true

155 A 105-year-old yogurt-eating male has an inguinal hernia. Which of the following statements is applicable to this case?

a) At this age an indirect hernia is more common than a direct
b) The hernia is liable to descend to the lower scrotum
c) The neck of the hernial sac lies medial to the inferior epigastric artery
d) The hernia begins between the lateral margin of the conjoint tendon and the internal ring
e) Strangulation is likely to occur

156 A femoral hernia often causes strangulation. In releasing the trapped intestine, the surgeon must remember that the lateral boundary of the femoral canal is the

a) linea alba
b) inguinal ligament
c) pectineal ligament
d) femoral artery
e) femoral vein

157 You are looking into the peritoneal cavity through a laparoscope. You notice fibrous cords radiating from the umbilicus. They are remnants of which embryological structures?

a) The urachus
b) The umbilical vein
c) The umbilical arteries
d) None of these
e) All of these

158 In the adult, an important anastomosis between the arterial supply of the original embryonic midgut and that of the embryonic hindgut may save the day when the main aortic branches to each are occluded: it is mediated via the

a) inferior pancreaticoduodenal artery
b) superior rectal artery
c) ileocolic artery
d) marginal artery of Drummond
e) left gastric artery

159 The thoracic duct receives lymph from

a) all abdominal viscera
b) all thoracic viscera
c) all viscera on the left side of the chest
d) the entire body except the head, neck and arms
e) the entire body except the head, neck and right arm

160 A cancer cell from the stomach may reach a supraclavicular lymph node (Virchow–Troisier's gland) via the

a) diaphragmatic nodes
b) right mediastinal lymph duct
c) mesenteric nodes
d) cisterna chyli
e) perisplenic lymph nodes

161 In removing an aneurysm of the upper abdominal or lower thoracic aorta, the upper lumbar and lower intercostal arteries are often temporarily occluded.

The structure most liable to suffer deleterious effects is the

a) adrenal gland
b) spinal cord
c) descending colon
d) diaphragm
e) psoas muscle

162 It may be necessary to devascularize the lesser curvature of the stomach for excessive bleeding. This is best accomplished by ligating the

a) short gastric arteries
b) left gastric artery
c) splenic artery
d) left gastroepiploic artery
e) gastroduodenal artery

163 The surgeon often has to remove a stone from the lower end of the common bile duct. This structure

a) enters the hepatic flexure of the colon
b) enters the middle of the second portion of the duodenum
c) enters the first portion of the duodenum
d) passes in front of the duodenum
e) enters the biliary pouch of the pancreas

164 The portal vein

a) contains many valves
b) is constituted by the splenic and superior mesenteric veins
c) lies in front of the neck of the pancreas
d) lies behind the inferior vena cava
e) drains all the gastrointestinal tract and its paired glands

165 A tumor cell originating in the sigmoid colon may reach the liver via the blood stream. Which one of the following veins would be *first* involved in this pathway?

a) Inferior vena cava
b) External iliac vein
c) Splenic vein
d) Ascending lumbar vein
e) Left renal vein

166 The pancreas

a) has two duct systems, developmentally
b) develops from the stomach
c) crosses behind the aorta and inferior vena cava
d) receives vessels from only the celiac trunk
e) crosses the lower half of the left kidney

167 The direct inguinal hernia

a) passes through a triangle formed by the lateral edge of rectus abdominis, inferior epigastric artery and inguinal ligament
b) follows the obliterated processus vaginalis down toward the scrotum
c) passes through the deep inguinal ring
d) passes posterior (deep to) to the inguinal ligament
e) is usually of congenital origin

168 The liver is often torn in a crush injury of the upper abdomen and the right lobe is particularly affected. The right lobe may require removal. The true line between the two lobes, across which the ducts, hepatic arteries and portal tributaries of each lobe do not pass, is one joining the

a) falciform ligament to the ligamentum venosum
b) quadrate lobe to the caudate lobe
c) gallbladder fossa to the inferior vena cava
d) gallbladder fossa to the ligamentum venosum
e) none of the above

169 A liver edge palpable 7.5 cm (3 in) below the costal margin may still be normal

a) in infants
b) in patients with large elongated lungs (emphysema-enlarged alveoli)
c) in people with congenitally especially-well-developed right lobes (Riedel's lobe)
d) all of these are true
e) only *b* and *c* are true

170 A young surgeon carrying out his first hernia operation is confused by the contents of the inguinal canal. He/she should expect to encounter all these structures except

a) the dartos muscle
b) the vas deferens
c) a persistent processus vaginalis
d) the genital branch of the genitofemoral nerve
e) the cremasteric muscle/fascia

171 The inguinal canal is a potentially weak area. It is true to say that

a) the superficial ring is reinforced posteriorly by the conjoint tendon
b) the deep ring is reinforced anteriorly by the fibers of origin of the internal oblique

c) between the two rings the posterior wall consists of fascia transversalis only, the weakest part of the canal

d) the canal is smaller in the female

e) all are true

172 The sigmoid (pelvic) colon may be very long and mobile. However, the surgeon should be able to distinguish it from pelvic-lying small intestine even through a small incision because

a) it is dark brown in color

b) it has a complete longitudinal muscle coat

c) it has no appendices epiploicae

d) it has much greater diameter than the jejunum

e) it has tenia coli

173 The surgeon may wish to mobilize the duodenojejunal flexure. He would have to sever the

a) falciform ligament

b) ligament of Treitz

c) hepatoduodenal ligament

d) greater omentum

e) lesser omentum

174 Portal vein obstruction due to liver disease results in an abnormal flow into the following portosystemic anastomosis:

a) esophageal branches of the left gastric and azygos veins

b) superior and inferior mesenteric veins

c) superior and inferior epigastric veins

d) superior rectal and lumbar veins

e) all of the above

175 A Meckel's diverticulum may cause an acute abdominal illness. It

a) presents in 22 percent of the population

b) represents remains of the urachus

c) is usually situated midway along the small intestine

d) averages 1 cm (0.4 in) in length

e) may be lined by more than one type of mucosa

176 Acute appendicitis is the commonest acute abdominal emergency. With regard to the appendix,

a) it does not have any lymphatic tissue

b) the position at the base is relatively inconstant

c) it opens into the first 2.5 cm (1 in) of the ascending colon

d) it averages 15 cm (6 in) in length

e) the tip lies most commonly in the retrocecal position

177 The posterior wall of the stomach is invaded by a tumor of one of the structures of the 'stomach bed'. This includes

a) the body of the pancreas

b) the spleen

c) the left suprarenal gland

d) none of the above

e) all of the above

178 A posterior ulcer of the first part of the duodenum may penetrate the wall and erode an artery running behind it, causing a severe internal hemorrhage. This artery is likely to be the

a) hepatic

b) right gastric

c) gastroduodenal

d) left gastric

e) splenic

179 Contributions to the sympathetic nerve supply of abdominal and/or pelvic viscera come from the

a) vagus nerves

b) lesser splanchnic nerves

c) pelvic splanchnic nerves

d) spinal cord segments L3 and 4

e) spinal cord segments T3 and 4

180 The rectum

a) is directly posterior to the cervix uteri

b) is a perfectly straight tube, as the name indicates

c) has tenia coli continued on to it

d) has a complete mesentery

e) is supplied by the termination of the superior mesenteric artery

181 Acid from a leaking gastric ulcer irritates the peritoneum covering the undersurface of the central tendon of the diaphragm, and may produce pain referred to the

a) inner side of the left forearm

b) umbilical region

c) ear

d) shoulder

e) sternal region

182   It is important to palpate the abdominal aorta; this is possible until its bifurcation

a)   at the plane across the highest points of the iliac crest
b)   at the transpyloric plane
c)   7.5 cm (3 in) below the umbilicus
d)   at an epigastric level
e)   at a hypogastric level

183   Posterior to the kidney are

a)   the crus of the diaphragm
b)   the quadratus lumborum
c)   the pleura sometimes
d)   the subcostal nerve
e)   all the above

184   The ureters

a)   are about 25 cm (10 in) long
b)   lie 5 cm (2 in) lateral to the tips of the lumbar transverse processes
c)   cross the bifurcation of the aorta
d)   are supplied mainly by the lumbar arteries
e)   are physiologically constricted halfway along their course

185   Amongst the structures that hold the uterus in position, the most effective are the

a)   cardinal or lateral ligaments
b)   round ligaments of the uterus
c)   broad ligaments
d)   parametrium
e)   uterovesical fold of peritoneum

186   With respect to the ovary,

a)   the vein from the left usually passes to the inferior vena cava
b)   its artery also supplies the cervix
c)   it is smoother and larger than the testis
d)   it can cause irritation of the femoral nerve if inflamed
e)   it can cause irritation of the obturator nerve if inflamed

187   When palpating the prostate from the rectum in a case of enlargement of the prostate, it is possible to feel the

a)   anterior lobe
b)   verumontanum
c)   median lobe
d)   lateral lobe
e)   uterus masculinus

188   The nervi erigentes (pelvic splanchnics) from S2, 3 and 4 can, on stimulation,

a)   empty the seminal vesicles
b)   empty the vas
c)   cause ejaculation
d)   cause erection
e)   reduce peristalsis in the sigmoid colon

189   Tumors of the testes often spread by lymphatics. The lymphatics from the testes initially drain to the

a)   superficial inguinal nodes
b)   deep inguinal nodes
c)   common iliac nodes
d)   external iliac nodes
e)   para-aortic nodes

190   The pudendal nerve is often anesthetized in obstetric and gynecologic procedures. It

a)   passes out of both greater and lesser sciatic foramina
b)   supplies the internal anal sphincter
c)   supplies the voluntary urethral sphincter
d)   travels in a canal close to the medial boundary of the ischioanal fossa
e)   supplies the ovary

191   Between the layers of the urogenital diaphragm lie the

a)   ischiocavernosus muscle of the male
b)   prostatic urethra
c)   bulbourethral glands in the male
d)   bulbourethral glands in the female
e)   crus clitoris

192   The bladder

a)   can never occupy the abdominal cavity even when overdistended in the adult
b)   in a boy of 2 years lies in the abdominal cavity
c)   is completely surrounded by peritoneum
d)   is lined with cuboidal epithelium
e)   is supplied by the first and second lumbar nerves

193   Lymphatics of which of the following areas or structures *do not* drain to the inguinal nodes?

a)   Skin of the buttock
b)   Testis
c)   Labia majora
d)   Sole
e)   Anal triangle of perineum

194 By performing a rectal examination in the male it is possible to palpate

a) the lateral lobe of the prostate
b) a distended bladder
c) a diseased seminal vesicle
d) the presacral fascial space
e) all of the above

195 Which of the following pairs are correctly associated?

a) Glans penis – corpus cavernosum
b) Voluntary urethral sphincter – superficial perineal 'pouch'
c) External anal sphincter – a voluntary sphincter
d) Greater vestibular glands (of Bartholin) – deep perineal 'pouch'
e) Valves of Houston – sigmoid colon

196 The anal canal

a) is not a boundary of the ischiorectal fossa
b) is surrounded by an involuntary external sphincter
c) has a very important angle to the rectum maintained by the puborectalis muscle
d) drains its lymph only to the common iliac nodes
e) is insensitive to pain

197 The prostate gland

a) is homologous with the uterus
b) contains no muscular fibers
c) is surrounded by a venous plexus draining into the vertebral plexus
d) has a median lobe that is palpable per rectum
e) is traversed by the vas deferens

Part VIII   Physical Examination

36 The anatomic basis for the physical examination of the living patient

Physicians of the past practiced their art without the benefit of modern diagnostic aids and were forced to develop expert clinical acumen. They followed a dictum that is still true today – make the diagnosis from the history and confirm it by physical examination. With regard to the history, it is convenient to point out here that referred pain, which is pain appreciated in an area remote from the tissue/organ actually responsible for the pain, is usually based on recognized anatomic nervous pathways. The commoner situations are illustrated in Fig. 36.1. Contrarily, spinal nerves and some cranial nerves (V, IX) refer pain to their dermatomal distribution and may be affected by diseases of the spine, spinal cord and canal, and the peripheral nerves and their roots. Examples of peripheral nerve involvements, known as entrapment neuropathies, are given in Fig. 36.1 and Tables 36.1 and 36.2.

Except for the examination of a limited number of orifices, including both ends of the alimentary tract, vagina, nasal cavity and external auditory meati, the internal structures of the body are accessible to examination only through the skin. Therefore BEFORE EMBARKING ON THE PHYSICAL EXAMINATION OF A PATIENT, IT IS STRONGLY URGED THAT THE STUDENT/RESIDENT BE THOROUGHLY FAMILIAR WITH THE SURFACE REPRESENTATION OF STRUCTURES LYING DEEP TO THE SKIN SURFACE.

Surface anatomy of the head and neck

Head

The nasion is the central point of the naso-frontal suture and is situated at the base of the nose. The inion or external occipital protuberance is a fairly large projection which can be felt on the occipital bone immediately above the furrow of the neck. A line joining these two points over the vertex of the skull marks the longitudinal fissure of the brain, the upper attached margin of the falx cerebri and the superior sagittal venous sinus. In part, the line also corresponds to the metopic suture, when present, and to the sagittal suture between the parietal bones. The pterion is situated 4.3 cm (1.75 in) behind the zygomatic process of the frontal bone and 3.7 cm (1.5 in) above the upper border of the zygomatic process of the temporal bone. This point represents the stem of the lateral cerebral sulcus, the middle cerebral artery, and is very close to the site of election for exposing the middle meningeal artery. The anterior meningeal point, the site for opening the skull to expose the injured meningeal vessels, is situated 5 cm (2 in) behind the zygomatic process of the frontal bone and 5 cm (2 in) above the same process of the temporal bone; at this point the artery lies more superficially, as it is exiting the

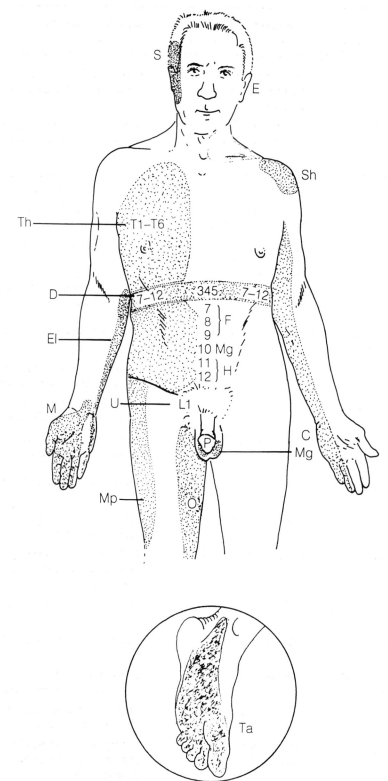

Fig. 36.1   Referred pain.

Site of pain	Nerves involved	Involved organ/tissue
Referred pain mediated by afferent sympathetic nerves and their spinal cord segments		
(F) Epigastrium	Spinal nerves T7–9 via sympathetic branches	Foregut and its derivatives, pancreas and biliary tract
(Mg) Periumbilicus	Spinal nerve T10 via sympathetic branches	Midgut—infra-ampullary duodenum to just short of the splenic flexure of colon, testes
(H) Hypogastrium	Spinal nerves T11, T12 via sympathetic branches	Hindgut—descending and pelvic colon, rectum
(U) Groin	Spinal nerve L1 via sympathetic branches	Ureter, especially with calculi (stones)
(C) Inner aspect of arm, forearm, hand	Spinal nerve T1 via sympathetic cardiac nerves	Heart, especially in angina pectoris
(P) Penis	Pudendal nerve via pelvic plexuses	Bladder base, especially with calculi (stones)
Referred pain mediated by afferent somatic nerves and appreciated as emanating from the roots or branch of the nerve		
(S) Side of face	V cranial nerve	Temporomandibular region, parotid gland, temporal artery
(E) External auditory meatus	V, X cranial nerve	Esophagus (hiatus hernia), oral cavity (especially tongue)
(Sh) Shoulder	Phrenic nerve (C3–5)	Pleural and peritoneal coverings of central diaphragm
Pain appreciated in the distal distribution of a somatic nerve due to a disturbance of the nerve at a proximal level		
(Th) Thorax	Intercostal nerves 1–6	Parietal pleura (upper), thoracic wall, thoracic spine, nerve roots
(D) Abdominal wall	Intercostal nerves 7–12	Parietal pleura (lower), pleural and peritoneal coverings of peripheral diaphragm
(O) Medial side of thigh and knee	Obturator nerve	Ovary, obturator hernia, hip joint
(El) Inner forearm and hand	Ulnar nerve	Ulnar nerve entrapment at elbow
(M) Lateral hand and 3½ fingers	Median nerve	Median nerve entrapment at wrist (carpal tunnel syndrome)
(Ta) Sole of foot	Posterior tibial nerve	Entrapment of posterior tibial nerve and its plantar branches
(Mp) Lateral thigh	Lateral cutaneous nerve of thigh	Entrapment of nerve at or above inguinal ligament (meralgia parasthetica)

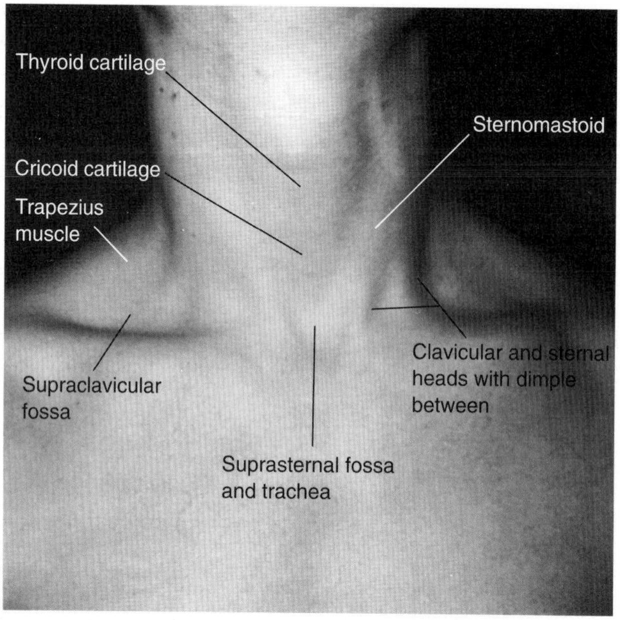
Thyroid cartilage
Sternomastoid
Cricoid cartilage
Trapezius muscle
Supraclavicular fossa
Clavicular and sternal heads with dimple between
Suprasternal fossa and trachea

Fig. 36.2 Surface anatomy of the head and neck.

osseous canal in which it is embedded inferiorly.

The tympanic antrum has been described in its relationship to MacEwen's suprameatal triangle (p. 185). The antrum usually lies about 1.25–1.90 cm (0.50–0.75 in) from the surface, except in the child where it is situated much more superficially.

Face

The parotid duct (Stensen) corresponds to the middle third of a line drawn from the tragus of the ear to a point midway between the ala of the nose and the red line of the upper lip. The duct can often be palpated on a masseter rendered tense by clenching the teeth. The latter act also defines the anterior border of the masseter muscle; anterior to this border lies the buccinator muscle. A line drawn parallel to and below the parotid duct from the lobule of the ear forward will mark the transparotid course of the facial nerve and the direction of its buccal branch.

At the junction of the inner and middle thirds of the supraorbital margin lies the supraorbital notch or foramen. A line drawn downward inferiorly from this point to the interval between the two lower premolars will connect the supraorbital, infraorbital and mental foramina. The infraorbital foramen lies about 8 mm (0.33 in) below the orbital margin, while the mental foramen lies midway between the alveolar and inferior borders of the mandible in the adult. As the teeth and gums disappear with age, the foramen begins to lie nearer the upper border of the mandible, and, when dentures replace teeth, pain and difficulty may occur as the nerves lie in more exposed positions.

Neck

The sternomastoid muscle is easily seen and felt. It separates an anterior triangle from a posterior triangle. Those who are geometrically inclined can go on subdividing each triangle into other triangles by using the muscles and tendons contained in these triangles.

Vessels

The course of the carotid artery has been mentioned previously (p. 225). The upper belly of the omohyoid muscle crosses the common carotid at the level of the cricoid cartilage, at which point the artery may be compressed against the anterior or carotid tubercle of the transverse process of the sixth cervical vertebra.

The three large anteriorly disposed branches of the external carotid arise in this area – the superior thyroid arises above the level of the upper border of the thyroid cartilage, the lingual opposite the great cornu of the hyoid bone and the facial a little above the cornu. These arteries may require ligation, and they can be easily found if these surface relationships are remembered. The facial artery becomes palpable where it curls around the inferior border of the mandible immediately anterior to the masseter muscle, about 3.75 cm (1.5 in) anterior to the angle of the mandible. The vessel then passes upward with an extremely tortuous course toward the inner canthus of the eye.

The superficial temporal artery can be palpated as it crosses the base of the zygomatic process of the temporal bone, immediately in front of the tragus of the ear, where it is accompanied by the auriculotemporal nerve.

The subclavian artery enters the neck opposite the sternoclavicular joint and can be represented by a curved line from this point to the midpoint of the clavicle; the highest point of the curve is a little less than 2.5 cm (1 in) above the clavicle.

The external jugular vein runs from its formation just behind the angle of the mandible to the middle of the clavicle. The internal jugular vein runs the same course as, but slightly external to, the internal and common carotid arteries.

Nerves

The vagus nerve passes down in, and the cervical sympathetic chain behind, the carotid sheath; they have the same surface relationship as the carotid artery and internal jugular vein. The positions of the three sympathetic ganglia have been mentioned previously (p. 241). The spinal accessory nerve crosses the transverse process of the atlas, which is a bony prominence felt immediately below and in front of the apex of the mastoid process. The nerve then enters the substance of the sternomastoid muscle at the junction of the upper and second quarters and emerges from the posterior border at the junction of the upper and middle thirds. Having left the muscle, it runs a downward and backward course to the anterior border of the trapezius; its course in the posterior triangle is somewhat variable, and this makes the nerve a somewhat elusive quarry during surgery. The site of emergence of the superficial nerves arising from the cervical plexus has been mentioned previously (p. 248).

The rima glottidis, the space between the true vocal cords, lies opposite the midpoint of the anterior border of the thyroid cartilage. In cases of asphyxiation due to either a foreign body impacting between the cords or swelling of the cords, it is necessary to make an opening into the airway below this point – this can be made through the cricothyroid membrane, which constitutes a laryngotomy, or below the cricoid cartilage, which constitutes a tracheostomy.

It is worth while revising the structures which lie in the middle line of the neck. From above downward lie the two mylohyoid muscles connecting the mandible to the body of the hyoid bone, the thyrohyoid membrane, the thyroid cartilage, the cricothyroid membrane, the cricoid cartilage, the first ring of the trachea, the isthmus of the thyroid gland, the trachea and the suprasternal notch flanked by the tendons of the sternomastoid muscles.

A finger on the cricoid cartilage will tell you that you are at the level of:

1 The sixth cervical vertebra.
2 The omohyoid as it crosses the common carotid artery.
3 The inferior thyroid artery as it swings medially posterior to the common carotid artery.

4 The middle cervical ganglion as it lies close to the inferior thyroid artery.

5 The end of the pharynx and the commencement of the esophagus.

6 The termination of the larynx and the beginning of the trachea.

7 The ligamentum arteriosum joins the left pulmonary artery to the aortic arch.

8 The azygos vein enters the superior vena cava.

Surface anatomy of the thorax

Extensive reference has been made to the surface anatomy of the thoracic wall, lungs, pleura, heart and major vessels. It is worth reiterating surface relationships of the large venous channels in the upper thorax, for they form the main route along which catheters are passed into the heart and beyond (e.g. pulmonary artery) to obtain important data. Furthermore, it is common practice to place catheters in the superior vena cava for feeding and other purposes, and this placement must be exact. The left brachiocephalic vein runs obliquely for 7.5 cm (3 in) behind the upper half of the manubrium from the left sternoclavicular joint to the lower border of the right first costal cartilage at its junction with the manubrium. Here it is joined by the right brachiocephalic vein, which runs obliquely downward immediately to the right of the manubrium. These two veins join to form the superior vena cava, which runs almost vertically downward, partly behind, but projecting beyond, the right margin of the manubrium. It enters the right atrium opposite the third costal cartilage, having been joined posteriorly by the azygos vein opposite the second costal cartilage.

For those who like to conserve mental energy, it is convenient to know that at the angle of Louis:

1 The superior mediastinum ends.

2 The second costal cartilage joins the manubrium.

3 The two pleural sacs meet.

4 The trachea bifurcates.

5 The thoracic duct crosses from right to left behind the esophagus.

6 The ascending aorta becomes the arch.

Surface anatomy of the vertebral column

The only parts of the vertebral column that can be visibly appreciated are the tips of some of the spinous processes. Except for the transverse processes of C1, 6 and 7, the spinous processes are also the only palpable parts of the vertebral column. Visible or palpable, the spinous processes lie at the bottom of a mid-line furrow which runs from the occiput to the sacrum.

Because of the forward convexity of the cervical spine, the only spinous process which normally is easily visible is that of C7, the vertebra prominens. The spinous processes of C1, 3, 4 and 5 are so small that they are only felt with difficulty; those of C2, 6 and 7 are usually palpable.

The spinous processes of the thoracic vertebrae are usually visible and palpable, especially in the stooping position. The spinous processes of the lumbar vertebrae are large and relatively easily palpable. The anterior convexity (lumbar lordosis), however, makes them difficult to see. That of the L4 lies at approximately the same level as the highest part of the crest of the ilium, an important relationship when a lumbar puncture is contemplated.

Surface anatomy of the abdomen

Reference has been made (p. 357) to vertical and horizontal planes, which divide the abdominal wall into nine areas. The value of the intercristal line in the performance of a lumbar puncture has also been mentioned (p. 294). The umbilicus is a rather unreliable landmark, as its position depends on many factors, including obesity and the abdominal protuberance of the aging; in the slender and fit, it lies on a level with the disc between the third and fourth lumbar vertebrae. The lineae semilunares are useful landmarks; each corresponds to the lateral border of the rectus muscle and extends from the pubic tubercle with a slight outward convexity to the tip of the ninth costal cartilage. A rather uncommon type of hernia (spigelian hernia) occurs along this line.

The inguinal canal, which measures 3.75 cm (1.5 in) in the adult, extends from the deep inguinal ring, situated 1.25 cm (0.5 in) above the midpoint of the inguinal ligament, to the superficial ring, whose base is the upper border of the pubic crest.

Gastrointestinal tract

The gastroesophageal junction of the stomach lies deeply behind the left seventh costal cartilage, about 1.25 cm (0.5 in) from the midline. The pyloric orifice lies opposite the first lumbar vertebra just to the right of the midline, and it is through this point that the transpyloric plane passes. While the lesser curvature is constant and can be represented by the curved line joining the two orifices described above, the greater curvature is subject to considerable variation. It is important, however, to remember that the fundus of the stomach corresponds to the level of the left dome of the diaphragm and therefore lies above and behind the apex of the heart.

The duodenum passes in a C-shaped manner from the pyloric orifice to reach the duodenojejunal flexure, which lies 2.5–3.75 cm (1–1.5 in) to the left of the second lumbar vertebra; the lowest point of the 'C' reaches the level of the third lumbar vertebra. It is convenient at this point to note that the head of the pancreas occupies the concavity of the duodenal loop, with the body crossing the midline at the level of the first and second lumbar vertebrae and the tail extending to the hilum of the spleen.

The mesenteric attachment to the small intestine extends from the duodenojejunal flexure to the junction of the right lateral vertical and transtubercular planes. This line curves obliquely downward, and a cyst occurring in the mesentery will tend to have the shape of a mass which runs obliquely downward and outward, and which can be moved at right angles to this line but not in the same plane.

With the ileocolic valve situated opposite the junction of the right lateral vertical and transtubercular planes, the cecum passes downward, forward and medially below the level of the transtubercular plane to occupy the right lower quadrant and the right half of the hypogastric region. It is important to note that although it is difficult to feel the cecum, it is commonly possible to compress it and elicit a gurgling noise. When the cecum becomes distended due to an obstruction more distally, its outline may be seen.

The appendix opens into the cecum just below and medial to the junction of the right lateral vertical and transtubercular plane at a point which does not coincide with the clinically important McBurney's point. The latter point, situated at the junction of the outer and middle thirds of a line joining the right anterior superior iliac spine to the umbilicus, is the usual site of maximal tenderness in patients with appendicitis. The marked tenderness usually limits the pressure applied by the palpating hand. Therefore, if one is able to exert sufficient pressure to gurgle the cecum with the examining hand, it is unlikely that there is an inflamed appendix lying beneath the hand. From the level of the transtubercular plane the ascending colon passes to the ninth right costal cartilage and then twists on itself at the hepatic flexure to become the transverse colon. This then extends to the left colic flexure which reaches the level of the left eighth costal cartilage. Between the two flexures, the position of the transverse colon is very variable and it may droop down as low as the false pelvis. The descending colon passes down to the iliac fossa, where it gains a mesentery and becomes the sigmoid colon. This convoluted loop of gut can often be palpated since it may contain solid stool and the mass may be mistaken for a tumor. An enema that empties the gut will dispel the 'tumor'. Furthermore, the mesentery of the pelvic colon may be long enough to allow the bowel to lie to the right of the midline. It is important to appreciate this, because inflammation of the sigmoid colon, as it lies in the right lower quadrant, often simulates an acute appendicitis. The sigmoid colon passes into the pelvis to end at the level of the third piece of the sacrum, which also indicates the commencement of the rectum.

The kidneys lie obliquely so that the upper poles are nearer the midline than the lower poles, with the left kidney at a higher level than the right. The hilum of the left kidney lies just medial to the anterior extremity of the ninth costal cartilage, with the hilum of the right kidney lying just below this level. As tenderness arising from a diseased kidney is usually appreciated posteriorly, it is worth knowing where these organs lie in relationship to the posterior abdominal wall. Each organ lies obliquely between the levels of the eleventh thoracic and third vertebral spinous processes, between 2.5 and 7.5 cm (1 and 3 in) from the middle of the back.

The abdominal aorta enters the abdomen at the level of the twelfth thoracic vertebra and terminates at the left side of the body of the fourth lumbar vertebra. It can be represented by a line joining a point about 2.5 cm (1 in) above the transpyloric plane and a point 1.25 cm (0.5 in) below and to the left of the umbilicus. The branches arise as follows: the celiac opposite the twelfth thoracic vertebra; the superior mesenteric opposite the level of the disc between the twelfth thoracic and first lumbar vertebra; the renals opposite the first lumbar vertebra; and the inferior mesenteric at the level of the third lumbar vertebra. The pulsation of the aorta, pushed anteriorly by the anterior convexity of the lumbar spine, can often be palpated, especially in the female. It is seldom that one can feel the common iliac artery; this vessel runs from the termination of the aorta to a point halfway between the anterior superior spine and the

pubic symphysis. The common iliac artery corresponds to the upper third of this line and the external iliac artery to the lower two-thirds. The inferior vena cava, formed at a lower level than the aorta, starts on the right side of the body of the fifth lumbar vertebra, about 2.5 cm (1 in) below and 1.25 cm (0.5 in) to the right of the umbilicus. The vein passes upward to the vena caval opening of the diaphragm at the level of the eighth thoracic vertebra.

The liver is a wedge-shaped structure with its apex lying at the same point as the apex of the heart, namely 8.25 cm (3.33 in) from the midline and in the fifth left intercostal space. Its lower edge then crosses the eighth left costal cartilage and the ninth right costal cartilage, after which it follows the lower limits of the right costal arch. It is important to recognize that its lower edge descends to halfway between the umbilicus and the xiphisternal joint, and therefore occupies the upper epigastrium. However, it is uncommon to be able to palpate the liver in this position as it lies behind the firm rectus sheaths and their contents. However, in lax abdominal walls, the organ can sometimes be palpated in this position. The upper limit of the liver runs to the right from the apex of the wedge on the left in a slightly upward direction to reach the sixth rib in the midaxillary line. It should be noted, therefore, that most of the liver lies behind the ribs and cartilages, a source of some comfort as these structures tend to prevent injury to this organ. For the causes of an easily palpable normal liver below the costal margin, see p. 394.

The gallbladder lies where the right linea semilunaris meets the costal margin — usually at the tip of the ninth costal cartilage.

The long axis of the spleen lies in the long axis of the ninth to eleventh ribs. The upper pole is only 3.75–5.0 cm (1.5–2 in) from the tenth thoracic spine, whereas the lower pole reaches as far anteriorly as the midaxillary line. It is not possible to feel a normal spleen, and it has to enlarge to at least twice its size before it can be palpated.

The transpyloric plane passes through the

following: the body of the first lumbar vertebra; the pylorus and body of the stomach; the ninth costal cartilages and therefore the hila of the kidney, the right just below and the left just above; the head, neck and body of the pancreas and therefore the splenic vein which lies posterior to the neck and body; the origin of the superior mesenteric artery; and the fundus of the gallbladder. The plane is slightly above the attachment of the transverse mesocolon and so separates the supracolic compartment with its contents, liver, spleen and fundus of the stomach, from the infracolic compartment with its contained small and large bowel.

Surface anatomy of the lower extremities

Hip region

The anterior and posterior creases of the upper thigh are related to the hip joint. In the neonatal child they should be equal and symmetrical; if this is not the case, suspect a congenital dislocation of the hip.

Posterolaterally the square-shaped greater trochanter can be easily palpated, especially in its lower part. Posteriorly the ischial tuberosity can be best appreciated sitting down even though it is covered by the lower part of the gluteus maximus muscle.

The central point of the saphenous opening, through which the long saphenous vein passes inward to join the femoral vein and through which a femoral hernia will emerge as it enlarges, can be consistently located 3.75 cm (1.5 in) below and 3.75 cm (1.5 in) lateral to the pubic tubercle. The femoral ring lies 1.25 cm (0.5 in) medial to the pulsating femoral artery immediately below the medial part of the inguinal ligament.

Thigh

With the thigh flexed, everted and slightly abducted, the femoral vessel corresponds to a line drawn from midway between the anterior superior iliac spine and the pubic symphysis to the adductor tubercle of the femur (see below). This line is very useful:

1 The upper 3.75 cm (1.5 in) corresponds to the common femoral artery.
2 The upper one-third indicates the common and superficial femoral vessels in the femoral triangle.
3 The upper two-thirds corresponds to the complete course of the femoral artery.
4 The middle third indicates the femoral artery in the subsartorial canal.

The long saphenous vein enters the thigh behind the medial femoral condyle and runs along a line from the adductor tubercle to the saphenous opening. This line represents a path well worn by vascular surgeons who harvest the vein at this level for arterial bypass surgery.

The femoral nerve runs a very short course before breaking up into its terminal branches. It can be located after it passes beneath the inguinal ligament, about halfway between the anterior superior iliac spine and the pubic tubercle. The nerve lies 1.25 cm (0.5 in) lateral to the femoral artery.

The sciatic nerve lies just to the medial side of the midpoint between the greater trochanter and the ischial tuberosity. From this point it can be traced superomedially in a gentle curve toward the greater sciatic foramen and inferiorly along a line to the middle of the popliteal fossa. At the junction of the middle and lower thirds of the thigh, the nerve divides into its two terminal branches. However, don't be surprised if, during dissection, the division is found to be as high as the greater sciatic foramen.

Knee region

The popliteal fossa is a lozenge-shaped hollow behind the knee joint. Above and laterally it is bounded by the easily palpable biceps tendon to whose medial side clings the common peroneal nerve. The tendon leads to the head of the fibula, which lies below, inferior and lateral to the lateral tuberosity of the tibia. Above and medially the popliteal space is bounded by the semitendinosus and semimembranosus tendons.

On the medial aspect of the knee, the gracilis and sartorius muscles form a well-marked prominence that is separated from the bulge of the vastus medialis by a depression. Deep in the depression the adductor magnus tendon lies under cover of the medial margin of the vastus medialis, and can be traced inferiorly to its insertion into the adductor tubercle. This tubercle also corresponds to the lower epiphyseal line of the femur.

Anteriorly the ligamentum patellae passes from the patella to the tibial tuberosity; both bony points are easily palpable.

The popliteal artery enters the upper angle of the popliteal space from the medial side and passes obliquely downward and outward to the midpoint of the space. From there it runs vertically downward to bifurcate opposite the level of the tibial tuberosity into its terminal branches. The pulsations are best felt in the lower part of the space in the flexed knee; the vessel is related more closely to the tibia than to the femur. The popliteal vein lies posterior to the artery and is joined by the short saphenous vein, which enters the popliteal fossa at its inferior angle.

The tibial nerve crosses the popliteal artery from without inward and continues into the leg. The common peroneal nerve hugs the biceps tendon before it leaves the popliteal space.

Leg

The medial surface and anterior border of the tibia are palpable throughout their length. Only the proximal and distal aspects of the fibula lie in a subcutaneous position, most of the shaft being embedded in muscles.

The bulk of the upper calf is formed by the gastrocnemius muscle, that of the lower calf by the soleus muscle. The tendo calcaneus is best appreciated in the lower third of the leg; it commences broadly and ends broadly, its middle portion, which lies about 3.75 cm (1.5 in) above its insertion, being its narrowest part.

The long saphenous vein enters the leg at a very constant point – the interval between the tendon of the tibialis anterior and the medial malleolus. The vein then ascends obliquely across the lower third of the tibia to reach the posteromedial border of the bone. Following this border it passes across the posterior aspect of the medial condyles of the tibia and femur to enter the thigh. In the leg the anteriorly lying saphenous nerve keeps the vein such close company that it is sometimes ligated or removed during operations on the vein.

The short saphenous vein lies behind the lateral malleolus. The vein ascends in a medial direction to reach the midline in the popliteal fossa where it usually ends by piercing the fascia and joining the popliteal vein. The sural nerve pierces the fascia a little below the middle of the leg and accompanies the vein into the foot. Those wishing to do a sural nerve biopsy must not be disappointed when they fail to find the nerve next to the vein in the upper half of the leg – the nerve is close to the vein but beneath the fascia.

The anterior tibial nerve lies deeply in the anterolateral compartment of the leg, accompanied by the anterior tibial artery. The posterior tibial nerve and blood vessels pass from the inferior angle of the popliteal space to a point about a thumb's breadth behind the medial malleolus. The superficial peroneal nerve can be seen, and often palpated, in the lower third of the leg when the ankle joint is plantar flexed. Passing over the neck of the fibula is the easily palpated common peroneal nerve.

Ankle and foot

The lateral malleolus lies 1.25 cm (1 in) below the medial malleolus and posterior to it; any alteration in this relationship implies a fracture with displacement (cf. lower ends of radius and ulna, Chapter 5). On the dorsum of the foot and anterior to the lateral malleolus is a soft mass, which is sometimes diagnosed as a fleshy tumor especially if it is well developed – it is the extensor digitorum brevis muscle. About 2.5 cm (1 in) below and in front of the medial malleolus the tuberosity of the navicular bone forms the most prominent bony point on the medial side of the foot; the tibialis posterior tendon can be traced to the site of its main insertion into this bony process. On the dorsal aspect of the navicular bone the tibialis anterior tendon can be palpated on its way to its insertion. Lying superficially in the angle between the tendons of the tibialis anterior and posterior (the 'tarsal snuffbox') is the origin of the long saphenous vein. Lying anterior to the navicular tuberosity are the medial cuneiform and first metatarsal bones; the tendon of the tibialis anterior can be palpated as it crosses the dorsal aspect of the navicular bone to insert into these bones.

The tendons around the ankle can be generally appreciated as they lie partly covered by retinacula and partly enclosed in synovial sheaths.

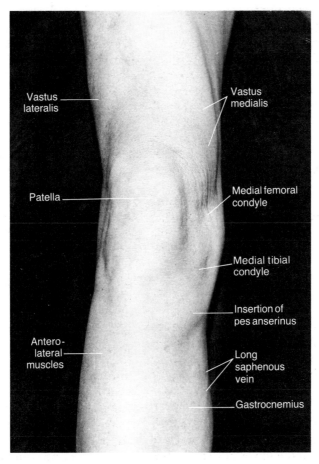

Fig. 36.3 Anterior view of the knee.

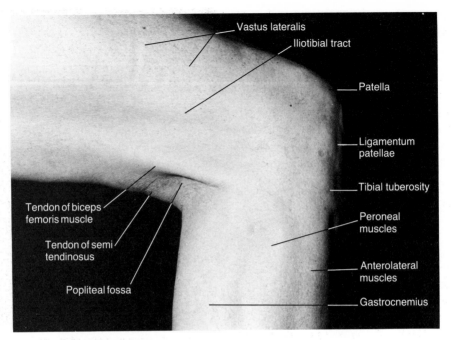

Fig. 36.4 The flexed knee and popliteal fossa.

Table 36.1 The clinical anatomy of nerve entrapments in the lower limb

Nerve	Site	Aetiology	Muscle affected	Sensory loss
Femoral	Lumbar plexus	Tumor or abscess in psoas major	Quadriceps + sartorius	Anteromedial thigh and leg towards big toe (saphenous nerve)
Lateral cutaneous nerve of thigh	Deep to inguinal ligament near anterosuperior iliac spine	Cyclist – meralgia paresthetica Weight loss	None	Outer thigh
Common peroneal	Neck of fibula	Pressure – e.g. plaster of Paris Trauma Tourniquet	Anterolateral compartments	Lateral shin + dorsum of foot
Deep peroneal	Upper shin	Edema from trauma Exercise	Anterior tibial compartment	Rarely, first web space
Posterior tibial	Tarsal tunnel	Trauma Rheumatoid arthritis	Intrinsic foot muscles	Sole

Surface anatomy of the upper extremities

Reference has been made throughout the preceding chapters on the upper extremity to the visible and palpable bony landmarks. In various sites, muscles, tendons and vessels can be delineated with varying degrees of clarity. Certain other specific relationships are described below.

Elbow region

With the forearm extended, the tip of the olecranon and the medial and lateral epicondyles lie in the same line. With the forearm flexed, the olecranon moves downward, and the lines joining the three bony points form a triangle. When the olecranon fractures and is pulled proximally by the triceps muscle, this triangle either disappears or its form is markedly altered (compare with normal elbow).

Wrist and hand

The anterior aspect of the wrist shows two conspicuous tendons, flexor carpi radialis and palmaris longus, and two tendons which

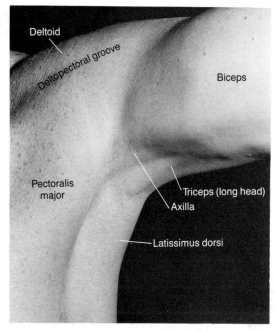

Fig. 36.6   Surface anatomy of the axilla.

are seen a little less clearly, flexor carpi ulnaris and flexor digitorum sublimis. The median nerve is located between the two conspicuous tendons, where it is easily anesthetized. In individuals not possessing a palmaris longus the nerve may be palpable.

Palmar and digital creases

The palmar aspect of the hand shows two transverse palmar creases. The distal of the two palmar creases is a useful landmark. It represents: (1) the proximal termination of the digital synovial sheaths of the second third and fourth fingers; (2) the site at which the digital nerves to the second, third and fourth web spaces bifurcate and where they may conveniently be anesthetized (please avoid the synovial sheaths); (3) the necks of the metacarpals of fingers two through five (see below); and (4) the distal end of a space between the medial flexor tendons and interossei, called the middle palmar space, sometimes a site of infection.

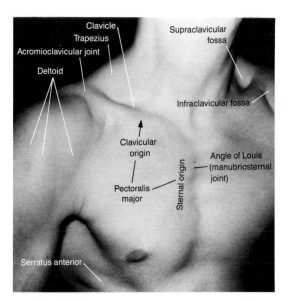

Fig. 36.5   Surface anatomy of the shoulder girdle.

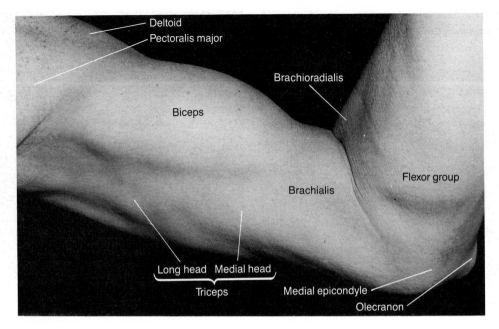

Fig. 36.7   Surface anatomy of the upper arm and elbow region.

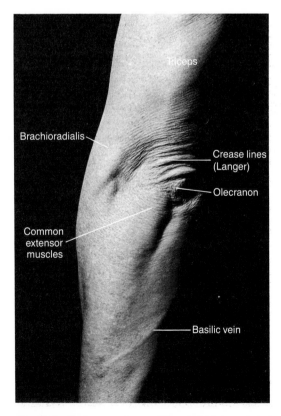

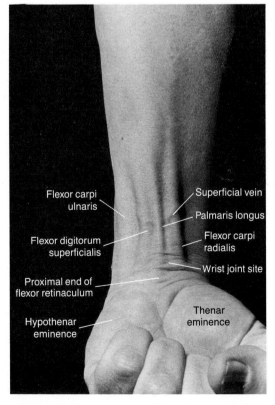

Fig. 36.8   Surface anatomy of the posterior aspect of the elbow.

Fig. 36.9   Surface anatomy of the anterior aspect of the wrist.

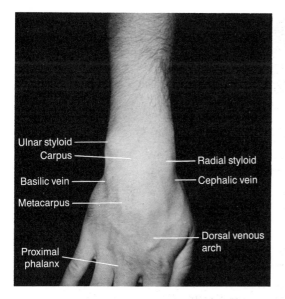

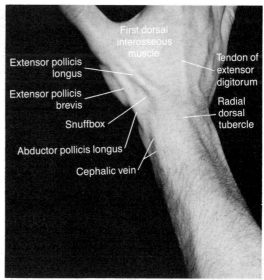

Fig. 36.10   Surface anatomy of the posterior aspect of the wrist.

The proximal palmar crease represents: (1) the distal end of a space between the lateral flexor tendons and interossei, called the thenar space – this is separated from the middle palmar space by a vertical septum, and may also be the site of an infective process; and (2) the distal end of the common synovial sheath for the four fingers.

Three digital creases lie in relation to the MCP and IP joints of the digits. These digital creases are rather sneaky. You would imagine that they represent the sites of the respective joints. Anyone who has tried to amputate a finger based on this belief will testify to the difficulty in finding the joint. In fact, the MCP joint lies 1.25 cm (0.5 in), the PIP joint 0.62 cm (0.25 in) and the DIP joint 0.31 cm (0.125 cm) proximal to the respective creases. In all three cases the prominence of the knuckle is formed by the head of the proximal bone of the joint.

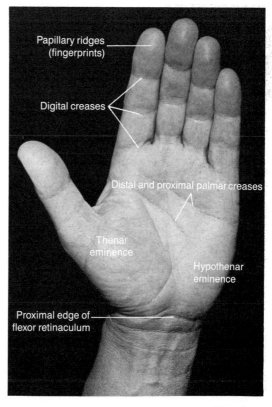

Fig. 36.11   The palmar creases of a normal hand.

Table 36.2   The clinical anatomy of nerve entrapments of the upper limb

Nerve	Site	Aetiology	Muscle principally affected	Sensory loss
Median	Carpal tunnel	Premenstrual Pregnancy Fracture Endocrine	Three thenar muscles: abductor pollicis brevis, flexor pollicis brevis, opponens pollicis	Lateral two and a half fingers + thumb
Anterior interosseous	Elbow	Fracture Edema	Flexor pollicis longus Flexor profundus of index + middle fingers	None
Ulnar	Medial epicondyle Between heads of flexor carpi ulnaris	Fracture Edema	Flexor carpi ulnaris Flexor profundus of ring + little fingers Hypothenar muscles Interossei	Medial one and a half, fingers
Ulnar deep branch	Carpus Hand	'Bottle opener's palsy' trauma Pressure fracture	Interossei only	None
Radial	Axilla	'Crutch palsy' or 'Saturday night palsy' Fracture of humerus	Elbow and wrist extensors	Back of hand, over first dorsal interosseous Brachioradialis reflex
T1	First rib	Cervical rib or fibrous band	All intrinsic hand muscles Superior tarsal (Horner's syndrome)	Medial border of arm
Posterior interosseous	Upper forearm within supinator	Elbow trauma Fibrous band	Wrist + finger extensors (Wrist drop)	None

Physical examination

The page numbers that appear in parentheses in the description of the physical examination that follows indicate page references to the underlying anatomic basis that pertains to the recognition of both normal and abnormal situations. The information is indicated on the actual pages by the emblem of a stethoscope in the margin adjacent to the relevant text or figure.

The student would be well advised to develop a method of physical examination that will lend itself to routine reproduction and therefore minimize errors of omission. A suggested method is to start at the top and finish at the bottom – top to toe. Because many physical signs are based on the anatomy of the area, such knowledge is the key to a thorough and fruitful clinical examination.

Head

Scalp

Examine the scalp for a swelling. A soft swelling of the scalp cannot be a lymph node (see p. 198) and is likely to be a sebaceous cyst (see p. 197).

A bony hard swelling is commonly a secondary tumor (see p. 480). In the case of an infant/child, palpate the fontanelles (see p. 194).

Eyes

Merely looking at the eyes will enable you to examine cranial nerves III, IV, VI (see p. 277) and part of VII (see p. 233), parasympathetic fibers of III (dilated pupil, see p. 277) and sympathetic fibers of spinal segment T1 (see p. 279).

Turning the lower eyelid inferiorly will expose the vascularized inner surface of the eyelid, allowing a rough estimate of the qualitative status of the blood (pallor indicates anemia); the sclera can also be inspected for yellow discoloration, usually an indication of jaundice.

The response to your questions will test cranial nerves VIII, hearing, as well as X and XII by virtue of normal speech mediated by intact laryngeal and tongue movements.

Ears

The external auditory meatus of the adult can only be examined by pulling the ear upward and backward (see p. 267). The external auditory meatus of the child is shorter (see p. 267).

The nerve supply of the internal meatus is V anteriorly and X posteriorly (see Fig. 17.1, p. 268). The tympanic membrane should be pearly gray.

Skull

Apply pressure from above downward to the three sinuses that are commonly the sites of inflammation; frontal, ethmoid and maxillary.

Nose

A deviated septum is common and mostly of little import. It is usually easy to see the inferior concha. Cranial nerve I should now be tested for the sense of smell – it is common for physicians to carry alcohol sponges in their pockets or there may be scented flowers in the room.

Oral cavity. Failure to examine the oral cavity is as serious a deficiency as omitting a rectal or vaginal examination.

Teeth and gums. The explanation for facial and neck pain as well as masses may be found in this area.

Tongue

Asking the patient to protrude the tongue will test the XII cranial nerve (see p. 283) and touching the protruded organ will evaluate the sensation and thereby nerve V (see Fig. 11.17, p. 209). Nerves V, IX, X and XI can be tested by observing the pharyngeal movements of the patient while swallowing on request.

The orifices of the parotid and submandibular ducts may be visualized if the history suggests such pathology.

Palatine tonsils. These should be inspected as they lie between the arches; they are often the site of pathology, especially in children.

Neck

Commence your examination by testing the flexion and extension of the cervical spine, except in patients with possible neck injuries. The presence of neck rigidity (spasm of the surrounding muscles) is very suggestive of irritation of the spinal canal by blood (subarachnoid hemorrhage) or pus (meningitis).

The commonest swelling of the neck is a lymph node/s (see p. 263). The second commonest swelling is the thyroid gland. Therefore swellings of these structures should first be excluded. Midline swellings of the neck (see p. 217). Initially the position of the normal midline structures of the neck should be checked starting with the trachea and esophagus. These midline swellings correlate to the anatomy of the part and are usually easily diagnosed.

Swellings other than those mentioned above are not common and are usually related to specific anatomical stuctures, such as the sternomastoid muscle (see p. 219), neurovascular structures in the carotid sheath (see p. 223) and the subclavian/axillary complex (see p. 244).

Next, palpate and listen to the carotid arteries (see p. 223).

Finally, test the last untested cranial nerve, XI. Ask the patient to shrug the shoulders, thereby putting the trapezius muscles into action.

Thorax

The time-honored sequence of INSPECTION, PALPATION, PERCUSSION and AUSCULTATION finds its most useful application in the thorax and abdomen. Start with the anterior chest.

Anterior chest

Inspection

Deviations from normal should be sought. At this time, observe the respiratory movements; rate and depth and area of diminished movement.

Palpation (breast)

The breast should be examined in the order of nipple, areola and parenchyma. It is a good routine to examine the axillary lymph nodes at this stage (see p. 68). Next, the respiratory movements are gauged bimanually. The apex beat is then visualized and palpated (see p. 153).

Percussion

The air-filled lungs are then percussed; they should reach the levels described on p. 141. Failure to do so usually means replacement of the air by fluid of some type, such as fluid, exudate, blood or resorption of the air. Anteriorly, only the upper lobes and the middle lobe and its counterpart, the lingula, are available for examination (see p. 141).

Auscultation

The breath sounds should be similar on the two sides, except for the clearer sounds at the right apex (see p. 163). and the encroachment by the heart on the left lung.

Cardiac sounds

This examination should cover the rate, rhythm, volume of the heart beat and the sounds of the closure of the four valves. With regard to the latter, the sounds in the living are best heard at areas that differ from their anatomic cadaveric location (see Fig. 9.10, p. 158).

Posterior chest

The patient then leans forward crossing the upper extremities in front of the chest. By thus displacing the scapulae anteriorly, the chest wall is exposed for examination.The same inspection, palpation, percussion and auscultation are now performed.

REMEMBER that, posteriorly, the lung and pleura reach their lowest levels and that effusions of fluid, a very common condition, gravitate to the lowest regions of the pleural sacs, where they are most easily detectable. Posteriorly, the lower lobes of the lung are dominant, with only a small portion of the upper lobe being represented above the third spinous process.

WHILE THE PATIENT IS IN THIS POSITION, examine the following: the vertebral column from neck to sacrum looking for deformities and points of tenderness over the spinous processes, the only accessible parts of the vertebrae; the erector spinae muscle group, commonly the site of pain and tenderness from muscle injuries; the area between the twelfth ribs and the iliac crests with special reference to tenderness in the loins indicating possible kidney disease.

Abdomen

The abdomen is examined with the patient lying comfortably on his/her back. Remember that the muscular abdominal wall is the only protection for the underlying viscera and a worthwhile abdominal examination is possible only with the muscles in a state of maximal relaxation. This is achieved by approximating the origins and insertions of the muscles by flexion of the hip joints and the spine; the latter is achieved by modestly flexing the neck and dorsal spine by the use of a pillow under the head or by raising the head of the examination table/bed.

Inspection

Look at the abdomen for movement of the abdominal wall, the presence of abnormalities, such as dilated vessels (see p. 364, Fig. 24.11), masses, distension and skin discoloration. The umbilicus may be the site of congenital abnormalities, dilated veins, metastatic deposits or a hernia.

Palpation

Structures that may normally be palpable: liver (see p. 394), right kidney (see p. 423), aortic pulsation (see p. 432), bladder of young child (see p. 446).

Structures that are normally impalpable. Gastrointestinal tract. The tract is normally impalpable; it needs significant pathologic changes before it is possible to appreciate a portion of the afflicted tract. It must be pointed out that a colon loaded with stool may be palpable; however, it is the stool that is palpable and not the colonic wall. Adrenals, pancreas, normal ovaries and tubes are also impalpable.

Solid Organs. The palpation of solid organs is often greatly assisted by using the descending diaphragm to push the upper abdominal organs against the examining hand; with the hand correctly placed, instruct the patient to take a deep breath.

Spleen. This organ is normally impalpable. When palpable below the left costal margin, the spleen is probably enlarged to twice its normal size. The organ is best felt by having the patient lie on the right side and taking a deep breath to allow the diaphragm to move it inferiorly.

Liver

The liver is mostly hidden behind the ribs and cartilages. The lower edge of the right lobe is, under favorable circumstances (a soft, lean abdominal wall) palpable at a finger breadth below the costal margin (see p. 393). Under similar conditions, the organ may be palpable in the epigastrium as it passes to the left hypochondrium.

Kidneys

The kidneys lie deeply against the posterior abdominal wall and the diaphragm. They may only be palpated through the anterior abdominal wall. In the normal patient, it may be possible to palpate only the lower pole of the right kidney on inspiration; the left kidney lies at a higher level and is not palpable (see p. 423). In people who have lost a considerable amount of weight, both kidneys may be palpable; the perirenal fat prevents this visceroptosis (see p. 426).

Intraperitoneal versus extraperitoneal mass. A palpable mass can lie inside or outside the peritoneal cavity. If the abdominal wall is rendered tense by asking the patient to attempt a sit-up and the mass can still be felt, the mass lies in the abdominal wall. To appreciate intraperitoneal masses, a relaxed abdominal wall is mandatory.

Tenderness and rigidity are the hallmarks of acute pathology in the abdominal wall or intraperitoneally. Remember that the visceral peritoneum evokes modest abdominal signs (autonomic nerve supply), whereas the parietal nerves supplying the parietal peritoneum and abdominal wall evoke a marked response (somatic nerve supply) (see p. 419). A useful tip in examining the acute abdomen is to start the examination in an area where the patient is not feeling pain and tenderness and rigidity are absent; positive findings elsewhere are then especially significant. This is important in a nervous patient whose whole abdomen may be tense and tender rendering the examination and the deductions based thereon far less significant. The area where pathology appears least is usually the left hypochrondrium, a good place to start the physical examination of the acute abdomen.

Lower abdomen

As the examination proceeds from the upper to the lower abdomen, the pelvic organs will now come under consideration.

Pelvic organs that may normally be palpated abdominally include: the bladder of the child who has not reached the age of eight, at about which time the organ commences its descent into the enlarging pelvis (see p. 446); the distended bladder of the adult (see p. 446); the pregnant uterus; the normal uterus if displaced anteriorly on bimanual examination. Bimanual examination, carried out with the bladder emptied (see p. 447), assesses the uterus as well as the ovaries and cul-de-sac (see p. 451 and p. 454).

All the above are dull to percussion as apposed to a distended tympanitic intestinal loop.

Inguinal region or groin

The examination now reaches the groin. The presence or absence of a hernia must now be established. This is usually accomplished by urging the patient to raise the intra-abdominal pressure by coughing. If a hernia is present, its anatomical type should be established, namely indirect or direct, inguinal or femoral. If the hernia passes from groin to scrotum, the path of the processus vaginalis, the hernia is indirect. If the hernia is confined to the groin, determine if it is above or below the inguinal ligament. If below the ligament, it is a femoral hernia, and if above the ligament, it is either a small indirect hernia where only the upper part of the processus vaginalis remains, or a direct hernia. The latter two are differentiated by blocking the internal ring (see p. 366) with two fingers. If a bulge cannot be appreciated on coughing, but can be when the fingers are removed an indirect hernia is present; the processus vaginalis has been temporarily closed by the fingers. If a bulge appears medial to your fingers, it is coming through Hesselbach's triangle (see p. 362, Fig. 24.9) and is a direct hernia.

Proceeding inferiorly, the male genital organs are examined next, the female organs being examined at a later stage. The scrotum is inspected and palpated, after which the testis, epididymis and spermatic cord are examined. The epididymis lies posterolateral to the testis (rarely this is reversed and is known as anteversion of the testis). Finally, the penis is examined.

Pelvis and perineum

At this stage the patient is turned into the left lateral position and the perineum is inspected.

Anus

This is examined with particular attention to the common abnormalities at the anal verge.

Anorectal examination. This very important examination is carried out with a well-lubricated glove. Much valuable information may be gained (see p. 456 and p. 457).

Vaginal examination

The vaginal examination precedes the rectal examination. The lateral position is convenient to inspect the cervix with the aid of a speculum. The patient is then turned to the supine position.

Lower extremities

Starting at the groin, this area is palpated for inguinal nodes (see p. 349). The limb is then inspected and palpated from thigh to toes. The peripheral pulses are then sought (see p. 344). The joints are then put through their range of motion, starting with the hips and passing inferiorly to the knees, ankles and tarsal joints. The two sides are compared with respect to range of motion and presence of pain.

The peripheral nerves are then assessed with respect to sensory and motor supply, as well as reflexes.

Upper extremities

These ere examined in the same way as the lower extremities. Finally, the blood pressure is taken on both arms. If raised, the patient is allowed to rest and the pressure estimation is repeated.

37 The anatomy of punctures

Venous punctures · Arterial punctures · Joint punctures · Nervous system punctures · Body cavity and organ punctures

The insertion of a needle into a vessel, space, body cavity or organ for diagnostic or therapeutic reasons is a common procedure.

Venous punctures

These are the most common punctures performed. For this reason, a few golden rules are in order.

All punctures may result in thrombosis, the incidence varying with the size and site of the vein, the size and bevel of the needle, the fluid instilled and the length of time the needle/cannula remains *in situ*. The upper extremity, especially the dorsum of the hand and the forearm, is the site of election because of the rarity of complications. The veins of the lower limb, on the other hand, because of the relative difficulty experienced in returning blood against gravity, are equipped with a large number of valves which are commonly the site of thrombotic incidents. In both extremities, the veins should be punctured in a disto–proximal sequence; a proximal puncture that leads to thrombosis may render the distal veins useless.

It is permissible to obtain blood from, but not instil solutions into, the superficial antecubital veins. This is because bending of the elbow is a common cause of needle or canula dislodgement; and splinting to avoid such movement is uncomfortable and often ineffective. Damage to the underlying median nerve and brachial artery is also possible as well as inadvertent injection of materials into the artery with serious sequela.

Veins of the upper limb

Cephalic vein (see p. 81 and Figs. 4.4 and 4.5)

Basilic vein (see p. 82 and Fig. 4.4)

Subclavian vein

The subclavian vein, together with the internal jugular vein, is often punctured for access to the large intrathoracic and cardiopulmonary structures. The course of this procedure is demonstrated in Figure 37.1. The subclavian vein reaches the upper border of the clavicle just medial to the midpoint, and medial to this it lies behind the clavicle before it passes posterior to the sternoclavicular joint.

Technique

- Trendelenberg position (head of bed 10–25° lower than the foot);
- position head to the contralateral side;
- place a rolled towel beneath and between the scapulae;
- insert needle below the midpoint of the clavicle and pass medially towards the suprasternal notch;
- stay parallel to the coronal plane; and
- stay close to the inferior border of the clavicle.

DANGERS: Puncture of pleura and/or subclavian artery.

Internal jugular vein (see also p. 225 and Figs. 12.10 and 12.11)

It is preferable to choose the vein on the right

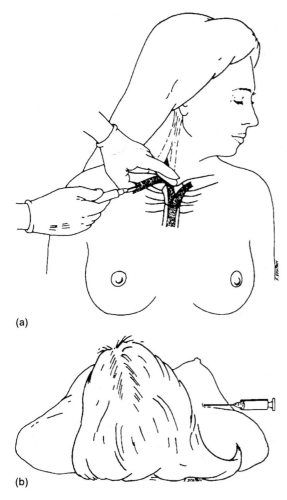

(a)

(b)

Fig. 37.1 Subclavian vein puncture. (a) The needle is passed deep into the clavicle with the thumb and middle finger indicating the landmarks. (b) The importance of keeping parallel to the coronal plane ('floor') is demonstrated.

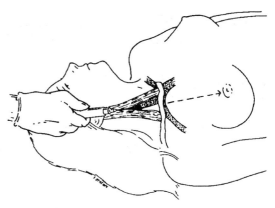

Fig. 37.2 Internal jugular vein puncture on the right side of the patient aiming for the nipple of the same side while staying lateral to and palpating the carotid artery with the non-dominant hand.

- palpate the carotid with the non-dominant hand and stay lateral to it; and
- do not insert needle more than 5 cm (2 in).

External jugular vein (see p. 216 and Fig. 12.6)

This vein is used mostly for access to the intrathoracic structures in children.

Technique

- place in supine position with the head rotated toward the opposite side, either in the Trendelenberg position or with a roll placed under the shoulder;
- raise the venous pressure by encouraging crying; and
- enter needle at the junction of the upper and middle thirds of a line passing from the angle of the mandible to the center of the clavicle.

Veins of the lower limb

Femoral vein (see p. 307 and Fig. 20.2)

This vein, lying in an area that is difficult to sterilize, is not a popular vessel for placing catheters because of the increased risk of infection. The femoral vein is the most superficial vessel in the groin crease and lies

side where it is usually larger and straighter and easier to cannulate.

Technique

- adopt a similar position to that described for the subclavian vein puncture;
- pass the needle at a 30° angle to the surface toward the apex formed by the manubrial and clavicular heads of the sternomastoid muscle;
- aim for the nipple of the same side (Fig. 37.2);

about two fingerbreadths inferior to the inguinal ligament.

Technique

- place finger of the non-dominant hand on the femoral pulse; and
- puncture skin at a point medial to the middle of the finger.

Long saphenous vein

This vein, lying in the 'anatomic snuffbox' of the foot (see p. 345 and Fig. 23.19), is often visible and can be made more so by applying a tourniquet to obstruct the veins.

Arterial punctures

Arteries that are commonly used for punctures include the radial, brachial and femoral arteries.

Arteries of the upper limb

Radial artery

The radial artery is the artery of choice for puncture because of its superficial location, good collateral supply, least painful site, and easy control of hemorrhage by applying

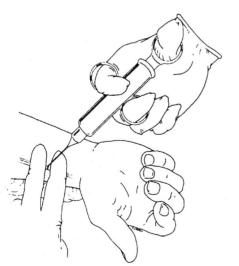

Fig. 37.3 Puncture of the radial artery. The lower forearm is supported by a small cushion.

pressure to the immediately subjacent bone (Fig. 37.3).

Brachial artery (see Figs. 3.23 and 5.7)

This vessel is palpable because of its subcutaneous position and is best punctured at a point about 2.5cm (1 in) below the midpoint of the arm. In this area, the ulnar nerve has left the brachial artery as it passes postero-medially, the profunda branch has already been given off and the median nerve still lies lateral to the vessel.

Arteries of the lower limb

Femoral artery (see Fig. 20.2)

Despite its size, the femoral artery is the least popular choice for arterial puncture. This is because closely related structures such as the femoral vein and nerve, the hip joint and the contents of a femoral hernia may be damaged. Furthermore, the groin is a difficult area to sterilize.

The artery can be palpated below the inguinal ligament at the center of a line drawn from the anterior superior iliac spine to the pubic symphysis.

Joint punctures

Shoulder joint

Lateral approach

Technique

- pass needle from immediately lateral to the acromial angle (junction of scapular spine and acromion) in a superior and medial direction (Fig. 37.4).

Medial approach

Technique

- pass needle from just lateral to the tip of

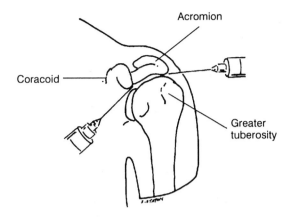

Fig. 37.4 Aspiration of the shoulder joint. Anterior and lateral routes are shown.

the coracoid process in a posterior, lateral and superior direction (Fig. 37.4).

Elbow joint

The most superficial part of the joint is the posterior aspect of the radiohumeral joint.

Technique

- flex elbow to a right angle with the arm in the midposition (semipronated); and
- pass needle immediately proximal to the easily palpable posterior radial head in an anterior direction (Fig. 37.5).

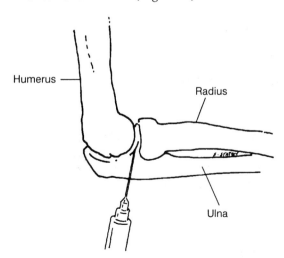

Fig. 37.5 Puncture of the elbow joint via the posterior aspect of the radiohumeral joint.

Wrist joint

Technique

- pass needle below and at right angles to the ulnar styloid process (Fig. 37.6).

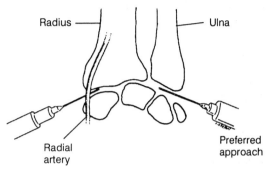

Fig. 37.6 Puncture of the wrist joint from either the ulnar or radial side. The ulnar approach is preferred as it avoids the radial artery which passes posteriorly in close relationship to the radial styloid.

Hip joint

Technique

- pass needle from the point just lateral to the upper border of the greater trochanter, medially and slightly superiorly parallel to the femoral neck (Fig. 37.7).

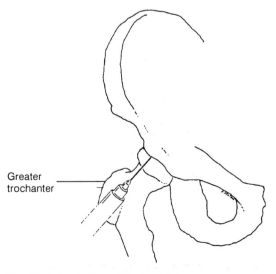

Fig. 37.7 Aspiration of the hip joint.

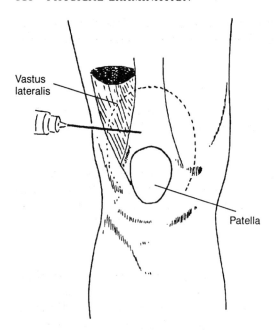

Fig. 37.8 Aspiration of the knee joint by the pre-ferred suprapatellar approach.

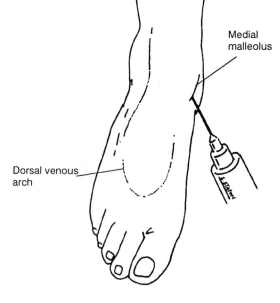

Fig. 37.9 Puncturing the ankle joint below the medial malleolus.

Knee joint

Technique

- pass needle into the suprapatellar bursa traversing the quadriceps muscle on the way (Fig. 37.8, see also Fig. 22.5).

Ankle joint

Technique

- pass needle from below the tip of either malleolus in a superior direction (Fig. 37.9).

Nervous system punctures

Lumbar puncture (Fig. 37.10)

Technique

- request the patient to sit or lie in the lateral position (sitting is preferred as the gravitational effect on the spinal fluid distends the arachnoid space)
- flex spine as much as possible to separate the spinous processes and laminae by bending the head forward if sitting or by approximating the head and knees if lying down (curling up);
- locate the point in the midline between the third and fourth lumbar spinous processes by palpation at the level of a line joining the highest points of the iliac crests;
- pass needle into this point;
- advance needle in the midline anteriorly for 5–7.5 cm (2–3 in) and between the spinous processes and through the interspinous ligament, dural and arach-noidal membranes.

Epidural puncture

Lying between the walls of the vertebral canal and the dura, the epidural space contains fat, venous plexuses, arteries, lymphatics and spinal nerve roots. Local anesthetic solution instilled into this space will produce effects on the nerve roots. In the lumbar region, this will produce anesthesia of the lower extremities. Similar results can be obtained

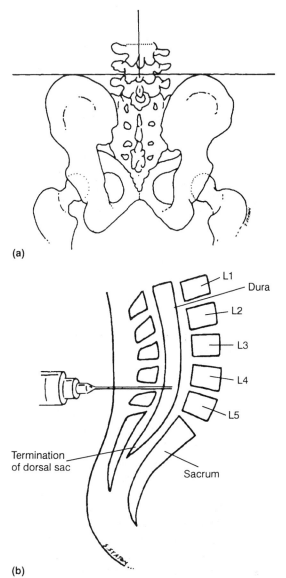

(a)

(b)

Fig. 37.10   (a) A line joining the highest point of the iliac crests cuts across the 4th lumbar spinous process. A lumbar puncture is carried out in the intervertebral space above or below this line. (b) Lumbar puncture through the 4th lumbar intervertebral space.

by placing the anesthetic solution in the subarachnoid space; the advantage, however, of the epidural approach is that the avoidance of the penetration of the membranes prevents leakage of the

cerebrospinal fluid, possible contamination of the nervous system and damage to the cauda equina.

Technique

- use the same approach as for the lumbar puncture but stop the needle short of the dura.

In the sacral region, anesthetic solution will affect the sacral nerves and for this reason an epidural injection is often used to alleviate the pain of delivery (Fig. 37.11).

Technique

- locate palpable gap at the lower end of the sacrum, the sacral hiatus (see Fig. 32.1);
- insert needle into this gap initially in an almost perpendicular direction; and
- advance needle in a vertical direction (Fig. 37.11).

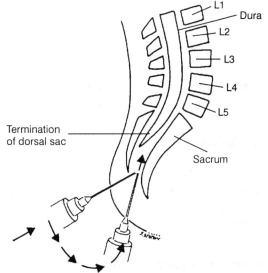

Fig. 37.11   Sacral epidural injection. The needle enters the sacral hiatus and deposits the anesthetic solution outside of the dural sac.

Peripheral nerves

Injections are usually made around peripheral nerves so as to render them temporarily anesthetic.

Head

The supra-and infraorbital, and the mental branches of the fifth nerves lie in the same vertical plane (see Fig. 11.7). The supraorbital nerve can be found at the junction of the inner and middle thirds of the upper margin of the orbit, where the supraorbital notch is often palpable. If the notch is impalpable because of its conversion into a foramen, the nerve may be located in a subcutaneous position at the superior orbital margin, 2.5 cm (1 in) from the midline.

The infraorbital nerve emerges from the infraorbital foramen which lies along the infraorbital margin, 0.6 cm ($^{1}/_{4}$ in) above and 0.6 cm ($^{1}/_{4}$ in) lateral to the nasal ala. A finger placed on the infraorbital margin prevents the injection entering the orbit.

The mental nerve emits from the mental foramen which lies opposite a point between the upper and lower borders of the mandible between the two premolar teeth (see Fig. 13.1).

The inferior dental and lingual nerves can both be anesthetized by an injection placed just above the large mandibular foramen. The needle is placed 0.6–1.25 cm ($^{1}/_{4}$–$^{1}/_{2}$ in) above the crown of the third molar tooth immediately medial to the ramus, aiming in a posterior and slightly lateral direction.

Upper limb

Performance of the more proximal nerve blocks, such as that of the brachial plexus, are the province of the anesthesiologist and not the neophyte physician. The more distal blocks, however, are extremely useful and well worth mastering, since injuries and infections of the hand are very common. Being served by the ulnar and median nerves (the radial nerve is a very minor contributor to either the sensory or motor function of the hand), it is these structures that need to be anesthetized.

ULNAR NERVE

In the case of the ulnar nerve, peripheral nerve block is better performed at the elbow because a block of this nerve at the wrist requires a ring block because of the dorsal branch given off above the level of the wrist. Additionally, injury to the ulnar artery is completely avoided at the elbow and, finally, the nerve is palpable at this level.

Technique

- rotate extremity internally at the shoulder and flex at both the shoulder and elbow joints (Fig. 37.12);
- locate the 'ulnar groove' between the tips of the olecranon and the medial epicondyle; and
- apply pressure in this area to palpate the nerve or if, for any reason, it is impalpable, parathesiae can be elicited.

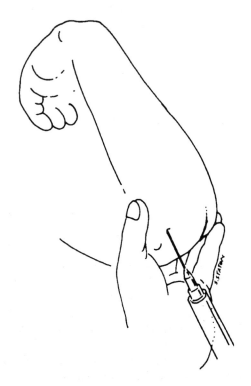

Fig. 37.12 The position of the elbow joint to facilitate location of the ulnar nerve.

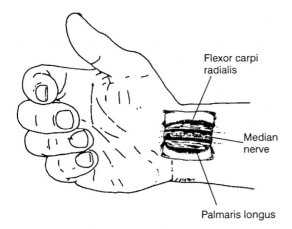

Fig. 37.13 The position of the median nerve with the hand closed and the wrist strongly flexed to render the tendons prominent.

MEDIAN NERVE

Technique

- supinate the forearm and draw a line across the anterior surface through the styloid process of the ulna;
- strongly flex the wrist to delineate the forearm tendons; and
- pass needle on the lateral side of the palmaris longus tendon or, if absent, the flexor digitorum superficialis (Fig. 37.13, see also Fig. 5.17).

DIGITAL BLOCK

This is a very frequently used block. It should be remembered that the palmar and plantar nerves also innervate the dorsal surfaces of the middle and distal phalanges. This means that a large part of the digit can be anesthetized by palmar or plantar injections. Ring (circumferential) blocks are, therefore, often unnecessary, thus limiting the volume of injected solution and thereby reducing the risk of inducing vascular compromise. To prevent this latter complication, the volume of solution injected should be less than 10 ml. It is important to remember that nail lesions are common and frequently require an interventional procedure which can be

performed with anesthesia by plantar/palmar injections (see p. 106).

Technique

- place punctures on the dorsolateral aspect of the interdigital folds where the skin is thin and nonadherent (Fig. 37.14);
- the palmar/plantar nerves lie close to and anterolateral to the phalanx;
- the needle should strike the bone, be withdrawn slightly and then passed 6–1 cm ($^{1}/_{4}$–$^{1}/_{2}$ in) anterolaterally; and
- to block the dorsal nerve, the needle is passed a similar distance in a dorsolateral direction.

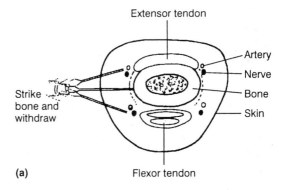

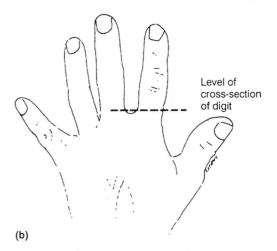

Fig. 37.14 Digital nerve block shown in cross-section (a) at level (b). The same technique applies to the toes.

Body cavity and organ punctures

Thorax

Pleural cavity (see Fig. 8.11)

Technique

• pass needle over the upper border of a rib through an intercostal space;
• the sites of choice are the anterior, middle or posterior axillary lines, where the muscular coverage is least (almost solely the serratus anterior); and
• the level of the tap depends on the circumstances for which the procedure is being carried out.

Pericardial cavity

Technique

• pass needle from slightly below and to the left of the xiphoid process toward the middle of the left scapula (approximately the same as the left shoulder); and
• the number 45 is a useful ally – the patient lies at a 45° angle, the needle is angled at 45° to the plane of the patient, and points 45° toward the left shoulder (Fig. 37.15).

Abdomen

Peritoneal cavity

Technique

• the site of choice is the midline at 2.5 cm below the umbilicus as it is likely to be avascular;
• this site presents little danger of the needle entering the bladder (which should be empty) or stomach (which may also be emptied by a nasogastric tube); and
• at this site, small bowel is likely to be encountered but, being freely mobile, is displaced without injury.

Liver (see also p. 393 and Fig 28.1).

The patient should be cooperative and able

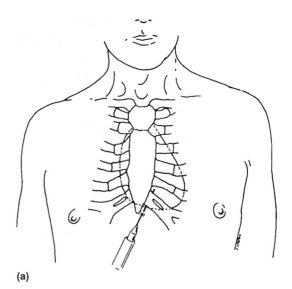

(a)

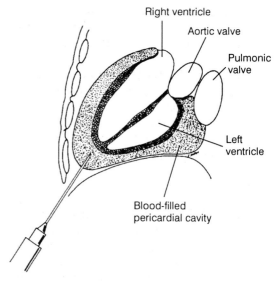

Right ventricle

Aortic valve

Pulmonic valve

Left ventricle

Blood-filled pericardial cavity

(b)

Fig 37.15 Anterior view (a) and sagittal view (b) of the technique for pericardial aspiration.

to suspend respiration on command so as to avoid movement of the biopsy needle in the liver that could result in a laceration.

Technique

• insert needle in the midaxillary line and over the superior border of the ribs of the 8th, 9th or 10th spaces;

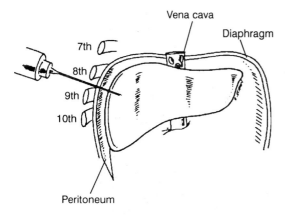

Fig. 37.16 Liver biopsy carried out through the 8th intercostal space.

- placing the needle at a higher level, it would necessarily pass through and possibly injure or infect the lung;
- the needle penetrates the intercostal space, costal and diaphragmatic pleura, diaphragm and parietal and visceral peritoneum; and
- before the needle enters the liver, the patient is instructed to hold their breath, and the biopsy is taken (Fig. 37.16).

Urinary bladder

In the child, the bladder is always an abdominal organ. In the adult, the bladder is a pelvic organ, becoming abdominal only when distended. In the latter case, the bladder's extraperitoneal inferolateral surfaces come into contact with the anterior abdominal wall, through which the bladder can be safely punctured for the placement of a catheter (Fig. 37.17).

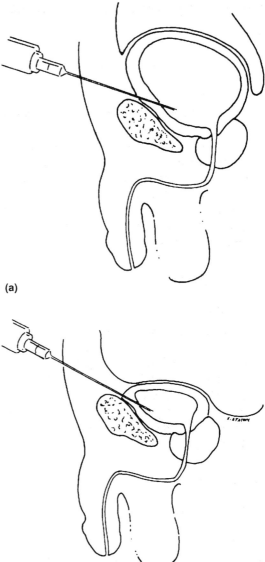

(a)

(b)

Fig. 37.17 Suprapubic aspiration of the (a) distended and (b) child's bladder.

Review questions on physical examination

(answers on p. 529)

198 **A 40-year-old male complains of intermittent pain in the right ear of several weeks' duration. Select the anatomic area and the pathology likely to be responsible for this symptom.**
a) The cardio-esophageal region hiatus hernia
b) A malignant tumor of the side of the tongue
c) An unerupted third lower right molar
d) A lesion of the glenoid lobule of the parotid gland – mumps (parotitis)
e) All the above are correct

199 **Which of the following surface markings are correct.**
a) The thyroid cartilage indicates the end of the pharynx and commencement of the esophagus
b) The duodeno-jejunal junction lies 2.5–3.75 cm (1–1.5 in) to the left of the second lumbar vertebra
c) The angle of Louis indicates the highest point reached by the normal aortic arch
d) The saphenous vein usually joins the femoral vein 7.5 cm (3 in) below the pubic tubercle
e) The level of the bifurcation of the popliteal artery lies 5 cm (2 in) below the tibial tuberosity

200 **Venous punctures with a view to placing catheters are everyday procedures. It is true to say that**
a) the femoral vein is the vein of choice as it lies most superficially
b) the subclavian vein is best punctured above the medial end of the clavicle as there is less muscle to penetrate
c) to puncture the internal jugular vein the needle should not penetrate more than 5 cm (2 in)
d) the superficial antecubital veins of the elbow are superficial and visible and therefore the veins of choice
e) because the veins of the leg are equipped with valves and can be rendered prominent by the use of a tourniquet they offer an ideal conduit

People in anatomy

ACHILLES
A mythical Greek warrior who was vulnerable only in the heel.
A' reflex – ankle jerk.
A' tendon – tendon of gastrocnemius and soleus (tendo calcaneus).

ALCOCK, Benjamin
1801–? In 1849 was appointed Professor of Anatomy in Queen's College, Cork, but was called upon to resign in 1853 due to a dispute concerning the working of the Anatomy Acts. Went to USA in 1855 and was heard of no more!
A's canal – for the internal pudendal vessels in the ischiorectal fossa.

BAKER, William M.
1839–1896 An English surgeon.
B's cyst – a cyst of the bursa of the popliteal fossa.

BALL, Sir Charles Bent
1851–1916 Professor of Surgery in Dublin.
B's valves – fusion remnants of the proctodeum and allantoic gut described by Morgagni 200 years earlier.

BARTHOLIN, Caspar (Secundus)
1655–1738 Succeeded his father, Thomas Bartholin, as Professor of Medicine, Anatomy and Physics in Copenhagen.
B's ducts – sublingual ducts that open into the submandibular duct.
B's glands – greater vestibular glands.

BATSON, Oscar Vivian
1894–1979 Professor of Anatomy, University of Pennsylvania.
B's plexus – vertebral venous plexus connecting skull to pelvis and thought to be an important route of metastatic spread.

BELL, Sir Charles
1774–1842 Surgeon, anatomist and artist.
Founder of The Middlesex Hospital Medical School. Later Professor of Surgery, Edinburgh.
B's nerve – long thoracic nerve to serratus anterior.
B's palsy – facial nerve paralysis of unknown etiology.

BIGELOW, Henry Jacob
1818–1890 Professor of Surgery at Harvard.
B's ligament – the iliofemoral ligament.

BOCHDALEK, Vincent A.
1801–1883 A Prague anatomist.
B's foramen – pleuroperitoneal canal.
B's gap – lumbocostal trigone.
B's gland – derived from primitive thyroglossal duct.

BOYDEN, Edward A.
1886–1978 An American anatomist.
B's meal – fatty meal used to test gallbladder evacuation.
B's sphincter – the separate sphincter of the common bile duct proximal to the sphincter of Oddi.

BRUNNER, Johann Conrad
1653–1727 Professor of Anatomy at Heidelberg and later in Strasbourg. He described the crypts of Lieberkühn a decade before they were attributed to L.

BUCK, Gurdon
1807–1877 A New York surgeon.
B's fascia – deep penile fascia.

BUDD, George
1808–1882 Professor of Medicine, King's College, London.

B–Chiari syndrome – hepatic vein occlusion leading to hepatomegaly and cirrhosis.

CAMPER, Pieter
1722–1789 Professor of Medicine, Anatomy, Surgery and Botany in Groningen, Holland, from 1763 to 1773.
C's fascia – fatty layer of the superficial fascia of the abdomen.

CLOQUET, Baron Jules Germain
1790–1883 Professor of Surgical Pathology, Hôpital St Louis, Paris.
C's gland – a lymph node found at the apex of the femoral canal belonging to the inguinal group.

COLLES, Abraham
1773–1843 Professor of Anatomy and Surgery in Dublin from 1804 to 1836.
C' fascia – perineal fascia and deep layer of the superficial fascia of the abdomen.
C' fracture – of the extremity of the radius.
C' ligament – reflected inguinal ligament.

COOPER, Sir Astley Paston
1768–1841 Surgeon, Guy's Hospital, London, and FRS in 1802.
C's fascia – cremasteric fascia derived from the internal oblique muscle.
C's ligament – the pectineal part of the inguinal ligament.
Ligaments of Astley C – suspensory ligaments of breast which account for dimpling in some carcinomata.

CORTI, Alfonso
1822–1888 An Italian anatomist.
Organ of C – specialized cells within the cochlear duct, consisting of a basilar membrane supporting hair cells which rub against the tectorial membrane.

COWPER, William
1666–1709 A London surgeon. FRS 1698.
C's glands – bulbourethral glands.

DENONVILLIERS, Charles Pierre
1808–1872 Professor of Anatomy and, later, Surgery in Paris.
D's fascia – rectovesical fascia.

DOUGLAS, James
1675–1742 Scottish anatomist and 'man-
midwife' who lived in London. Physician to the Queen. FRS 1706.
Line of D – the arcuate line of the posterior layer of the sheath of rectus abdominis.
Pouch of D – rectouterine peritoneal pouch.

DUPUYTREN, Guillaume
1777–1835 A French surgeon and pathologist.
D's contracture – fibrosis of the palmar fascia causing fixed flexion deformity usually of the ring and little fingers.

EDINGER, Ludwig
1855–1918 A German anatomist.
E–Westphal nucleus – midbrain group of cells forming part of the oculomotor nerve from which originate the preganglionic parasympathetic fibers to the eye.

ERB, Wilhelm H.
1840–1921 A German neurologist.
E's palsy – damage to the upper trunk of the brachial plexus.

EUSTACHIO (EUSTACHI or EUSTACHIUS), Bartolomeo
1520–1574 Professor of Anatomy in Rome, and Physician to the Pope.
Eustachian tube – the auditory tube (strictly, its cartilaginous part).
E valve – valve of the inferior vena cava.

FALLOPIUS (FALLOPIO), Gabriele
1523–1563 Professor of Anatomy and Surgery in Padua.
F' canal – facial nerve canal.
Fallopian tube – uterine tube.

FRIEDREICH, Nikolas
1825–1882 A German neurologist.
F's ataxia – hereditary spinal ataxia.

FROMENT, Jules
1878–1946 Professor of Medicine, Lyons.
F's sign – seen in ulnar palsy when the distal phalanx of the thumb flexes whilst trying to hold a piece of paper against the index finger. This due to adductor pollicis weakness permitting overaction of flexor pollicis longus.

GALEN, Claudius
c. 130–200 Second century physician and gatherer of medical information. Many of his

erroneous ideas were regarded as gospel until the fifteenth and sixteenth centuries.
Great vein of G – internal cerebral vein.

GARTNER, Hermann Treschow
1785–1827   A Copenhagen anatomist.
G's canal or duct – the vestige of the mesonephric duct found in the broad ligament of the uterus and sometimes extending in the wall of the vagina to the vulva.

GEROTA, Dumitru
1867–1939   Professor of Surgery, Bucharest. He also studied lymphatic injection techniques in Berlin.
Fascia of G – perinephric fascia.

GIMBERNAT Y ARBOS, Manuel Louise Antonio de
1734–1816   Catalan surgeon and Professor of Anatomy in Barcelona; surgeon to King Carlos III.
G's ligament – lacunar ligament.

GLISSON, Francis
1597–1677   Regius Professor of Physics, Cambridge. He was a founder member of The Royal Society and described the sphincter of Oddi 200 years before Oddi.
G's capsule – fibrous capsule of the liver.

HARTMANN, Henri Albert
1860–1952   Professor of Surgery, Paris.
H's pouch – dilatation at the neck of the gallbladder, sometimes the site of gallstone impaction.

HEISTER, Lorenz
1683–1758   Professor of Surgery at Altorf, Nuremberg, and later also of Anatomy at Helmstadt.
H's valve – 'spiral valve' due to mucosal folds in the cystic duct.

HESSELBACH, Franz Kaspar
1759–1816   A German surgeon.
H's fascia – cribriform fascia.
H's triangle – inguinal triangle.

HIGHMORE, Nathaniel
1613–1685   An English anatomist.
H's antrum – maxillary antrum.

HILTON, John
1804–1878   Surgeon at Guy's Hospital, London, from 1849 to 1871.

H's law – a nerve which crosses a joint, gives a branch to that joint.

HORNER, Johann F.
1831–1886   A Zürich ophthalmologist.
H's syndrome – facial and ophthalmic paresis, due to cervical sympathetic interruption.

HOUSTON, John
1802–1845   Lecturer in Surgery in Dublin, and Physician to the City Hospital.
Valves or folds of H – transverse rectal folds.

HUNT, James R.
1872–1937   An American neurologist.
H's syndrome – facial paralysis, otalgia, aural herpes due to geniculate ganglion herpes zoster.

HUNTER, John
1728–1793   A Scottish anatomist and surgeon who worked in London. Founder of the Hunterian Museum, now in the custody of the Royal College of Surgeons of England.
H's canal – adductor canal or subsartorial canal of the thigh.

KIESSELBACH, Wilhelm
1839–1902   German laryngologist.
K's area – anterior portion of nasal septum, often the site of epistaxis. (syn: Little's area).

KILLIAN, Gustav
1860–1921   Berlin Professor of Otolaryngology.
K's bundle – lowest fibers of inferior constrictor of pharynx.
K's dehiscence – weakness between fibers of inferior pharyngeal constrictor.

KLUMPKE, Augusta Déjérine-
1859–1927   A French neurologist.
K's paralysis – paralysis of the intrinsic muscles of the hand, often due to birth injury.

LATARJET, Andre
1877–1947   A French anatomist.
L's nerve – a branch of the anterior vagus to the upper border of the pylorus and antrum of the stomach.

LOUIS, Antoine
1723–1792   A surgeon and physiologist of Paris whom American authorities cite as noting the

'angle' but this is not mentioned in his works.
Angle of L – the manubriosternal junction.

MacEWEN, Sir William
1848–1924   Scottish surgeon who trained under Lister.
M's triangle – suprameatal triangle which is the surface marking of the mastoid antrum.

MACKENRODT, Alwin K.
1859–1925   Professor of Gynecology in Berlin.
M's ligament – cardinal (lateral cervical) ligament of the uterus.

McBURNEY, Charles
1845–1914   A New York surgeon.
M's point – between 3.75 and 5 cm (1.5 and 2 in) from the right anterior superior iliac spine upon a line to the umbilicus. The given site of maximum tenderness with appendicitis.

MECKEL, Johann Friedrich
1714–1774   Professor of Anatomy, Botany and Gynecology in Berlin.
M's space/cave – the trigeminal cavum. The subarachnoid space around cranial nerve V as it lies in the middle cranial fossa.

MECKEL, Johann Friedrich the younger
1781–1833   Grandson of the preceding.
A German comparative anatomist and embryologist.
M's cartilage – cartilage of the first branchial arch.
M's diverticulum – a diverticulum of the ileum; a persistent proximal part of the vitellointestinal duct.

MEIBOMIUS (MEIBOM), Hendrik
1638–1700   Professor of Medicine at Helmstadt and later Professor of Poetic Art.
Meibomian cyst – chalazion due to infection of a tarsal gland.
M' gland – tarsal gland.

MONTGOMERY, William F.
1797–1859   Professor of Midwifery in Dublin.
M's tubercle – elevated reddish areolar glands easily seen in pregnancy.

MORGAGNI, Giovanni Battista
1682–1771   A Padua anatomist and pathologist.
Appendix or hydatid of M – appendix testis or vesicular appendix of epoöphoron.
Columns of M – anal columns.

Foramen of M – foramen cecum of the tongue.
Lacunae/Fossa of M –fossa navicularis.

MÜLLER, Heinrich
1820–1864   A German anatomist.
M's muscle – superior tarsal muscle of the eyelids, or circular fibers of the iris.

MÜLLER, Johannes Peter
1801–1858   Professor of Anatomy and Physiology in Berlin.
M's duct – primordial female genital duct – the paramesonephric duct.

NEWTON, Sir Isaac
1643–1727   English physicist who discovered the laws of gravity.

ODDI, Ruggero
1864–1913   An Italian physician.
Sphincter of O – sphincteric muscle fibers around the termination of the bile duct and main pancreatic duct.

OSGOOD, Robert Bayley
1873–1956   Orthopedic surgeon at Massachusetts General Hospital, Boston.
O–Schlatter's disease – osteochondritis of tibial tubercle.

PAGET, Sir James
1814–1899   An English surgeon.
P's disease of nipple – malignant change in the skin surrounding the nipple.

PANCOAST, Henry Khunrath
1875–1939   Professor of Radiology in Philadelphia, Pennsylvania.
P's syndrome – apical carcinoma of the lung associated with Horner's syndrome, brachial plexus palsy, pain down the arm and distension of neck veins.

PETIT, Jean Louis
1674–1750   A Paris surgeon.
P's triangle – lumbar triangle.

PEYER, Johan Konrad
1653–1712   Professor of Logic, Rhetoric and Medicine in Schaffhausen, Switzerland.
P's patches – aggregated lymphoid follicles in the lower ileum.

PFANNENSTIEL, Hermann Johann
1862–1909   A German gynecologist.
P's incision – transverse suprapubic incision used for pelvic surgery.

POTT, Percival
1713–1788   An English surgeon.
P's disease – tuberculosis of the spine.
P's fracture – ankle fracture involving the malleoli and ligaments.

POUPART, François
1661–1709   Surgeon to the Hôtel Dieu, Paris.
P's ligament – inguinal ligament.

RAYNAUD, Maurice
1834–1881   A Parisian physician.
R's disease – peripheral vascular disturbance consisting of spasmodic contractions of the digital arteries.

REISSNER, Ernst
1824–1878   A German anatomist.
R's membrane – vestibular membrane of the inner ear, dividing scala vestibuli and scala media.

RETZIUS, Andreas Adolf
1796–1860   Professor of Anatomy and Physiology at the Karolinska Institute, Stockholm, Sweden.
Cave of R – prevesical space.

RIEDEL, Bernhard M.C.L.
1846–1916   A German surgeon.
R's lobe – a tongue-like extension of the right lobe of the liver.

SANTORINI, Giovanni D.
1681–1737   An Italian anatomist.
Cartilage of S – corniculate cartilage.
Duct of S – accessory pancreatic duct.

SCARPA, Antonio
1747–1832   A Venetian anatomist, orthopedist and ophthalmologist.
S's fascia – deep layer of the superficial fascia of the lower abdomen.
S's nerve – nasopalatine nerve.
S's triangle – femoral triangle.

SCHLATTER, Carl
1864–1934   Professor of Surgery, Zurich.
Osgood–S disease – osteochondritis of tibial tubercle.

SCHLEMM, Friedrich S.
1795–1858   A German anatomist.
Canal of S – venous sinus of the sclera found at the corneoscleral junction.

SIBSON, Francis
1814–1876   An English anatomist.
S's fascia – the suprapleural membrane.

SPIGELIUS, Adrian (van der Speighel)
1578–1625   A Belgian anatomist in Padua.
Spigelian hernia – herniation along the linea semilunaris of the abdomen.
S' lobe – caudate lobe of liver.

STENSEN, Niels or Nicholas
1638–1686   Danish anatomist, who later became a minister and a bishop.
S's duct – duct of the parotid gland lying across the masseter muscle.

TENON, Jacques René
1724–1816   Professor of Pathology in the Academy of Sciences, Paris, and Chief Surgeon at the Salpetrier.
T's capsule or fascia – the fascial sheath of the eyeball.

THEBESIUS, Adam C.
1686–1732   A German physician.
T' valve – valve of the coronary sinus.
Thebesian veins – small veins draining directly into the cardiac chambers.

TREITZ, Wenzel
1819–1872   Austrian who was Professor of Pathological Anatomy in Cracow, and later Professor of Pathology in Prague.
Ligament of T – suspensory ligament of the duodenum.

TRENDELENBURG, Freidrich
1844–1924   A German surgeon.
T position – operating position with head down and pelvis elevated, as used in the treatment of shock.
T sign – dipping of the pelvis in congenital hip dislocation.

TROISIER, Charles Emile
1844–1919   Professor of Pathology, Paris.
T's sign – enlargement of supraclavicular node usually due to upper gastrointestinal tract carcinoma.

VATER, Abraham
1684–1751   Professor of Anatomy, Botany, Pathology and Therapeutics in Wittenberg, Germany.
Ampulla of V – hepatopancreatic ampulla.

VENUS
Mythical Roman goddess of love.
V's dimples – site of posterior superior iliac spine and the middle of the sacroiliac joint. These are best seen on well-rounded ladies.

VIDIUS (Guido Guidi)
1500–1569   Italian anatomist and physician to Francis I of France; from 1548, Professor of Medicine at Pisa.
V canal – pterygoid canal.
V nerve – nerve of the pterygoid canal.

VIRCHOW, Rudolf
1821–1902   German pathologist who emphasized the cell as the fundamental unit in pathology.
V's node – supraclavicular node. (*See also* Troisier.)

VOLKMANN, Richard von
1830–1889   German Professor of Surgery.
V's contracture – tissue loss, especially muscular, due to ischemia.
V's splint – guttered splint for lower extremity fractures.

WALDEYER, Heinrich W. G. von
1836–1921   Professor of Pathological Anatomy in Breslau, and later in Berlin.

W's ring – lymphatic ring in the pharynx.
W's fascia – posterior pelvic wall fascia.

WESTPHAL, Karl F. O.
1833–1890   A German neurologist.
Edinger–W nucleus – group of cells in the midbrain, forming part of the oculomotor nerve from which preganglionic parasympathetic fibers to the eye are derived.

WHARTON, Thomas
1614–1673   An English anatomist.
W's duct – main submandibular duct opening at the frenulum of the tongue.
W's jelly – connective tissue (mucous) of the umbilical cord.

WINSLOW, Jacob Benignus
1669–1760   A Danish anatomist who, at the age of 74, was appointed Professor of Anatomy, Physics and Surgery in Paris.
Foramen of W – epiploic foramen.
Ligament of W – oblique popliteal ligament.
Pancreas of W – uncinate process of the pancreas.

WIRSUNG, Johann G.
1600–1643   A German anatomist in Padua.
W's duct – the main pancreatic duct.

WOLFF, Kaspar Friedrich
1733–1794   Professor of Anatomy and Physiology in St Petersburg. One of the founders of modern embryology.
W'ian duct – mesonephric duct.
W'ian body – mesonephros.

Answers to review questions

Part I

1 a	(15)	2 a	(45)	3 c	(15)	4 c	(16)	5 d	(22)	6 e	(23)
7 d	(26)	8 b	(27)	9 a	(28)	10 e	(28,9)	11 e	(30,1)	12 c	(35)
13 b	(39)	14 e	(39)	15 c	(52)	16 e	(41,43)	17 b	(39)	18 c	(45)
19 c	(46)	20 d	(47)								

Part II

21 c	(71)	22 d	(73)	23 b	(106)	24 c	(93)	25 b	(94)	26 b	(79)
27 e	(92)	28 c	(73)	29 e	(103)	30 c	(106)	31 d	(81)	32 d	(95)
33 a	(76)	34 c	(66)	35 b	(97)	36 d	(107)	37 e	(79,92)	38 c	(102)
39 a	(102)	40 e	(106)	41 a	(62)	42 e	(73)	43 d	(76)	44 c	(60)
45 a	(94)										

Part III

46 a	(120)	47 b	(122)	48 c	(125)	49 a	(115)	50 d	(126)	51 a	(117)

Part IV

52 d	(139)	53 e	(159)	54 d	(162)	55 c	(153)	56 d	(156)	57 d	(152)
58 a	(160)	59 b	(168)	60 c	(168)	61 b	(372)	62 d	(162)	63 b	(242)
64 c	(141)	65 b	(145)	66 c	(157)	67 b	(159)	68 c	(145)	69 b	(159)
70 d	(374)	71 e	(246)	72 b	(140)	73 a	(364)	74 b	(137)	75 e	(160)
76 a	(158)	77 e	(165)	78 e	(164)	79 c	(152)	80 d	(160,246)	81 b	(150)
82 e	(163)	83 c	(166)	84 b	(131)	85 b	(159)				

Part V

86 c	(264)	87 e	(216)	88 d	(231)	89 a	(232)	90 d	(234)	91 c	(228,256)
92 c	(226)	93 e	(248)	94 b	(260)	95 d	(285)	96 d	(244)	97 d	(224)
98 b	(211)	99 b	(232)	100 c	(183)	101 d	(200)	102 a	(205)	103 a	(276)
104 d	(187)	105 b	(267)	106 c	(232)	107 b	(193)	108 c	(195)	109 c	(200)
110 d	(205)	111 a	(254)	112 c	(223)	113 b	(199)	114 d	(223)	115 c	(223)
116 b	(218)	117 c	(219)	118 a	(242)	119 e	(234)	120 c	(66)	121 d	(244)
122 c	(241)	123 e	(240)	124 b	(233)	125 b	(248)	126 d	(252)		

Part VI

127 b	(305)	128 b	(304)	129 a	(324)	130 d	(339)	131 e	(340)	132 c	(308)
133 c	(325)	134 d	(306)	135 b	(343)	136 c	(136)	137 b	(333)	138 a	(303)
139 d	(349)	140 b	(310)	141 e	(332)	142 d	(335/6)	143 c	(336)	144 c	(342)
145 c	(325)	146 a	(339)	147 e	(349)	148 e	(334)	149 c	(315 348)	150 b	(298)
151 a	(339)	152 b	(320)	153 d	(437)						

Part VII

154 e	(388)	155 c	(368)	156 e	(370)	157 e	(395,447)	158 d	(417)	159 a	(419)
160 d	(419)	161 b	(434)	162 b	(407)	163 b	(402)	164 b	(409)	165 c	(409)
166 a	(399)	167 a	(368)	168 c	(395)	169 d	(393/4)	170 a	(370)	171 e	(366/7)
172 e	(388)	173 b	(386)	174 a	(414)	175 e	(388)	176 e	(390)	177 e	(383)
178 c	(405)	179 b	(420)	180 a	(457)	181 d	(374)	182 a	(432)	183 e	(424)
184 a	(426)	185 a	(454)	186 e	(455)	187 d	(460)	188 d	(43)	189 e	(468)
190 c	(471)	191 c	(465)	192 b	(446)	193 b	(468)	194 e	(457)	195 c	(459)
196 c	(456)	197 c	(480)								

Part VIII

198 e	(493)	199 b	(497)	200 c	(513)

Index

531